Ecological Studies, Vol. 124

Analysis and Synthesis

Ecological Studies

Volumes published since 1989 are listed at the end of this book.

Springer
New York
Berlin
Heidelberg
Barcelona
Budapest
Hong Kong
London
Milan
Paris
Santa Clara
Singapore
Tokyo

Walter C. Oechel   Terry Callaghan
Tagir Gilmanov   Jarle I. Holten   Barrie Maxwell
Ulf Molau   Bjartmar Sveinbjörnsson
Editors

Global Change and Arctic Terrestrial Ecosystems

With 132 Illustrations

Springer

Walter C. Oechel
Global Change Research Group
San Diego State University
San Diego, CA 92182
USA

Library of Congress Cataloging-in-Publication Data
Oechel, W.C.
Global change and arctic terrestrial ecosystems/Walter C. Oechel.
p. cm.—(Ecological studies; v. 124)
Includes bibliographical references and index.

ISBN-13: 978-1-4612-7468-1 e-ISBN-13: 978-1-4612-2240-8
DOI:10.1007/ 978-1-4612-2240-8

1. Climatic changes—Arctic regions. 2. Land use—Environmental aspects—Arctic regions. 3. Ecology—Environmental aspects—Arctic regions. I. Title. II. Series.
QC981.8.C5033 1996
574.5'2621—dc20 96-11797

Printed on acid-free paper.

Softcover reprint of the hardcover 1st edition 1997

Production coordinated by Chernow Editorial Services, Inc., and managed by Karen Phillips; manufacturing supervised by Jeffrey Taub.
Typeset by Best-set Typesetter Ltd., Hong Kong.

9 8 7 6 5 4 3 2 1

SPIN 10475697

Preface: Global Change and Terrestrial Ecosystems: The GCTE Research Programme for the Arctic

Compared with most other biomes, arctic terrestrial ecosystems are relatively constant over time. They are strongly dominated by temperature which varies less from year to year than does precipitation—a dominant influence on the dynamics of other systems. Given that global warming is expected to result in relatively faster and greater changes in high latitudes, global change will have a particularly strong impact on arctic terrestrial ecosystems. The arctic will therefore be a most important region for detecting global change (the signal-to-noise ratio will be strong, because the signal will be bigger and the background variation is relatively low). It will also be important in terms of feedback effects to further atmospheric and climate change, because arctic ecosystems are potentially important sources of trace gases, and induced changes in biological composition will have significant effects on albedo and evapotranspiration. All of this calls for a well-focused, priority research program. The aim of this introductory paper is to provide some pointers as to what this research effort might comprise.

Global change, in the context of the International Geosphere–Biosphere Program (IGBP), is defined as the individual and interactive effects of changes in atmospheric composition, UV radiation, climate, and land use. The Global Change and Terrestrial Ecosystem Project of the IGBP addresses the ways in which these global changes will affect terrestrial ecosystems at local, landscape, regional, and continental scales, and the ways in which these induced changes will result in further changes in the atmosphere and climate. Specifically, its objectives are:

1. To predict the effects of changes in climate, atmospheric composition, and land use on terrestrial ecosystems, including key production systems.
2. To determine how these effects lead to feedbacks to the atmosphere and the physical climate system.

The GCTE research programme provides a useful coordinating mechanism for regional studies related to global change and terrestrial ecosystems, such as that in the arctic. GCTE's Operational Plan (IGBP Report No. 21, Steffen et al., 1992) gives a framework for coordinating the research and for placing it in a global context. Figure I.1, from the Operational Plan, shows the structure of the project, including integrating facilities and activities, some of the products of the research programme, and some of the major groups with which GCTE interacts. Although the four Foci (F1 to F4 in Figure I.1) are presented as separate boxes, and the leaders are indeed based in different institutions, there is considerable overlap and collaboration. The major transects, the long-term ecological modeling activity (LEMA), the elevated CO_2 network, and the crop networks act as integrating facilities that bring scientists from the different Foci and Activities together.

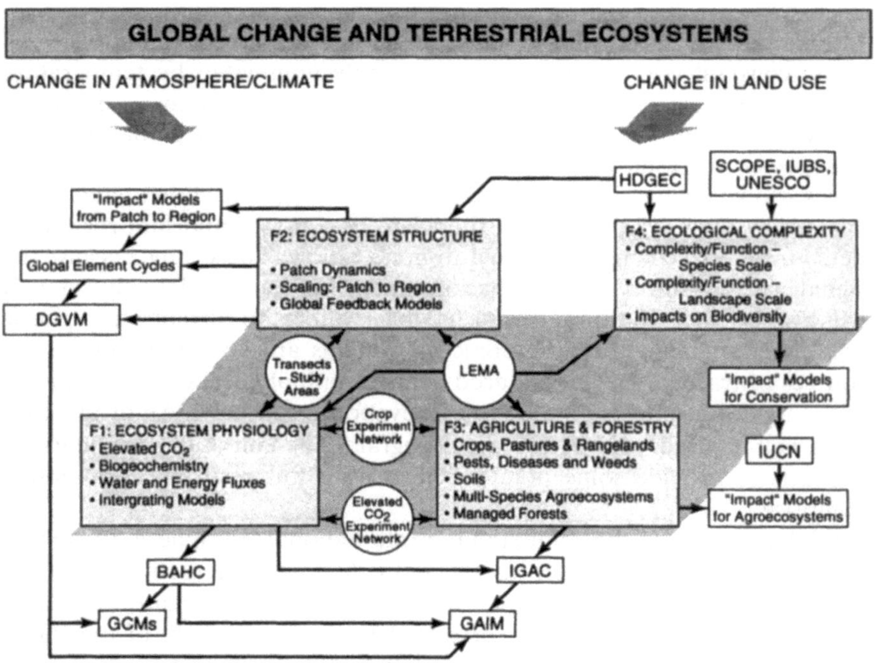

Figure I.1. GCTE's Operational Plan.

Our aim in jointly sponsoring this symposium was to strongly encourage arctic ecology researchers to consider the questions posed by the GCTE plan and to orientate their research towards addressing these issues. The task is beyond any one research group and given the increasing difficulty in obtaining the significant funds required for complex research beyond the interests of single nations, there is a real need for a collaborative effort.

Until very recently the questions ecologists have been expected to answer have been only local or regional in scale, and the driving forces and outflows from ecosystems could be taken as given, or as sinks. But the rest of the science community and the policy analysts are looking to ecologists to provide answers to very important questions about the ways in which ecosystems interact with their environments, and how these interactions may change as the structure and composition of the ecosystems change. We cannot afford a disparate set of individually inspired projects. If we are to provide the answers expected of us, we must focus our efforts in a well-orchestrated, complementary set of research projects that together address the priority issues. The global change problem has forced ecology into the arena of big science. Our effort will not be in the nature of a single, major experiment or expedition, but the required effort is just as large and complex; and because it is necessarily spatially dispersed, it is more difficult to execute.

I present below a brief account of the research issues that arctic ecologists are expected to resolve, in the format of the GCTE Operational Plan, followed by some suggestions for implementation.

The GCTE Research Programme

Focus 1: Ecosystem Physiology

The primary aim of Focus 1 is to understand and model the effect of global change on primary ecosystem processes, such as the exchange of energy, water, and trace gases with the atmosphere, element cycling and storage, and biomass accumulation or loss. A central thesis of Focus 1 is that the ways in which ecosystems function—their physiology—will be strongly affected by the combined and interactive suite of changes in atmospheric CO_2, land-use practices, and the likely changes in the means and extremes of temperature and rainfall. In the arctic, the predicted rise in temperature will have direct effects on plants and animals and will indirectly affect productivity via effects on mineralization rates. Nutrient supply is a major limitation to primary productivity in many high latitude regions. These effects will interact with enhanced UV-B and atmospheric CO_2 levels and changed precipitation patterns to have a significant effect on the performance of ecosystems.

The GCTE research programme calls for a set of manipulative experiments at the scale of whole plant communities or ecosystems. The experiments should vary combinations of temperature and CO_2 and determine the effects on system processes, such as soil carbon accumulation and decomposition; competitive relationships between plants; changes to nutrient fluxes and pools; plant tissue quality; energy and water balance; and plant phenology and architecture. A number of experiment techniques, such as glasshouses, open-top chambers, and ecocosms have been suggested. In view of the (expected) relatively greater significance of CO_2 effects in other biomes, the development of FACE experiments has not been recommended as a priority issue for arctic systems. However, this will need to be reviewed.

The experimental work will be accompanied by the development of integrated models of ecosystem physiology. The models will be designed to predict how global change will affect the physiology of arctic terrestrial ecosystems in the longer (decadal) time frames.

A critical ecosystem process in the arctic context is the movement of carbon compounds through the terrestrial biosphere. Tundra and boreal ecosystems presently store large amounts of soil carbon, and there is a possibility that globally significant amounts of this stored carbon will be released to the atmosphere as temperatures increase. Thus, a major emphasis in GCTE's research in the arctic will concentrate on determining the size and extent of major terrestrial carbon pools and the impact of global change on these pools. Particular attention needs to be given to ecosystem controls on the fluxes of important greenhouse gases such as CH_4, CO_2, and N_2O.

In addition to the individual manipulative experiments, GCTE is using a gradient approach to studying changes in the terrestrial regulation of regional and global biogeochemistry. In this approach, experiments and observations are distributed along a gradient of a controlling environmental parameter, such as temperature or precipitation. The high latitude northern hemisphere region has been identified as critical, and GCTE intends to establish at least three major N–S transects in tundra/boreal zones in the circumpolar region. Measurements to be made at the experimental sites along the transect include productivity, grazing and other disturbance effects, tissue quality, decomposition, soil moisture, soil nutrients and their dynamics, root dynamics, and trace gas fluxes. The relatively slow dynamics of high latitude systems means that a long-term commitment (10+ years) will be required for this research. The transects will also provide an organizing framework for more extensive measurements of vegetation, biogeochemistry, and land use on tundra/boreal regions.

Changes in both arctic ecosystem structure (see below) and function will likely have important implications for direct feedback to climate as well as to global biogeochemical cycles. In the arctic, potential changes to ecosystem structure could have a significant impact on regional albedo, considering the possible movement northwards of boreal forest at the expense of tundra. Changes in both ecosystem structure and function could also lead to changes in bulk surface conductance with implications for the surface energy balance.

Focus 2: Change in Ecosystem Structure

Coupled with their effects on ecosystem physiology, the driving forces of global change will lead directly and indirectly to changes in the distributions of plant and animal species and the species composition of ecosystems. Changes in ecosystem composition will, in turn, lead to further changes in ecosystem function. The goal of Focus 2 is to model this complex suite of impacts and responses, so that the pattern of change in ecosystem composition and structure can be predicted.

The northward spread of the boreal forest is the major predicted change in ecosystem structure in the tundra. These predictions, however, are based on "equilibrium" vegetation models that do not take transient dynamics into account. When the processes and rates by which ecosystems actually change are taken into account, the short- to medium-term results can be rather different. For example, if the predicted warming in high latitudes is accompanied by seasonal drying, large areas of present-day forest may be subject to more frequent fires, and this will lead to more grassy or heath-like vegetation. Such a change would, of course, lead to releases of greenhouse gases to the atmosphere. Other factors, such as increases in storm frequencies and intensities and in outbreaks of insect infestations and diseases, could also have significant impacts on the composition of high latitude systems.

Much of our present knowledge about these systems is based on experiments, observations, and models at the patch scale, usually some tens or hundreds of meters in extent. At larger scales, new processes become important, such as the disturbances described above. Thus, the scaling of ecosystem dynamics from the patch scale to landscapes (hundreds of meters to a few kilometers) and eventually up to regional scales is an area of central concern to the development of predictive models.

To facilitate the modeling of ecosystem change at a variety of scales, GCTE is establishing a set of intensive study areas in key biomes around the world. The aim is to build up a coherent system of small-scale experimental and observational sites, larger scale observations and remotely sensed data to support the development of a nested set of models. In most cases the GCTE study areas will be coincident with the biogeochemical transects, which will facilitate the linkage of ecosystem dynamics models to simulations of ecosystem function. For both the transects and study areas, close collaboration will be required between experimentalists and modelers.

GCTE hopes to establish two or three study areas in the tundra/boreal zone in the near future. Already there are preliminary plans for two projects that could serve as study areas: one is the Northern European Terrestrial Ecology Profile (NETEP), which extends from Svalbard down through Scandinavia to northern Europe; and the other is an extension of the present BOREAS experiment and is being developed as part of the Canadian N-BIOME programme.

Focus 3: Global Change Impact on Agriculture and Forestry

A significant fraction of the Earth's land cover is now converted to intensively managed and highly modified systems devoted to production of food and fiber. Many of these systems are already threatened by damage to soil and water resources through poor technology and management practices, and will be further impacted by global change. Through Focus 3, GCTE aims to improve our capability to predict global change impacts on key agronomic species and on managed forests, pastures, and rangelands.

As far as the arctic is concerned, Focus 3 is probably the least relevant of the four GCTE Foci. There is relatively little production of food and fiber in the region compared to the temperate and tropical zones.

However, global change may alter that situation significantly. The northward spread of boreal forests coupled with increasing economic pressures on several countries in the region (Russia in particular) would lead to rapidly increasing use of these forests for timber production. GCTE is developing an international research program on global change and managed forests. The programme will be based on networks of experimental sites in tropical, temperate, and boreal regions designed to explore the climatic limits and tolerances of important tree species. Modeling groups will synthesize the experimental and observation data and develop a predictive capability for the responses of managed forests to changing climate and atmospheric composition. This effort will be crucial in projecting future patterns of production forestry around the world, including the potential for the expansion and northward movement of the boreal forest-based industry.

Focus 4: Global Change and Ecological Complexity

Within the mounting concern about the loss of biological diversity, a crucial question is how the change in ecological complexity will affect ecosystem function. The aim of GCTE's Focus 4 is, therefore, to determine the impact of global change on ecological complexity and on the relationship between ecological complexity and ecosystem function. We have used the term "ecological complexity" rather than "biodiversity," because it involves much more than just the numbers of species. It includes the spatial patterns of species and communities, because these also influence population dynamics and landscape and regional scale processes.

In the arctic, the relationship between ecosystem complexity and ecosystem function is little understood. Even less is known about how this relationship will be affected by rapid environmental change. Arctic terrestrial ecosystems appear to be relatively stable (i.e., constant) over time and may have little resilience to disturbance effects. It is not known how a relatively small change in complexity will influence ecosystem function.

A major thrust of GCTE's Focus 4 is to undertake experiments and observational studies to determine the impacts of various kinds and combinations of global change on complexity and the consequent effects on function. One approach is to undertake reciprocal experiments, in which biological composition (structure) and abiotic environment (factors affecting physiological function) are separately altered, and the effects on each other measured. The details of such experiments for arctic ecosystems have yet to be determined.

Freshwater Ecosystems

The role of arctic freshwater ecosystems in global change is particularly important. The role of peatlands and northern wetlands as potential carbon sources or sinks has already been discussed. In addition, the movement of nutrients and organic matter via groundwater, streams, and rivers plays a major role in regional biogeochemistry.

The impacts of global change on freshwater ecosystems and their role in biogeochemical cycles has not been emphasized in the IGBP program thus far. This situation is, however, changing. The subsurface and surface flow of water and transport of nutrients is now part of the BAHC (Biospheric Aspects of the Hydrological Cycle) core project. GCTE will not be developing a special freshwater ecosystems section, but it is clear that the arctic research community will need to address this issue.

Implementation Strategy

The implementation of a coordinated research program on global change and terrestrial ecosystems in the arctic requires three elements: (1) sufficient scientific staff with an interest in undertaking global change research, (2) adequate resources to support the required research, and (3) mechanisms and resources to achieve the coordination.

The first of these does not seem to be a problem. There appears to be a significant number of ecologically oriented researchers interested in global change research in the arctic region, as evidenced by the strong response to this conference. The Scandinavian and North American scientific communities have traditionally been active in international research, and it is particularly pleasing to see a growing contribution from the Russian scientific community to the global change effort. Russia has much to offer in the tundra/boreal region, so a strong Russian contribution will add significantly to arctic global change research.

With regard to the second requirement, resources for global change research is not an easy task, particularly in these difficult economic times when funds for scientific research in general are limited. However, there are sources of funds and experience has shown that when scientists can develop coordinated national and regional research programs, resources can be obtained. Which brings us to the third requirement, and the main reason for this symposium.

The prospects for the development of a coordinated global change research program on arctic terrestrial ecosystems appear particularly good. There are already a number of organizations poised to play a prominent role in coordination. The International Arctic Science Committee (IASC) has developed an excellent framework for global change research in the region (IASC, 1994), which includes a strong component on terrestrial and marine ecosystems. The MAB Northern Science Network is also a well-established coordinating mechanism. The IASC priorities include the effects of elevated CO_2, UV-B, changed climate and nutrients; the effects on pools and fluxes of carbon (both these correspond to GCTE Focus 1 priorities); impacts on distribution; composition and structure of ecosystems (corresponding to GCTE Focus 2); impacts on diversity and its relation to ecosystem function, genetic variability, and related issues (GCTE Focus 4). Their other priorities on water and waterborne transport relate jointly to GCTE, BAHC, and LOICZ.

What is required now is some rationalization and cooperation among these groups to ensure that the task of coordinating global change research in the arctic is done without duplication and wastage of resources. It we can develop a coherent, prioritized, and demonstrably collaborative research effort we will maximize support for the program and we will also maximize the likelihood of supplying the answers to the questions about global change and arctic ecosystems.

Brian H. Walker

References

International Arctic Science Committee. 1994. *Scientific Plan for a Regional Research Programme in the Arctic on Global Change.* Proc. of a Workshop at Reykjavik, Iceland, 22–25 April 1992. National Academy Press, Washington DC.

Steffen, W.L., Walker, B.H., Ingram, J.S., and Koch, G.W. 1992. *Global Change and Terrestrial Ecosystems: The Operational Plan.* IGBP Report No. 21. International Geosphere Biosphere Secretariat, Royal Swedish Academy of Sciences, Stockholm.

Contents

Contributors

Allan N.D. Auclair	Science and Policy Associates, Inc., Washington, DC 20005, USA
Julie A. Bedford	Science and Policy Associates, Inc., Washington, DC 20005, USA
Ellen Bielawski	Arctic Institute of North America, University of Calgary, Alberta T2N IN4, Canada
William Dwight Billings	Department of Botany, Duke University, Durham, NC 27708-0338, USA
Hugh Boyd	Environment Canada, Canadian Wildlife Service, Ottawa, Ontario K1A 0H3, Canada
Terry V. Callaghan	Department of Animal and Plant Sciences, Sheffield Centre for Arctic Ecology, University of Sheffield, Sheffield S1D 5BR, UK
Bengt Å. Carlsson	Department of Ecology, Plant Ecology, University of Lund, S-223 62 Lund, Sweden

F. Stuart Chapin III	Department of Integrative Biology, University of California, Berkeley, CA 94720, USA
Wolfgang Cramer	Potsdam Institute for Climate Impact Research, D-144 12 Potsdam, Germany
Anne E. Giblin	The Ecosystem Center, Marine Biological Laboratory, Woods Hole, MA 02543, USA
Tagir G. Gilmanov	Department of Vertebrate Zoology and General Ecology, Moscow State University, Moscow, Russia
Anne Gunn	Department of Renewable Resources, Government of Northwest Territories, Yellowknife, Northwest Territories, Canada
Sarah E. Hobbie	The Ecosystem Center, Marine Biological Laboratory, Woods Hole, MA 02543, USA
Brian Huntley	Environmental Research Centre, University of Durham, Durham DH1 3LE, UK
Sven Jonasson	Department of Plant Ecology, University of Copenhagen, 2D, DK-1353 Copenhagen, Denmark
Douglas L. Kane	Water Resources Center, University of Alaska, Fairbanks, AK 99775, USA
Claude Lavoie	Centre d'études nordiques and Départment de biologie, Université Laval, Sainte-Foy, Québec G1K 7P4, Canada
Glen M. MacDonald	Department of Geography, McMaster University, Hamilton, Ontario L8S 4K1, Canada
Robert B. McKane	The Ecosystem Center, Marine Biological Laboratory, Woods Hole, MA 02543, USA
Jesper Madsen	National Environmental Research Institute, Ronde, Denmark
Barrie Maxwell	Arctic Adaptation Division, Canadian Climate Centre, Downsview, Ontario, Canada

Ross W. Mein — Department of Forest Science, University of Alberta, Edmonton T6G 2H1, Canada

Ulf Molau — Department of Systematic Botany, University of Göteborg, S-413 19 Göteborg, Sweden

Knute J. Nadelhoffer — The Ecosystem Center, Marine Biological Laboratory, Woods Hole, MA 02543, USA

Walter C. Oechel — Global Change Research Group, San Diego State University, San Diego, CA 92182, USA

Nikolai S. Panikov — Institute of Microbiology, Russian Academy of Sciences, Moscow 117811, Russia

Edward B. Rastetter — The Ecosystem Center, Marine Biological Laboratory, Woods Hole, MA 02543, USA

Peter R. Rowntree — Meteorological Office, Hadley Centre for Climate Prediction and Research, Bracknell, Berkshire RG12 2SY, UK

Brit Salbu — Laboratory for Analytical Chemistry, Agricultural University of Norway, N-1432 Aas, Norway

Peter A. Scott — Atmospheric Environment Services, Centre for Atmospheric Research Experiments, Egbert, Ontario L0L 1N0, Canada

Leonid R. Serebryanny — Institute of Geography, Russian Academy of Sciences, Moscow 117811, Russia

Gaius R. Shaver — The Ecosystem Center, Marine Biological Laboratory, Woods Hole, MA 02543, USA

Terje Skogland — Norwegian Institute for Nature Research, Trondheim, Norway

Mats Sonesson — Abisko Scientific Research Station, Abisko, Sweden, and Institute of Plant Ecology, University of Lund S-223 62 Lund, Sweden

Per Strand — Norwegian Radiation Protection Authority, N-1345 Østeraas, Norway

Bjartmar Sveinbjörnsson	Institute of Biology, University of Iceland, Reykjavík, Iceland, and Department of Biological Sciences University of Alaska, Anchorage, AK 99775, USA
Arcady A. Tishkov	Institute of Geography, Russian Academy of Sciences, Moscow 117811, Russia
George L. Vourlitis	Global Change Research Group, San Diego State University, San Diego, CA 92182, USA
Brian H. Walker	Global Change and Terrestrial Ecosystem, Core Project Office, Canberra, Australia
Ross W. Wein	Department of Forest Science, University of Alberta, Edmonton, Alberta T6G 2H1, Canada
Boris A. Yurtsev	Komarov Botanical Institute, Russian Academy of Sciences, St. Petersburg, Russia

Introduction
Challenges for the Future: Arctic and Alpine Ecosystems in a Changing World

William Dwight Billings

Beyond and above timberline, the cold-adapted low vegetations of arctic tundras and alpine fellfields are open to great changes in biotic structure: environmentally, floristically, faunistically, metabolically, and sociologically. The Arctic has been predicted to warm more rapidly and to a greater extent than the rest of the biosphere (Maxwell, 1992). Also, these once pristine and relatively isolated places are now subject to considerable disturbance, due mostly to increasing human populations and the use of industrial equipment and vehicles (Billings, 1973, 1979a, 1979b).

The plants dominating tundra vegetation in all its aspects differ in one fundamental way from all other plants on earth: they can carry on metabolic and reproductive processes at growing season temperatures only slightly above or even below 0°C. If growing seasons become longer and warmer, how will these species and populations fare?

Many environmental and ecosystemic problems are apparent in a changing Arctic. These represent challenges for attention now and in the future. If arctic and alpine ecosystems are to be sustainable, these problems must be attacked, solved, and such knowledge put to work in a practical way.

As a start, I have listed several of these arctic–alpine challenges in order of my own opinions as to their relative urgency:

1. The need for continuous monitoring of vegetation, soils, and environment on a number of scales.

2. The preservation of permafrost and prevention of thermokarst.
3. The problem of increases in greenhouse gases in the atmosphere.
4. Loss of stratospheric ozone, and resultant increase in ultraviolet-B radiation.
5. Tundra ecosystem carbon balance: relationships between atmospheric carbon dioxide, plant metabolism, soil carbon, and soil nitrogen.
6. Biotic migrational rates, distances, and invasions into arctic and alpine ecosystems.
7. Acclimation potentials of ecotypic plant populations.
8. The future of snow in arctic and alpine regions: a problem of water resources.
9. Last, but not least, there is the challenge, in the next century, of a longer, snow-free, and warmer growing season for tundra vegetation on cold soils that may be carbon rich but nutrient poor. Daylength and sun angles are the only constants through this long arctic summer season; but in future climates, solar radiation may actually decrease because of cloudiness while air temperatures increase. Will native tundras be replaced by trees or even agricultural crops?

The Need for Continuous Monitoring of Vegetation, Soils, and Environment

Of major importance to understanding the ecology of change in the biosphere is the use of improved and continuous instrumental monitoring of whole ecosystems at all levels, both physical and biological. Monitoring is necessary to provide advance warning, in terms of years and decades, of the effects of disturbances both biological and physical. Such disturbances are often the triggers that set off future ecosystemic change in tundras. Twentieth century science has been surprised far too often by the delayed effects of changes, such as increases in radioactivity, atmospheric carbon dioxide, destruction of stratospheric ozone with concomitant increases in ultraviolet-B (UV-B) radiation, and biological invasions that have occurred when nobody was ecologically aware of what was happening or expecting the consequences. Monitoring should include remote sensing and in situ measurements of the structure and fluxes within these ecosystems and the component biomes of the biosphere. Such monitoring should include mapping the extent of green vegetation and the amounts of chlorophyll in real time through the year.

Among the techniques, including remote and in situ sensing by appropriate instruments, should be ground mapping, repeat photography, and biological censusing at the community and population levels, both floristically and faunistically. Such techniques are informative of changes in community structure, biological invasions, and biodiversity. Urgently needed at the physiological level are more experimental and automated measurements of photosynthesis and respiration rates of arctic and alpine plants and their water-use efficiencies under increasing atmospheric carbon dioxide both in the field and under controlled conditions.

Permafrost Warming and Thermokarst

Most of the arctic tundra is underlain with permafrost. Such permanently frozen ground has existed for thousands of years, long before Holocene warming began. In the coastal wet tundra on marine sediments, the permafrost is also "wet" and is characterized by white ice lenses or ice wedges below shallow, peaty, wet soils and vegetation that insulate the permafrost during the tundra summers. The "active" layer of soil above the permafrost thaws briefly during the short arctic summer to depths of ca. 20 to 50 cm. It is in the active layer where plant root growth and nutrient uptake occur. The roots of some plants, notably various species of *Eriophorum*, follow the thaw down to the top of the permafrost table where they grow horizontally along the frozen surface at temperatures of 0°C or even lower (Billings et al., 1978).

Permafrost is a fundamental part of arctic tundra ecosystems, it is the "glue" that holds these ecosystems together. In the wet coastal tundra ecosystem of the North Slope of Alaska, large white ice wedges embedded in permafrost have formed extensive polygonal patterns during the thousands of years since the land rose above the sea. As tundra plants invaded these patterned and very cold soils, they formed vegetations that are themselves patterned on the ice wedge polygoned substratum. Such polygoned vegetation changes through several thousand years in a cyclic sequence (the thaw-lake cycle) that never reaches any stage that is completely stable (Billings and Peterson, 1980, 1992). This cycle of shallow lakes and terrestrial polygons is dependent on permafrost and its patterned ice wedges. If these ice wedges warm to temperatures above freezing; by increasing soil temperatures, by exposure to warmer air and flowing water because of climatic change, or disturbance by vehicles; the result is accelerated erosion (thermokarst) due to melting of the permafrost. It is significant that Lachenbruch and Marshall (1986) have reported increasing permafrost temperatures in the Alaskan arctic tundra. Recently, Oechel et al. (1993), based on weather data from Barrow and Prudhoe Bay, have concluded that summer air temperatures over the North Slope tundra have increased through the last 25 years. This does not bode well for the maintenance of permafrost and low soil temperatures as a carbon sink.

Monitoring of arctic soil and permafrost temperature is urgently necessary and of very high priority in understanding possible global warming. If, indeed, the Arctic is the geographic region in which temperature changes will be most pronounced in the next few decades, such temperatures will provide the earliest warning of global climatic change. Preservation of the permafrost and thereby protection of the overlying vegetation, is necessary if thermokarst and loss of the wet coastal tundra is to be prevented.

The problems created by thermokarst and the loss of permafrost in the wet coastal tundra have been shown by Billings and Peterson (1992) along Footprint Creek near Barrow, Alaska. Accelerated thermokarst along this creek is an example of what only slightly warmer waters can do to a pristine tundra ecosystem, the "Voth Tundra." Much of this previously undisturbed wet tundra of ca. 14 ha has been destroyed by thermokarst during the last 30 to 40 years, due primarily to the artificial drainage of water still flowing by ditch into Footprint Creek from four thaw-lakes artificially drained in 1950 to protect some natural gas wells east of the village. Billings and Peterson (1992) measured the loss of 1050 metric tons of soil, ground ice, and vegetation that has been washed out into the Chukchi Sea from the Voth Tundra through Footprint Creek during the 11 summers between 1977 and 1988. This is only a small sample of what can happen when such tundra ecosystems and their permafrost are disturbed: the Voth Tundra is gone. If the upper 2 m of permafrost are lost by melting for whatever reason, the coastal tundra is lost. Tundra vegetation and permafrost have been inseparable for thousands of years. If the insulation provided by such vegetation is damaged or destroyed by any disturbance, the underlying permafrost is doomed.

The Problem of Increases in Greenhouse Gases in the Atmosphere

The various natural "greenhouse" gases in the atmosphere (carbon dioxide, methane, water vapor, nitrous oxide, etc.) have fluctuated in concentrations during the Pleistocene and Holocene as climates and vegetation have changed (Dansgaard and Oeschger, 1989). Principal among the increasers is carbon dioxide. Analyses of ice core bubbles in Greenland and Antarctica indicate CO_2 concentrations varied during the glacials and interglacials of the Pleistocene from ca. 190 to ca. 300 ppmv. The preindustrial benchmark of atmospheric CO_2 during the 18th and early 19th centuries, as measured directly and in ice cores, was near 275 to 280 ppmv (Siegenthaler and Oeschger, 1987). In the last 200 years, atmospheric CO_2 concentrations have risen at a steadily increasing rate.

The evidence for this CO_2 increase is from staightforward monitoring. Such continuous monitoring began on the barren summit of Mauna Loa on the island of Hawaii in the mid-Pacific in 1958 by Keeling and his coworkers (1982). The notable Mauna Loa curve has been the standard for 35 years. In 1958, the CO_2 in the atmosphere was at ca. 316 ppmv; it is presently at ca. 355 ppmv. The Mauna Loa measurements have been augmented now by those from many other CO_2 monitoring stations scattered over the surface of the earth, including one near our research area at Barrow, Alaska. The basis for the global warming hypothesis rests on these CO_2 data obtained from years of steady monitoring at Mauna Loa and the other stations. Such careful and long-term measurements of CO_2 and the other greenhouse gases must be continued.

Loss of Stratospheric Ozone and Resultant Increase in Ultraviolet Radiation

Similar monitoring of the synthetic chlorofluorocarbons should also continue. The chlorofluorocarbons are two-edged swords because they are not only greenhouse gases but also break down the stratospheric ozone layer that shields the earth from much of the biologically damaging solar ultraviolet-B radiation (Cicerone, 1987; Rowland, 1989).

Such radiation is not only a serious problem for human populations but also for plants and vegetation (Caldwell et al., 1982; Billings, 1984). Caldwell, Robberecht, and Billings (1980) measured solar UV-B radiation along the arctic–alpine life zone beyond and above timberline from Barrow, Alaska, down the Rocky Mountains of North America to the Andes of Venezuela and the central Andes of Peru. Total daily shortwave solar radiation varied along this gradient only by a factor of less than 15% for the entire length of the gradient. In contrast, total daily effective UV-B radiation varied by sevenfold from low values in the Barrow tundra to the paramo vegetation of the equatorial Andes. As of the late 1970s, then, UV-B irradiance in that part of the Arctic was quite low compared to that received by alpine plants in the Rocky Mountains, the tropical Andes, and on the summit of the volcano Haleakala on Maui in Hawaii, where we also measured incoming solar UV and plant response. All measurements were made at each location at the time of highest noon sun in the year 1977.

Native plants on all these high mountains in these middle latitudes and the tropics have evolved reflectance and absorption barriers to the penetration of UV-B radiation to the leaf chlorophyllous tissues. Some arctic plant populations, when put in higher UV-B environments in the laboratory or phytotron, were much more sensitive to damage of the photosynthetic tissues and processes (Robberecht et al., 1980; Caldwell et al., 1982). One of these species was the widespread arctic–alpine plant *Oxyria digyna* in which photosynthetic inhibition was significantly greater in the arctic ecotypes than in those from the high alpine sites in Colorado and California. The challenging problem in these cases lies in knowing what will happen to arctic plants if the stratospheric ozone hole enlarges. Will the arctic plant populations be severely injured or will they acclimate in various ways to higher UV-B irradiation? We don't know the answers yet.

The increases in CO_2, methane, water vapor, and chlorofluorocarbons are at the root of the estimates of global warming and the role of temperature in biospheric change. One does not have to emphasize the importance of globally monitoring the atmospheric concentrations of these gases through time, and of the inputs of the gases into and through the atmosphere. More research must be initiated and continued on what can be done to lower input rates, and also how to augment the arctic and alpine ecosystemic sinks.

Tundra Ecosystem Carbon Balance: Relationships Between Atmospheric Carbon Dioxide, Plant Metabolism, Soil Carbon, and Soil Nitrogen

Simply put, the carbon balance of an ecosystem is the difference between carbon capture as CO_2 in photosynthesis by the ecosystem's green vegetation at the primary producer trophic level and the respiratory loss of CO_2 at all trophic levels, including green plants, herbivores, carnivores, and decomposer non-green plants and monerans. The efflux of methane is also part of the equation, as are the inorganic import of carbon and mass export of carbon compounds in water or wind.

The measurement of an ecosystem's carbon balance with any precision is far from simple, and in most cases is very difficult. Oechel and Lawrence (1985) have stated it succinctly: "Calculating the carbon balance for a single plant is difficult; attempting to estimate the carbon balance for an ecosystem such as the taiga is rudimentary at best." Obtaining a rough carbon budget for any ecosystem requires, as a minimum, quantitative data on the following parameters: (1) gross and net photosynthesis of the whole plant community, (2) the loss of carbon by dark respiration and photorespiration of the whole biotic community, including the green vegetation, its leaves, stems, flowers, fruits, and seeds with their export rates and decomposition, all belowground roots, rhizomes, bulbs, tubers, and other vascular plant carbon storage sinks. To this must be added the carbon metabolism of all the cryptogams: algae, bryophytes, lichens, (3) the fluxes and sinks of all other trophic levels, including decomposers and pathogens such as bacteria and fungi, herbivores and carnivores, including mosquitos!

We attempted such a carbon balance for all trophic levels in the wet coastal tundra ecosystem at Barrow, Alaska, during the IBP days of the early 1970s (Chapin et al., 1980). At that time, the measurements showed a carbon budget where photosynthetic carbon capture in $g\,C\,m^{-2}\,yr^{-1}$ was +214 g Cm yr, respiratory and export carbon loss was $-105\,g\,C\,m^{-2}\,yr^{-1}$, and carbon balance was positive into the system at $+109\,g\,C\,m^{-2}\,yr^{-1}$ (see Billings, 1987). Essentially, the arctic tundra is a "belowground ecosystem" when one considers the large biomass of storage organs such as roots, stembases, and rhizomes. So, most of this positive balance of carbon went into the cold soil as live and dead roots, rhizomes, and litter. Live roots, rhizomes, and stembases contain high quantities of stored carbohydrates. These are harvested and stored by lemmings that eat them under the snow during the winter. Eskimo (Inupiat) children used to gather and eat these "mouse nuts." Do they still?

Most of the carbon in the wet, cold soil eventually becomes peat in the "active layer." In the past, the insulating effect of the dead litter and live vegetation has allowed the permafrost table to rise into the peat and preserve the organic carbon for centuries until sporadic warming initiates thermokarst. Then decomposition is renewed aerobically with an efflux of CO_2 (Peterson and Billings, 1975; Billings and Peterson, 1992).

In the geologic and historic past, the Alaskan arctic tundra has been a net sink for CO_2 and biologically produced carbon compounds (Chapin et al., 1980; Billings, 1987). A large taiga bog of black spruce (*Picea mariana*) and *Sphagnum* moss in central Alaska also was a net sink for carbon through thousands of years and as recently as the early 1980s (Billings, 1987). The result is a carbon-rich peat 3 to 4 m deep throughout a bog ecosystem with an area of ca. 20 km^2. This peat has been maintained for several millennia in the frozen soils of this taiga wetland near Fairbanks. In estimates of global soil carbon pools, the combined arctic tundra–boreal forest global soil carbon pool estimates range from 21% (Raich and Schlesinger, 1992) to ca. 27% (Post et al., 1982) of the total global soil carbon pool, or approximately one quarter of the earth's soil carbon. In a warming earth, this represents a ticking time bomb of potential atmospheric CO_2 that has been stored frozen in tundra and muskeg, and out of atmospheric circulation, for several thousand years (Billings, 1987). Billings et al. (1982, 1983, 1984), in prediction experiments with intact frozen tundra cores from Barrow, Alaska, have shown that the postulated increase in summer temperatures at Barrow will turn that wet tundra from a sink to a source of CO_2. Also, if the depth of thaw increases, the water table in that tundra will drop from the surface to depths of 10 to 20 cm or more. As we have shown, if this happens, efflux of decompositional CO_2 will increase dramatically (Billings et al., 1983). This may be happening now in the tundra of the North Slope, and perhaps elsewhere. Oechel et al. (1983) have reported recent warming of this tundra region, with concomitant loss of CO_2 to the atmosphere from the tundra ecosystem. Their results suggest that this may be acting as a positive feedback on the concentration of atmospheric CO_2 and increased warming, as we predicted that it might (Billings et al., 1982).

In a third microcosm experiment in the phytotron using a group of our frozen tundra cores from Barrow, we found that the addition of nitrogen to the thawing microcosms (nitrogen being the present limiting soil nutrient factor in that tundra) had far more effect on ecosystem net carbon gain than increasing the atmospheric CO_2 concentration over the cores (Billings et al., 1984).

Our experimental microcosm results under controlled conditions have been confirmed under field conditions in tussock tundra at Toolik Lake north of the Brooks Range, Alaska, by Tissue and Oechel (1987). They conclude that growth of the principal dominant graminoid there

(*Eriophorum vaginatum*) is more limited by soil nutrient supply than by carbon availability for photosynthesis. The tundra plants growing in the field at Toolik Lake under small "greenhouses" at elevated CO_2 concentrations and those at ambient CO_2 levels had similar photosynthetic rates at their respective growth CO_2 concentrations within the natural tundra vegetations. The reduction of photosynthetic capacity for plants growing at elevated CO_2 levels did not appear to be due to stomatal closure or end-product inhibition.

Shaver et al. (1992) also conclude that changes in atmospheric CO_2 concentrations have relatively little long-term effect on whole-system carbon budgets and productivity of wet and moist arctic tundras. They present a conceptual model of carbon/nutrient interactions in terrestrial ecosystems using carbon and nitrogen. Such a model should be useful in studying the effects of climatic change on other terrestrial nutrient-limited ecosystems.

For these arctic tundras, perhaps the most important factor is the potential for redistribution of the large pool of nitrogen in soils with low C/N ratios and its uptake by vegetation with higher C/N ratios. If climatic warming should be the cause of soil CO_2 efflux in northern ecosystems, those carbon losses may be offset by the release of soil nutrients by increasing depth of thaw resulting in plant carbon gains in photosynthesis (Shaver et al., 1992; Schimel et al., 1990).

In summary, in regard to arctic tundra carbon balance now and in the future, the most probable effects of higher CO_2 levels on the Alaskan tundra will be through its indirect effects on temperature: lowering water table levels, thus allowing oxygenation of the peat and increasing soil organic matter decomposition. Carbon dioxide is not now limiting to tundra ecosystem productivity, but soil nutrients and length of snow-free growing season can be limiting there and will likely continue to limit productivity of this cold arctic ecosystem. This whole carbon balance problem is extremely complex. It is a real challenge that needs much research attention.

Biotic Migrational Rates, Distances, and Invasions Into Arctic and Alpine Ecosystems

At present, the most readily available information for northward migration rates of species of plants in a warming climate is based on Holocene paleoecological macrofossils and microfossils including pollen grains and spores as discussed in several publications (Overpeck, Bartlein, and Webb, 1991; Davis, 1989; Davis and Zabinski, 1992; Jacobson, Webb, and Grimm, 1987). This kind of research is extremely valuable as an analogue to what may happen in the next century or two if the climates of the Northern Hemisphere warm even more dramatically than they did during parts of the Holocene. Such paleoecological research is needed and must be continued. Additionally, however, migrational rates, distances, directions, and resistances based on quantitative measurements of present living plant populations in biotic communities are necessary. Such hard data would use aerial and/or space photography and radar. But quantitative "ground truth" data based on plainly marked permanent transects, plots, and mapping also will be required from vegetation ecologists who are competent in plant identification under field conditions. Also, I am a firm believer in carefully done repeat photography from surveyed points toward targets on the ground; such photographs have been very useful in my own work for many years.

Of comparable importance to migration rates of species of plants and animals is the other side of the coin: the death and possible extinction of local populations of species in biological communities that are changing floristically or disintegrating ecologically due to climatic change, rise in sea level, loss of permafrost, and loss of soil by thermokarst erosion. Simultaneously with climatic change, there often is increased competition from the northward migrations of native and/or exotic weedy species of plants and animals as ecosystems become warmer. At the present time, the tundra around Barrow, Alaska, still has no adventive weedy plant species derived from central Alaska. But the road to Prudhoe Bay, with considerable truck traffic from Fairbanks, provides ready access to plant "hitchhikers" and probably has a few adventive species invading the Arctic by now. It is surprising that the airplane flights into Barrow have not brought in more adventive propagules. Maybe they do, but germination and establishment are not successful because of low soil temperatures. At Barrow, we have not yet observed such adventives from the taiga biome of central Alaska.

The upward migration of timberlines in the western North American Cordillera could eventually result in smaller and less continuous mountain peak and high ridge "islands." Such vegetational change upward from the subalpine zone would decrease the opportunities for migration of the truly alpine biota either north or south along the crest of the Rocky Mountains. In colder climates of the past, these alpine Cordilleran routes were open and available to many cold-tolerant species of plants, a situation that has already been reduced in the last two or three centuries. Some alpine species will become extinct, locally at least, but others migrating up from below may take their places and thus change the composition and functioning of the ecosystem.

As an example of this latter situation, which is occurring in the Alps, Hofer (1992) tabulated lists of vascular plant species on 14 peaks in the Bernina region of the Swiss Alps between 1905 and 1985. On 12 out of the 14 peaks, in the subnival and nival zones, significant immigration of species from below has occurred in that 80-year period. All the adventive species now present at the end of those 80 years are typically montane, alpine, or arctic taxa with seeds mostly distributed by wind. The average number of species per peak has increased from 16 to 28. Hofer attributes this increase to the rise in air temperature on these mountains in this century and the retreat of glacial and firn ice. The rise in temperatures between 1911 and 1990 in the higher zones of the Alps is ca. 0.6 to 0.8°C. Rübel (1912), reporting on his 1905 survey, which is the initial basis of Hofer's comparison, listed 70 vascular plant species above timberline on 20 Bernina peaks. Hofer, using only 14 of these 20 peaks, found 108 species. Only one peak, Piz Languard at 3261 m, has fewer species than found by Rübel in his 1905 census. Apparently, in the Bernina Alps, immigration from below has dominated considerably over extinction during this century.

Very recently, Grabherr et al. (1994), in the Austrian and eastern Swiss Alps, using 26 alpine summits exceeding elevations of 3000 m, found a similar situation of plant species migrating upward. They measured vascular plant cover, abundance, and species richness on these peaks and summits in 1992 and compared these data with historical records of similar data earlier in this century. From the very precise historical records, compared to present distributions of nine typical nival plant species, Grabherr et al. calculated that maximum rates of upward migration on these 26 peaks were close to 4 m of elevation per decade. However, most rates were near 1 m. Within the 20th century, the Austrian meteorological stations show that the mean annual air temperature of this alpine region has increased by 0.7°C. These authors ascribe the upward migration of these plant species to this moderate warming trend. They agree with my hypothesis (Billings, 1993, and as stated above for the North American Cordillera) that if this warming continues there could be multiple extinctions of alpine plant species on the higher peaks worldwide.

Present models of how plant species could move northward or upward on mountains in a CO_2-rich world of the next century are rather simplistic. Most models assume that there are almost no resistances in the north to invasions or migrations from the south. Actually, there are a number of such resistances. These include:

- Glacially scoured rocks or thin soils.
- Acid podzols with low supplies of nutrients.
- Cold soils underlain with permafrost.
- Boggy and wet peatlands.
- Existing boreal forests that shade out seedlings.
- The lack of appropriate bird or mammal seed vectors.
- Daylength problems with flowering, leaf, and bud dormancy mechanisms; latitudinally controlled daylength and solar angle parameters are tied in with plant genetic structure.
- Increased chance of severe frost during the growing season.
- Lack of sufficient seed sources from local ecotypes.

Such resistance information and quantitative estimates need to be incorporated into the models.

The present vegetation of Canada has been produced only during the last several thousand years by plant migrations from southern refugia since deglaciation of the Wisconsinan and other ice sheets. Even after these many thousands of years, the floristic richness of eastern Canada is relatively low compared to the rich floras of the unglaciated southeastern quarter of the North American continent. These past migrations northward from refugia far to the south, however, have been natural ones under relatively low atmospheric CO_2 concentrations and have occurred during relatively steady and slow climatic changes. The next two centuries will probably see much faster warming and certainly much higher levels of atmospheric CO_2. The migrations of plant populations and species will probably not keep up with such warming in which the only constant will be the unchanged length of summer days unique to each latitude. This lack of photoperiodic, temperature, and soil synchrony may lead to sparsely vegetated areas where the old original vegetation has died out but more southerly species have not yet arrived. One other prospect is that adventive species from other continents, including herbaceous weeds, may occupy the floristically poor gaps or disturbed areas. It is quite likely that wholly new vegetation types will result in such places. Billings (1990) has described such a situation after the invasion of the European annual grass *Bromus tectorum* into the cold mountains of the Great Basin in western North America. These postulated new vegetation types possibly could be poor habitats for the native birds and mammals of arctic ecosystems. Thus, the problems posed by migrations and invasions call for a great deal of thought, broad-based ecological research, and planning.

Acclimation Potentials of Ecotypic Plant Populations

Oxyria digyna, alpine sorrel, is a widespread perennial arctic–alpine plant species in the Polygonaceae. My research group started working with it in 1958 to find out why one species could be so successful in growth and reproduction over such a wide range of latitudes (from 28°N to 83°N in Pearyland, Greenland, and from sea level in the Arctic to elevations of over 3800 m in the middle latitude alpine regions) (Mooney and Billings, 1961; Billings et al., 1965; Billings et al., 1971).

Using controlled conditions in the Duke Phytotron, in which we grew ca. 35 widely distributed arctic and alpine populations of *Oxyria* from seed to flowering and fruiting status, we have, over the years, come to the following conclusions concerning its ability to prosper in such a widespread array of cold summer climates. In brief:

1. The higher the latitude of origin, the longer the photoperiod needed for flowering.
2. Gradual reduction of photoperiod over several weeks results in production of a perennating bud containing a pre-formed inflorescence that will emerge at the start of summer in the following year. Arctic ecotypes

go into this dormancy at longer photoperiods than do alpine populations from the middle latitudes.

3. Plants of alpine ecotypes grow faster at higher temperatures whereas arctic ecotypes have lower optimum temperatures for growth and flowering.
4. Dark respiration rates at any experimental temperature in the laboratory are increased by cold acclimation in all ecotypes. But actual respiration rates in the field may not be higher than in warm-acclimated plants due to the colder temperatures that govern metabolism under field conditions. Thus, dark respiration homeostasis is almost ideal in all populations and shows no ecotypic difference. This assures the provision of metabolic energy to cells even under very cold conditions (Körner and Larcher, 1988).
5. Maximum net photosynthesis rates are lower after warm acclimation in all arctic populations but in only a few alpine populations.
6. The temperature of maximum net photosynthesis is increased by warm acclimation.
7. Maximum net photosynthesis rates are higher in arctic ecotypes than in alpine ones; this difference is increased by cold acclimation.
8. Alpine ecotypes show ideal homeostasis in net photosynthesis whereas arctic ecotypes show only a low degree of partial homeostatic adjustment in net photosynthesis.
9. Efficiency of chlorophyll is higher in alpine ecotypes.
10. Cold acclimation increases Hill reaction activity by isolated chloroplasts from alpine ecotypes when the plants were grown under a 14-h photoperiod; the results are less clear at longer daylengths.
11. Mitochondrial oxidative activity is increased by cold acclimation in the laboratory with such rates being higher in arctic ecotypes. As with acclimation of dark respiration, field temperatures determine actual mitochondrial activity of acclimated plants under natural conditions.

We conclude that in *Oxyria digyna* acclimation is under genetic control and that "acclimation ecotypes" exist. The most plastic phenotypes in regard to acclimation occur in the alpine ecotypes. In a warming Arctic, more research of this type with other species, especially the dominants, is very much needed.

The Future of Snow in Arctic and Alpine Regions: A Problem of Water Resources

People in all biomes are dependent on sufficient liquid water supplies for domestic supplies and agriculture. This applies as well to the Arctic, particularly during the winter when most of the water is frozen. More research is needed on ecosystem water balance and the roles of snow and ice in energy balance in those cold climates that may become warmer with simultaneous increase in vegetational cover and total leaf area. The usual water source of local precipitation, runoff from regional mountains, lakes, aquifers, and other groundwaters are likely to be subject to considerable stress and change in a warming climate, maybe for the better—or maybe not.

The roles of vegetational patterns involved in the catchment, short-term maintenance, and hydrology of arctic, alpine, and subalpine snowdrifts, snowpacks, their meltwater yield, and contents cannot be ignored in the consideration of any scenarios of regional warming effects in the middle and high latitude mountains (Billings, 1969, 1973, 1993; Walker et al., 1993; Walker et al., 1996). This is of particular importance in those mountain ranges bordering regions that are dependent on mountain runoff as their principal water source. Research on these snow ecology problems is still very much needed, in spite of the excellent record of snow hydrologists throughout this century in Europe and North America. I would be remiss not to mention the classic pioneer research on snow ecology and snow surveys by Dr. J.E. Church in the Sierra Nevada of California and Nevada more than 90 years ago. Church set the stage for our present knowledge of the interactions between snow and mountain vegetation.

Conclusions: Attempting to Predict Arctic and Alpine Futures

The key word in this discussion of challenges for research on arctic ecosystems in an unstable, changing biosphere is "change." I shall attempt to paraphrase the distinguished ecologist William S. Cooper (1926): "Change in vegetation and ecosystems is like a braided glacial river flowing through time." The complexities of the separations, rejoinings, and new combinations of plant and animal species are almost limitless and filled with surprises.

One cannot view the biosphere provincially either in space or time. There are no crosswalls or real barriers to biotic migrations or environmental changes on the present earth. This is even more true of regional ecosystems or biomes. Limiting factors to the successful growth and reproduction of biotic systems exist and are unique to specific ecotypic populations. And, within ecosystems, there is an eventual "carrying capacity" for people and animals. This is also true globally. Carrying capacity is determined, of course, by primary productivity of the vegetation and the nutrient availability of the soil, and the length of the growing season.

Any change in a limiting factor may set off long-lasting chain reactions in an ecosystem or in the biosphere. The initiator of such a chain reaction is a "trigger factor" (Billings, 1952). A trigger factor can be either a changed limiting factor, for example, sudden warming of the Arctic, or it can be an entirely new environmental factor, a "surprise." An example of the latter is the destruction of stratospheric ozone by CFCs, a man-caused surprise. The ultimate future of surprises and possible chain reaction results are generally still unknown. Such are real ecological challenges.

The biosphere has changed continually during the past in both space and time, and will continue to do so. Some of these changes have been very slow and over long time periods. Other changes have been quite rapid, for example, the onset and termination of the Younger Dryas in the Arctic and North Atlantic region, as shown by both marine and ice core data. Both onset and termination of this late-glacial event occurred within elapsed times of 10 to 20 years from near-glacial conditions to relatively warm and vice versa (Mayewski et al., 1993). Such sudden changes in temperature and atmosphere in mid-Holocene were natural and, of course, preindustrial. Many natural events are episodic and present major problems: earhquakes, floods, droughts, and volcanic eruptions. An example: Mt. Pinatubo's eruption in the Philippines in 1991 changed temperatures in the northern middle latitudes and also detrimentally affected the stratospheric ozone layer in those same latitudes. Global average total ozone in 1992 was 2 to 3% lower than in any previous year (Gleason et al., 1993). The atmospheric ash and dust from Pinatubo is a main suspect.

In the last two to three centuries, within the span of the Industrial Revolution, there has been a "speed up" of environmental and biospheric change that, in many cases, is almost exponential in progression. Much of this acceleration has been caused by well-documented changes in composition and temperature of the atmosphere. Extinction and migration of the biota are also increasing, much of this due to deforestation. As a result, these latter centuries have been filled with biospheric surprises that were not expected, not predicted, and that have changed our world.

In the last decade, much attention has been given to the question, "Are ecosystems and the biosphere sustainable?" (Clark and Munn, 1986; Lubchenco et al., 1991). This is not a trivial question. The future of the earth as a habitable place depends upon the answer. Lubchenco and her committee propose three research priorities: (1) ecological causes and consequences of global change in all its aspects, (2) ecological determinants and consequences of biological diversity and the effects of global and regional changes on such diversity, and (3) the management of sustainable ecological systems and the interface between ecological processes and human social systems. The implementation of this very important and complex research program on earth ecology is spelled out in detail in Lubchenco's committee report.

The most succinct and to the point approach to the question of the principles of ecosystem sustainability is that of Chapin (1993). He defines "a sustainable ecosystem as one . . . that maintains its characteristic diversity of major functional groups, productivity, and rates of biogeochemical cycling in the face of some disturbance." Such a system occurs when negative feedbacks tend to maintain the current characteristics of an ecosystem; positive feedbacks hasten change away from stability and the status quo.

Can future biospheric characteristics and interactions be predicted? At this period in time, estimates of approximate values and fluxes can be attempted by monitoring changes of all kinds on a global basis, by experiments both in the field and under controlled conditions using microcosms of native species, and by mathematical modeling. Facts are necessary to all these approaches. We must conserve all sources of these facts ("information") before they are lost forever: genetic, environmental, populational, communities, ecosystemic, and people, all in the broadest sense. And we must be ready for surprises.

Beyond all this, cooperation is needed between and among all the sciences, including ecology and economics, between the sciences and politics ("governments"), and among nations. All of these problems concern the future of a livable earth. The "braided stream" flows on, and ever faster.

Acknowledgment. I thank Professor Dr. Christian Körner for his critical reading of an earlier version of this manuscript and his helpful suggestions. I also thank an unknown reviewer for comments that have made this a better contribution. And, most of all, to my wife, Shirley, for reading and suggesting improvements in the manuscript's many versions.

References

Billings, W.D. 1952. The environmental complex in relation to plant growth and distribution. *Q. Rev. Biol.* 27:251–265.

Billings, W.D. 1969. Vegetational pattern near alpine timberline as affected by fire–snowdrift interactions. *Vegetatio* 19:192–207.

Billings, W.D. 1973. Arctic and alpine vegetations: Similarities, differences, and susceptibility to disturbance. *BioScience* 23:697–704.

Billings, W.D. 1979a. High mountain ecosystems: Evolution, structure, operation, and maintenance. In: Webber, P.J. (ed.), *High Altitude Geoecology* (Chap. 6, pp. 97–125). AAAS Selected Symposium 12. Boulder, CO: Westview Press.

Billings, W.D. 1979b. Alpine ecosystems of western North America. In: Johnson, D.A. (ed.), *Special Management Needs of Alpine Tundra Ecosystems* (pp. 6–21). Denver, CO: Range Management Society.

Billings, W.D. 1984. Effects of UV-B radiation on plants and vegetation as ecosystem components. In: National Research Council (ed.), *Causes and Effects of Changes in Stratospheric Ozone: Update 1983* (Chap. 13, pp. 206–217). Washington, DC: National Academy Press.

Billings, W.D. 1987. Carbon balance of Alaskan tundra and taiga ecosystems: Past, present, and future. *Quat. Sci. Rev.* 6:165–177.

Billings, W.D. 1990. *Bromus tectorum*, a biotic cause of ecosystem impoverishment in the Great Basin. In: Woodwell, G.M. (ed.), *The Earth in Transition: Patterns and Processes of Biotic Impoverishment* (Chap. 15, pp. 300–322). New York: Cambridge University Press.

Billings, W.D. 1993. The effects of global and regional environmental changes on mountain ecosystems. In: Despain, D.G. (ed.), *The Greater Yellowstone Ecosystem.* Denver, CO: U.S. National Park Service.

Billings, W.D., and Peterson, K.M. 1980. Vegetational change and ice-wedge polygons through the thaw-lake cycle in arctic Alaska. *Arctic Alpine Res.* 12:413–432.

Billings, W.D., and Peterson, K.M. 1992. Some possible effects of climatic warming on arctic tundra ecosystems of the Alaskan North Slope. In: Peters, R., and Lovejoy, T.E. (eds.), *Climatic Warming and Biological Diversity* (Chap. 18, pp. 233–243). New Haven: Yale University Press.

Billings, W.D., Godfrey, P.G., and Hillier, R.D. 1965. Photoperiodic and temperature effects on growth, flowering, and dormancy of widely distributed populations of *Oxyria*. *Bull. Ecol. Soc. Am.* 46:189 (abstract).

Billings, W.D., Peterson, K.M., and Shaver, G.R. 1978. Growth, turnover, and respiration rates of roots and tillers in tundra graminoids. In: Tieszen, L.L. (ed.), *Vegetation and Production Ecology of an Alaskan Arctic Tundra* (Chap. 18, pp. 415–434). Ecological Studies 29. New York: Springer-Verlag.

Billings, W.D., Godfrey, P.J., Chabot, B.F., and Bourque, D.P. 1971. Metabolic acclimation to temperature in arctic and alpine ecotypes of *Oxyria digyna*. *Arctic Alpine Res.* 3:277–289.

Billings, W.D., Luken, J.O., Mortensen, D.A., and Peterson, K.M. 1982. Arctic tundra: A source or sink for atmospheric carbon dioxide in a changing environment? *Oecologia* 53:7–11.

Billings, W.D., Luken, J.O., Mortensen, D.A., and Peterson, K.M. 1983. Increasing atmospheric carbon dioxide: Possible effects on arctic tundra. *Oecologia* 58:286–289.

Billings, W.D., Peterson, K.M., Luken, J.O., and Mortensen, D.A. 1984. Interaction of increasing atmospheric carbon dioxide and soil nitrogen on the carbon balance of tundra microcosms. *Oecologia* 65:26–29.

Caldwell, M.M., Robberecht, R., and Billings, W.D. 1980. A steep latitudinal gradient of solar ultraviolet-B radiation in the arctic-alpine life zone. *Ecology* 61:600–611.

Caldwell, M.M., Robberecht, R., Nowak, R.S., and Billings, W.D. 1982. Differential photosynthetic inhibition by ultraviolet radiation in species from the arctic-alpine life zone. *Arctic Alpine Res.* 14:195–202.

Chapin, F.S., III. 1993. Principles of ecosystem sustainability. *Bull. Ecol. Soc. Am.* 74(2):189–190 (abstract).

Chapin, F.S., III, Miller, P.C., Billings, W.D., and Coyne, P.I. 1980. Carbon and nutrient budgets and their control in coastal tundra. In: Brown, J., Miller, P.C., Tieszen, L.L., and Bunnell, F.L. (eds.), *An Arctic Ecosystem; The Coastal Tundra at Barrow, Alaska* (Chap. 12, pp. 458–482). Stroudsburg, PA: Dowden Hutchinson, Ross, Inc.

Cicerone, R.J. 1987. Changes in stratospheric ozone. *Science* 237:35–42.

Clark, W.C., and Munn, R.E. (eds.). 1986. *Sustainable Development of the Biosphere*. New York: IIASA and Cambridge University Press.

Cooper, W.S. 1926. The fundamentals of vegetational change. *Ecology* 7:391–413.

Dansgaard, W., and Oeschger, H. 1989. Past environmental long-term records from the Arctic. In: Oeschger, H., and Langway, C.C., Jr. (eds.), *The Environmental Record in Glaciers and Ice Sheets* (pp. 287–318). New York: John Wiley and Sons, Ltd.

Davis, M.B. 1989. Lags in vegetation response to greenhouse warming. *Climatic Change* 15:75–82.

Davis, M.B., and Zabinski, C. 1992. Changes in geographical range resulting from greenhouse warming: Effects on biodiversity in forests. In: Peters, R.L., and Lovejoy, T.E. (eds.), *Global Warming and Biological Diversity* (Chap. 22, pp. 297–308). New Haven: Yale University Press.

Gleason, J.F., et al. 1993. Record low ozone in 1992. *Science* 260:523–526.

Grabherr, G., Gottfried, M., and Pauli, H. 1994. Climate effects on mountain plants. *Nature* 369:448.

Hofer, H.R. 1992. Veranderungen in der Vegetation von 14 Gipfeln des Berninagebietes zwischen 1905 und 1985. *Berichte Geobot. Inst.* ETH Rübel, Zürich 58:39–54.

Jacobson, G.L., Jr., Webb, T., III, and Grimm, E.C. 1987. Patterns and rates of vegetation change during the deglaciation of eastern North America. In: Ruddiman, W.F., and Wright, H.E., Jr. (eds.), *North America and Adjacent Oceans During the Last Deglaciation. The Geology of North America* (Vol. K-3, Chap. 13, pp. 277–288, 2 map plates). Boulder, CO: The Geological Society of America.

Keeling, C.D., Bacastow, R.B., and Whorf, T.P. 1982. Measurements of the concentration of carbon dioxide at Mauna Loa Observatory, Hawaii. In: Clark, W.C. (ed.), *Carbon Dioxide Review: 1982* (pp. 377–385). New York: Oxford University Press.

Körner, Ch., and Larcher, W. 1988. Plant life in cold climates. In: Long, S.F., and Woodward, F.I. (eds.), *Plants and Temperature* (Vol. 42, pp. 25–57). Cambridge, UK: Society for Experimental Biology.

Lachenbrach, A.H., and Marshall, B.V. 1986. Changing climate: Geothermal evidence from permafrost in the Alaskan Arctic. *Science* 234:689–696.

Lubchenco, J., et al. 1991. The sustainable biosphere initiative: An ecological research agenda. *Ecology* 72:371–412.

Maxwell, B. 1992. Arctic climate: Potential for change under global warming. In: Chapin, F.S., III, Jefferies, R.L., Reynolds, J.F., Shaver, G.R., and Svoboda J. (eds.), *Arctic Ecosystems in a Changing Climate: An Ecophysiological Perspective* (Chap. 2, pp. 11–34). San Diego: Academic Press, Inc.

Mayewski, P.A., et al. 1993. The atmosphere during the Younger Dryas. *Science* 261:195–197.

Mooney, H.A., and Billings, W.D. 1961. Comparative physiological ecology of arctic and alpine populations of *Oxyria digyna*. *Ecol. Monogr.* 31:1–29.

Oechel, W.C., and Lawrence, W.T. 1985. Taiga. In: Chabot, B.F., and Mooney, H.A. (eds.), *Physiological Ecology of North American Plant Communities* (Chap. 4, pp. 66–94). New York: Chapman and Hall.

Oechel, W.C., Hastings, S.J., Vouritis, G., Jenkins, M., Riechers, G., and Grulke, N. 1993. Recent change of arctic tundra ecosystems from a net carbon dioxide sink to a source. *Nature* 361:520–523.

Overpeck, J.T., Bartlein, P.J., and Webb, T., III, 1991. Potential magnitude of future vegetation change in eastern North America: Comparisons with the past. *Science* 254:692–695.

Peterson, K.M., and Billings, W.D. 1975. Carbon dioxide flux from tundra soils and vegetation as related to temperature at Barrow, Alaska. *Am. Midland Nat.* 94:88–98.

Post, W.M., Emanuel, W.R., Zinke, P.J., and Stangenberger, A.G. 1982. Soil carbon pools and world life zones. *Nature* 298:156–159.

Raich, J.W., and Schlesinger, W.H. 1992. The global carbon dioxide flux in soil respiration and its relationship to vegetation and climate. *Tellus* 44B:81–99.

Robberecht, R., Caldwell, M.M., and Billings, W.D. 1980. Leaf ultraviolet optical properties along a latitudinal gradient in the arctic-alpine life zone. *Ecology* 61:612–619.

Rowland, F.S. 1989. Chlorofluorocarbons and the depletion of stratospheric ozone. *Am. Sci.* 77:36–45.

Rübel, E. 1912. *Pflanzengeographische Monogaphie des Berninagebietes*. Leipzig: Engelmann.

Schimel, D.S., Parton, W.J., Kittel, T.G.F., Ojima, D.S., and Cole, C.V. 1990. Grassland biogeochemistry: Links to atmospheric processes. *Climatic Change* 17:13–25.

Shaver, G.R., Billings, W.D., Chapin, F.S., III, Giblin, A.E., Nadelhoffer, K.J., Oechel, W.C., and Rastetter, E.B. 1992. Global change and the carbon balance of arctic ecosystems. *BioScience* 42:433–441.
Siegenthaler, U., and Oeschger, H. 1987. Biospheric CO_2 emissions during the past 200 years reconstructed by deconvolution of ice core data. *Tellus* 39B:140–154.
Tissue, D.T., and Oechel, W.C. 1987. Response of *Eriophorum vaginatum* to elevated CO_2 and temperature in the Alaskan tussock tundra. *Ecology* 68:401–410.
Walker, D.A., Billings, W.D., and DeMolenaar, J.G. 1996. Snow-vegetation interactions in arctic and alpine environments. In: Jones, H.G., Hoham, R.W., Pomeroy, J.W., and Walker, D.A. (eds.), *Snow Ecology.* Cambridge: Cambridge University Press (In press).
Walker, D.A., Halfpenny, J.C., Walker, M.D., and Wessman, C.A. 1993. Long-term studies of snow-vegetation interactions. *BioScience* 43:287–301.

1. Context for Global Change

1. Recent Climate Patterns in the Arctic

Barrie Maxwell

During the course of the past century, there has been an overall increase in global temperature of some 0.5°C. This is fairly uniformly reflected in both the Northern and Southern Hemispheres, as is the temporal pattern of temperature change. Thus, in the Northern Hemisphere, the coolest period of the past 100 years occurred in the early part of the 20th century; this was followed by a fairly abrupt warming during the 1920s, which culminated around 1940. Cooling occurred thereafter until the early 1970s, followed by another abrupt warming into the 1980s. The past decade has, in fact, been the warmest of the past 100 years with the five warmest individual years having been recorded during it.

Much of the interest in this trend in temperature has resulted from the issue of global warming. There is a general consensus (IPCC, 1990) that the natural greenhouse effect, which keeps the earth warmer than it would otherwise be, is being enhanced by human activities that are substantially increasing the atmospheric concentrations of the greenhouse gases (carbon dioxide, methane, chlorofluorocarbons, and nitrous oxide). This will result, on average, in an additional warming of the earth's surface. The marked warming of the past decade is certainly in agreement with the enhanced greenhouse gas effect, although it cannot be attributed directly to it as yet, as observed values are still within the range of historic variability.

Global circulation models (GCMs) have been used to develop climate scenarios under increased atmospheric concentrations of greenhouse gases.

These models suggest an increase in global mean temperature of 3°C before the end of the next century; they also indicate that high northern latitudes will warm more than the global mean in winter. This potential polar amplification of greenhouse-gas-induced global warming suggests that greenhouse effects may start to show up sooner in the Arctic than in other areas of the Northern Hemisphere. In this sense, the Arctic may well serve as a bellwether (Walsh, 1991). Therefore, careful monitoring of current arctic climate patterns will be judicious.

Data

This review focuses mainly on data covering the 1960 to 1990 period. Most of the material presented is drawn from the current literature and reference should be made to the individual papers for details of the data sets used. The global data set compiled by the Climate Research Unit at East Anglia, UK, is used for the regional analyses of temperature and precipitation included in this chapter. In most cases, information is presented in terms of anomalies relative to the 1951 to 1980 mean values.

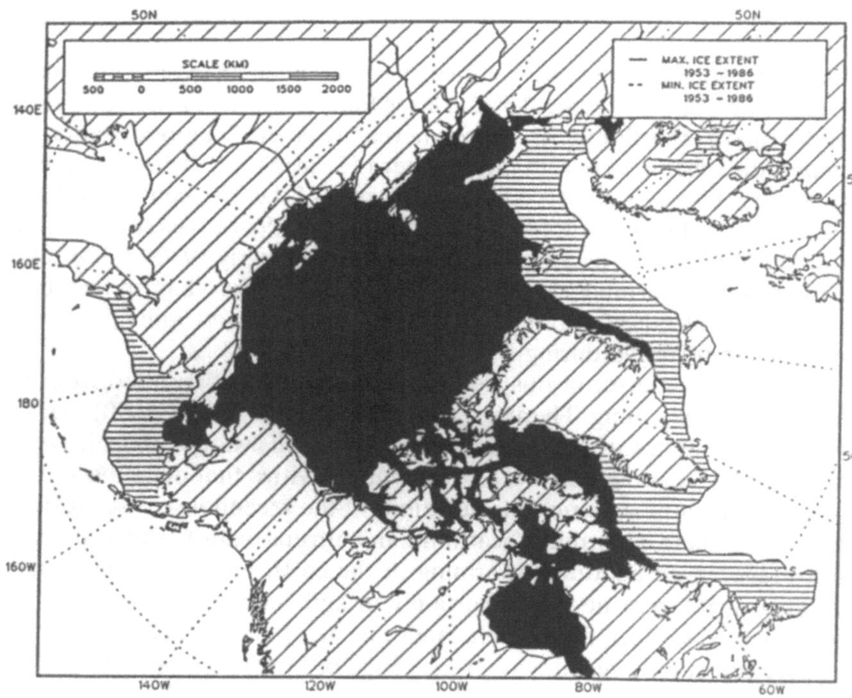

Figure 1.1. Composite sea-ice extent for winter: minimum is shaded; maximum is shaded and horizontally hatched. From Agnew (1993).

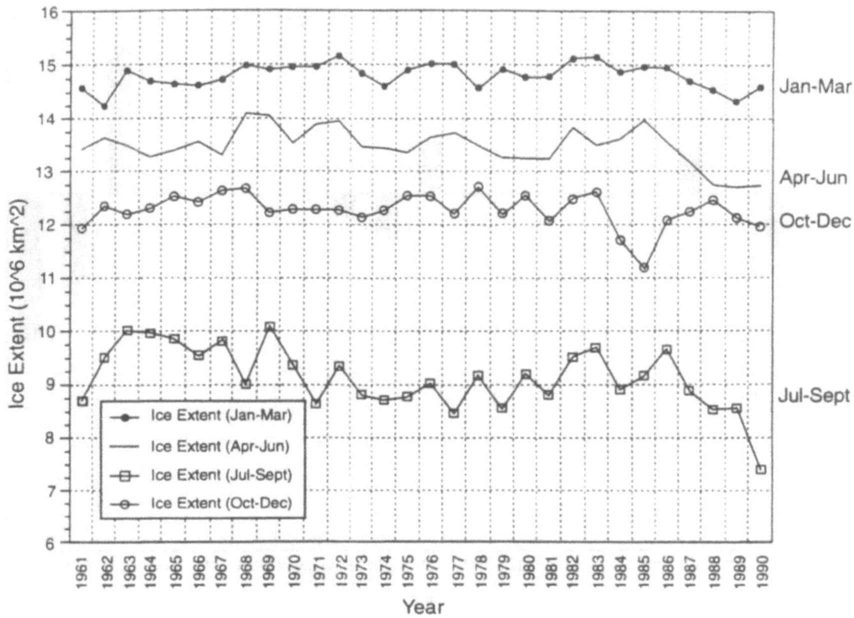

Figure 1.2. Time series of end-of-month arctic sea ice extent averaged over 3-month ice seasons: January to March, April to June, July to September, and October to December. From Chapman and Walsh (1993).

Recent Patterns

Sea Ice

The extreme limits of winter sea-ice coverage in the Arctic are shown in Figure 1.1. These are composite maps utilizing data from the period 1953 to 1986. Although geography is an obvious constraining factor, there is regardless a marked latitudinal variation in sea ice extent around the pole. Off the Canadian eastern seaboard, for example, the maximum extent is south of 50°N. Near northern Scandinavia, the most southerly position of maximum extent is near 75°N. And in the Bering Sea, the maximum southern position is near 55°N. This very regional character to the sea-ice pattern is a feature that is seen repeatedly in other features of arctic climate.

The mean area of arctic sea-ice coverage is approximately 11 million km^2. Chapman and Walsh (1993) examined such data on a year-by-year basis for the period 1961 to 1990. Figure 1.2 illustrates the changes in arctic sea-ice extent as a whole over that 30-year period for four different ice seasons: winter (January to March), spring (April to June), summer (July to September), and autumn (October to December). These suggest a recent decrease in sea ice extent, which is most noticeable in summer. Particularly noteworthy is the 1990 value, which was the lowest of the period by a wide margin.

To provide a regional perspective on this, Figure 1.3 shows the percentages by which mean arctic sea ice extent changed from 1961–75 to 1976–90

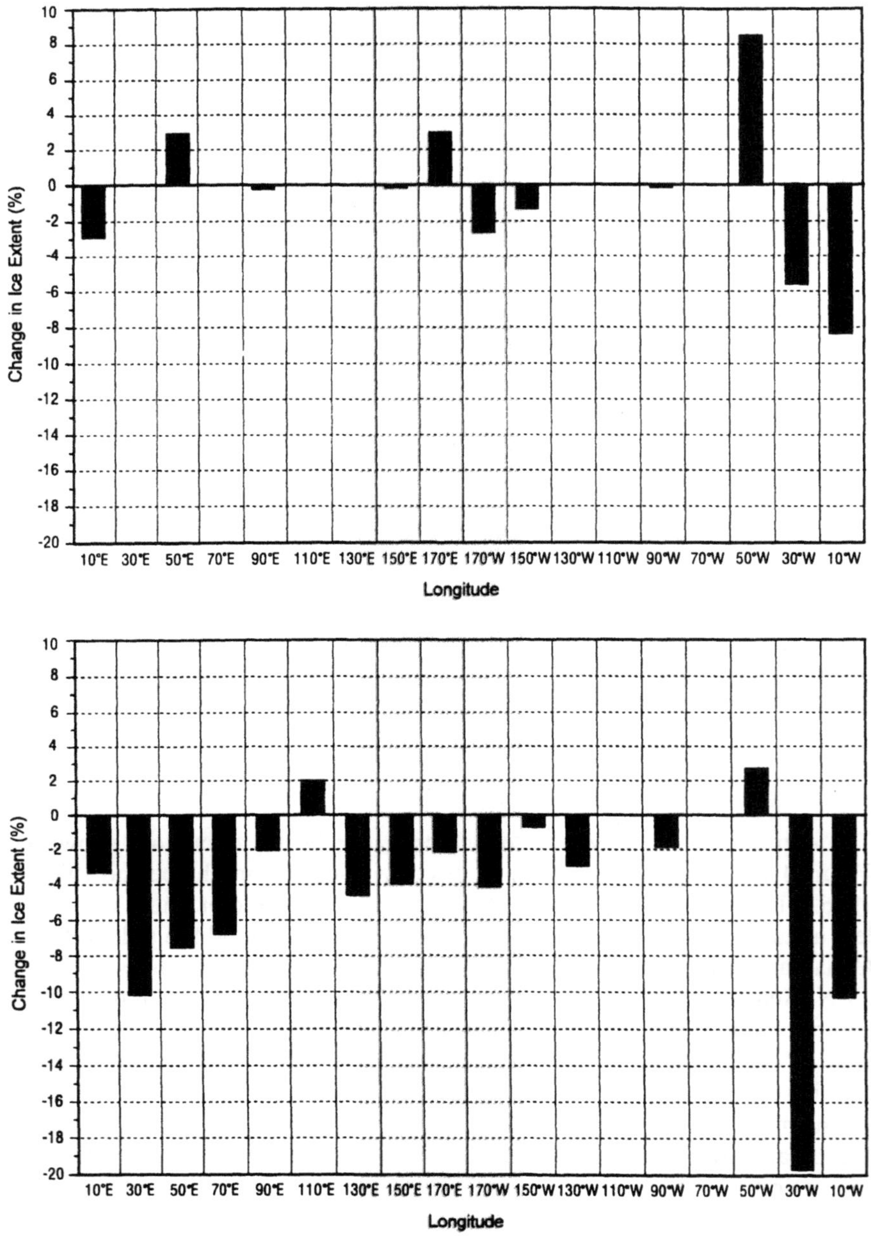

Figure 1.3. Percentages by which mean arctic sea ice extent changed from 1961–75 to 1976–90. Changes are plotted by 20° longitudinal sector for (upper) winter (December to February) and (lower) summer (June to August). From Chapman and Walsh (1993).

as a function of longitude. Results for both winter (December to February) and summer (June to August) are given. In winter, changes are generally fairly small, with two exceptions. In the eastern North Atlantic (0–40°W), decreases of about 5 to 8% occurred; farther west in the Davis Strait and Labrador Sea areas (40–60°W), an increase of about 8% occurred. The changes in summer are much more dramatic. In most sectors, they are negative and in some instances they exceed 10%, including the Greenland Sea (0–40°W) and the Barents Sea (20–40°E).

Possible changes in thickness are another aspect of the sea-ice regime in the Arctic. Thickness data are not so readily available on a wide areal basis as are those of extent. McLaren et al. (1992) analyzed subsurface ice thickness data gathered during the course of six submarine cruises carried out in six different years during the period 1977 to 1990. Their results suggested a trend towards thinner ice by some 15% over the time period, but the interannual variability in the data limited the confidence that could be placed in such a trend. One complicating factor is the role of snow cover in limiting ice thickness. Brown and Coté (1992) studied ice thickness measured at four fast-ice sites in the Canadian High Arctic over the period 1950 to 1989. They found no evidence of systematic ice thinning. Recent thinning and thickening trends at two of the sites, however, were consistent with changes in the average depth of snow covering the ice.

Atmospheric Circulation

Figures 1.4 and 1.5 show the mean patterns at the 50-kPa (~5000 m ASL) and 100-kPa (near surface) levels in the atmosphere. These illustrate an important feature that dominates the atmospheric circulation of the Arctic, namely, a deep, cold, low-pressure area known as the circumpolar vortex, which extends through the middle and upper troposphere and the lower stratosphere (Maxwell, 1992). This feature appears on the 50-kPa maps as a pronounced area of low heights extending over the North Pole with three distinct troughs, particularly in winter. The two most intense of these are situated over eastern Canada and the Sea of Okhotsk, with the third, somewhat less pronounced, lying over the northwestern portion of Siberia. The two strong troughs are reflected on the surface patterns in the Icelandic and Aleutian lows, respectively.

During the past 30 years with the increased frequency of lighter sea-ice years, there has been a change in intensity of these two surface features. Agnew (1993) found that for the entire Arctic, the Icelandic and Aleutian winter lows were generally stronger for light ice years as opposed to heavy ice years. This confirms an earlier evaluation of surface pressure by Trenberth (1990), who looked at changes between the 1924–76 and 1977–88 November-to-March periods. The more intense circulation around the lows as a result would tend to increase northward meridional atmospheric trans-

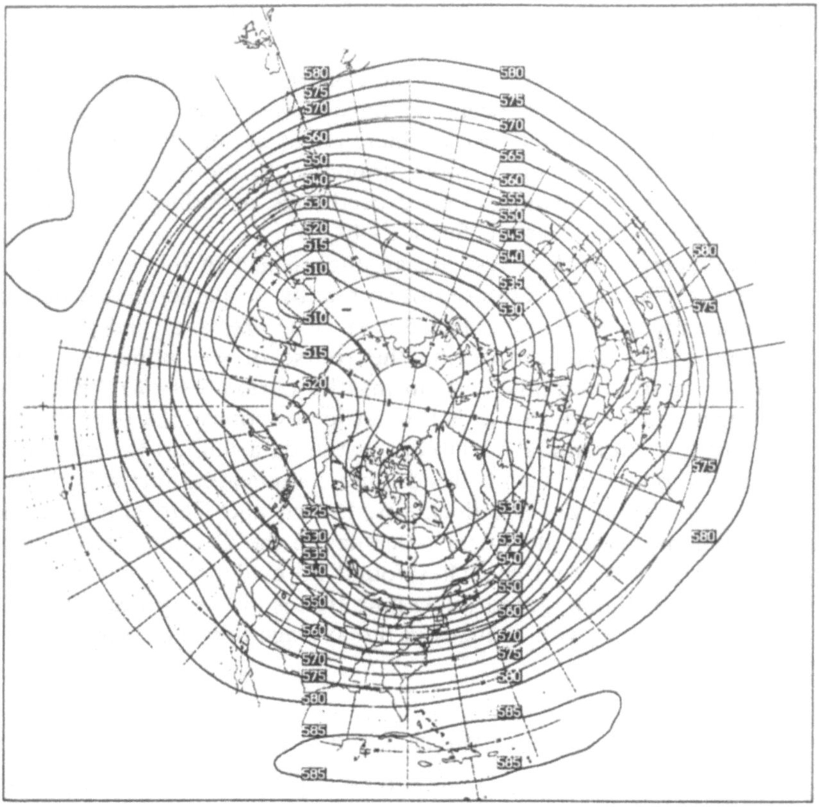

(a)

Figure 1.4. (a) Circumpolar 50-kPa mean height pattern for January based on the 1949 to 1978 time period. Heights, given in decameters (dm) and contoured at 5-dm intervals, represent the elevation above the earth's surface at which a pressure of 50-kPa occurs. From Harley (1980).

port in the lower layers and also increase ocean heat transport through increased wind stress and vorticity.

The 50- to 100-kPa thickness anomaly field for light ice for the Arctic as a whole in winter is shown in Figure 1.6. Changes in the thickness of this layer of the atmosphere reflect warming (increased thickness) or cooling (decreased thickness) of the lower part of the troposphere, including the surface. In the figure, there are marked regions of increased thickness over the Alaska, Yukon, and Mackenzie areas of Canada, and over much of

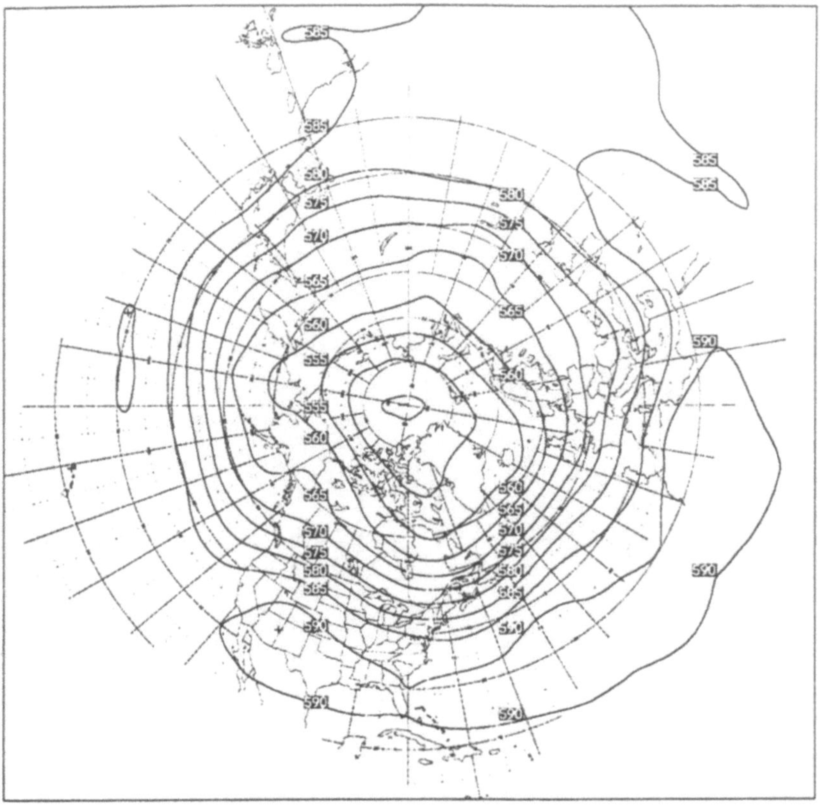

(b)

Figure 1.4. *Continued* (b) Circumpolar 50-kPa mean height pattern for July based on the 1949 to 1978 time period. Heights, given in decameters (dm) and contoured at 5-dm intervals, represent the elevation above the earth's surface at which a pressure of 50-kPa occurs. From Harley (1980).

northern Russia. There are strong thickness decreases over a region extending from the Davis-Labrador region of Canada eastward over Iceland, much of Scandinavia, and Novaya Zemlya.

Temperature

The normal surface air temperature patterns of the Arctic (Figure 1.7) reveal marked differences with longitude in all seasons of the year. For example, at 70°N, mean January temperature varies from near freezing over the Norwegian Sea to as low as –42°C over Greenland and northern Siberia. In summer, the range is somewhat smaller, but still appreciable—

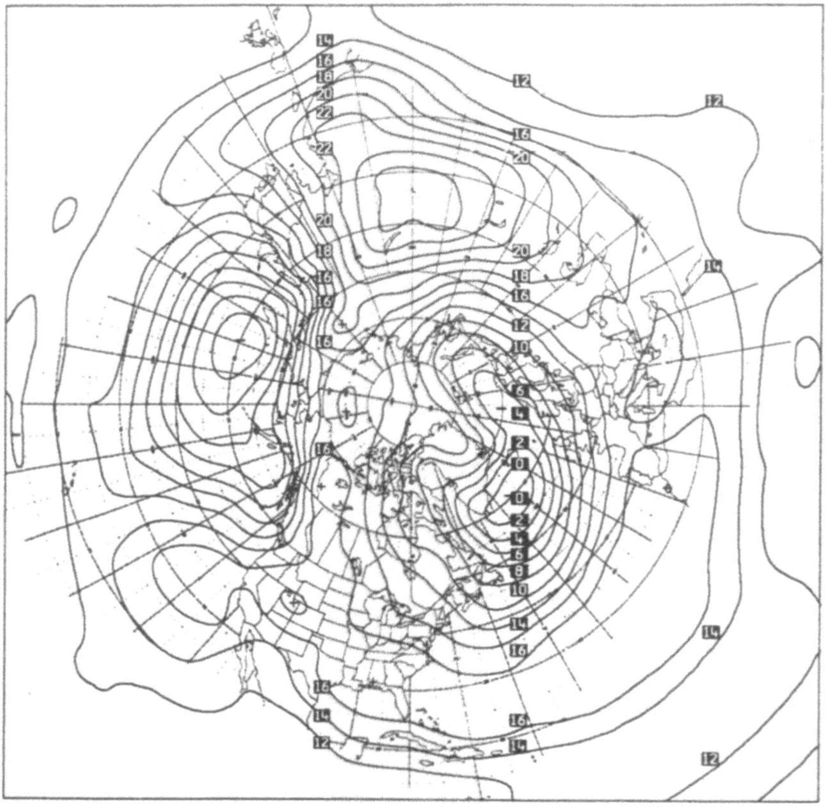

(a)

Figure 1.5. (a) Circumpolar 100-kPa mean height pattern for January based on the 1949 to 1978 time period. Heights, given in decameters (dm) and contoured at 2-dm intervals, represent the elevation above the earth's surface at which a pressure of 100-kPa occurs. The 100-kPa pattern is very much indicative of the mean sea level pressure pattern. From Harley (1980).

from as high as +12°C over northern Norway to −12°C over central Greenland. These variations are a function of the major climate controls, including atmospheric circulation, solar radiation input, physical geography, and the nature of the underlying surfaces.

The temperature record for the 1961 to 1990 period reveals a similarly regional character to the circumpolar pattern. Figure 1.8 shows best-fit linear trends, both the annual pattern and seasonal ones based on 3-month means: December to February (winter), March to May (spring), June to August (summer), and September to November (autumn). These are essen-

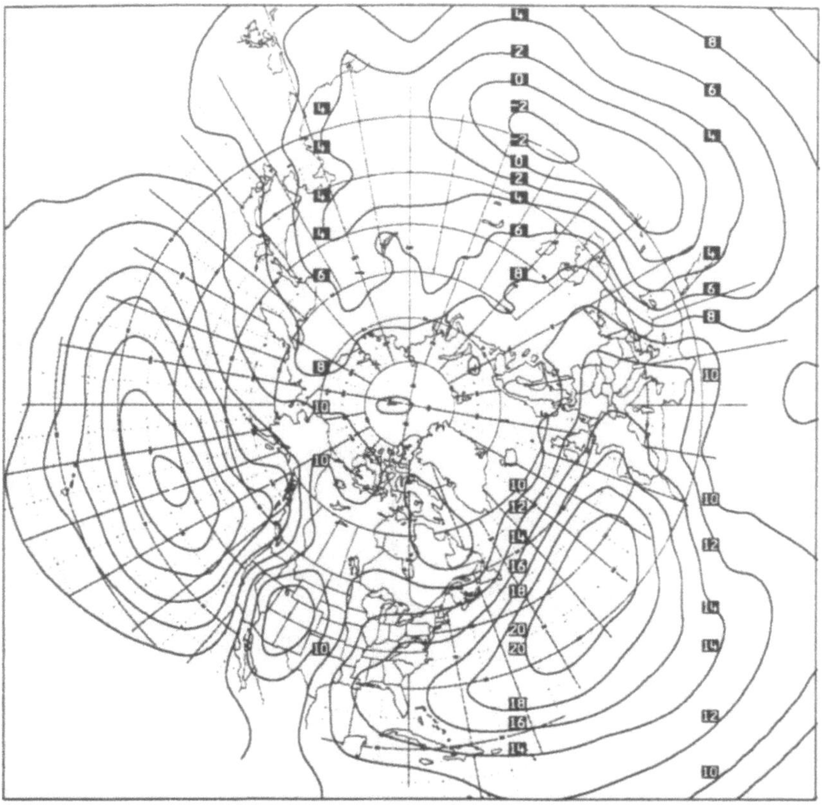

(b)

Figure 1.5. *Continued* (b) Circumpolar 100-kPa mean height pattern for July based on the 1949 to 1978 time period. Heights, given in decameters (dm) and contoured at 2-dm intervals, represent the elevation above the earth's surface at which a pressure of 100-kPa occurs. The 100-kPa pattern is very much indicative of the mean sea level pressure pattern. From Harley (1980).

tially the same as the maps published by Chapman and Walsh (1993) and illustrate the same main features, notably:

(a) winter and spring are the seasons when warming dominates;
(b) warming is greatest over the land areas of the Mackenzie and Yukon regions of Canada, Alaska, and northern Russia;
(c) cooling persists throughout the year over a region extending from the Baffin-Davis area of Canada through the southern half of Greenland to Iceland; and

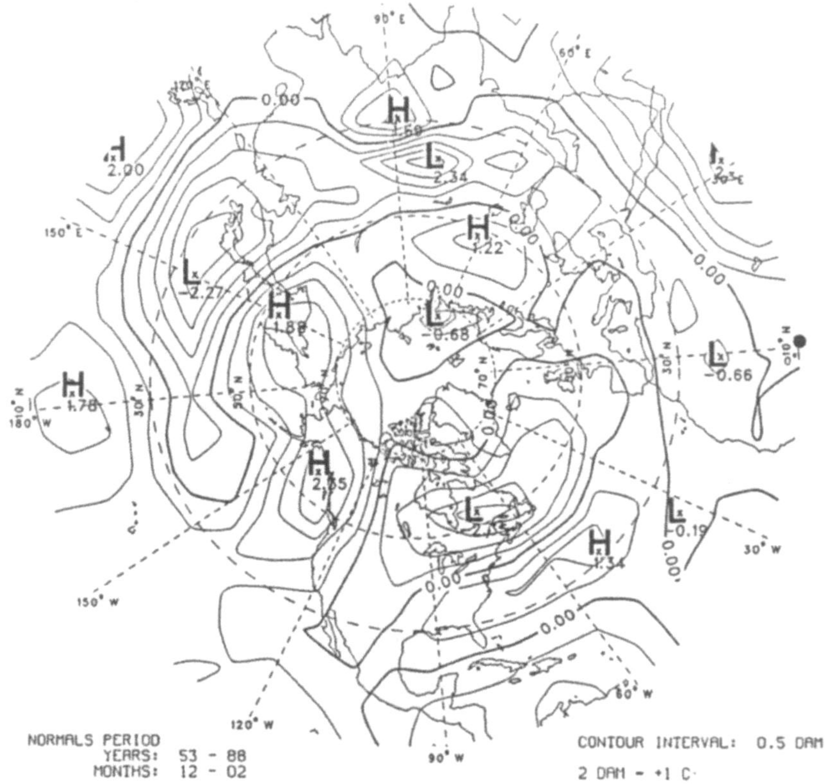

Figure 1.6. Composite 50- to 100-kPa thickness anomaly field for light ice for the Arctic as a whole. From Agnew (1993).

(d) the summer pattern suggests an areally averaged trend that is nearly zero.

The recent warming may be compared with the record of the past 100 years, both in terms of the whole Northern Hemisphere (Figure 1.9) and the Arctic (60 to 90°N) alone (Figure 1.10). These figures, which actually show yearly anomalies from the 1951 to 1980 average, indicate warming during the first 40 years of this century followed by slight cooling through the late 1960s and then warming during the past 20 years. The arctic warming earlier in the century is much more marked than the most recent warming, in contrast to the situation for the hemisphere as a whole. Chapman and Walsh (1993) pointed out that the earlier warming was concentrated in the 60–75°N area, whereas the most recent warming is greatest in the

50–60°N zone and apparently decreases slightly as one goes farther north.

Despite the recent overall summer warming being little different from zero, Figure 1.8(d) does show a marked pattern of alternating areas of warming and cooling around the area south of the Arctic Basin. Due to the importance of mean June and July temperature for various aspects of the arctic terrestrial environment, the trends are presented in detail for those months for five areas: Northwestern Canada and Alaska; Eastern Canadian Arctic and Southern Greenland; Scandinavia; North Central Russia (Yamal); and Northeastern Russia. Both year-to-year values and a best-fit linear trend are shown (Figure 1.11). Table 1.1 summarizes the trends illustrated.

The strong winter warming that has been observed in the Arctic suggests that changes may also be occurring in the characteristics of surface-based inversions as they are closely rooted to actual surface temperature. Such inversions are typically 400 to 850m in thickness and occur on more than 70% of winter days (Bradley et al., 1992). Bradley et al. (1993) examined winter inversion data for the period 1966 to 1990 and found that a systematic reduction in midwinter surface-based inversion depths occurred during that period for North American stations (Figure 1.12) located within a wide area extending from 62 to 162°W (Alert to Kotzebue) and from 70 to 83°N (Inuvik to Alert).

Examination of some Russian data suggested that this pattern of decreasing inversion thickness may extend to 104°E (Cape Chelyuskin). These results are generally in accord with those of Kahl et al. (1993b), who found a warming trend in the 85- to 70-kPa layer (~1500 to 3000m) of the troposphere for the same geographical area of the Arctic. On the other hand, a cooling trend was recorded for that layer over northwestern Russia, Scandinavia, and Greenland.

Precipitation

The time series of annual precipitation anomalies (using the 1951 to 1980 normal value as a base) covering the period 1890 to 1991 for the Arctic is shown in Figure 1.13. Precipitation gradually increased until the early 1920s, followed by a levelling-off or slight decline until the early 1950s. Since then there has been a fairly consistent increase, to the point where values at present are nearly 10% above the 1951 to 1980 normals.

As with temperature, there is very much a regional character to the arctic precipitation pattern. In Canada, for example, much of the exposed eastern coasts of eastern arctic islands such as Baffin and Devon receive annual precipitation totals in excess of 300mm; at comparable latitudes to the west over Victoria and Prince of Wales Islands, totals are only half or less of that

(a)

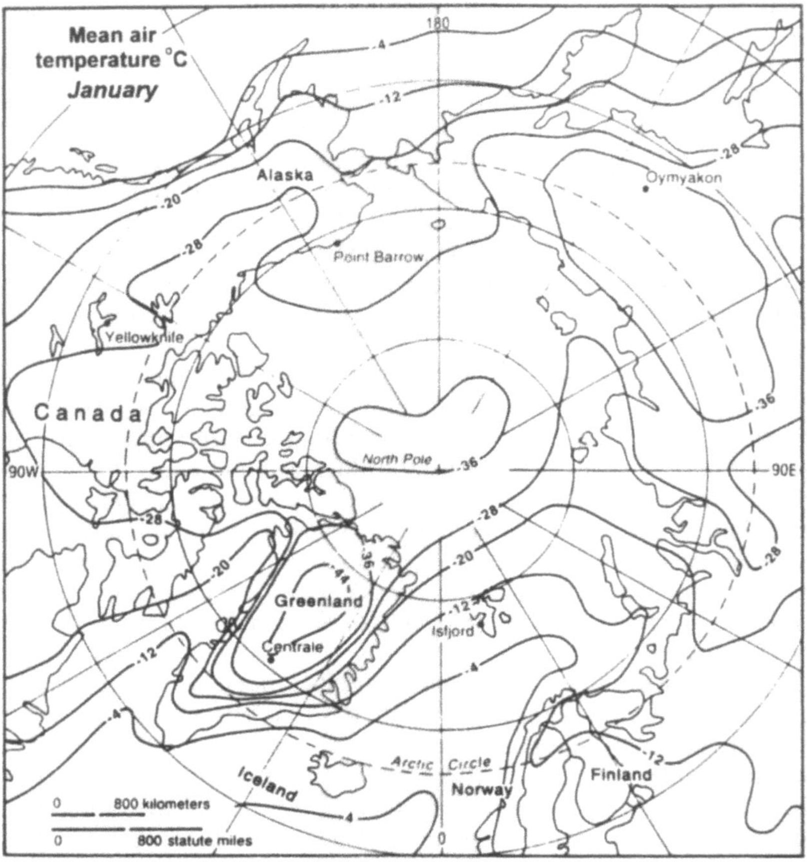

Figure 1.7. Circumpolar mean surface air temperature for (a) January.

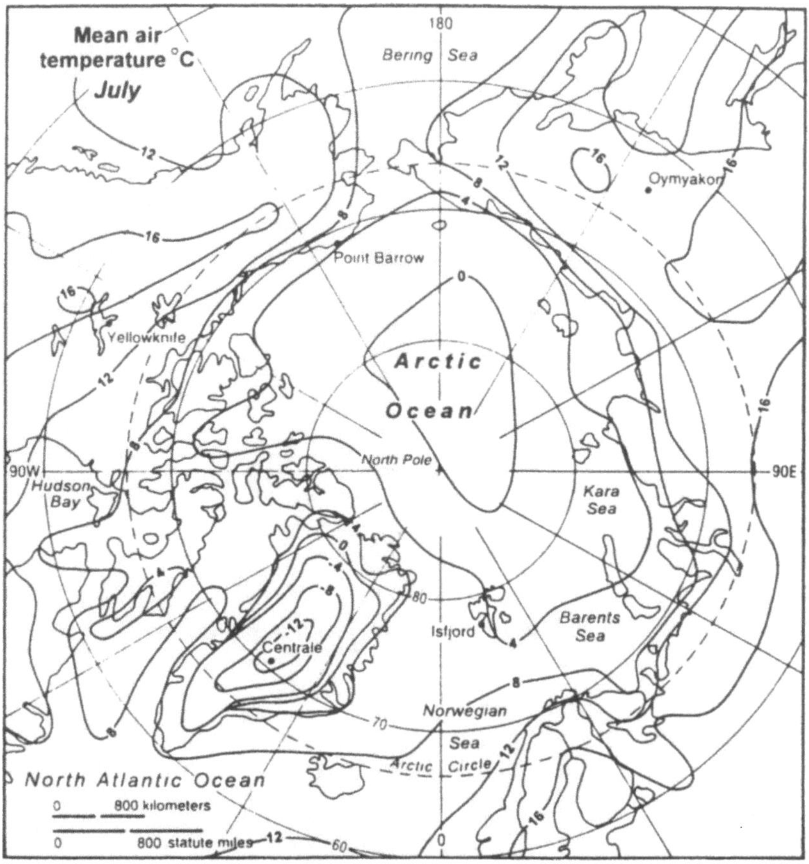

(b)

Figure 1.7. *Continued* (b) July. From CIA (1978).

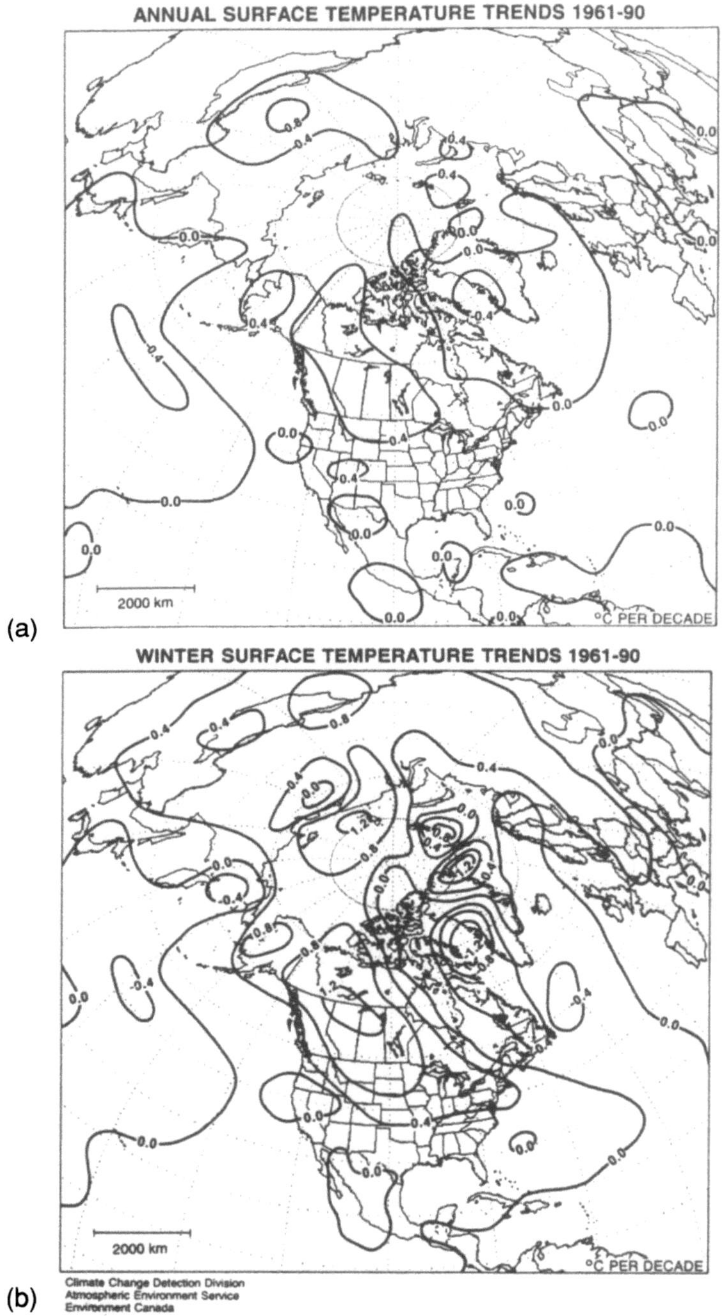

Figure 1.8. Circumpolar best-fit linear trends of mean surface air temperature for the period 1961 to 1990. Annual and seasonal patterns are shown: (a) annual; (b) winter (December to February).

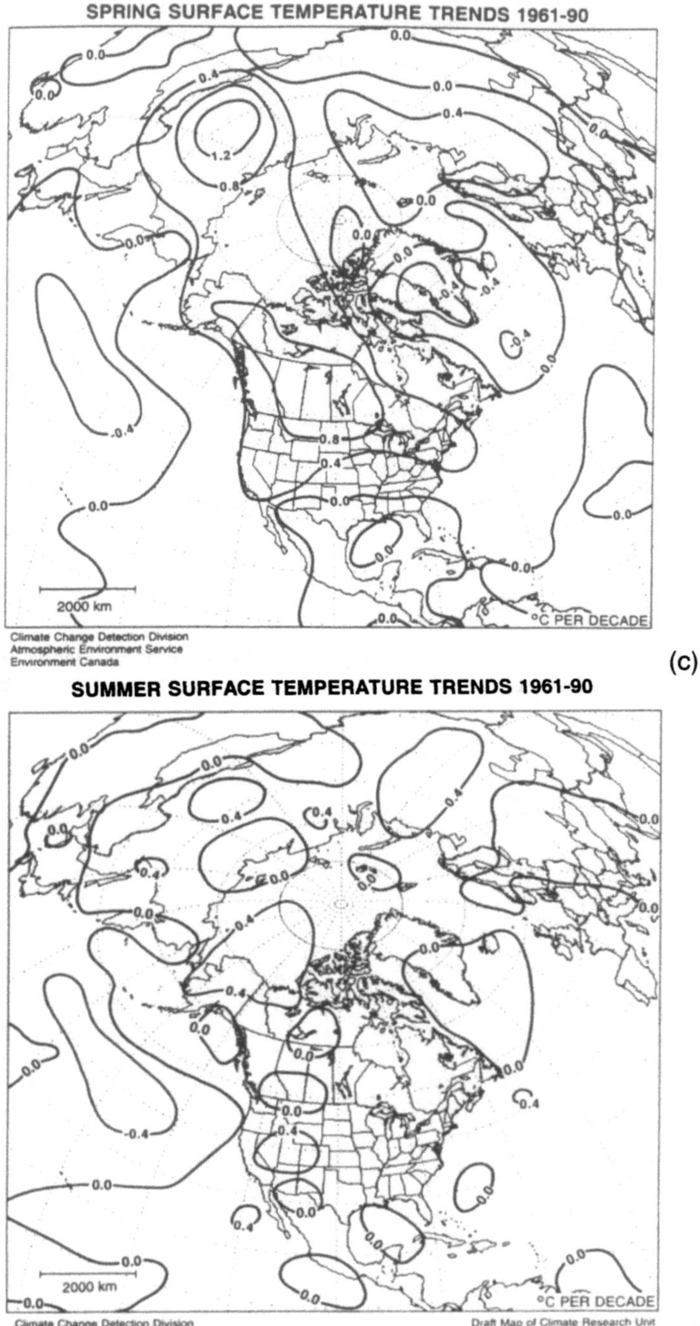

Figure 1.8. *Continued* (c) spring (March to May); (d) summer (June to August).

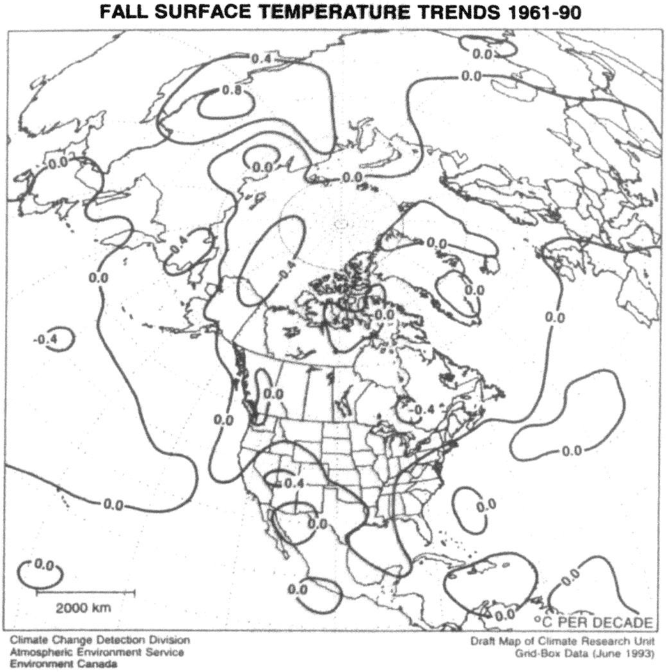

(e)

Figure 1.8. *Continued* (e) autumn (September to November). Maps courtesy of Climate Change Detection Division, Canadian Climate Centre (Reycraft, personal communication).

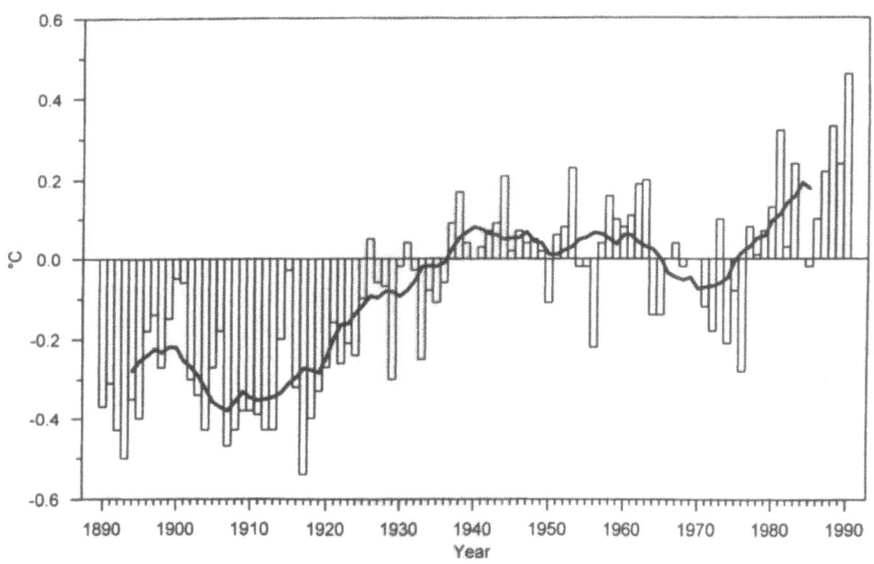

Figure 1.9. Time series of mean annual surface air temperature over the past 100 years for the Northern Hemisphere. Shown are anomalies from the 1951 to 1980 normal period, on a year-to-year basis (bars) and as a running mean (solid line).

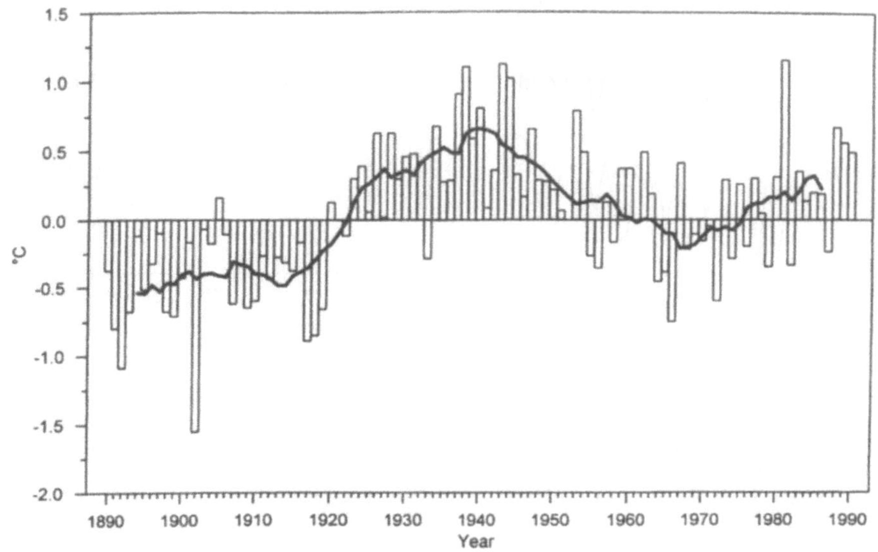

Figure 1.10. Time series of mean annual surface air temperature over the past 100 years for the Arctic (60–90°N). Shown are anomalies from the 1951 to 1980 normal period, on a year-to-year basis (bars) and as a running mean (solid line).

amount. Figure 1.14 illustrates precipitation variations during the 1961 to 1991 period, both annually and for the month of July, for the same five arctic regions discussed above under temperature. Table 1.2 summarizes the best-fit linear trends.

Snow Cover

Weekly snow cover maps produced from visible satellite imagery are available since 1972. An analysis of data from the 1972 to 1989 period has been carried out by Robinson and Dewey (1990). Their results indicate that, for the Northern Hemisphere, snow cover was at its lowest extent for the 18-year period during 1988 and 1989—some 8 to 10% below the period mean of 25.7 million km^2. The 1981 to 1989 period generally had less extensive cover than the previous 9 years.

Figure 1.15 illustrates the year-to-year variation in seasonal snow cover extent for both the North American and Eurasian components of the hemispheric total. The Eurasian component dominates the hemispheric signal. Decreases in snow cover extent are evident in the 1980s, during all seasons over Eurasia and in spring over North America. Hemispheric seasonal means were 3.7 to 8.4% lower for the 1981 to 1989 period than for the first half of the record. An update of this analysis (Robinson et al., 1993) suggests continued reduction of hemispheric snow cover through 1992, particularly in spring.

The date of snow cover disappearance on the arctic tundra has been studied by Foster (1989), who looked at records from a number of North

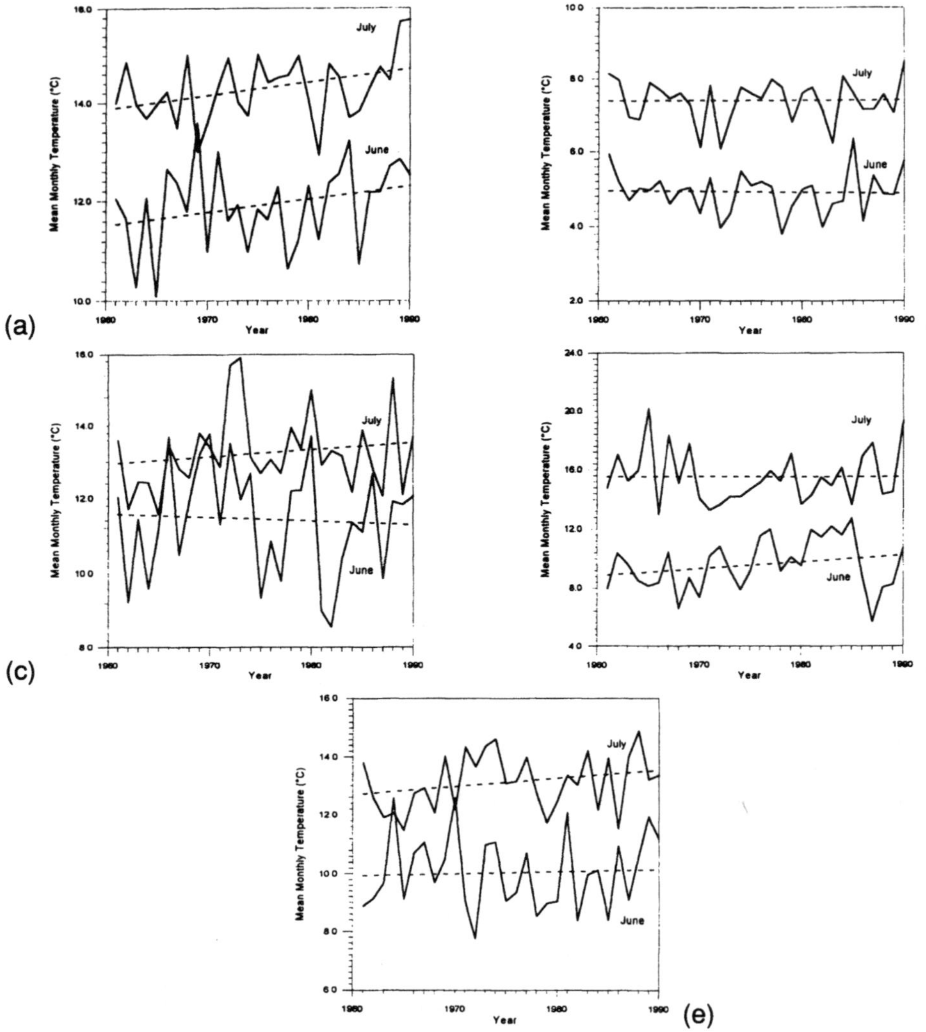

Figure 1.11. June and July mean surface air temperature in the Arctic during 1961 to 1990. Year-to-year values (solid line) and best-fit linear trends (dashed line) are shown for five arctic regions, clockwise from the upper left: (a) Northwestern Canada and Alaska, (b) Eastern Canadian Arctic and Southern Greenland, (c) Scandinavia, (d) North Central Russia (Yamal), and (e) extreme Northeast Russia.

Table 1.1. Surface Air Temperature Trends (°C/Decade) During 1961–1990

Region	June	July
Northwestern Canada and Alaska	+0.25	+0.28
Eastern Canadian Arctic and Southern Greenland	−0.02	0.00
Scandinavia	−0.12	+0.23
North Central Russia (Yamal)	+0.47	−0.02
Extreme Northeast Russia	+0.07	+0.30

Figure 1.12. Mean year-to-year winter December to March) surface-based temperature inversion depths for eight Canadian and Alaskan stations during the period 1967 to 1990. From Bradley et al. (1993).

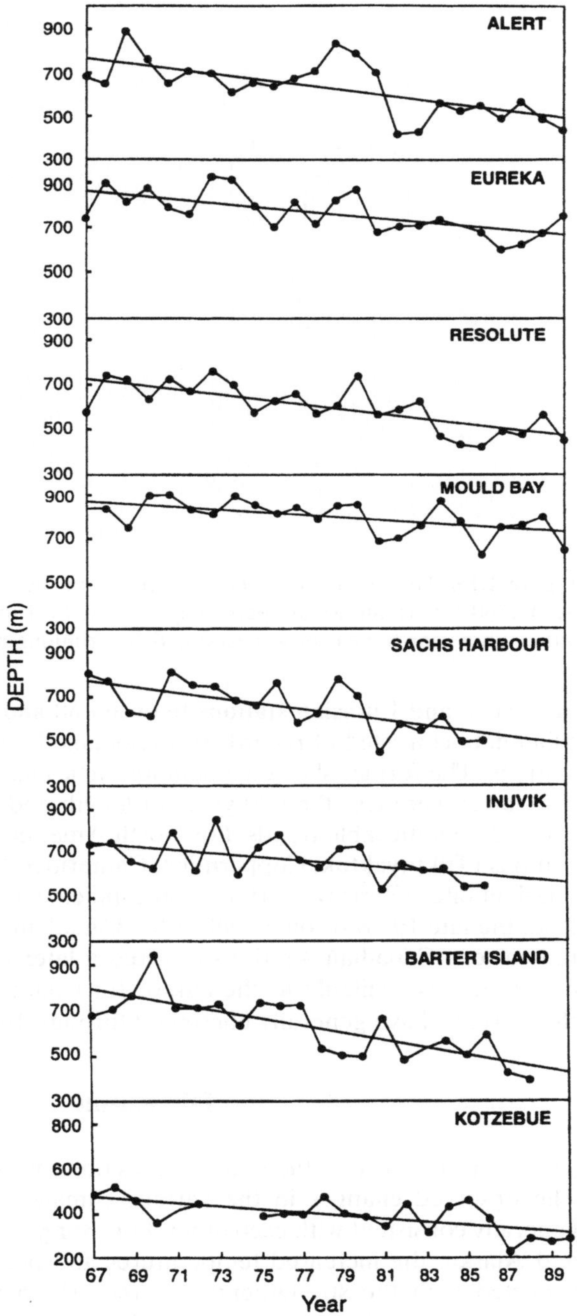

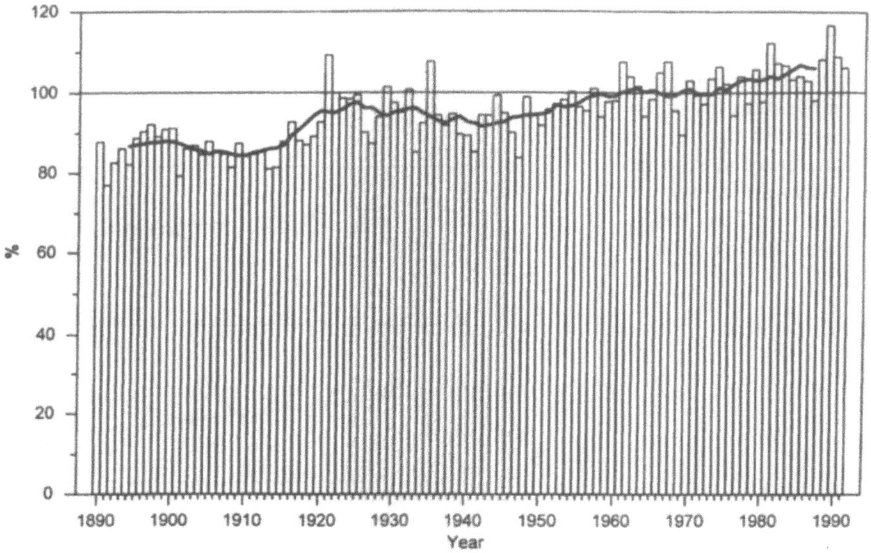

Figure 1.13. Time series of annual precipitation over the past 100 years for the Arctic (60–90°N). Shown are percentages of the 1951 to 1980 normal period annual value, on a year-to-year basis (bars) and as a running mean (solid line).

American and Eurasian stations between 68 and 74°N latitude. The Eurasian data consisted of records from Sodankyla, Finland, and five Russian stations. The former showed a definite cyclic character, but no clear trend. The latter were less than 20 years in length ending in the 1950s and also showed no noticeable trends. For North America, however, four Canadian stations (Tuktoyaktuk, Coppermine, Cambridge Bay, Hall Beach) and two Alaskan ones (Barrow, Barter Island) indicate a trend to earlier snowmelt from the late 1960s through 1983. One Canadian station (Clyde, located in the eastern Canadian Arctic) suggested a later snowmelt. Updated snow cover records available to the author for Canadian stations indicate that these trends have generally continued through 1992.

Discussion

The trends outlined in the preceding sections are summarized in Table 1.3. The observed changes in the various climate and related elements are generally consistent with each other. For example, in Northwestern Canada and Alaska, the increased temperatures and precipitation are both to be expected with the strengthening of the Aleutian low and the resulting increased southerly to southwesterly flow of air over the area. With spring being the season of greatest warming, the trends to decreased spring snow cover extent and earlier dates of snow cover disappearance are also to be

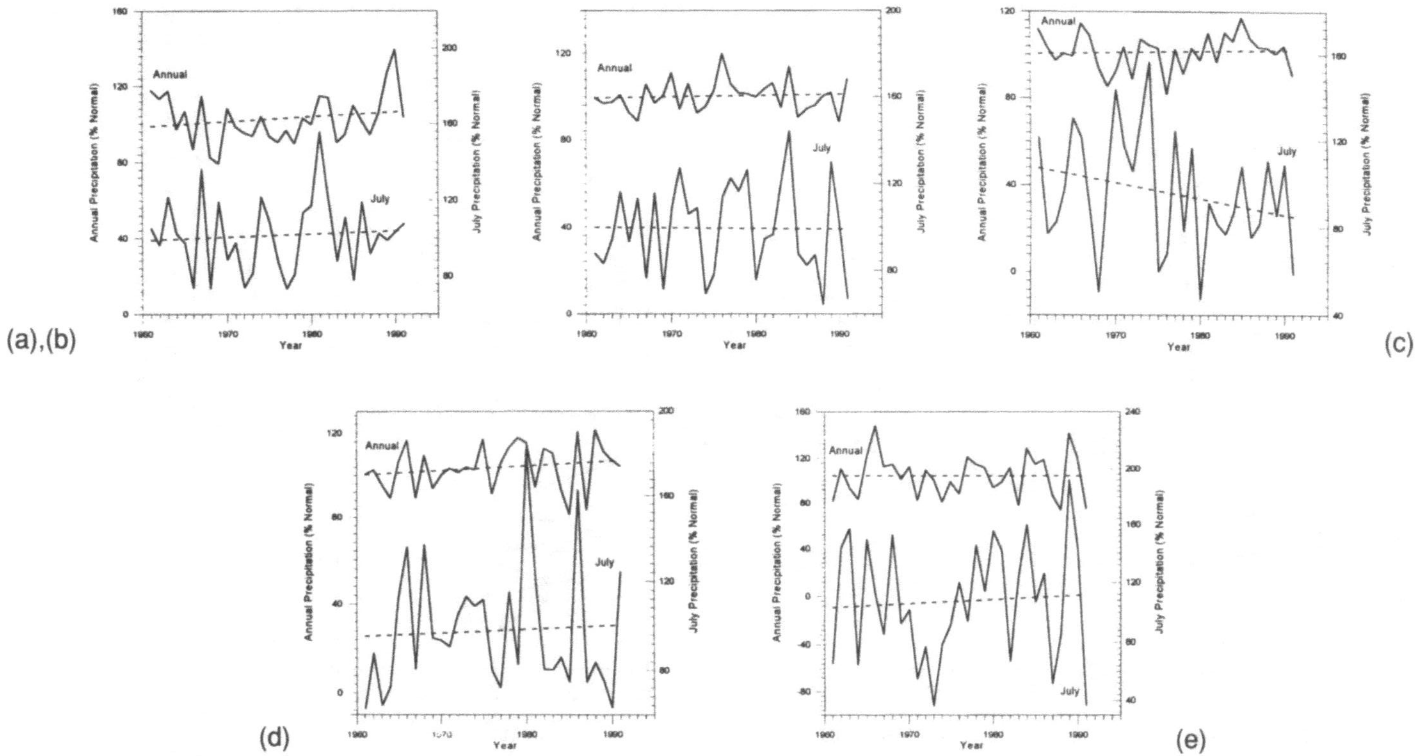

Figure 1.14. Annual and July precipitation in the Arctic during 1961 to 1991. Shown are percentages of the 1951 to 1980 normal values, on a year-to-year basis (solid line) and as a best-fit linear trend (dashed line), for five arctic regions, clockwise from the upper left: (a) Northwestern Canada and Alaska, (b) Eastern Canadian Arctic and Southern Greenland, (c) Scandinavia, (d) North Central Russia (Yamal), and (e) extreme Northeast Russia.

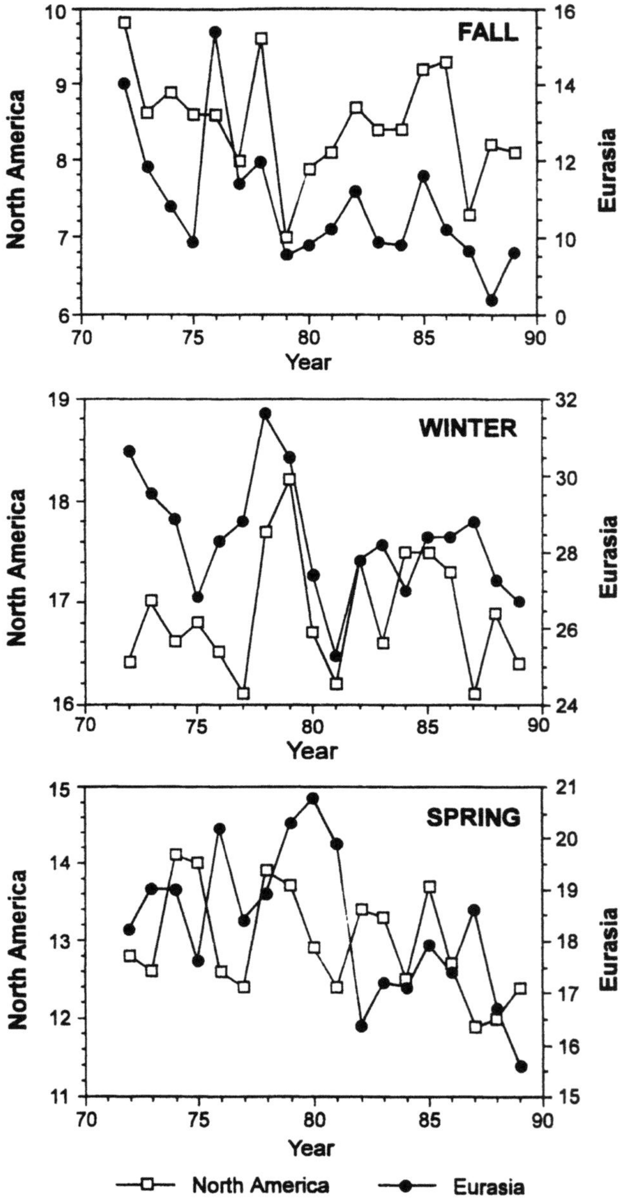

Figure 1.15. Seasonal snow cover (million km²) over Eurasia and North America from winter 1971–72 to autumn 1989. Autumn (September to November) winter (December to February) spring (March to May). From Robinson and Dewey (1990).

Table 1.2. Precipitation Trends (%/Decade) During 1961–91

Region	July	Annual
Northwestern Canada and Alaska	+1.7	+2.7
Eastern Canadian Arctic and Southern Greenland	−0.3	+0.7
Scandinavia	−7.3	+0.7
North Central Russia (Yamal)	+1.7	+2.3
Extreme Northeast Russia	+3.0	0.0

expected. Conversely, in the eastern Canadian Arctic and Greenland, strengthening of the Icelandic low to the southeast has resulted in increased northerly to northwesterly flow over the area. This is reflected in generally lower temperatures and a trend to increased sea ice extent. Precipitation, although somewhat increased, is substantially less so than in Northwestern Canada and Alaska or the Yamal area of Russia—the two areas of greatest warming in the Arctic.

The importance of these apparent trends may be partly judged by the degree to which they are statistically significant. The bold entries in Table 1.3 indicate those trends that have been found by their respective authors, at least at the 5% level. The implication of this result is the simple fact that, although compelling, most observed recent arctic climate trends are still generally within the range of historic variability. As with other observed climate trends outside the Arctic, they are consistent with an enhanced greenhouse gas effect but not yet directly attributable to it.

Table 1.3. Climate and Related Element Trends by Region

	Northwestern Canada and Alaska	Eastern Canadian Arctic and Greenland	Scandinavia	North Central Russia (Yamal)	Extreme Northeast Russia	Arctic Ocean
Surface Temperature						
Annual	**Increase**[a]	**Decrease**	No trend	**Increase**	No trend	Decrease
January	**Increase**	**Decrease**	No trend	Increase	Decrease	Decrease[b]
July	**Increase**	No trend	Increase	**Increase**	Increase	No trend[b]
Precipitation						
Annual	Increase	Increase	Increase	Increase	No trend	
July	Increase	Decrease	Decrease	Increase	Increase	
Snow Cover Extent	North America as a whole			Eurasia as a whole		
Fall	No Trend			**Decrease**		
Winter	No Trend			**Decrease**		
Spring	Decrease			**Decrease**		
Snow Cover Data of Disappearance	Earlier	No trend	No trend			
Sea Ice						
Extent	**Decrease**	Increase		**Decrease**	**Decrease**	No trend
Thickness						Decrease
Atmospheric Circulation	Strengthening of the Icelandic and Aleutian lows					

[a] Temperature statistical significance results from Skinner (personal communication).
[b] Trend information from Kahl et al. (1993a).

Table 1.4. GCM Results Under a 2 × CO_2 Scenario (Compared to 1 × CO_2) for Various Arctic Regions in Winter

	Northwestern Canada and Alaska	Eastern Canadian Arctic and Greenland	Scandinavia	North Central Russia (Yamal)	Extreme Northeast Russia	Arctic Ocean
Temperature						
CCC	Increase	Increase	Increase	Increase	Increase	Increase
NCAR	Increase	Decrease	Decrease	No change	Increase	Decrease
Precipitation						
CCC	Increase	Increase	Increase	Increase	Increase	Increase
NCAR						
Snow Cover	North America as a whole			Eurasia as a whole		
CCC	Decrease			Decrease		
NCAR						
Sea Ice						
CCC	Decrease	Decrease		Decrease	Decrease	Decrease
NCAR	Decrease	Increase		No change	Decrease	No change
Sea Level Pressure						
CCC	Decrease	Decrease	Decrease	Decrease	Decrease	Decrease
NCAR	Decrease	Increase	Decrease	Decrease	Decrease	Decrease

It is, however, useful to compare these trends with GCM-generated climate patterns under a scenario of doubling of the atmospheric concentration of carbon dioxide. Two GCM results are considered—the Canadian Climate Centre (CCC) high resolution model for a doubled CO_2 equilibrium experiment (Boer et al., 1992) and the National Centre for Atmospheric Research (NCAR) community climate model coupled to an ocean general circulation model with CO_2 increasing linearly at 1% per year (Washington and Meehl, 1989). The model results are summarized in Tables 1.4 and 1.5 for both winter (December to February) and summer

Table 1.5. GCM Results Under a 2 × CO_2 Scenario (Compared to 1 × CO_2) for Various Arctic Regions in Summer

	Northwestern Canada and Alaska	Eastern Canadian Arctic and Greenland	Scandinavia	North Central Russia (Yamal)	Extreme Northest Russia	Arctic Ocean
Temperature						
CCC	Increase	Increase	Increase	Increase	Increase	No change
NCAR	Decrease	Decrease	Increase	Increase	Decrease	No change
Precipitation						
CCC	Increase	No change	Increase	Increase	Increase	No change
NCAR						
Sea Ice						
CCC	Decrease	Decrease		Decrease	Decrease	Decrease
NCAR						
Sea Level Pressure						
CCC	Decrease	No change	Decrease	Decrease	Decrease	Increase
NCAR						

(June to August). Only temperature information was readily available for the summer from the NCAR model.

A comparison of these Tables 1.4 and 1.5 shows general agreement between the results of the models. The region where the most apparent differences occur is that of the Eastern Canadian Arctic and Southern Greenland. In winter there, the CCC model suggests a temperature increase, sea-ice extent decrease, and sea level pressure decrease, whereas the NCAR results are opposite for all three elements. The same contrast in temperature trends is also evident in summer.

In comparing the model results with the currently observed trends in the Arctic, those regions where the models are in agreement are showing current trends that are, for the most part, consistent with the direction of changes that might be expected under an enhanced greenhouse gas effect. This shows up for all the climate and related elements considered here. In the Eastern Canadian Arctic and Southern Greenland region where the model results differ, currently observed trends tend to agree with the direction of changes suggested by the NCAR model. The fact that the NCAR model results presented here are from a transient run and that the model involves a coupled ocean circulation component may suggest that its outputs will be closer to future reality (although there are other deficiencies in the model, Washington and Meehl, 1989). This suggests that the Eastern Canadian Arctic and Southern Greenland trends may well be a sign of things to come in that region. At the very least, close attention to ongoing conditions in the region is warranted.

Summary

The short review of recent literature on climate patterns in the Arctic during the past three decades and the regional analyses of temperature and precipitation data drawn from the East Anglia world climate data set presented in this chapter (and summarized in Table 1.3) have shown a consistent pattern in three different senses. Firstly, there are regional differences among the recent trends observed for any given climate or related element. This is consistent with the normal pattern of any climate element over a vast area such as the circumpolar Arctic. From this point of view, the fact that a large area such as the Arctic, which is experiencing an overall warming, might have one or two regions within it that are undergoing the opposite trend is not unusual or inconsistent at all. Secondly, the recent trends observed for each element are generally consistent with each other. Thus, higher temperature is generally associated with less sea-ice and snow cover extent. Thirdly, the recent trends observed are generally consistent with those projected by the global circulation models under increased atmospheric CO_2 concentrations.

This level of consistency provides a measure of confidence in the validity of the observed trends as an early indicator of global warming. It must be stressed, however, that the observed trends are still generally within the range of historic variability so that it cannot yet be stated with certainty that an enhanced greenhouse-gas effect is being observed.

References

Agnew, T. 1993. Simultaneous winter sea-ice and atmospheric circulation anomaly patterns. *Atmos. Ocean* 31:259–280.

Boer, G.J., McFarlane, N.A., and Lazare, M. 1992. Greenhouse gas-induced climate change simulated with the CCC second-generation general circulation model. *J. Climate* 5:1045–1077.

Bradley, R.S., Keimig, F.T., and Diaz, H.F. 1992. Climatology of surface-based inversions in the North American Arctic. *J. Geophys. Res.* 97 (No. D14):15699–15712.

Bradley, R.S., Keimig, F.T., and Diaz, H.F. 1993. Recent changes in the North American Arctic boundary layer in winter. *J. Geophys. Res.* 98 (No. D5):8851–8858.

Brown, R.D., and Coté, P. 1992. Interannual variability of landfast ice thickness in the Canadian High Arctic, 1950–89. *Arctic* 45:273–284.

Central Intelligence Agency (CIA). 1978. *Polar Regions Atlas*. Document GC78-10040. Washington, DC: U.S. Government Printing Office. 66 pp.

Chapman, W.L., and Walsh, J.E. 1993. Recent variations of sea ice and air temperature in high latitudes. *Bull. Am. Meteorol. Soc.* 74:33–47.

Foster, J.L. 1989. The significance of the date of snow disappearance on the arctic tundra as a possible indicator of climate change. *Arctic Alpine Res.* 21:60–70.

Harley, W.S. 1980. Northern Hemisphere monthly mean 50-kPa and 100-kPa height charts. Rep. CLI 1-80, Atmospheric Environment Service, Downsview, Ontario.

IPCC. 1990. *Climate Change: The IPCC Scientific Assessment*. Cambridge: Cambridge University Press.

Kahl, J.D., Charlevoix, D.J., Zaitseva, N.A., Schnell, R.C., and Serreze, M.C. 1993a. Absence of evidence for greenhouse warming over the Arctic Ocean in the past 40 years. *Nature* 361:335–337.

Kahl, J.D., Serreze, M.C., Stone, R.S., Shiotani, S., Kisley, M., and Schnell, R.C. 1993b. Tropospheric temperature trends in the Arctic: 1958–1986. *J. Geophys. Res.* 98 (No. D7):12825–12838.

Maxwell, B. 1992. Arctic climate: Potential for change under global warming. In: Chapin, F.S., III, Jefferies, R.L., Reynolds, J.F., Shaver, G.R., and Svoboda, J. (eds.), *Arctic Ecosystems in a Changing Climate: An Ecophysiological Perspective* (pp. 11–34). San Diego: Academic Press.

McLaren, A.S., Walsh, J.E., Bourke, R.H., Weaver, R.L., and Wittmann, W. 1992. Variability in sea-ice thickness over the North Pole from 1977 to 1990. *Nature* 358:224–226.

Robinson, D.A., and Dewey, K.F. 1990. Recent secular variations in the extent of Northern Hemisphere snow cover. *Geophys. Res. Lett.* 17:1557–1560.

Robinson, D.A., Dewey, K.F., and Heim, R.R., Jr. 1993. Global snow cover monitoring: An update. *Bull. Am. Meteorol. Soc.* 74:1689–1696.

Trenberth, K.E. 1990. Recent observed interdecadal climate changes in the Northern Hemisphere. *Bull. Am. Meteorol. Soc.* 71:988–993.

Walsh, J.E. 1991. The Arctic as a bellwether. *Nature* 352:19–20.

Washington, W.M., and Meehl, G.A. 1989. Climate sensitivity due to increased CO_2: Experiments with a coupled atmosphere and ocean general circulation model. *Climate Dynam.* 4:1–38.

2. Quaternary Environmental Changes and Ecosystems of the European Arctic

Leonid R. Serebryanny and Arcady A. Tishkov

The analysis of the state and dynamics of modern ecosystems has to be reconciled with palaeoecological investigations, providing a reliable retrospective base for forecasting future trends of impacts. Such an approach is extremely useful for the realization of international scientific programs like IGBP in any regions, including high latitude ones.

Unfortunately, the knowledge of high latitude environmental changes is still inadequate, and abstract hypotheses often replace concrete real information, even nowadays. The ecosystems of islands and archipelagoes of the European Arctic are usually considered as primitive and ecologically immature. Such appraisal is treated as a consequence of their young age. It is curious to note in this connection the hypothesis on the existence of a huge Panarctic ice shield that covered all the region during Quaternary glaciations, including the last one. When it is taken into account, one is forced to accept that the term "biosphere" had been replaced by glaciosphere at the considerable area of the Northern Hemisphere during hundreds of thousands of years, and the formation of high latitude ecosystems took place only during the last several thousand years, since the Holocene climatic optimum.

The authors of the Panarctic ice shield hypothesis (Denton and Hughles, 1981) unfortunately did not take into account the development of biota in the European Arctic. Until very recently, this information was rather scanty but it is now more abundant due to the progress of geological surveys and study of key sections. In this chapter we invoke some selected pollen analytical data on the Quaternary in Novaya Zemlya and Spitsbergen; more extensive descriptions are presented in other publications (Serebryanny and Malyasova, 1993; Serebryanny et al., 1983, 1993).

Our approach was based upon the comparative analysis of palaeobotanical data, including pollen, spores, and macroscopic plant fossils. It should be realized that the formation of local vegetation is required to be reconstructed, mainly with reference to macroscopic remains and then subsequent identification of local elements in pollen spectras. But these spectras give an integral reflection of the vegetation of wetlands and their surroundings, together with some allochtonous components. The thorough separation of all these contributions in successive time-span series helps to reconstruct vegetational and environmental changes in the Arctic.

Natural Background and Zonation

The areas under study are the European Arctic islands and archipelagoes, including Novaya Zemlya, Svalbard, and Franz Josef Land. They are attached to continental shelves with depths less than 100 m (Figure 2.1). Distances between the main archipelagoes are 200 to 250 km, but during glaciations they were much smaller due to lower sea levels.

The relief of Arctic archipelagoes is elevated and rugged. Dissected plateaus and low mountains with the highest summits at 1000 to 1200 m above sea level occupy large areas; coastal plains are usually not so extensive and are almost lacking in some regions. Glaciers cover 56% of the total area in Svalbard, 80% in Franz Josef Land, and 25% in Novaya Zemlya.

Ice-free areas in the European Arctic belong to the natural zones of polar deserts and tundras. The boundaries of these zones are shown in Figure 2.2. In spite of high latitude position, the vegetation of polar deserts and tundras is relatively rich in species and has a complicated pattern of spatial organization. In Franz Josef Land 36 species of vascular plants are identified (Odasz, 1992), which confirms effective capacity for plant survival, even in ice-free refugia. The most favorable environments with rich vegetation are attached to bird cliffs. In general, Franz Josef Land is a typical polar desert area.

Aleksandrova (1977) treats Svalbard (without North-Eastern Land) as an independent district of the arctic tundra subregion. The basic guidelines for such a viewpoint were total absence of Siberian species, presence of Amphi-Atlantic elements, and specific cryohumid nature of vegetation.

In the wake of Young (1971), Yurtsev et al. (1978) proposed a more differential plant–geographical subdivision of Svalbard. They referred its western coastlands, including Hornsund, Bellsund, Konigsfjord, and Prince Karl Forland, to the arctic tundra subzone and other parts of the archi-

Figure 2.1. The geographical position of islands and archipelagoes in the European Arctic. The isobath of 100 m is shown: I = Svalbard; II = Novaya Zemlya; III = Franz Josef Land.

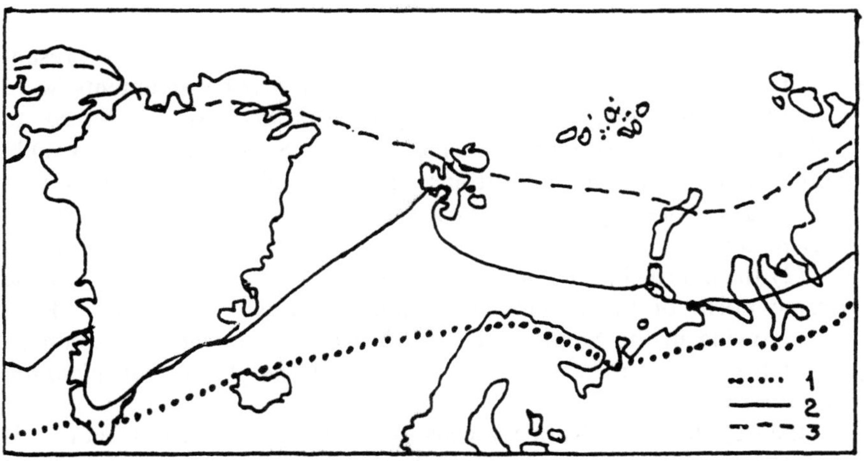

Figure 2.2. The natural floristic zones and subzones of the European Arctic: 1 = subarctic tundra; 2 = arctic tundra; 3 = polar desert.

pelago to the periglacial humid variant of the high-arctic tundra subzone. Similar viewpoints were held by Ronning (1970).

The high latitude position of Svalbard, present glaciation, and the impact of warm West Spitsbergen current cause the meridional microzonality of vegetational cover (Summerhayes and Elton, 1928). From west to east, subarctic communities of Isfjord coastlands (participation of *Betula nana*, *Vaccinium uliginosum*, *Rubus chamaemorus* in plant cover) are followed by arctic tundras and polar deserts. Rather elevated and rugged terrain provides the broad distribution of oro-polar deserts and oro-tundra communities.

The northern bulge and elevated areas of Novaya Zemlya are considered polar deserts and other parts of this archipelago are arctic tundras (Aleksandrova, 1977). The boundary between these two main zones is traced between Krestovaya and Arkhangelskaya gubas (bays) at the Barents Sea coast. It is worth noting that the assignment of the most part of Novaya Zemlya to arctic tundras is doubtful, because its southwestern coastlands are very distinct as concerns the appearance of plant cover and floristic contents. Our observation at the Rusanova Peninsula and Malye Karmakuly area indicates a rather broad distribution of bush and dwarf bush communities with the participation of Salix species and *Betula nana*, the presence of swamps with flat elevations and Sphagnum mosses, and appearance of meadow tundras with herbs and grasses. These peculiarities permit assigning western coastlands of Novaya Zemlya to subarctic tundras. Hence, the set of zonal subdivisions becomes larger than what was taken into account in palaeoecological reconstructions.

Pollen Data

Pollen analysis provides clear evidence of very old origin of the present vegetation in Novaya Zemlya. In three sites marine sediments were revealed in the socles of Holocene terraces and were referred to Eopleistocene due to microfaunistic data (V.S. Zarkhidze and O.F. Baranovskaya, oral communications). Pollen analysis helped to reveal three groups of plant species: (1) those growing now in the studied localities (*Salix polaris*, *Ranunculus pygmaeus*, *Draba sp*., *Minuartia sp*., *Cerastium sp*.), (2) those growing in southern areas, for example, at the site in the extreme north of the archipelago (Spokoinaya River valley), and (3) those growing in the Southern Island (*Valeriana capitata*) and southern part of the Northern Island (*Rubus chamaemorus*, *Thalictrum alpinum*, *Vaccinium sp*.).

In the key site at the Kara Sea coast of the Southern Island (Figure 2.3) deposits with Asteraceae (up to 40%) and Caryophyllaceae (10 to 30%) were found. Fossil pollen flora contains not only Arctic (*Salix polaris*), Arcto-Alpine (*Dryas octopetala*), and Hypoarctic (*Betula nana*, *Empetrum*

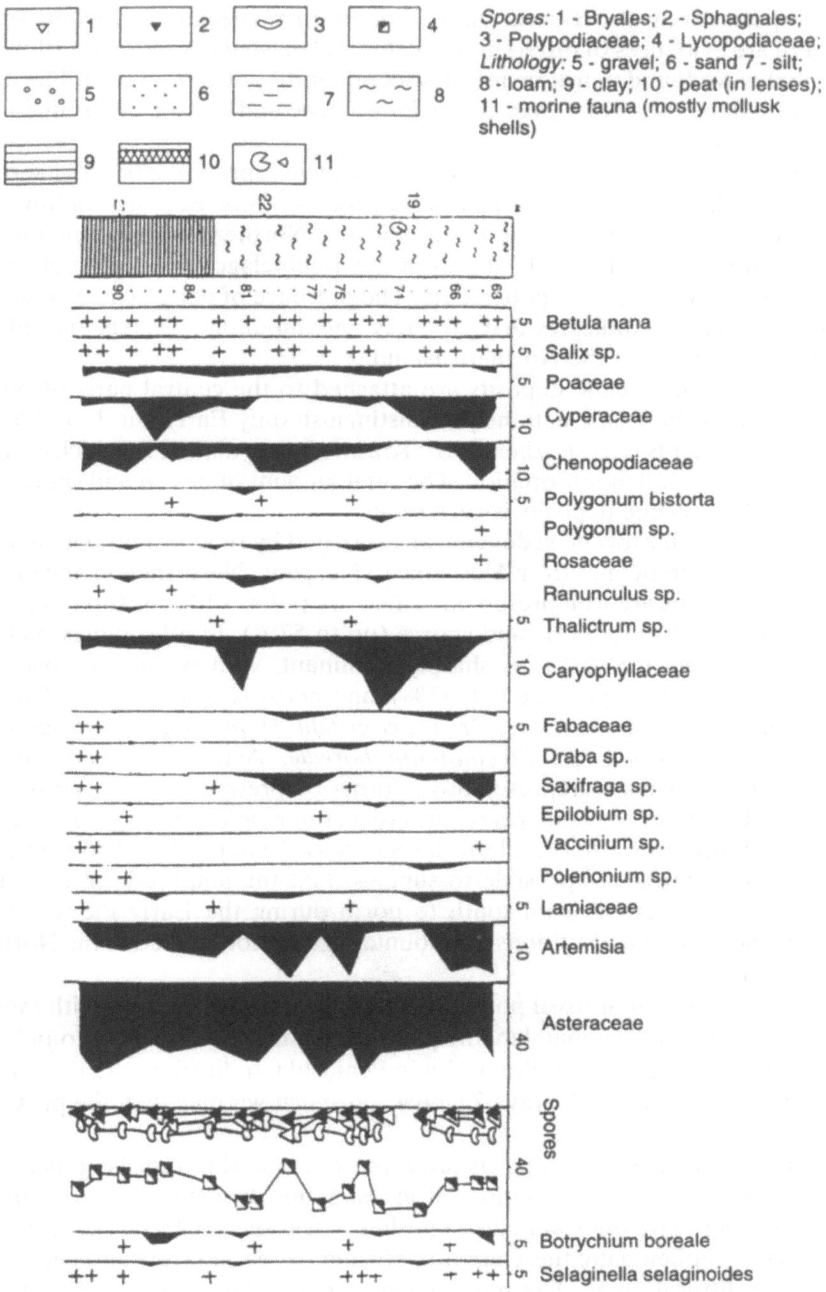

Figure 2.3. Pollen diagram of Eopleistocene maritime clays and silts at the mouth of the Listvenichnaya River, Southern Island, 40 m a.s.l.

sp.) species, but also Boreal ones (*Eriophorum vaginatum*, *Ledum sp.*). There are evidences of different life forms that existed at tundra interfluves. Probably we found maritime meadow communities of Boreal type. The age of the sediments is older than classic Pleistocene and may be determined as Eopleistocene.

The oldest loose sediments of Novaya Zemlya represent a broader series of zonal plant communities than present ones—from Boreal meadows in the Southern Island to arctic tundras in the Northern Island. But forest communities were definitely absent in the archipelago. The ice sheet existence is not confirmed by pollen data. The presence of periglacial elements, such as Ephedra, indicates that probably only mountain glaciers existed at higher elevations of the Northern Island.

The true Pleistocene deposits are attached to the central parts of both main islands. Faunistic data help to distinguish only Early and Late Pleistocene sediments (Krasnozhen, 1982; Krasnozhen et al., 1982), which correspond well to pollen information. The total amount of pollen and spores in Middle Pleistocene deposits is very small.

The Early Pleistocene sediments are enriched by redeposited pollen and spores of different ages from Mesozoic to Neogene. Nevertheless, the Quaternary remains are well preserved. There are pollen of herbs (up to 42%), dwarf bushes (up to 35%) and spores (up to 57%). In all samples pollen of *Betula nana* (58–62%) is sharply dominant, with pollen of Poaceae (16–67%), Caryophyllaceae (14–32%), and herbs (Rosaceae, incl. *Potentilla sp.*, *Rubus chamaemorus*, *Dryas octopetala*, *D. punctata*, *Saxifraga sp.*, Primulaceae, *Draba sp.*, *Polemonium boreale*, Asteraceae, *Polygonum viviparum*). So many representative forms of present tundra plants occurred. Taking into consideration differences in heights of the studied sites (400 to 500m above sea level on the Southern Island, 800 to 900m in the Northern Island), it is possible to suppose that the amplitude of tectonic movements increased from south to north during the Early Pleistocene. The most favorable conditions for mountain glaciation existed in the Northern Island.

The comparison of fossil pollen spectra from Novaya Zemlya with modern pollen rain at the mainland helps to determine the similarity to pollen spectra of the southern tundra of Kanin Peninsula. Indications are the Early Pleistocene climate of Novaya Zemlya was much warmer than the present climate.

The Late Pleistocene deposits are represented at different heights from 60 to 300m a.s.l. During geological investigation two transgressive series and two regressive ones were distinguished. Clear signs of facial changes are marked by pollen data: there are no pollen in deep-sea layers and very few grains in littoral areas. The most pollen-enriched layers are in the upper sublittoral, attached to terraces 170 to 180m a.s.l. They contain typical pollen spectras of the Kumzha, a period that corresponds to Riss-Wurm. There are two zones in pollen spectras. The lower zone contains forest-type pollen spectra with a dominance of pine, including *Pinus sylvestris* and *P. siberica* (64%), and the presence of spruce, *Picea excelsa* (10–12%), alder, and birch; only singular grains of fir (Abies), broad-leaved species (Ulmus, Carpinus), and hazel (Corylus) were identified. The upper zone is characterized by the prevalence of spores (67%) and arboreal pollen consisting mainly of birch (tree, bush, and dwarf bush forms [30%]). There are singular grains of Ulmus (2%) and Corylus (3%). Certainly arboreal pollen in both zones was transported a long distance from the mainland, but it is evidence of forest penetration to the Arctic ocean coasts during this interglacial period.

The most representative section is demonstrated in Figure 2.4. It is noteworthly to mark high contents of *Betula nana* (up to 30%), and the presence of Salix and arctic herbs, *Artemisia sp.sp.*, maximum in the lower part of the diagram. The upper part is characterized by higher and richer contents of herbaceous pollen (including Poacea and Cyperaceae, with a maximum of Bryales and Asteraceae), presence of typical arctic tundra plants (*Dryas octopetala*, *Rumex arcticum*, *Sagina nivalis*, *Epilobium arcticum*), and lesser contents of *Betula nana* and Salix. Long-transported pollen is absent due to the low elevation of the site.

We can trace two stages in vegetation development during Kumzha (Riss-Wurm) Interglacial. At the western coastland of Southern Island: firstly, southern dwarf bush tundra and secondly, typical tundra (herb moss or dwarf bush moss communities). In the Northern Island pollen spectra of Kumzha Interglacial reveal southern variants of arctic tundras. Glaciation was very similar to the present situation.

There is evidence for the existence of a warm period in the middle of Late Pleistocene. This period, called the Krestyakhino Interval, was a bit cooler than Kumzha Interglacial. Typical pollen diagrams with tundra spectra were obtained for several sites attached to the terraces at 180 to 200 m a.s.l. and mouths of palaeorivers in the Southern Island (Serebryanny and Malyasova, 1993).

The oldest ^{14}C-dated deposits are situated at the Barentz Sea coast of Novaya Zemlya. Peat buried in lacustrine deposits at the north of Gusinaya Zemlya, Southern Island, is dated at 15,310 ± 610 BP (LU 188). There is a dominance of Bryales spores, with small pollen amounts of Poaceae, Cyperaceae, and herbs (*Polemonium boreale*, *Taraxacum sp.*, *Cirsium sp.*, *Aster sp.*, *Artemisia sp.*). It resembles typical tundra spectra.

Buried peat from Mityushikha Bay is dated at 15,120 ± 120 BP (LU 1446). Pollen spectras are similar to the above-mentioned ones, but with a great variety of herbs (Figure 2.5). This locality is very close to present day glaciers, and probably it was a refugium during late glacial time when glaciation was more extensive.

Nevertheless, evidently even during the greatest cooling the ice sheet of Novaya Zemlya had no connections with Northern Uralian ice body, and outlet glaciers advanced into the shelf area only in bounded areas. Accord-

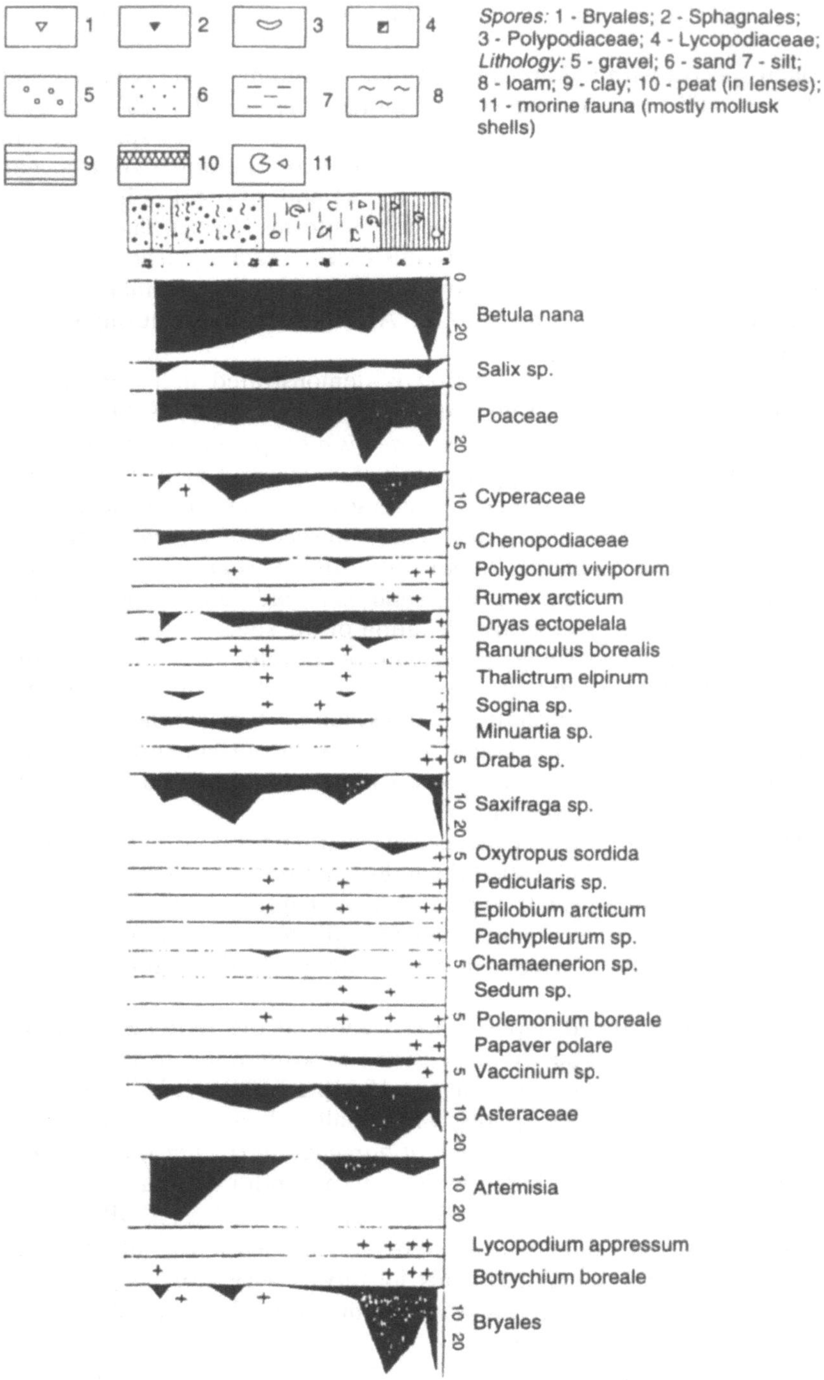

Figure 2.4. Pollen diagram of Kumzha (Riss-Wurm) marine loams at the middle course of the Rogacheva River valley, Southern Island (71°45′N).

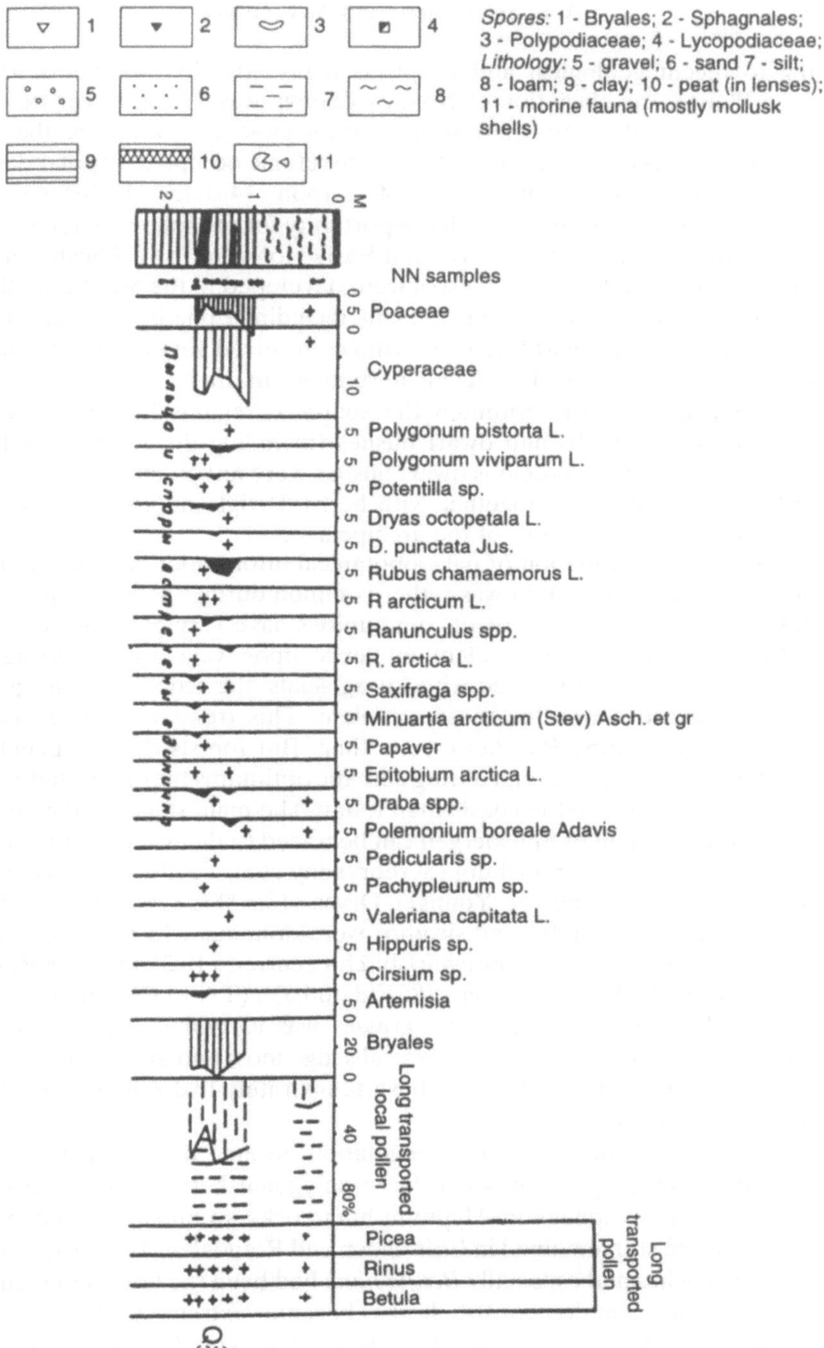

Figure 2.5. Pollen diagram of late glacial peat and clays from Mityushikha Bay, Northern Island.

ing to recent geological and geophysical investigations (Pavlidis, 1992), Late Pleistocene glaciation of Novaya Zemlya was a little larger than the present area and the Northern Island was the most ice covered. At the same time numerous coastal areas were ice-free refugia occupied by polar deserts and arctic tundras. The study of carbon 14-dated Holocene sites (Serebryanny, 1993) provided the opportunity for the following reconstructions. During the Younger Dryas and Early Holocene, dwarf bush communities with herbs and green mosses were developed in the Southern Island. Many periglacial elements were present, including Ephedra, Artemisia, and Chenopodiaceae. Dwarf birch and willows occurred very often. The northern areas were occupied by Arctic herb moss tundras.

During the climatic optimum the southern regions had an increased participation of herbs and dwarf bushes (including dwarf birch, willows, Ericales). In northern regions dwarf bushes were not abundant; sedges and grasses were dominant together with herbs. Periglacial elements become less noticeable in the flora of the archipelago.

To make an estimation of palaeobotanical information it is necessary to note that such pollen flora was rather common during the recent period of development and present plant communities have very old origins.

Climatic changes had evident influence upon vegetation. During the warmest intervals, compared with interglacials, the scope of plant groups and communities was the most complete. This trend was distinctly expressed, even during Pre-Pleistocene time. But forests did not penetrate into the archipelago, though during climatic optimums they migrated northwards much closer to the coasts than today. The main stages in the vegetational development of Spitsbergen can be traced to the recently studied site in Grondalen near Barentsburg (Serebryanny et al., 1993). The basal layers were deposited during the Younger Dryas (10,590 ± 120 BP) (TA 1572), when sedge moss tundra and swamps were dominant. In the neighboring Reindalen more grasses and dwarf birch occurred (10,360 ± 260 BP) (Tln 270). During the Early Holocene (8510 ± 100 BP) (TA 1537), pollen of *Salix polaris*, dwarf birch, sedges, and grasses was identified. Vegetation was rather abundant and variegated: typical sedge moss and grass moss mineral swamps intermingled with dwarf bush temperately hydrophylic communities in drained sites.

A similar mosaic structure of vegetation became dominant during the peak of climatic optimum, when large polygonal herb moss swamps and typical subarctic Sphagnum-Hypnum hummocky swamps occurred. Some relics of them still remained in Colesdalen and Reindalen. The participation of dwarf bushes and especially *Betula nana* had been the largest one during the Atlantic. Simultaneously, herbs became extremely differentiated (Ranunculaceae, Polygonaceae, Vacciniaceae, *Oxyria digyna*). The highest quantity of long-transported pollen (Pinus, Alnus, Betula sect. Albae, Ulmus, Carpinus) was also found during this period. The favorable environment for northwestern airflows from the temperate zone appeared due to earlier deglaciation of North America.

The subboreal was rather warm in Svalbard due to increased pollen occurrence of Caryophyllaceae, Asteraceae, Brassicaceae, and Bryales compared with subrecent spectra. The pollen contents of Poaceae also was high, because these plants were disseminated in herb/moss swamps and wet meadows (*Dupontia fisherii*, *Alopecurus alpinus* a.o.). But the share of *Betula nana* and Cyperaceae became smaller.

The cooler subatlantic environments were marked by loss of thermophilic Sphagnum mosses and Drepanocladus sendtneri, pollen contents of *Betula nana* sharply declined, but we marked the increase of *Salix polaris* and distribution of Ranunculaceae and Saxifragaceae, typical also for present conditions.

Botanical Evidence

Palaeobotanical results confirm the continuous development history of floristic complex and plant cover in the European Arctic. This process had no essential time–space interruptions, and constant exchange occurred with the mainland and between archipelagoes. The main evidence for such conclusions were obtained on examination of specific features of present flora and vegetation. The analysis of floristic connections and character of plant cover allows coming to the following conclusions.

1. The floras of European Arctic islands are extremely rich in species when considering their geographical position (see Table 2.1).

Table 2.1. Comparison of Species Abundance (Species Number) of Different Families in Vascular Plant Floras of Western Coastlands of Spitsbergen and Some Areas of the Russian Arctic[a]

Families	Polar Deserts		Arctic Tundras		Typical Tundras
	Franz Josef Land (80°N)	Taimyr Peninsula Cape Chelyuskin (77°N)	Taimyr Peninsula, Mary Pronchishcheva Bay (75°N)	Svalbard, Isfjord (78°N)	Taimyr Peninsula, Tareya (72°N)
Poaceae	16	8	15	28	28
Cyperaceae	1	2	15	13	21
Juncaceae	3	3	3	6	8
Polygonaceae	2	2	2	3	7
Caryophyllaceae	8	6	11	13	19
Ranunculaceae	3	4	7	10	14
Brassicaceae	10	8	16	16	23
Saxifragaceae	10	10	14	13	13
Rosaceae	1	3	3	6	8
Asteraceae	—	2	4	4	18

[a] Tishkov (1986).

To estimate floristic peculiarities of the western coastland of Spitsbergen we compared our data on the Isfjord area with the material from other Arctic regions. Evidently the representation of species in the main families is rather high. The leading position belongs to Poaceae and Brassicaceae, which determine similarity of Svalbard flora with typical tundra. But there are lesser indices for Cyperaceae and Asteraceae, with no Leguminosae or other families at all. It is connected with recent evolution history of Arctic floras. Changes in local floras—from 50 species in Franz Jozef Land to 200 in the Southern Island of Novaya Zemlya—are illustrated in Figure 2.6.

2. The local floras are very similar in their core parts: Svalbard and Novaya Zemlya have 63 common species, all 50 species in Franz Jozef Land occur in Svalbard, and there are 105 common species in Svalbard and Eastern Greenland coastland (Figure 2.7). The weak local endemisin and similarity of floras are evidence of continuous common development and constant exchange.

3. The key importance for the Quaternary palaeoecology of Arcto-Atlantic relates to the problem of maturity of floristic structure and plant communities compared with adjacent regional floras. Yurtsev et al. (1978) set apart the European-Western Siberian Province from the Arctic floristic region and marked its boundary position and formation in the contact zone

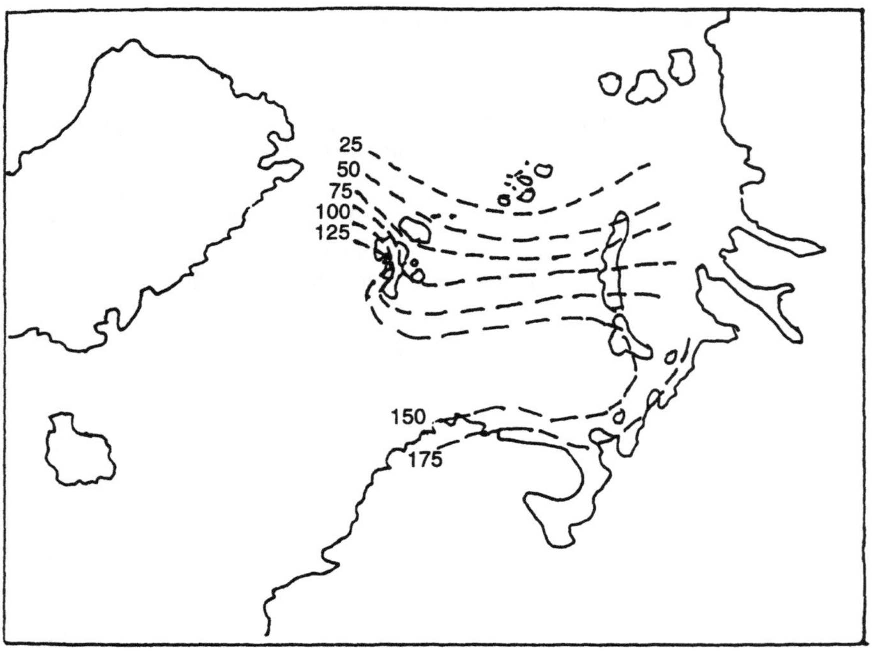

Figure 2.6. Species abundance in local floras of European Arctic.

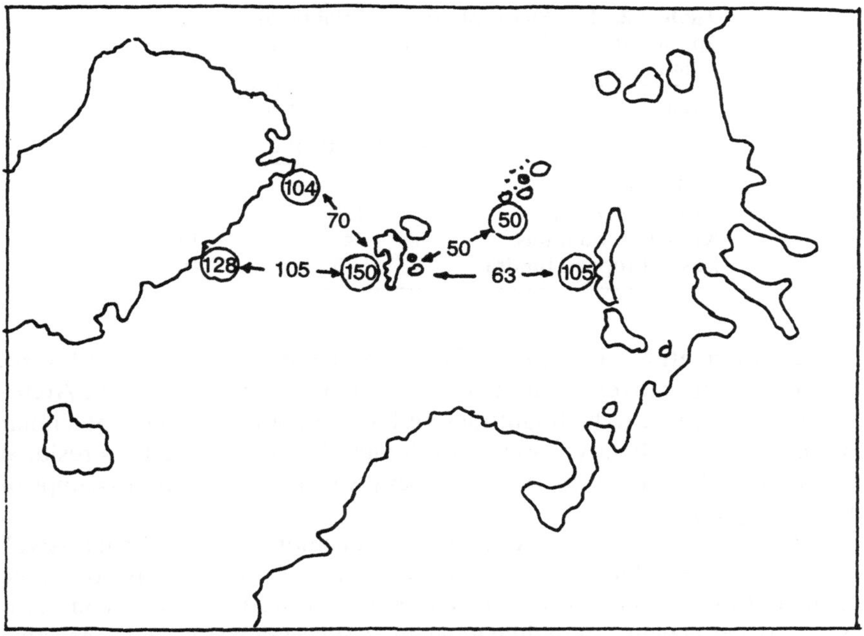

Figure 2.7. Indices of common plant species between the European Arctic archipelagoes. Total number of species is shown in circles.

of "eastern" and "western" migration routes. We believe that formation of an individual succession system without "Siberian" and "American" species in climax group in the above province bears witness to a long duration of this process. The Holocene of the Arctic is equal only to two to three normal "succession cycles," and it would be too difficult to create a succession system providing the formation and restoration of vegetation on different grounds during such a short time interval.

4. The "key" event in floristic connections and vegetational development of the Arcto-Atlantic region may be indicated by the conduction of Amphi-Atlantic species (Table 2.2).

The presence of Amphi-Atlantic elements to floras of the region was used as a differential index of floristic division. Eastern boundaries of most species of this group are very abrupt. But the main factor is connected with their importance for the composition of plant communities: for example, in Novaya Zemlya only *Silene acaulis* has an essential coenotic role among all the Amphi-Atlantic species, but the Siberian plant *Carex ensifolia* ssp. arctosibirica participates actively in the structure of climax communities. The importance of Amphi-Atlantic species (*Betula nana*, *Deschampsia brevifolia*, *Carex parallela* a.o.) is much greater in communities.

Table 2.2. The Share of Amphi-Atlantic and Siberian Species in the Floras on some Areas of the Arcto-Atlantic

Areas	Species	
	Amphi-Atlantic	Siberian
Svalbard	18	—
Novaya Zemlya	12	8
Khibiny Mountains	4	14
East-European tundras	—	16

The participation of Amphi-Atlantic species is diminished southward. Probably comparatively continuous formation of floristic complex Arcto-Atlantic took place in high latitudes and in a climatic environment similar to the present or slightly warmer conditions (it is evidenced by a residual distribution of *Betula nana* and hummocky Sphagnum-Hypnum swamps of subarctic type).

5. The main obstacle for eastward penetration of many "aggressive" species (Siberian) during Late Pleistocene-Early Holocene was not an ice shield, but the existing succession system. It appeared to be very conservative as concerns plants of middle and climax succession stages. At the same time possibilities for the distribution and naturalization of pioneer species (e.g., those setting on primary grounds) always existed. This group of plants makes up the core of circumpolar species. As concerns such species, the laws of island biogeography operated during Late Pleistocene and Holocene, when rates of invasion and occupation were determined by the distance from sources of biota, sizes of islands, directions of sea currents, and sea level changes.

Only ecological niches of "pioneer species" were vacant permanently. Oropolar, petrophylic, swamp, Arctotundra, and Subarctotundra communities were filled up, inaccessible for the invasion of new species, in Svalbard, Novaya Zemlya a.o., even in the Late Pleistocene.

6. In different periods of Holocene and Late Pleistocene Svalbard had an impoverishing influence on its flora. During colder periods the share of Subarctic and Arcto-Boreal species diminished as well as the share of Arctic and Arcto-Alpine species during warmer periods. But great ruggedness of terrain, considerable latitudinal differences, variegated bedrock and loose deposits, and asymmetry of glaciation preserved the floristic core and the succession system, at least during the Late Pleistocene and Holocene and probably during all the Pliocene and Quaternary, when there were variegated ice-free areas. In various high mountains such refugia existed even during severe glaciations.

The comparison of palaei- and recent data on the development of Arctic floras and vegetation as well as evolution of biota and ecosystems helps to make objective judgments on total environmental changes and to make correct ecological forecastings. Such an approach based on the search of past analogues forms the key for understanding present ecological situations. We hope that our attempt will be useful for the solution of similar problems under IGBP program.

Summary

1. There is close correlation of Quaternary and especially Holocene events in Novaya Zemlya, Franz Josef Land, and Svalbard.
2. The active exchange of island biotas took place between archipelagoes and the mainland before the Oldest Dryas. Probably the second activization maximum occurred during temporary marine regressions in the Atlantic-late Subboreal.
3. The climatogene reconstructions of island biotas were realized due to their own genopool. Typical island processes of extinction and immigration had no great importance in arctic successions.
4. In the Late Pleistocene large but not continuous glaciation nodules existed in the Arcto-Atlantic islands together with numerous coastal, valley, and nunatak refugia.
5. At the same period succession system of vegetation was formed and it provided the following rapid colonization of ice- and water-free surfaces.
6. The modern trends of climate changes in the Arcto-Atlantic resemble the situation at the beginning of Holocene climatic optimum and other periods. The response of Arctic vegetation upon global changes may be well modeled on the basis of Holocene events.
7. The economic use of Arcto-Atlantic modified the succession system in some areas (e.g., increased areas of disturbed lands and plant deficiency of the early stages of successions), but it did not lead to irreversible consequences, because we still observe high biodiversity, plasticity, and dynamics of biota.

References

Aleksandrova, V.D. 1977. *The Geobotanical Regionalization of Arctic and Antarctic*. Leningrad: Nauka (in Russian).

Krasnoshen, A.S. 1982. The history of recent development of southern part of Novaya Zemlya. In: *Geology of the Southern Island of Nivaya Zemlya* (pp. 100–109). Leningrad: Sevmorgeologiya.

Krasnozhen, A.S., Baranovskaya, O.F., Zarkhide, V.S. et al. 1982. The Upper Quaternary deposits of the Southern Island of Novaya Zemlya. In: *Late Cenozoic Stratigraphy and Palaeogeography of the Arctic* (pp. 40–52). Leningrad: Sevmorgeologiya.

Odasz, A.M. 1992. Vascular plant distribution in Franz Josef Land. In: Ggertz, I., Morkved, B. (eds.), *Environmental Studies From Franz Josef Land, With Emphasis on Tikhaia Bay, Hooker Island* (pp. 83–130). Oslo: Worsk Polazinstitutt.

Pavlidis, Yu.A. 1992. The extent of last glaciation in the Arctic Basin. *Oceanology*, 32(5):525–521 (in Russian, English summary).

Ronning, O.I. 1970. *Synopsis of the Flora of Svalbard* (p. 1). Oslo: Norsk Polarinstitutt Arbok.

Serebryanny, L.R., and Malyasova, E.S. 1993. The natural history of Novaya Zemlya. In: Boyarsky, P.V. (ed.), *Novaya Zemlya* (Vol. 2). Russian Institute for Cultural and Natural Heritage (in Russian).

Serebryanny, L.R., Tishkov, A.A., Malyasova, E.S., Solomina, O.N., and Ilves, E.O. 1983. The reconstruction of vegetational development in the high latitudes of Arctic. *Izv. Akad. Nauk SSSR. Ser. Geogr.*, 6:75–84 (in Russian). (Translated in: *Polar Geogr. Geol.* 9(4):308–320, 1985.)

Serebryanny, L.R., Tishkov, A.A., Malyasova, E.S., Solomina, O.N., and Ilves, E.O. 1993. Holocene palaeoecology of the Arcto-Atlantic. *Izv. Akad. Nauk Rossii. Ser. Geogr.* 3:39–52 (in Russian).

Summerhayes, V.S., and Elton, C.S. 1928. Further contributions to the ecology of Spitsbergen. *J. Ecol.* 16:193.

Tishkov, A.A. 1985. The primary successions of Arctic tundras of the western coastlands of Spitsbergen (Svalbard). *Izv. Acad. Nauk SSSR. Ser. Geogr.* 3:99–105.

Young, S.B. 1971. The vascular flora of St. Lawrence Island with special reference to floristic zonation in the Arctic.

Yurtsev, B.A., Tolmachev, A.I., and Rebristaya, O.V. 1978. Floristic limitions and subdivision of the Arctic. In: *The Arctic Floristic Region* (p. 3). Leningrad: Nauka.

3. The Impact of Hydrologic Perturbations on Arctic Ecosystems Induced by Climate Change

Douglas L. Kane

The physical environment of the polar regions of the world go through large seasonal swings, such as air temperature, duration and intensity of solar radiation, surface albedo, and surface energy balance, as well as the magnitude of many of the hydrologic processes. These fluctuations affect both the animal and plant biota, particularly in the Low Arctic where they are more predominant. Long-term alterations induced by climatic change, superimposed on the seasonal changes, could further influence the ecosystem. In the last 20 years or so we have begun to study how ecosystems work in the Arctic and how the various physical, chemical, and biological processes are interwoven and sustain today's environment.

Hydrology plays an important role in the world's climate because of the water fluxes and the energy associated with these fluxes. In cold regions of the world, precipitation as snow can significantly alter the surface radiation balance and so the duration of the snowcover is important. To successfully predict future climates globally, it is imperative that we understand and incorporate into general circulation models (GCMs) the hydrologic processes of the high latitudes.

Hydrology is closely interrelated to meteorology and in cold regions this is accentuated by phase change back and forth from liquids to solids (as well as associated latent heat): freezing and thawing of soils, snowmelt, aggradation and degradation of ice covers on lakes and streams and the formation of rime, etc. As one progresses farther north, the seasons become dominated by snow and ice. Biologically, this has important implications to the plants and animals that have adapted to and inhabit these areas.

The Intergovernment Panel on Climate Change (IPCC, 1990) made a scientific assessment of climate change over the next few decades. There is absolutely no doubt that greenhouse gases [carbon dioxide, methane, nitrous oxides, and chlorofluorocarbons (CFCs)] are increasing dramatically. The same panel estimated in 1992 (IPCC, 1992a) that under a scenario of a doubling of CO_2 the global mean surface temperature is unlikely to lie outside the range of +1.5° to 4.5°C. This conclusion was reached assuming a scenario of "business-as-usual" (IPCC, 1990). It is also acknowledged that the mean surface air temperature has increased by 0.3 to 0.6°C over the last 100 years. Coincidentally, this is of the same magnitude as the natural climate variability. This is one reason why they concluded that it will be at least a decade or more before any detection of the greenhouse enhancement of our climate is made. Finally, they point out that there are many uncertainties in our predictions in regards to timing, magnitude, and regional patterns of climate change.

The main differences between the 1990 assessment and the 1992 assessment are that the latter one incorporates results from transient models (rather than equilibrium models) with ocean–atmosphere components (our understanding of climate feedback processes is improving, and models continue to be more physically realistic). Also, more attention is being directed toward the role of gases and aerosols (basically ozone and sulfur-derived airborne particles).

GCMs have generally predicted that there will be global warming, that the greatest warming will occur in the arctic regions of the world, and that most of this warming will be seasonal during the winter months. From these results it is simple to draw the following conclusions:

1. The depth of soil freezing will be less or the amount of heat extracted from the ground during the winter months will be less.
2. The snow-free season should be longer. How the distribution will be partitioned between spring and fall seasons is not clear; however, the extent of the snow-free season will be greater in the fall than spring.
3. More rainfall should occur, with less snowfall. This issue is not totally clear because the GCMs also predict changes in precipitation, with some areas having an increase and others experiencing a decrease.
4. The southern limit of permafrost in the Northern Hemisphere should migrate northward and there should be a general warming of the permafrost.
5. The probability of early melt periods and the formation of ice lenses in the snowpack should increase.
6. The spatial and temporal distribution of snowcover in the world should decrease.

In the following sections the impact of climate change on the hydrologic processes within the Arctic and the biological ramifications of this altered hydrologic cycle are discussed.

Climate Change in the Arctic

Why is the Arctic important in the climate change scenario and what role does hydrology play in this process? These two questions are the essence of this chapter. In terms of impact on people and the food-growing industry, arctic regions of the world are the least impacted. However, in terms of the world climate, the arctic regions are just as important as other areas, and changes in the Arctic can significantly influence the climate elsewhere. Although GCMs all have their limitations, one common prediction is that greater warming will occur in the Arctic than elsewhere and this difference could be two to three times greater than elsewhere. It is also predicted that the greatest warming will occur during the winter months. In an area where much of the near-surface soils are at temperatures just below the freezing point of water, this is quite significant.

Warming of the surface soils may substantially alter the permafrost distribution as we know it today. Lachenbruch and Marshall (1986) have found that in the past few decades near the permafrost table there has been a warming of 2 to 4°C for northern Alaska. In the near-surface considerable carbon is tied up in the soils. The decomposition of these soils could produce considerable carbon dioxide and methane, which would have additional global consequences.

There are very few places in the Arctic where sufficient, high-quality hydrologic data exist and the hydrologic processes are understood well enough so that estimates can be made of how changing climate is going to affect the hydrology and what this means to the ecosystems that prevail there. Imnavait Creek in the Alaskan Arctic is the most intensively studied watershed, and therefore it provides an opportunity to model a wide range of present-day conditions, as well as making some climate change-induced predictions (Hinzman and Kane, 1992).

Arctic Hydrology

Although the hydrology of the Arctic is important globally, until recently this largely unpopulated region has received little attention. The concern of global change has acted to catalyze new interest in arctic hydrologic processes. As part of the world's climate engine, it has long been recognized as an area that sheds excess energy from the earth's surface. But it also represents an area that has in the past stored large quantities of carbon in the surface soils. The decomposition of these organic soils, which may accelerate in a warmer climate, could produce additional quantities of either carbon dioxide or methane, depending upon soil moisture conditions in the active layer. This, of course, would amplify the global warming problem.

One of the more important aspects of climatic change, if it is warming, is that the structure of arctic watersheds will change (Kane et al., 1992). Both continuous and discontinuous permafrost will be impacted. For continuous permafrost, it means that the active layer thickness may change; this is important because of the amount of soil water storage and the area available for flow down the hillslope will increase with a thicker active layer. In discontinuous permafrost, a warmer climate could completely eliminate shallow permafrost and allow recharge to the deeper subpermafrost groundwater. Kane et al. (1991) found for Imnavait Creek, in the Alaskan Arctic, that warming of 2°C increased the present active layer thickness of 50 cm to almost 70 cm over a 50-year period. An average annual temperature increase of 4°C almost doubled the thickness of the active layer. This assumed that the makeup of the active layer remains the same in terms of the soil constituents. They also assumed that the temperature changed uniformly over the year. In a subsequent article, Hinzman and Kane (1992) showed that the distribution of warming over the year had little effect on the final depth of thaw after 50 years of warming. They also found that if the warming was greater than 4°C, the active layer failed to freeze back completely and a talik formed. This has repercussions in that subsurface drainage could continue throughout the winter in the active layer, a process that presently does not take place in areas of continuous permafrost. The complete melting of permafrost would take considerable time, approximately the same amount of time it took to form.

Phase change is a dominant process in arctic regions. The freezing and thawing of ponds and streams, the formation and decay of *aufeis*, freezing and thawing of the active layer, melting of snow, condensation, sublimation, evaporation, and transpiration are all hydrologic processes that are active in the Arctic. Therefore, rates at which hydrologic processes proceed are closely tied to the magnitude of energy fluxes. Although sensible heat flux is important in such processes as warming the air, soil, and snowpack, latent heat at both ends of the liquid spectrum is crucial to arctic hydrologic processes.

The permafrost hydrology of North America is summarized in an article by Woo (1986), where he emphasizes the importance of the various processes and stresses the need to examine both the heat and mass fluxes associated with many of the processes. A more succinct presentation of arctic hydrology is included in Kane et al. (1992) along with tables summarizing data on watershed water balances for both the Low and High Arctic in North America.

The impact of climate change on the hydrology of the Arctic is mostly qualitative (Woo, 1990). Data bases for this area of the world are quite

Table 3.1. Spring and Annual Water Balances for Imnavait Creek Watershed, Alaskan Arctic

Year	Snowpack Water (cm)	Summer Precip. (cm)	Total Precip. (cm)	Snowmelt Runoff (cm)	Summer Runoff (cm)	Total Runoff (cm)	Evapo-trans. (cm)	Pan Evap. (cm)
1985	10.2	25.1	35.3	6.6	*	*	*	*
1986	10.9	16.3	27.2	5.7	6.2	11.9	15.3	31.0
1987	10.8	27.2	38.0	7.1	17.9	25.0	13.0	32.0
1988	7.8	25.2	33.0	3.9	7.2	11.1	21.9	33.2
1989	15.5	25.7	41.2	9.4	7.8	17.2	24.0	42.0
1990	10.6	16.3	26.9	6.4	2.8	9.2	17.7	39.4
1991	8.2	24.9	33.1	5.6	4.3	9.9	24.2	37.7
1992	18.1	24.1	42.2	14.4	6.3	20.7	21.5	32.8
1993	12.5	20.8	33.3	7.9	14.6	22.6	10.8	32.1

*Data unavailable.

sparse and scientific studies are lacking (Kane, 1993). Output from GCMs is questionable for the polar regions because of the poor quality of data used in the models. Many of the induced hydrologic changes due to climate change are going to be subtle and therefore difficult to identify without good-quality field measurements. Presently, we have very little good-quality data; in addition, there is considerable variation from year to year in many of the processes. For example, snowmelt may be earlier, as predicted by Hinzman and Kane (1992), but the onset of snowmelt can in any year occur from early May until the end of May for the watershed they studied. Other hydrologic processes vary also, as illustrated in Table 3.1 for Imnavait Creek in the Alaskan Arctic.

The most important hydrologic indicators of climate change in the Arctic are the seasonal extent of snowcover, soil moisture in the active layer, evapotranspiration, and runoff. Each of these components of the arctic hydrologic cycle will be discussed relative to precipitation inputs.

Precipitation

Both the short duration and the small number of precipitation stations in the Arctic have contibuted to the fact that we have a poor understanding of spatial distribution and quantity of precipitation in this region. The problem is made worse because for long periods of time we have used gauges that catch less than the true amount of precipitation, particularly during snowfall (Benson, 1982; Clagett, 1988; Woo et al., 1982). In the Alaskan Arctic north of the Brooks Range, an area greater than 200,000 km^2, there have only been two stations with any length of record, but the data are unreliable because of the windy conditions that prevail during the winter snowfall.

The annual variability in both snow on the ground just prior to ablation and summer precipitation, along with total annual precipitation, is illustrated in Table 3.1. These data also illustrate the seasonal variation both within and between years. The snowpack water content data do not include

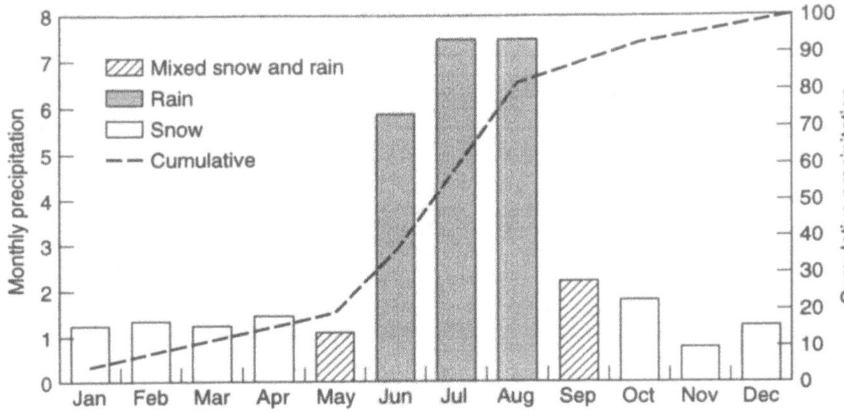

Figure 3.1. The mean monthly precipitation (cm) and the cumulative annual precipitation (%) for Imnavait Creek Watershed, Alaskan Arctic (Jan. 1984 to Sept. 1992).

water that may have been lost to sublimation; therefore, the actual annual precipitation would be greater than reported here. Because we are using these data to make hydrologic predictions of runoff from snowmelt, we are only interested in what is on the ground at the onset of melt. Hydrologically, what is more important is redistribution of the snowpack by wind. In all arctic environments, wind is an important factor and it is not necessary that it accompany the snowfall. Wind associated with the event results in an undercatch of snowfall precipitation and redistribution, whereas wind during and after an event can result in redistribution only.

Precipitation as snow can occur during any month of the year, but it is unusual for rainfall to occur during the winter months. This is probably less important hydrologically than biologically. The formation of dense layers that develop in a snowpack after rainfall early in the winter would force those animals that need to reach a food source on the ground to expend more energy over a longer period of time than if it occurs late in the winter. Most arctic plants have adapted to the harsh environment; how severely they are impacted depends on the duration and air temperatures associated with summer snowfall events.

How precipitation amounts and intensities would change in a warmer climate in the Arctic is difficult to predict at this time. GCMs in general have predicted some modest increases in precipitation. With higher rates of evapotranspiration the amount of precipitation has to increase because of limited atmospheric storage. It is the spatial distribution of this precipitation that is unkown. More important may be that ice cover in the seas of the Arctic will decrease and there will be greater source areas for moisture, a phenomenon unique to the Arctic.

Global warming with longer summers and shorter winters would result in more of the annual precipitation falling as rainfall. Presently, for the Alaskan Arctic, 40% (on average) of the annual precipitation comes as snow (Figure 3.1). With no additional precipitation, this would result in shallower snowpacks. Increased precipitation would help offset this loss. This is an area where there are so many factors to consider that predictions have to be utilized with care. Ecosystem response is clearly going to be controlled by the amounts of evapotranspiration and precipitation.

Seasonal Snowcover

Snow plays an important role in both the surface energy and hydrologic balances. The albedo of arctic snow is about 0.8 if the vegetation is buried; during the summer months without snow the albedo is 0.2 (Figure 3.2). Wind-blown dust can decrease the albedo of the snow cover. This is a common phenomenon along the major river drainages and such man-made structures as the haul road in Alaska that runs from Prudhoe Bay to the south. In the Alaskan Arctic, snow can occur at any time, but from mid-May to mid-September it usually melts within a few days. Once the snowcover is permanent in September, it continues to accumulate until sometime in May

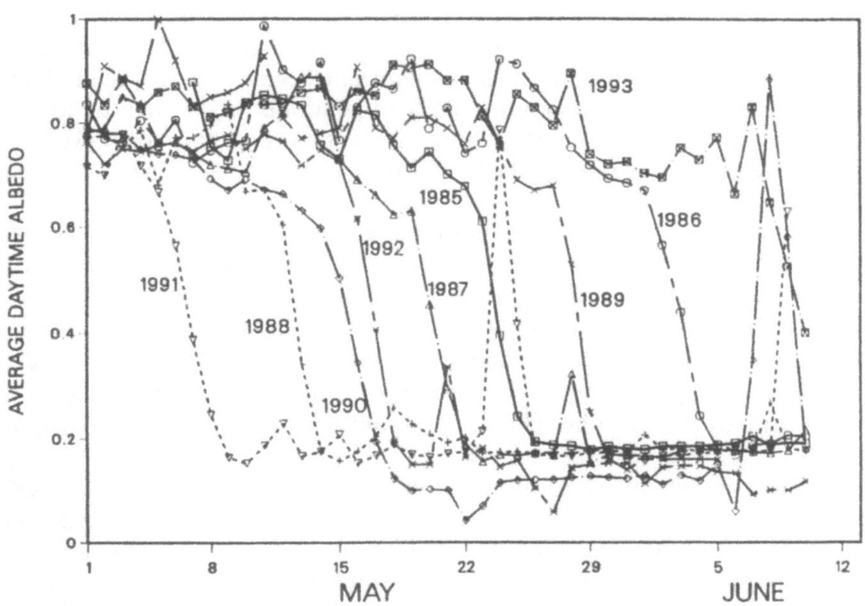

Figure 3.2. Surface albedo for nine snowmelt and summer seasons, Imnavait Creek Watershed.

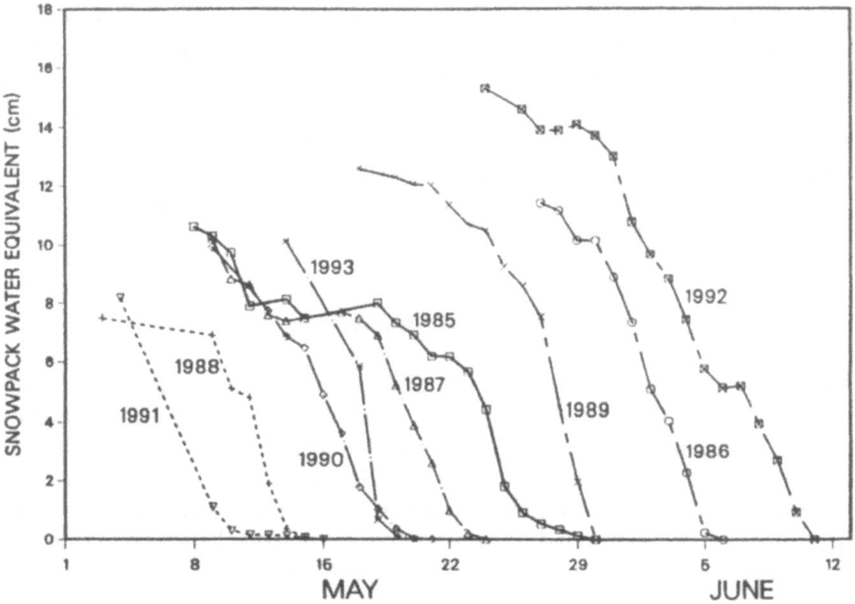

Figure 3.3. Snowpack water equivalent (cm) during ablation for the period 1985 to 1993, Imnavait Creek Watershed, Alaskan Arctic.

when snowmelt is initiated. During the winter months redistribution by wind occurs. Hydrologically this is significant because snow accumulates along drainages where it is more likely to generate runoff than be lost to evaporation during ablation. Those areas that are bare or have less snowcover with vegetation sticking through are going to have a lower albedo and melt first.

At Imnavait Creek, average basin water content of the snowpack has varied from a low of 7.8 cm to a high of 18.1 cm (Table 3.1, Figure 3.3), measured just prior to ablation. This watershed is about 950 m above sea level and receives more precipitation than the coastal areas. The greater the snowpack water content the greater the volume of runoff. Limited storage in the active layer is responsible for this outcome. Ablation is relatively short, usually 7 to 10 days (Figure 3.3), but it can be interrupted by cold spells.

Climate change in the form of warming will shorten the seasonal extent of snowcover. In the spring, the slope of the mean daily temperatures is very steep upward, but in the fall the same slope during cooling is much flatter. This would imply that in the spring warming would occur in a relatively short time, whereas the same temperature change in the fall would take much longer. Therefore, snow will not disappear that much earlier in the spring, and in the fall the date that snow is sustained will be delayed considerably for a climate scenario of warming. This is in fact what Hinzman and Kane (1992) reported for Imnavait Creek: a 4°C warming scenario advanced snowmelt by 1 week in the spring and delayed snow accumulation in the fall by about 3 weeks. In total, this represents almost 1-month reduction in snowcover. This obviously has repercussions on the surface energy balance. As mentioned, snowmelt can occur over a 1-month window and plants obviously take advantage of early snowmelt. If snow disappears on the average of 1 week earlier each year, plants will respond to this change. In the fall, the question is whether there is enough light for the plants to continue to function or will they eventually be replaced by other plants?

Runoff

Runoff generation is a result of either snowmelt or rainfall. Snowmelt rate is controlled by the amount of available energy, and rainfall intensity is controlled by atmospheric conditions. It can, at any time with temperatures above freezing, rain at a rate greater than the ablation rate. This means that, with the limited storage in the active layer, greater floods will occur from rainfall than snowmelt. Nevertheless, every spring significant streamflows occur in arctic watersheds. Part of this is due to the volume of water in the snowpack and the fact that snow damming produces a higher peak runoff response. Kane et al. (1991) found that 60 to 70% of the snowpack melted in the Imnavait Creek watershed before significant runoff was produced. Peak flow generally occurs rapidly as a slush flow cuts a channel through the snowpack in the valley bottom and the accumulated snowmelt adjacent to this channel drains. Hinzman and Kane (1991) modeled these processes with good results.

Redistribution of snow in a watershed by wind is responsible for the snow-damming effect in the valley bottom. Snow also accumulates in depressions on the hillslope, referred to as water tracks. These small, steep gradient drainages are efficient at removing water off the side slopes, even though deeper snowpacks exist there. However, in the valley bottoms, where the stream gradients are much less, the meltwater accumulates in the snowpack until it is able to cut a channel through the snowpack. The process of snow accumulation along drainages results in a higher ratio of runoff than would occur if the snow was uniformly distributed over the watershed.

Timing of ablation in the Alaskan Arctic can vary from year to year, with shallow snowpacks melting in early May and thicker snowpacks melting in late May as a general rule. With thicker snowpacks, the albedo remains higher for longer periods of time and therefore delays snowmelt (Figures 3.2 and 3.3). It is not unusual for snowmelt to be interrupted by at least one cold spell lasting several days.

Summer runoff is closely tied to rainfall. After snowmelt, flows generally recede significantly and will remain low until sufficient rainfall occurs to both rewet the active layer and produce runoff. The summer months of June, July, and August are the wettest in the Alaskan Arctic (Figure 3.1). Runoff response is rapid when the rainfall amounts reach a certain threshold. Kane et al. (1989) found that little or no runoff was produced following relatively dry periods when rainfall quantities were less than 12 to 15 mm. Figure 3.4 shows the variability in the distribution of snowmelt and rainfall runoff for several years at Imnavait Creek. This variability is closely tied to the precipitation patterns.

During climatic warming, runoff volume will likely decrease at the expense of additional evapotranspiration. The exception to this would be if precipitation increases sufficiently to offset the additional evapotranspiration. With an earlier snowmelt, the runoff season for the small drainages will be initiated earlier in the year and will continue longer into the fall. If warming is sufficient to develop a talik between the seasonal frost and permafrost table, many of the headwater streams may eventually have small quantities of flow throughout the winter months.

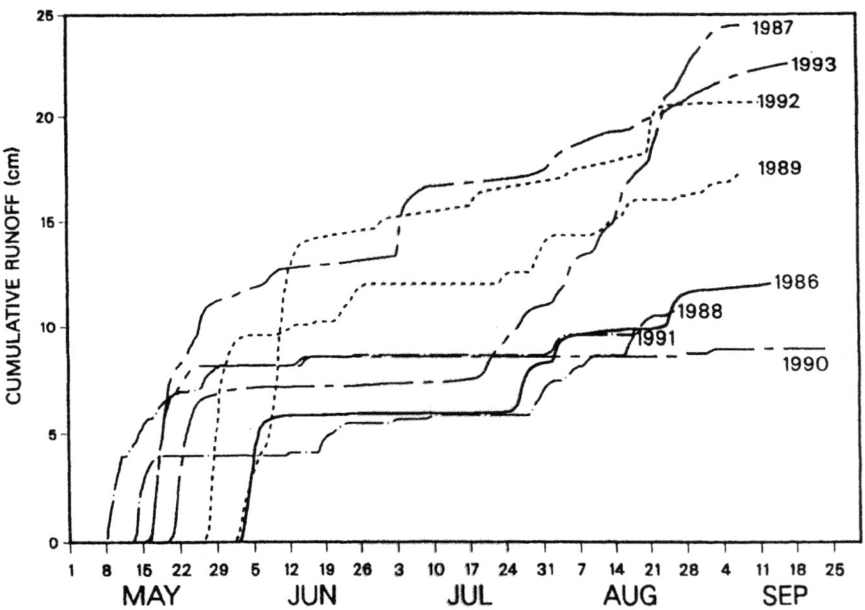

Figure 3.4. Cumulative runoff for eight snowmelt and summer seasons, Imnavait Creek Watershed.

Evapotranspiration

Most years, more water leaves arctic watersheds as evapotranspiration than as both rainfall- and snowmelt-generated runoff combined (Table 3.1). There is also a wide range of evapotranspiration from year to year; for an 8-year period at Imnavait Creek it varied from 13 to 24.2 cm. Generally, wet years have higher evapotranspiration than dry years, but this is not always the case. The distribution of the rainfall is critical; summers with high precipitation but few storms are conducive to more runoff than evapotranspiration. Likewise, numerous light rainfalls enhance evapotranspiration.

There have been numerous evaporation-related studies in the North American Arctic and the results have been summarized by Kane et al. (1990). Several techniques have been used to calculate fluxes returning moisture to the atmosphere for environments where evaporation dominates, such as lakes, to the watershed scale where transpiration is more important. Mean daily rates of evapotranspiration varied from 0.7 to 4.6 mm/day, with rates generally decreasing at the higher latitudes. In Table 3.1, evapotranspiration as determined by the water balance of a small watershed can be compared to point evaporation from a pan. The ratio of calculated evapotranspiration to the measured pan values vary between 0.4 and 0.7.

With a warmer climate there is more available energy for evapotranspiration. It is predicted that evapotranspiration will increase in most parts of the world, although it is site specific. With an average annual increase in air temperature of 4°C, Hinzman and Kane (1992) foresee an increase in evapotranspiration, both with no change in precipitation and a 15% increase in precipitation. With a 5% decrease in precipitation, which has not been predicted in any GCMs, they show a decrease of a few percent.

Soil Moisture

In those northern latitudes where continuous permafrost prevails, soil water of the active layer is essentially the only subsurface flow of any consequence. Also, the amount of storage is severely limited by the shallow depth of the active layer. Large quantities of groundwater are stored as ice in the permafrost, but hydrologically they are not important. Even in the winter time, downslope water movement in the active layer ceases once it is frozen. In northern Alaska, the active layer is usually completely refrozen by December; in warmer areas where permafrost is near 0°C the active layer may not completely refreeze until late March when the minimum soil temperatures occur.

Typical soil conditions in Arctic Alaska are shallow organic soils (5 to 25 cm) over mineral soils. Except during periods of drought and where well-drained soils exist because of topography, the mineral soils are near saturation, but the moisture content of the organic soils can vary substantially (Hinzman et al., 1991). Late summer rainfall prior to freeze-up guarantees that the soils are wet prior to freezing. However, the large pores in the organic soils allow rapid gravity drainage before freeze-up and some vertical migration of moisture into the overlying snowpack in response to an upward heat flux during the winter months. During snowmelt, the organic surface soils provide some storage; Kane et al. (1991) estimated from laboratory and field measurements that this value averaged over the entire Imnavait Creek watershed was about 15 mm. For the 9 years of data for Imnavait Creek (Table 3.1), this represents from 8 to 20% of the snowpack water content. Similar estimates of active layer storage were reported for summer conditions (Kane et al., 1989), where no significant increase in runoff was observed for rainfall events of less than 15 mm unless it had rained very recently.

After snowmelt, the active layer is near saturation; hillslope drainage and the onset of transpiration reduce soil moisture levels with time. Precipitation in May is low and in many years this trend carries over into June. Sustained low precipitation results in very dry organic soils and even some drying of the upper mineral soils. In the Alaskan Arctic, about 60% of the annual precipitation falls during the summer months of June, July, and August (Figure 3.1); because of the limited soil storage, moisture contents of the active layer over the summer months are closely tied to the precipitation pattern.

Predictions of soil moisture change in response to a warmer climate by GCMs indicate that those areas that are relatively dry now, such as the Great Plains in the United States, will become drier. However, one should be cautious in using GCM results on regional scales. Physically, it makes sense that a warmer climate would increase evapotranspiration; this may result in increased precipitation, but not necessarily at this site. Hinzman and Kane (1992) found for Imnavait Creek that predicted soil moisture levels in the active layer rose earlier in response to earlier snowmelt. The lack of spatially distributed field measurements of active layer soil moisture makes it difficult to adequately assess what may happen in a changing climate. This, coupled with the fact that the active layer will physically change if the surface energy balance is altered, increases the complexity.

Animal Response

Numerous four-legged animals, from those small creatures that live under the snow to the large polar bear that roams the arctic ice, will be impacted by changes in the climate and hydrology. Because of their nonaggressive behavior, large numbers, size, and extent of range, caribou (*Rangifer tarandus*) have been studied in more depth than most other arctic animals.

There are several hydrologic changes that could severely impact their lifestyles, such as the temporal and spatial extent of snowcover, midwinter snowmelt events, and depth and density of snow. Caribou spend a large part of the year traveling over the snowcover, foraging for food under the snowcover, and calving in areas where the snowcover lingers.

The energy expended by caribou in reaching their major food source under the snowcover can vary depending on both the density and depth of the snowpack. Fancy and White (1985) found that caribou expended twice the energy in cratering for food in a dense snowpack with a thin hard crust than for the same activity in light, uncrusted snow. It is not surprising that in years of heavy snow animals must expend more energy to obtain food and that body fat and protein may be utilized to offset this added energy output. For the Porcupine caribou herd in Northwestern Canada, Russel et al. (1993) found that in years of heavy snow regionally, caribou exploited over-wintering sites where the snowpack was the shallowest. The depth of the snowpack in the Arctic varies considerably from year to year (Hinzman et al., 1996); under a scenario of climate warming the period of snowmelt accumulation will definitely be shortened (Hinzman and Kane, 1992) but the trends in the amount of snowfall are less certain. Although relatively rare now, midwinter snowmelt events would become more frequent with a warmer climate. Refreezing of this meltwater in the cold snowpack would result in a layer(s) of ice developing within the snowpack that could provide a significant energy barrier to those animals digging or cratering in the snow. Presently, the density of the snowpack is governed by the wind conditions that prevail; average density of a windblown snowpack is about 0.3 to $0.35\,g/cm^3$, generally sufficient for a person to walk on top. The density of ice layers could approach $1\,g/cm^3$, but the thickness could be limited to the amount of ablation or rainfall. In 10 years of studying the hydrology at Imnavait Creek, we have observed only two events where ice layers developed in the snowpack. They both occurred late in the winter season, a few weeks before ablation, and were about 1 to 2 cm thick and located near the surface of the snow.

For those animals traveling over a snowcover, a denser snowpack is advantageous in terms of energy expenditure. It has been reported (Klein et al., 1987) that leg length in caribou is related to nutritional constraints, energy efficiency of foraging through snow of various depths, the efficiency of locomotion where long migrations occur, and fleetness related to predator avoidance. For caribou, longer legs are desirable for locomotion and fleetness, but likely undesirable for foraging at ground level.

During calving, snow appears to be an important factor for caribou. Although there are some conflicting data, Eastland et al. (1989) report that female caribou prefer to calve in those areas where the snowcover is the greatest but the snowpack is mottled and cotton grass (*Eriophorum vaginatum*) is available (Eastland and White, 1991). The need to avoid predators and nutritional needs of lactating females are the likely reasons for the selection of these calving areas. A much earlier ablation period could dramatically affect caribou population dynamics (Eastland and White, 1991). However, Hinzman and Kane (1992) found that snowmelt would only be about 1 week earlier in the Alaskan Arctic and that the window of snowmelt from year to year was about 1 month. More significantly, they predicted that snow accumulation in the fall would be delayed by about 3 weeks; nutritionally this could be important for the caribou.

The response of other animals in the Actic to climate change would be similar to the caribou for the same activity. What effect the development of ice layers in the snowpack could have on those small animals living beneath the snowpack is unknown. If the active layer continues through the winter months to release carbon dioxide and other gases, and they become entrapped beneath a relatively impermeable layer, this could be detrimental to those animals.

Vegetation Response

Even with the present climate, there are ongoing changes occurring with arctic vegetation. The difference between what has been happening and what may happen is that the rate of change might be much greater with additional greenhouse gases in our atmosphere. There are many processes that will influence what happens; soil mineralization, nutrient availability, soil moisture, atmospheric deposition, and biomass decomposition are all interrelated and will affect plant production. The established plants in the Arctic are unique because they survive and propagate where the soil temperatures are just above the freezing point of water and at any time during the summer the air temperature can drop below 0°C and be accompanied by snow.

The greatest impact of climatic change to vegetation may come from soil warming. Figure 3.5 shows the present thermal regime of the active layer and how it would respond to a gradual warming of 4°C over 50 years. The simulated present temperatures compare well with measured field values. Warmer soils in the root zone [Figure 3.5(c)] and a deeper active layer would be the two outcomes that influence vegetation the most. A deeper active layer allows drainage to occur later in the winter and delays freezing of the active layer by about 3 months.

If precipitation does not increase, active layer moisture contents will be reduced by increased evaporation (Hinzman and Kane, 1992). With a longer summer season and a deeper active layer, more drainage from the active layer will occur; this will also produce a drier active layer. Maxwell (1992) predicts that annual global precipitation could increase 15% with the Arctic getting an additional increase of 5 to 10% over the annual global precipitation. This is about the same magnitude as the increase in evapotranspiration reported by Hinzman and Kane (1992). It has been conjec-

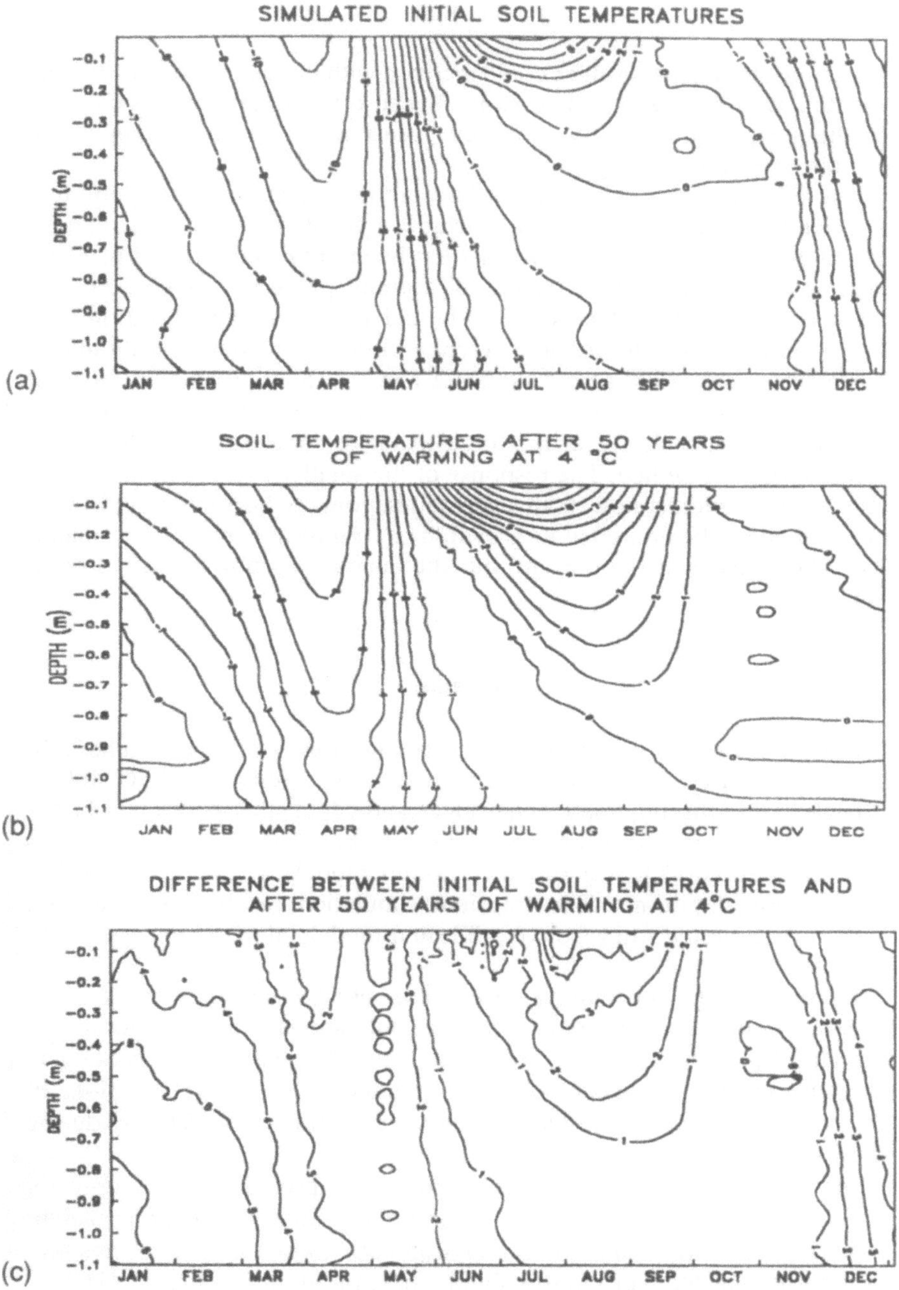

Figure 3.5. (a) Active layer and upper permafrost temperatures prior to climatic warming, Imnavait Creek Watershed, Alaskan Arctic. (b) Active layer and upper permafrost temperatures after 4°C over a period of 50 years for Imnavait Creek Watershed. (c) Differences in active layer and upper permafrost temperatures over a 50-year period, Imnavait Creek Watershed.

tured that stomatal resistance could increase with increasing carbon dioxide and this would reduce transpiration. Because the water table is so close to the surface most of the time, evaporation is a large component of evapotranspiration.

There is much speculation about how arctic vegetation will respond to a warmer climate. Billings (1992) concludes that plants have never experienced the rate of climate change that has been predicted; therefore, we do not know if plant migration and the soil-forming processes can keep up with the change in climate. The southern boundary of permafrost and the tree line would shift northward in a warmer climate. In Alaska, a mountain barrier runs east–west across the state and this may prevent, or at least delay, the migration of the tree line. Sveinbjörnsson (1992) concluded that it was highly speculative at this time to make conclusions of the future location of the tree line. The response of the smaller existing flora is probably of more interest now. Is this vegetation going to become more robust, or is it going to be limited in its growth by nutrients and moisture availability? These issues are important, but the answers do not appear to be at hand.

Summary

Climate change is a worldwide phenomenon; however, early results indicate that greater change may occur in the arctic regions (IPCC, 1992b; Manabe and Stouffer, 1980; Schlesinger and Mitchell, 1985). Only in the last couple of years have we moved away from GCM equilibrium models with the double CO_2 scenario to transient models with coupled ocean–atmosphere algorithms. The transient simulation results are consistent with earlier quoted rates from equilibrium models of 0.3°C per decade of warming.

Increases in the concentration of greenhouse gases in the earth's atmosphere will produce differential changes, both spatially and temporally, in the climate. Although some areas may actually experience a cooling in response to greater concentrations of greenhouse gases, arctic regions of the world are predicted to have the greatest warming, particularly during the winter season.

Incremental increases in radiative fluxes will provide more energy for both sensible warming of soil and atmosphere and hydrologic processes requiring latent heat. In a warmer environment, evaporation and transpiration during the summer months should increase in the Arctic.

Hydrologic changes may be difficult to quantify because of the large annual variations that we see now. Snowmelt events in the wintertime are rare. A warmer environment may produce more of these events. Although this would enhance surface travel by animals, it would make it difficult for animals to break through this layer to forage and may also trap gases that are emanating from the ground and produce an environment unsuitable for those animals living below the snowcover. Presently, snow in the Arctic during the summer months is quite common. The occurrence of these events in a warming scenario would decrease. Earlier ablation in the spring would occur, but it now appears that this change would not be that substantial; however, the long-term cumulative effect may be important to the flora and fauna.

The hillslope hydrology of the active layer could be significantly impacted by climate change. With the existing soil conditions we now have, it is predicted that the active layer would increase in thickness. This is not surprising: if you go to a warmer setting you find a deeper active layer. Hydrologically this will mean potentially greater subsurface storage and more water movement downslope. How this would impact the soil moisture levels and subsequently the vegetation is unknown. It is not easy to make statements about how the wildlife may be affected because for some it may be advantageous to have a drier active layer and for others. It may not be beneficial.

Two unkowns that complicate the future predictions of climate change in the Arctic are precipitation and cloud cover. Warming predicted by general circulation models has been shown to be very sensitive to cloud cover. Hydrologically, greater runoff would result from greater precipitation. Also, more snowfall would produce later snowmelt, offsetting the impact of warming. We need to perform two tasks if we are going to be able to identify hydrologic alterations induced by climate change. First, we need hydrologic studies of all the processes at various watershed scales and, second, the collection of good-quality data that will allow us to detect small process changes. This is not easy in a harsh, remote environment.

References

Benson, C.S. 1982. Reassessment of winter precipitation on Alaska's Arctic Slope and measurements on the flux of wind blown snow. Rep. UAG R-288, University of Alaska Fairbanks, Geophysical Institute.

Billings, W.D. 1992. Phytogeographic and evolutionary potential of arctic flora and vegetation in a changing climate. In: Chapin, F.S., III, Jefferies, R.L., Reynolds, J.F., Shaver, G.R., and Svoboda, J. (eds.), *Arctic Ecosystems in a Changing Climate, an Ecophysiological Perspective* (pp. 91–109). New York: Academic Press, Inc.

Clagett, G.P. 1988. The Wyoming windshield—an evaluation after 12 years of use in Alaska. Proc. of Western Snow Conference, pp. 113–123.

Eastland, W.G., and White, R.G. 1991. Potential effects of global warming on calving caribou. In: Weller, G., Wilson, C.L., and Severin, B.A.B. (eds.), International Conference on the Role of Polar Regions in Global Change, Geophysical Institute and Center for Global Change and Arctic System Science, pp. 460–464.

Eastland, W.G., Bowyer, R.T., and Fancy, S.G. 1989. Effects of snow cover on the selection of calving sites by caribou. *J. Mamm.* 70(4):824–828.

Fancy, S.G., and White, R.G. 1985. Energy expenditures by caribou while cratering in snow. *J. Wildl. Manage.* 49(4):987–993.

Hinzman, L.D., and Kane, D.L. 1991. Snow hydrology of a headwater arctic basin 2. Conceptual analysis and computer modeling. *Water Resources Res.* 27(6):1111–1121.

Hinzman, L.D., and Kane, D.L. 1992. Potential response of an arctic watershed during a period of global warming. *J. Geophys. Res.* 97(D3):2811–2820.

Hinzman, L.D., Kane, D.L., Gieck, R.E., and Everett, K.R. 1991. Hydrologic and thermal properties of the active layer in the Alaskan Arctic. *Cold Regions Sci. Technol.* 19:95–110.

Hinzman, L.D., Kane, D.L., Benson, C.S., and Everett, K.R. 1996. Energy balance and hydrological processes in an arctic watershed. In: Reynolds, J.F., and Tenhunen, J.D. (eds.), *Landscape function and disturbance in arctic tundra* (pp. 131–154). Ecological Studies, Vol. 120, New York, Springer.

Intergovernmental Panel on Climate Change (IPCC). 1990. *Climate Change, the IPCC Scientific Assessment* (Houghton, J.T., Jenkins, G.J., and Ephraums, J.J., eds.). Cambridge: Cambridge University Press.

Intergovernmental Panel on Climate Change (IPCC). 1992a. *Scientific Assessment of Climate Change*. Cambridge: Cambridge University Press.

Intergovernmental Panel on Climate Change (IPCC). 1992b. *Climate change 1992, The Supplementary Report to the IPCC Scientific Assessment* (Houghton, J.T., Callander, B.A., and Varney, S.K., eds.). Cambridge: Cambridge University Press.

Kane, D.L. 1993. Meteorological and hydrologic studies in the Alaskan Arctic in support of long-term ecological research. In: Redmond, K.T., and Tharp, V.L., (eds.), Proc. of the Ninth Annual Pacific Climate (PACLIM) Workshop, April 21–24, 1992. California Dept. of Water Resources, Interagency Ecological Studies Program, *Tech. Rep.* 34:13–21.

Kane, D.L., Gieck, R.E., and Hinzman, L.D. 1990. Evapotranspiration from a small Alaskan Arctic watershed. *Nordic Hydrol.* 21:253–272.

Kane, D.L., Hinzman, L.D., Benson, C.S., and Everett, K.R. 1989. Hydrology of Imnavait Creek, an Arctic watershed. *Holarctic Ecol.* 12:262–269.

Kane, D.L., Hinzman, L.D., Benson, C.S., and Liston, G.E. 1991. Snow hydrology of a headwater arctic basin 1. Physical measurements and process studies. *Water Resources Res.* 27(6):1099–1109.

Kane, D.L., Hinzman, L.D., Woo, M.K., and Everett, K.R. 1992. Arctic hydrology and climate change. In: Chapin, F.S., III, Jefferies, R.L., Reynolds, J.F., Shaver, G.R., and Svoboda, J. (eds.), *Arctic Ecosystems in a Changing Climate, an Ecophysiological Perspective* (pp. 35–57). New York: Academic Press, Inc.

Klein, D.R., Melgaard, M., and Fancy, S.G. 1987. Factors determining leg length in *Rangifer tarandus*. *J. Mamm.* 68(3):642–655.

Lachenbruch, A.H., and Marshall, B.V. 1986. Changing climate: Geothermal evidence from permafrost in the Alaskan Arctic. *Science* 234:689–696.

Manabe, S., and Stouffer, R.J. 1980. Sensitivity of a global climate model to an increase in CO_2 concentration in the atmosphere. *J. Geophys. Res.* 85:5529–5554.

Maxwell, B. 1992. Arctic climate: Potential for change under global warming. In: Chapin, F.S., III, Jefferies, R.L., Reynolds, J.F., Shaver, G.R., and Svoboda, J. (eds.), *Arctic Ecosystems in a Changing Climate, an Ecophysiological Perspective* (pp. 11–34). New York: Academic Press, Inc.

Russel, D.E., Martell, A.M., and Nixon, W.A.C. 1993. Range Ecology of the Porcupine Caribou Herd in Canada. Nordic Council for Reindeer Research, Harstad, Norway. *Rangifer Special Issue*. 8.

Schlesinger, M.E., and Mitchell, J.F.B. 1985. Climate model simulations of the equilibrium climatic response to increased carbon dioxide. *Rev. Geophys.* 25:760–798.

Sveinbjörnsson, B. 1992. Arctic tree line in a changing environment. In: Chapin, F.S., III, Jefferies, R.L., Reynolds, J.F., Shaver, G.R., and Svoboda, J. (eds.), *Arctic Ecosystems in a Changing Climate, an Ecophysiological Perspective* (pp. 239–256). New York: Academic Press, Inc.

Woo, M.K. 1990. Consequences of climate change for hydrology in permafrost zones. *Journal of Cold Regions Engineering* 4(1):15–20.

Woo, M.K., Heron, R., and Marsh, P. 1982. Basal ice in High Arctic snowpacks. *Arctic Alpine Res.* 14(3):251–260.

Woo, M.K. 1986. Permafrost hydrology in North America. *Atmos. Ocean* 24(3):201–234.

4. Global and Regional Patterns of Climate Change: Recent Predictions for the Arctic

Peter R. Rowntree

The Greenhouse Effect

Certain gases in the atmosphere, commonly called greenhouse gases, absorb and emit radiation at wavelengths at which the Earth's surface emits. Consequently, some of the radiation emitted to space through the top of the atmosphere emanates not from the surface but from within the atmosphere. As the intensity of radiation is proportional to the fourth power of the temperature of the emitter, and the atmosphere is mostly colder than the surface, these gases reduce the radiative loss of energy to space. This is called the "greenhouse effect."

The greenhouse effect is not new. It has been keeping the Earth warm since its formation. In geologically recent times the most important greenhouse gases have been water vapor and carbon dioxide (CO_2). In the absence of greenhouse gases and clouds, equilibrium requires that the heating of the surface by incoming solar radiation must be balanced by longwave radiation emitted from the earth's surface. If we know the reflectivity of the earth for solar radiation—it is about 30% at present—we can calculate the temperature of the earth's surface needed to give this balance to be about −18°C or 255 K. With greenhouse gases present, the level from which radiation is effectively emitted to space is raised to several kilometers above the surface, where the temperature is about 30 K (or degrees Celsius) lower than that near the surface. Consequently, the surface temperature can be higher by about this amount and the earth will emit the same amount of radiation to space as in the absence of greenhouse gases.

The concentrations of several greenhouse gases are observed to be increasing. CO_2, methane, nitrous oxide, and several chlorofluorocarbons (CFCs) have all increased since the start of the industrial era. However, the atmospheric concentration of sulphate aerosols, generated by the burning of coal and oil, has also increased, enhancing the atmosphere's reflectivity for solar radiation. Also, stratospheric ozone, a greenhouse gas, has been depleted over the last 20 years or so due to the increases in CFCs. Its cooling effect is thought to be perhaps as large as the warming due to the CFCs (Isaksen et al., 1992).

Measurements of air in ice cores, supported by late 19th century observations, suggest that since the start of the industrial era in about 1750, the largest contribution, about $1.5\,W\,m^{-2}$ so far in terms of its effect on longwave radiation at the top of the troposphere, has come from CO_2. Methane and nitrous oxide have together contributed another $0.5\,W\,m^{-2}$. A "best estimate" of the opposite effect of aerosols is also about 0.5 to $1\,W\,m^{-2}$, partly by direct reflection of solar radiation by the aerosols, and partly because increased numbers of condensation nuclei lead to more, smaller cloud droplets and so to higher cloud albedos. Because the aerosols are rapidly washed out, there are large geographical variations in their effects with the maximum effects near the main source regions. This is in contrast to the longer-lived trace gases, which vary in concentration by only a few percent through the atmosphere. Volcanic eruptions, such as that of Mt. Pinatubo in 1991, are thought to have significant short-term effects on global mean temperature; however, they are not expected to induce long-term change unless there is a major change in their frequency or intensity.

The direct effect of a doubling of CO_2 would be to warm the surface temperature by about 1.3 K. However, a number of feedback factors act to modify this. One of the most important found in models is the increase in a warmer atmosphere of water vapor, which is the largest contributor to the natural greenhouse effect. Another is the positive albedo feedback due to snow and ice: as the climate warms, snow and ice extents decrease, reducing the amount of solar radiation that is reflected back to space and so providng a further increase in the radiative forcing. Clouds are also found to be important, but different models indicate widely varying feedbacks, some positive, some negative. Consequently, the temperature response found in climate models to a doubling of CO_2 varies between about 2 and 5 K.

Modeling Climate Change Due to the Greenhouse Effect

Climate Prediction

To predict the effects of increasing CO_2 and other trace gases on climate, it is necessary to use a global model of climate based on the relevant physical equations. To make a climate simulation, the procedure is much as in a numerical forecast, the equations representing the atmospheric and oceanic physics being integrated forward in time. After 5 days the accuracy of prediction of individual atmospheric events is deteriorating; however, if the model is run further, its anticyclones and depressions should continue to develop and move in a realistic fashion, and realistic oceanic fronts and circulations should be maintained. By running the model on for many years a simulation of climate can be obtained—the realism of this will be discussed shortly. To estimate the effect of a change in forcing, the relevant parameter (e.g., the amount of CO_2) is changed and the simulation is repeated with this change.

Equilibrium and Transient Response Experiments

Results obtained from climate models over the last decade have been from two types of experiment, commonly called equilibrium experiments and transient response experiments. The equilibrium experiments are intended to estimate the equilibrium response to an increase (usually a doubling) of CO_2. Because of the large amount of computing that would be needed to take a model with a deep ocean to equilibrium, they usually use a "slab" ocean model consisting of a mixed layer 50 m thick with no currents but sufficient thermal inertia to represent seasonal variations. The advection of heat in this slab model is represented by a prescribed seasonally varying "flux correction" by which energy is input to the ocean to maintain realistic temperatures and sea-ice distributions in a long-term climatological sense. In any perturbation experiment, for example with CO_2 doubled, the same flux correction as calculated for the control experiment is applied.

The "transient response" experiments use a model with a deep ocean to predict the evolving response to gradually increasing greenhouse gases. Thus, ocean currents are represented, including vertical motion on the gridscale and subgridscale convective and diffusive mixing, which disperse the effects of surface fluxes of heat, water, and momentum throughout the ocean. The simulations so far published that were obtained in this way have suffered from errors in ocean surface temperatures (e.g., Murphy, 1995; Washington and Meehl, 1989); these are sufficiently large to cause major errors in the atmospheric circulation—a serious problem given the quite realistic atmospheric simulations that can be obtained with the higher resolution models when climatological sea surface temperatures are used (Gates et al., 1990). Coupled models thus are likely to require substantial improvement to reduce this "climate drift" to an acceptable level.

To avoid an unacceptable climate drift many models retain the flux correction discussed above for the slab model. There is an assumption implicit in the use of the flux correction technique that the errors in the simulations of the two climates are similar. This is a useful first approximation to the errors, though it is likely to become increasingly invalid as the differences between the ocean circulations in the control experiment and the perturbation experiment increase. A major thrust of effort on coupled model development is consequently to eliminate the need for flux corrections.

Models and Experiments

As discussed by Gates et al. (1992) in the 1992 IPCC (Intergovernmental Panel for Climate Change) Report, four coupled models of the atmosphere and deep ocean have been used to study the transient response to increasing greenhouse gases. These are from the Geophysical Fluid Dynamics Laboratory (GFDL) at Princeton, NJ (Manabe et al., 1992), the National Center for Atmospheric Research (NCAR) at Boulder, CO (Meehl et al., 1993), the Max Planck Institute (MPI) at Hamburg, Germany (Cubasch et al., 1992), and the Hadley Centre (HC) at the Meteorological Office, Bracknell, UK (Murphy and Mitchell, 1995). Three of these models include flux corrections to optimize the realism of the control simulation, the exception being the NCAR model.

The structure of these models is indicated in Table 4.1. Note that three of the models use a spheric harmonical (or "spectral") representation of the atmosphere for advection. The atmospheric physics is computed on a 4.5° by 7.5° grid for the R15 models, and a 5.6° by 5.6° grid for the T21 spectral resolution. The models were run both with constant CO_2 (the control experiments) and with gradually increasing CO_2. Except for the NCAR model all were run to the time of doubling of CO_2 or beyond. Meehl et al. (1992) report results for the NCAR model to year 60.

Except as discussed above for the atmospheric advection in the spectral models, the temperatures, winds or currents, and other variables in the

Table 4.1. Structure of Four Models Used for Transient Response Experiments

Model	GFDL	NCAR	MPI	HC
Atmosphere				
Horizontal resolution[a]	R15	R15	T21	2.5°, 3.75°
Layers	9	9	19	11
Ocean				
Horizontal resolution[a]	4.5°, 3.75°	5°	4°	2.5°, 3.75°
Layers	12	4	11	17
Greenhouse gas scenario				
CO_2 increase (yr)[b]	1%	1%	IPCCA	1%
CO_2 doubling time (yr)	70	100	60	70

[a] R15 and T21 indicate spectral resolutions (see text) (R = rhomboidal and T = triangular). For the nonspectral representations, the first figure is the latitudinal, the second the longitudinal resolution—only one figure is given if both are the same.

[b] 1% compound for GFDL and HC, 1% simple for NCAR. "IPCCA" refers to the Scenario A (also called "Business-as-usual") used in the 1990 IPCC Report (Houghton et al., 1990).

atmosphere and ocean are represented for a three-dimensional array of boxes. In the horizontal, the gridbox array follows lines of constant latitude and longitude, whereas in the vertical there are typically 10 to 20 layers in both atmosphere and ocean. Because of the large computing requirements, the horizontal grid length used in CO_2 experiments to date has been limited to about 2.5° of latitude; Gates et al. (1990) found that, with this resolution, quite a realistic simulation of the atmospheric circulation was obtained. With coarser resolution, major errors were evident in the simulated circulation patterns (see Figure 4.1 in Gates et al.). However, the ocean simulation at a resolution of 2.5° latitude is quite poor. The typical horizontal scale of important circulation features in the ocean, such as the Gulf Stream, is less than that of features such as jet streams and baroclinic mid-latitude cyclones in the atmosphere; it is clear that in the future the ocean model will need to be on a finer mesh.

Atmosphere and ocean gridpoint models require a high east–west resolution near the poles to reduce the errors associated with the curvature of the latitude circles. To avoid an unacceptably short timestep and even more massive computing resources, some kind of filtering (e.g., Fourier filtering) is used to control shortwave features.

The increase of 1%/yr used in these experiments (the IPCC Scenario A has a rate of increase of radiative forcing similar to that with a 1%/yr increase in CO_2) is faster than the observed past rate of increase for CO_2 which has risen from about 0.3%/year in the 1960s to nearly 0.5%/year in the 1980s. Earlier increases starting from about 1750 were slower though not negligible. As discussed earlier, the globally averaged changes in radiative forcing due to other trace gases and aerosols may have been roughly in balance over recent decades, though their regional effects probably have not, due to the uneven distribution of sulphate aerosols. It is thus not possible to relate the changes with a 1%/year increase in CO_2 to the real world in any precise way.

Most of the results discussed here will be from the Hadley Centre model, which is described in more detail below, with some comparisons to results from other coupled models. Space does not allow extensive comparisons, and maps from other models are mostly not available on projections appropriate for the Arctic land areas on which the discussion is focussed. However, as noted earlier, the relatively low horizontal resolution provided by R15 or T21 spectral models does not allow as realistic simulations of the circulation as can be obtained with the Hadley Centre model, especially in high latitudes, so this model may provide the best guidance on circulation changes.

The Hadley Centre (HC) Model

The Hadley Centre (HC) atmospheric model used in the experiments discussed here is a development of the UKHI model assessed in the 1990 IPCC

Report by Gates et al. (1990). As indicated in Table 4.1, it represents the surface pressure, and the temperature, humidity, and wind components on 11 layers on a 2.5° by 3.75° latitude–longitude mesh. The deep ocean model has temperature, salinity, and currents on 17 layers and the same horizontal mesh as the atmospheric model. The land surface is represented by a model with the same horizontal mesh as the atmosphere. Fourteen surface and subsurface characteristics are defined as a function of the soil and vegetation types for each gridbox, which are specified from the 1° data sets constructed by Wilson and Henderson-Sellers (1985). The soil is represented by four layers for thermal processes and one layer plus a canopy water store for hydrologic processes. Sea-ice is modeled using the "zero-layer" thermodynamic approach of Semtner (1976), but with an explicit representation of the fraction of the gridbox covered by leads due to Hibler (1979). The atmospheric model calculates separate surface turbulent and radiative fluxes for the ice and lead fractions of the gridbox.

Validation of HC Control Experiment

As shown by Gates et al. (1990), the HC (Meteorological Office) model simulates the atmospheric simulation in a generally realistic manner. For sea level pressure (PMSL, pressure at mean sea level) (Figure 4.1), the major observed features are present in winter: the Aleutian and Icelandic lows and a high over Siberia with a ridge across the east Eurasian Arctic Ocean to link with a north–south ridge over Canada. However, the trough along the north coast of western Eurasia is weaker than observed. In summer, the observed pattern shows a weak circumpolar trough near 60°N, with minima near Labrador and the east Atlantic, whereas over Siberia this trough becomes an extension of the Asian summer low. There is a weak high near the pole. The model captures these features though there is no clear east Atlantic low. However, the high over the Arctic Ocean is too strong, being 6 to 8 hPa above the minima near 60°N compared with about 4 hPa for the observed.

The simulated precipitation (not shown) at years 66 to 75 of the control experiment is generally similar to that shown for the UKHI simulation by Gates et al. (1990) and includes the main observed features. In winter, precipitation north of 60°N is less than 1 mm/day except over the southern half of Alaska and adjacent parts of northwest Canada, and over Scandinavia. In summer, precipitation exceeds 1 mm/day over all Arctic land in the HC model and, except some northern coastal areas, in the observed. The model has an area exceeding 2 mm/day extending north to 65 to 68°N over Eurasia and to near 60°N over North America. This is more extensive than in the UKHI model shown in Gates et al., making it more realistic over much of Eurasia, but less so over North America.

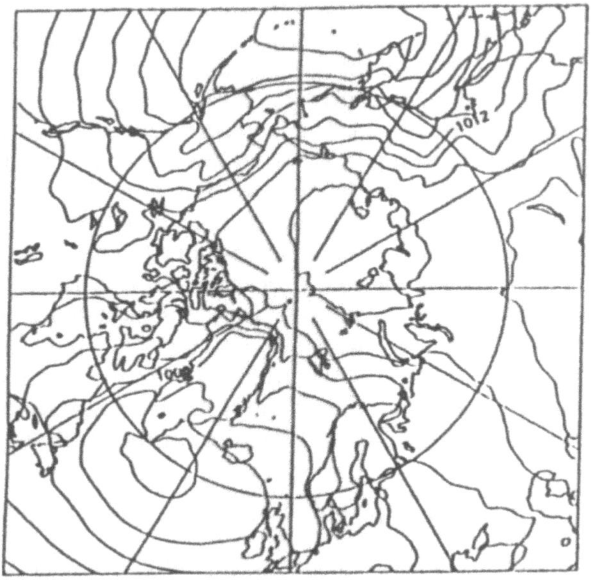

(a)

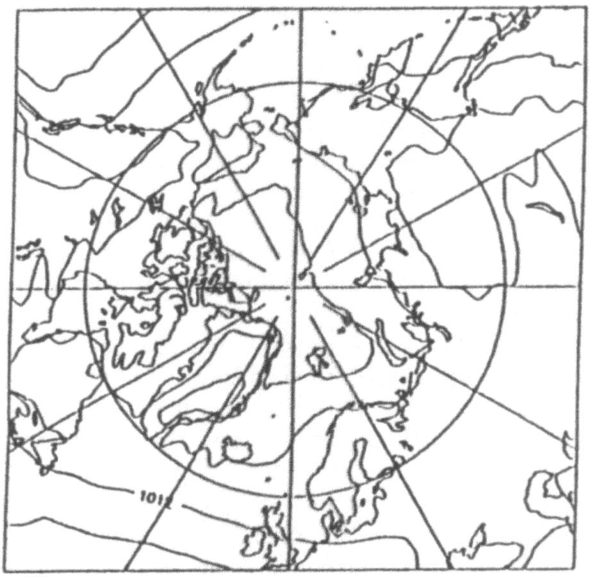

(b)

Figure 4.1. Sea level pressure (PMSL) for (a) Winter (December to February) and (b) Summer (June to August) averaged over years 66 to 75 of the Hadley Centre transient experiment. Contour interval: 4 hPa.

Results from the Hadley Centre Transient Experiment

Temperature Changes

The annually averaged temperature changes in transient response experiments have been shown by Gates et al. (1992). The responses obtained from these experiments at the time of doubling of CO_2 are similar in many ways to those from equilibrium experiments (Mitchell et al., 1990) but there are important differences, in part due to the inclusion of the deep ocean and ocean currents; this allows representation of the vertical mixing, which slows the response of the ocean, so that by the time of doubling at 1%/yr (typical of the rates to be expected in the next few decades) the global mean warming is only about 60 to 70% of the equilibrium figure, and the warming of the oceans is generally less than that of the land. Parts of the ocean with deep mixing in winter, notably the Antarctic Ocean and the North Atlantic Ocean, are particularly slow to warm. The upward mixing of heat in the high latitude oceans may be reduced by increases in precipitation, which freshen the near-surface water and so stabilize the vertical density structure. In the HC experiment (Figure 4.2), some small regions of cooling in both areas still persist at years 66 to 75 (when CO_2 is doubled) when the mean land surface warming is about 3 K (or degree C) compared with 1.5 K for the ocean. In contrast, in equilibrium experiments, the high latitude oceans could display some of the largest temperature rises (e.g., Bretherton et al., 1990, Figure 6.5(b)).

The typical magnitudes of warming for different surfaces in the HC experiment may be ordered from large to small in each season, as shown in Table 4.2. In summer the greatest warming occurs over those land areas that have less snow (e.g., the Canadian Archipelago) as a result of the climate change. Warming is almost as large over those parts of the land that are drier as a result of the increase in CO_2 and are so dry as to limit evaporation (and so the evaporative cooling); mostly, such areas are south of 60°N in the model. In winter, regions with sea-ice show the largest warming, but in summer sea-ice regions are second only to the more deeply mixed oceans for their lack of warming. This is because the sea-ice surface is already at melting point most of the summer in the control experiment, and though it absorbs more heat due to a longer melt season with increased CO_2, it cannot warm further till the ice is melted; even then the large thermal inertia of the ocean allows only slow warming. The effects of the extra heat gained in summer on surface temperature manifest in autumn and winter when ice is thinner and less extensive, with the largest warming over areas where the ice extent is much reduced with increased CO_2, such as the Barents Sea, Bering Strait, and Sea of Okhotsk (Cattle and Thomson, 1993). Thus, there is a large positive ice albedo feedback on the warming at high latitudes, but it reveals itself in the temperature changes in seasons when the solar radiation is weak or absent.

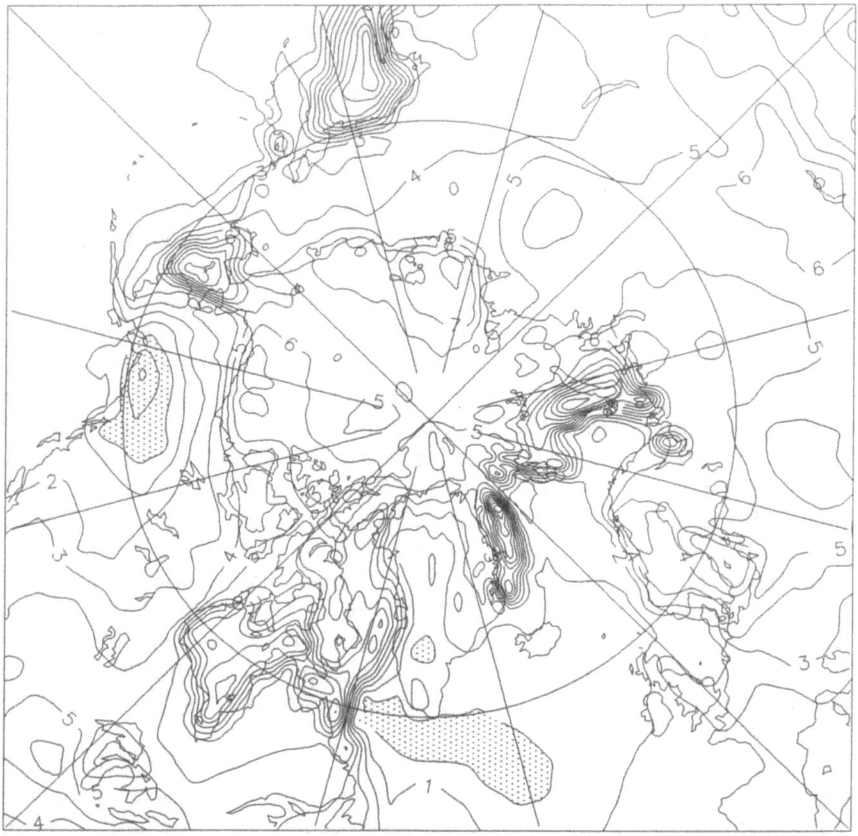

(a)

Figure 4.2. Differences in surface temperature (K) at years 66 to 75 of the Hadley Centre transient experiment (increased CO_2—Control). Contours at 1 K intervals. Shading where negative. (a) December to February.

Interactions Between Temperature and Circulation Changes

The atmospheric circulation can be substantially modified by spatial variations of the warming (Figure 4.3). One example is a general reduction of pressure over the continents; this corresponds to the greater warming of the continents than of the oceans and, in summer, represents an intensification of the continental lows associated with the summer monsoon. A less obvious example pointed out by Hall et al. (1994) is that the reduced baroclinic gradient over the west Atlantic associated with the large winter warming over eastern Canada weakens the storm track in this region, while the cooling near southern Greenland enhances baroclinicity further east. This, together with the effects of a moister atmosphere increasing the potential

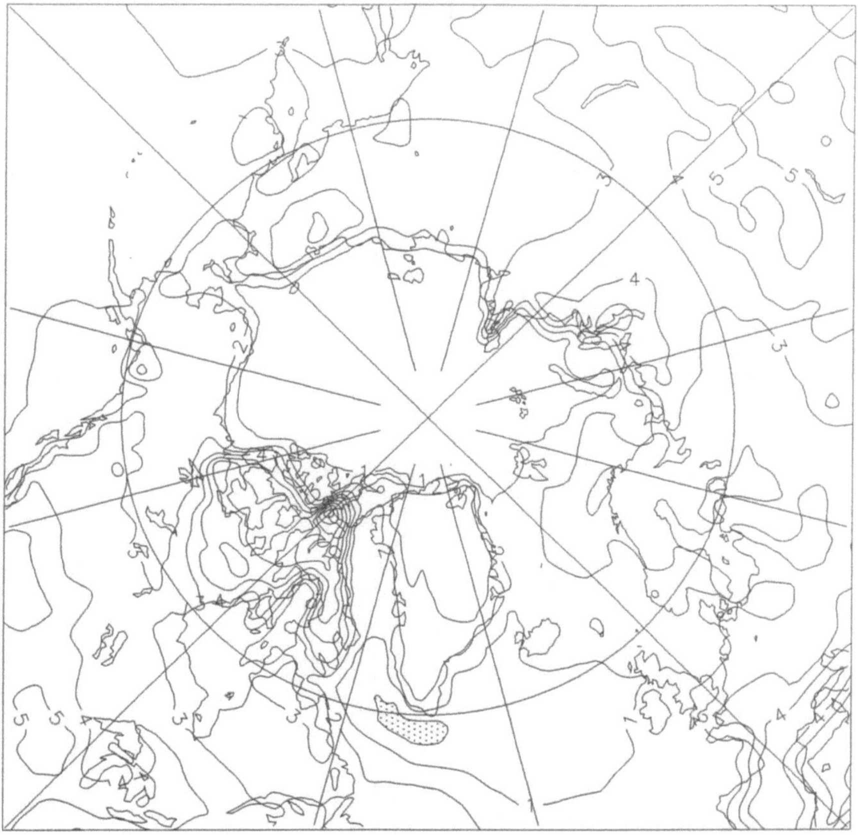

Figure 4.2. *Continued* (b) June to August.

Table 4.2. Ordering of Typical Temperature Changes from Highest to Lowest with Typical Changes (K) at Years 66 to 75 of the Hadley Centre Experiment in Middle and High Northern Latitudes

	Typical Change
For winter	
Melted sea-ice	about 10
Other sea-ice	+4–+8
Melted snow	+4–+8
Most land	+3–+6
Land with ocean/ice to windward	−1–+4
Most ocean	+1–+2
Ocean with deep mixing	−1–0
For summer	
Melted snow	+4–+7
Dried land	+3–+6
Most land	+2–+4
Land with ocean/ice to windward	+1–+3
Most ocean	+1–+3
Sea-ice	+0–+1
Ocean with deep mixing	−1–+1

(a)

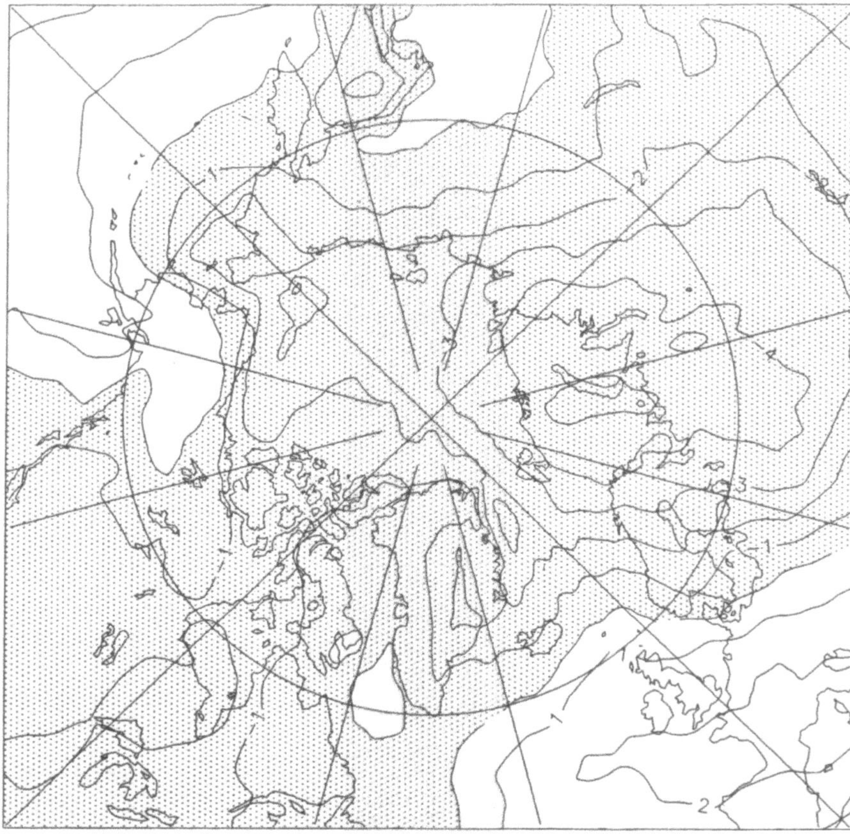

Figure 4.3. Differences in pressure at mean sea level (hPa) at years 66 to 75 of the Hadley Centre transient experiment (increased CO_2—Control). Contours at 1 hPa intervals. Shading where negative. (a) December to February.

for latent heat release in storms, leads to a downstream intensification of the storm tracks and associated westerly winds. In the HC model, this generates an increased westerly flow in autumn and winter over the North Atlantic sector and northern Eurasia south of a fall of pressure over the Arctic.

Some aspects of the climate change are thus of a regional nature and can be attributed to changes in circulation patterns. For example, Figure 4.2 shows a warming of up to 5 K over Scandinavia in winter, which is likely to be due in part to an increased westerly wind component from the warm ocean to the west. On the other hand, over the Pacific coasts of eastern Alaska and western Canada, where a reduction of northward flow weakens warm air advection, there is cooling or only slight warming, though

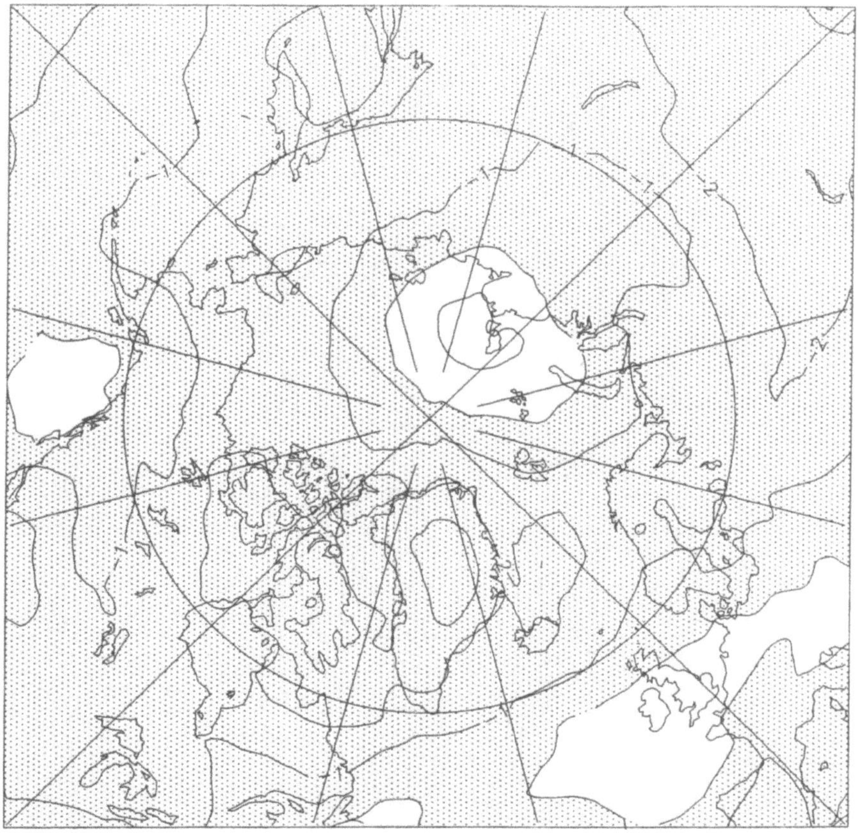

(b)

Figure 4.3. *Continued* (b) June to August.

part of this may be due to the small warming of the ocean to windward (SW). This feature is particularly strong in this decade; averages for years 31 to 70 show a warming of 0 to 1 K and PMSL changes only half as large though of similar pattern. In summer, the largest warming over northern Asia is in the sector (60 to 100°E) affected by increased southerly flow from warmer regions around an increase in pressure near Severnaya Zemlya. These circulation changes, although not likely to be correct in detail, are indicative of the type of regional effects to be expected with greenhouse warming.

Precipitation Changes

With enhanced CO_2, modeled precipitation increases generally in winter over land north of 45°N and over most of the Arctic land areas in summer (Figure 4.4). One of the few exceptions north of 60°N at years 66 to 75 is in

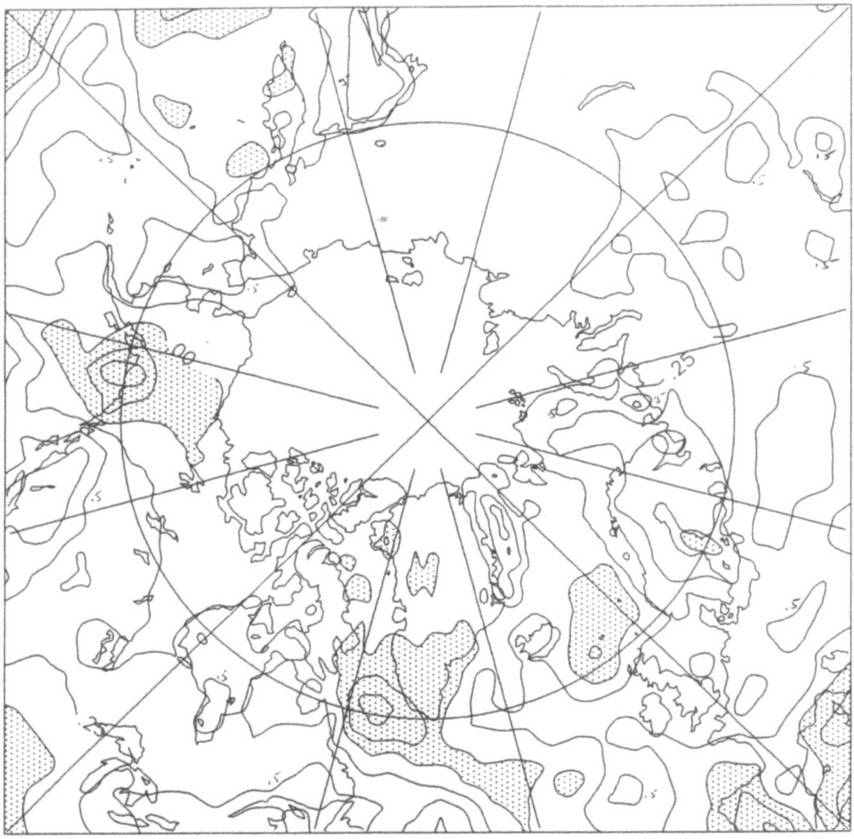

(a)

Figure 4.4. Differences in precipitation (mm/day) at years 66 to 75 of the Hadley Centre transient experiment (increased CO_2—Control). Contours at 0.25 mm/day intervals. Shading where negative. (a) December to February.

the Ob-Yenisei basins in summer; the long period (years, 31 to 70) mean shows a similar but weaker decrease in this region, with a break near 65°N. As discussed above, the PMSL difference charts show this to be a region with either higher pressure or increased flow from the arid regions to the south; this feature is also less pronounced in the 40-year mean. Precipitation is also decreased over the Pacific coastal areas of eastern Alaska and northwest Canada and over southern Greenland. These regions are adjacent to seas where warming is particularly limited by the deep mixing of the oceans and, at least in the Atlantic, the weakening of the thermohaline circulation, which transports warm water northward. The general rise in atmospheric water vapor content, which is mainly responsible for the increases in precipitation, is likely to be weak or absent in these regions. The

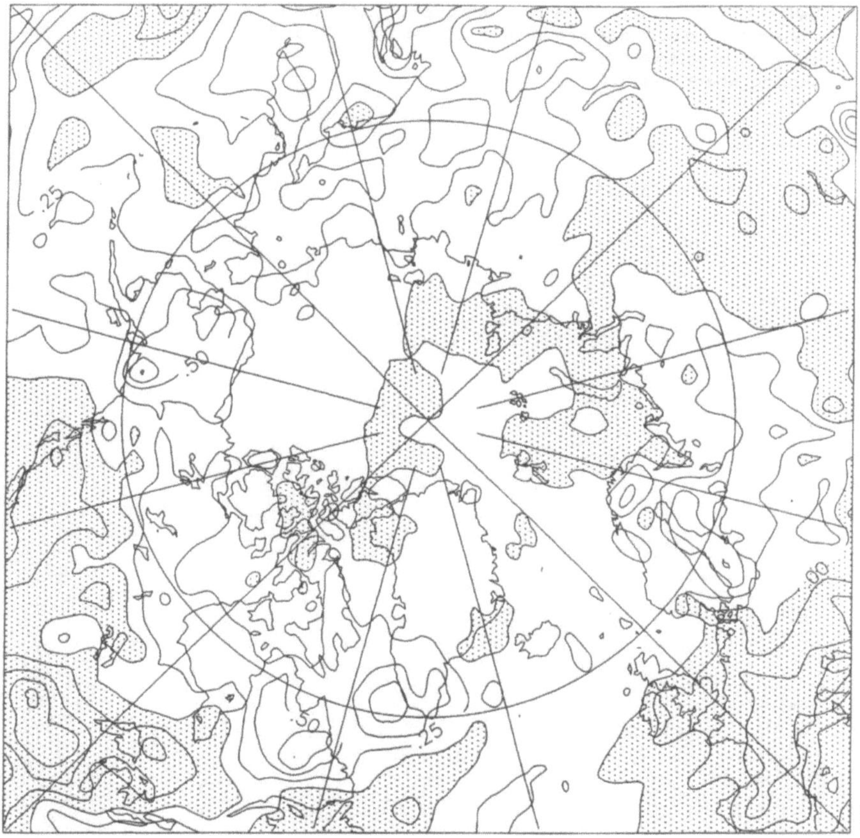

(b)

Figure 4.4. *Continued* (b) June to August.

increases in precipitation in summer near Hudson's Bay, Scandinavia, and east and west of the Bering Strait are all associated with falls of pressure of 1 hPa or more.

Soil Moisture Changes

With such general increases in precipitation, one might expect a widespread increase in soil moisture. However, equilibrium experiments have shown rather consistently (Mitchell et al., 1990, Figure 5.8) that soil moisture is predicted to decrease over middle and high latitudes in summer. The main reason for this, at least where the rainfall changes are small, is that evaporation is predicted to increase from spring onwards due to earlier snowmelt and to the effect of higher temperatures on the evaporation through the Clausius-Clapeyron relation (e.g., see Rowntree, 1991). In the HC transient response experiment (Figure 4.5), summer soil moisture north of 60°N is

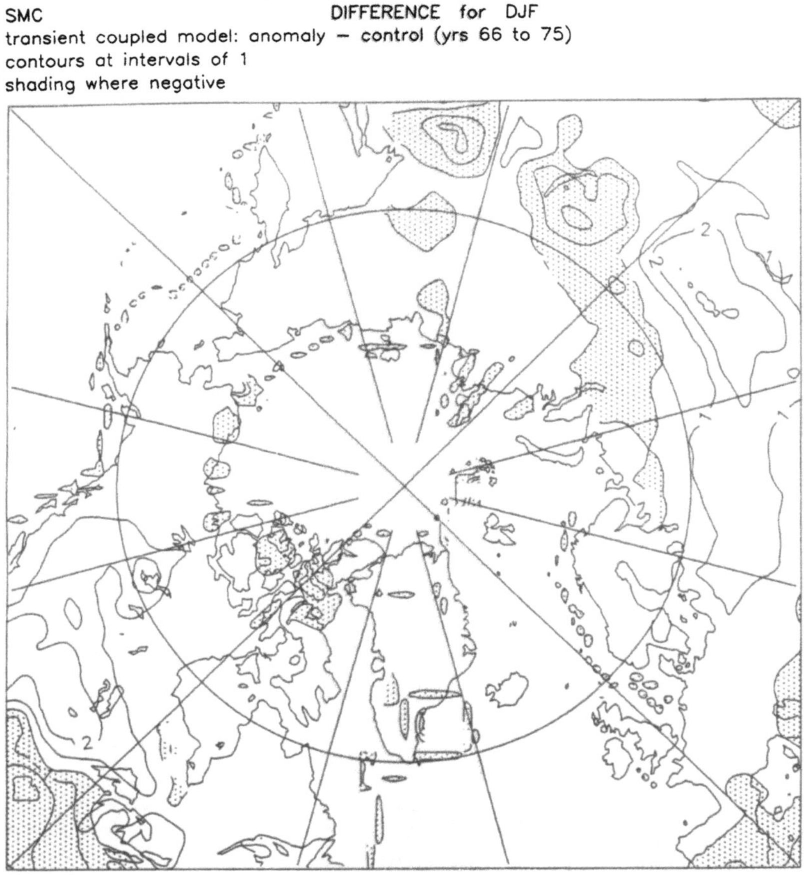

(a)

Figure 4.5. Differences in soil moisture content (cm) at years 66 to 75 of the Hadley Centre transient experiment (increased CO_2—Control). Contours at 1 cm intervals. Shading where negative. (a) December to February.

increased only where rainfall is increased by more than 0.2 mm/day. This is consistent with the changes in evaporation (not shown), which are typically an increase of about 0.25 mm/day in the June to August mean. Elsewhere there are decreases that are mostly less than 1 cm north of 60°N but reach 2 cm where rainfall is decreased in the Ob basin in this decade; this peak decrease is only 1 cm in the years 31 to 70 mean.

There is little consistent increase in the soil moisture differences after years 31 to 40; the same applies to the changes in precipitation and PMSL. The explanation is believed to be the similarity of the land–ocean temperature contrasts through this period. This is discussed further in the section on "Comparison of Transient Experiment Results." The years 31 to 70 seasonal mean soil moisture differences (not shown) reveal that soil moisture is increased over most regions north of 60°N in autumn and spring as well as winter due to increased rainfall throughout the year.

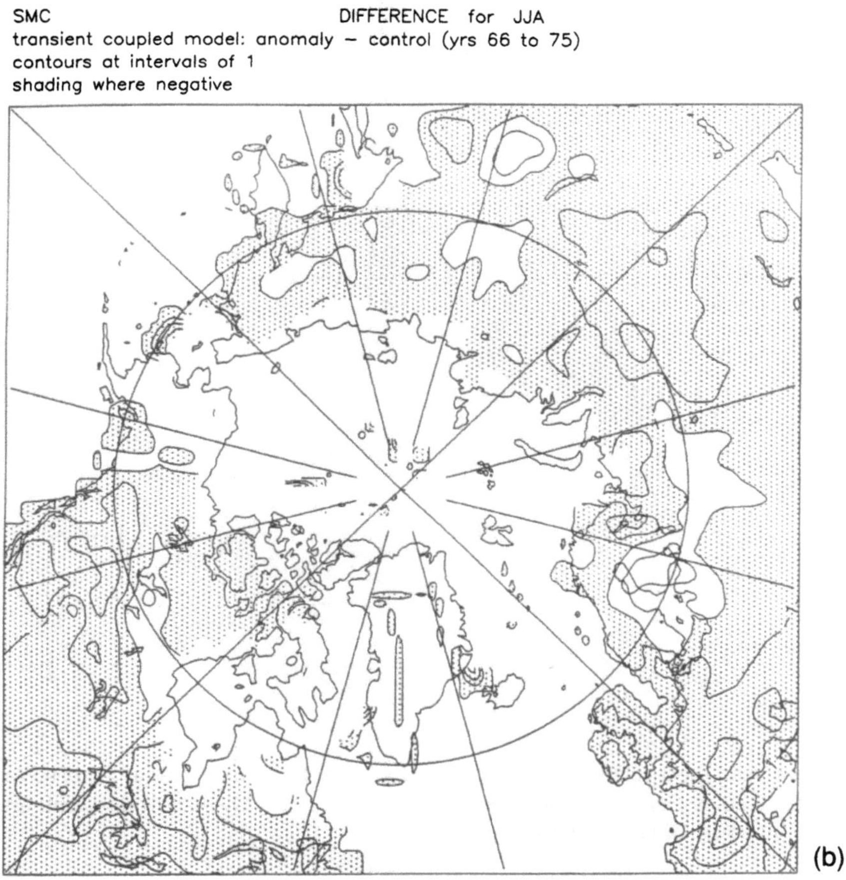

Figure 4.5. *Continued* (b) June to August.

Changes in Intensity of Precipitation

Analysis of the results over Europe (Gregory and Mitchell, 1995) from the equilibrium experiment UKHI used in Mitchell et al. (1990) has shown that in the warmed CO_2 climate, for a given average rainfall, the rainfall tends to be more intense with, for example, the rainfall on the wettest day of the season generally increasing more (or decreasing less) than the mean rainfall. This is presumably because of the greater water content of warmer air. With more rain on the wet days, the percentage of dry days commonly increases for a given mean rainfall. The year-to-year variability (seasonal SD) also generally increases more than the mean. The original analysis was confined to midlatitude regions of Europe. However, the analysis of dry day frequency has been repeated for the summer season (June to August) of the last 10 years of the transient experiment, using a criterion for a dry day of

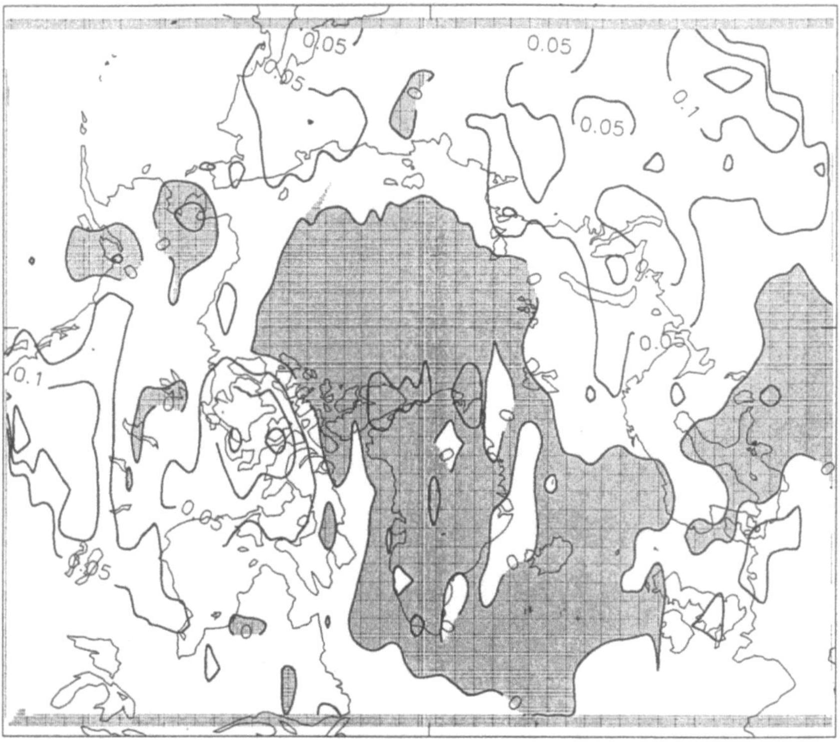

Figure 4.6. Increase in probability of a dry day in June to August at years 66 to 75 of the Hadley Centre transient experiment (increased CO_2—Control). Contours at 0.05 intervals. Shading where negative.

less than 0.05 mm. Figure 4.6 (Gregory, personal communication) shows that most Arctic land areas have more dry days, including much of the region with more rainfall.

Snowfall and Snowmass Changes

The modeled snowcover is generally realistic in the model. Gates et al. (1990) show this for North America with the 1-cm liquid water equivalent contour close to the observed 5-cm snowdepth contour, indicating correct snowdepth if the snow density is the 0.2 g/cm^3 typical of seasonal snowcover. They noted that results were similar over Europe and western Asia, but that snow was too extensive over eastern Asia. In the HC transient response experiment (not shown), the 1-cm snowmass contour over North America in the December to February mean is similarly located. Over coastal eastern Asia, it extends south to about 40°N, which appears to be realistic because Matson et al. (1986) show continuous cover north of about 45°N in winter and 25% frequency at about 40°N, where snowfalls are heavy when they do occur.

Over northern Asia, the modeled snowmass exceeds 50 kg/m² in winter. In the June to August mean, the 10 kg/m² contour is mostly near 70°N except at 60 to 80°E, where it is near 65°N. South of here, the snow-free season must end by June 1 or very soon after. Lydolph (1977) shows a map of date of disappearance of snowcover; the June 1 contour is near 70°N from 100 to 160°E, with a lower latitude further west reaching about 66°N at 60 to 80°E. This comparison indicates that the summer snowcover has realistic extent in the model over the Asian Arctic. A similar comparison with the corresponding map for North America in Hare and Hay (1974) indicates comparable quality in that region, the model capturing the NW–SE alignment of the early June snow margin from northern Alaska to central Hudson's Bay.

The increase in CO_2 reduces winter snowmass (Figure 4.7) over most regions south of 60°N, and also over regions of Scandinavia and southern Alaska north of this latitude, providing a positive snow albedo feedback. Most of the rest of the Arctic land has increased snowmass in this season, with increases exceeding 20 kg/m² northwest of Hudson's Bay and in the northern parts of the Ob and Yenisei basins. In these regions of increased snowmass the increase in precipitation has more effect on the snow budget than the later start to the snow season. The differences in snowfall rate in winter (Figure 4.8) show that a much more extensive area has increased snowfall in association with the increased total precipitation. This includes most of the land areas with a persistent winter snowcover, except the Pacific coast regions of Alaska and western Canada and southern Greenland, which have less total precipitation, as discussed earlier.

In summer, snowfall and snowmass are decreased over virtually all Northern Hemisphere land, again providing a positive snow albedo feedback; the only exceptions are the central and eastern parts of Greenland. The snowmass changes over Greenland are much larger than elsewhere because, being permanent snow, the changes have accumulated over 70 years from the start of the experiment. It is interesting that the regions of Greenland that are warmed least also have less snowfall and so lose snowmass, whereas the regions that warm most gain snowmass. The same happens over Antarctica.

Comparison of Transient Experiment Results

As discussed earlier, four experiments were compared by Gates et al. (1992), run using the NCAR, GFDL, MPI, and HC models.

(a)

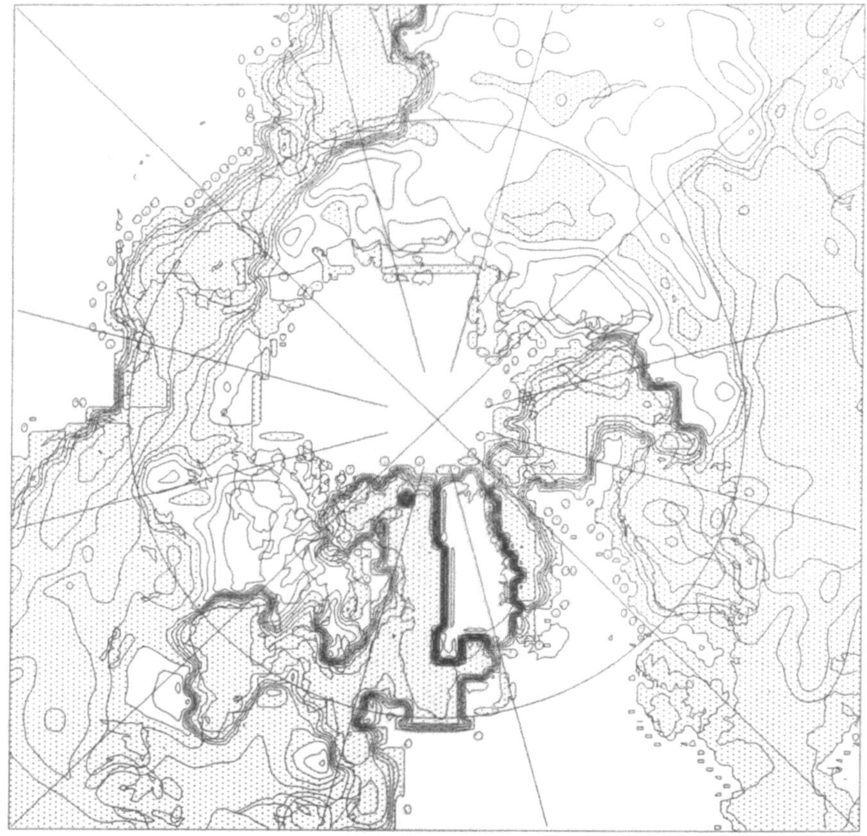

Figure 4.7. Differences in snow mass (kg/m^2) at years 66 to 75 of the Hadley Centre transient experiment (increased CO_2—Control). Contours at 0. ±5, 10, 20, 50, 100 kg/m^2. Shading where negative. (a) December to February.

Averages were shown for about the time of doubling, except that the NCAR run was for about a 45% increase, when the equilibrium warming would be just over half that at the time of doubling (*over* half because of the logarithmic relation between temperature increase and CO_2 increase).

Only annual mean temperature changes were shown for the experiments, but there is additional information available, mostly in the articles on the results referenced earlier.

Table 4.3 shows estimated annual mean temperature changes for selected regions and differences between the regions. The NCAR differences have

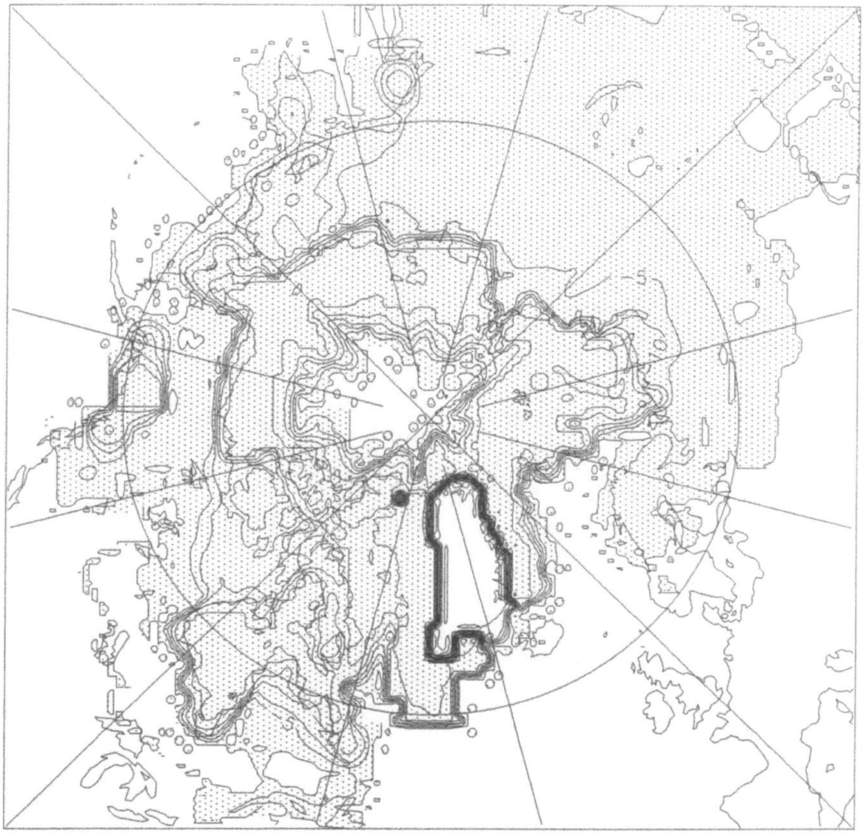

(b)

Figure 4.7. *Continued* (b) June to August.

been doubled to allow for the smaller CO_2 increase. There are large differences between the patterns of warming in the different experiments. The NCAR model is quite distinct, with little difference between the warming over the Arctic Ocean and that at 60°N. However, although the other models have a similar overall gradient between the Arctic Ocean and the equator (3.2 to 4.7 K), they differ markedly in its distribution. For example, comparing the HC and GFDL models, they have similar gradients between the Arctic Ocean and 60°N over land, but the HC model has a much smaller warming over the ocean at this latitude, and so a larger north–south gradient between there and the Arctic Ocean. The MPI model has a gradient of this size or greater over both ocean and land.

The ocean–land contrast at 60°N is much larger in the HC model than in the others. This is likely to be important in generating ocean–land contrasts

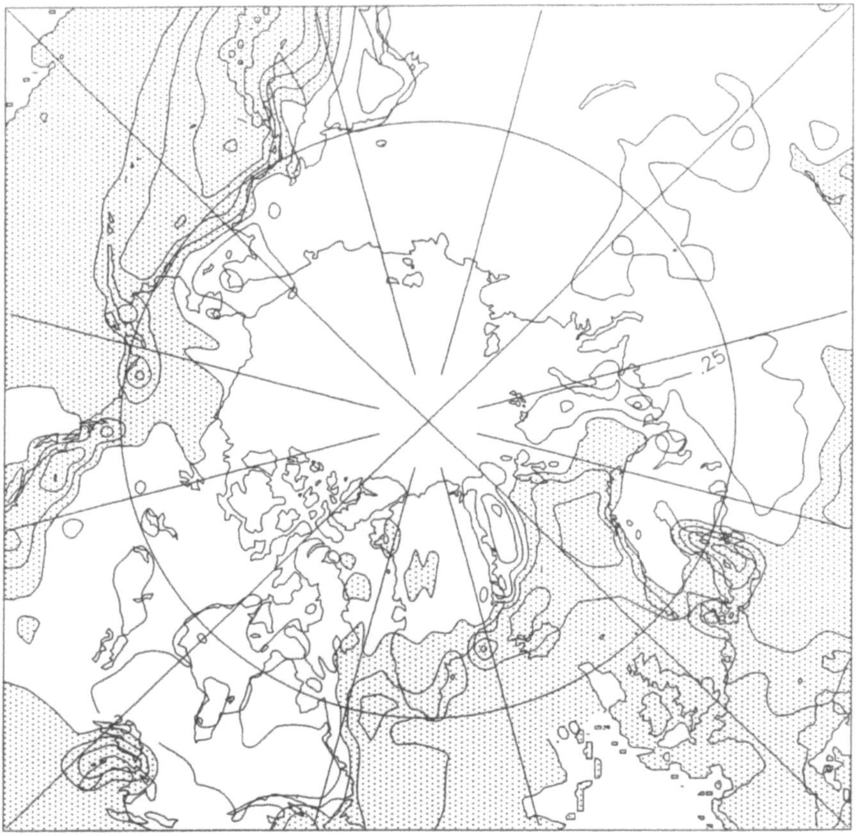

(a)

Figure 4.8. Differences in snowfall (kg/m²/day) at years 66 to 75 of the Hadley Centre transient experiment (increased CO_2—Control). Contours at 0.25 kg/m²/day intervals. Shading where negative. (a) December to February.

in atmospheric variables. An interesting feature of the HC results is that after the first 30 years clear patterns of precipitation and circulation change have developed in the HC model, which then persist with little further amplification through the next 40 years. These features appear to be associated with the development of the ocean–land contrasts, which show a similar temporal evolution. The weakness of these ocean–land contrasts in the other models may explain the fact that similar temporal behavior is not obvious; for example, the GFDL soil moisture differences shown by Gates et al. (1992) and Manabe et al. (1992) increase rather steadily from years 40 to 60 and 80 to 100. The reasons for the large ocean–land contrast in the HC

(b)

Figure 4.8. *Continued* (b) June to August.

model are not fully known, though Murphy and Mitchell (1995) have suggested a role for cloud feedbacks over land, with the warming reducing the cloudiness there and so intensifying the warming.

The soil moisture differences at about the time of doubling of CO_2 can also be compared. These are similar in the GFDL model to those already discussed for the HC model, with decreases predominant in summer and increases in winter over Arctic land. The MPI model has a similar but smaller seasonal variation in the effects of CO_2 on soil moisture. In particular, the summer drying is much less widespread. All the models tend to have more precipitation in the Arctic in winter; in summer patterns are patchy, with the HC model showing the most consistent increases over Arctic land.

Table 4.3. Range and Estimated Average (Ave.) of Annual Mean Temperature (K) Differences[a] (NCAR Values Doubled, See Text)

	GFDL Range	Ave.	MPI Range	Ave.	NCAR Range	Ave.	HC Range	Ave.
Arctic Ocean	4–6	5	4–8	6	1–5	3	3–7	5
Land 60°N	2–5	3.7	1–3	1.9	1–4	2.4	1–5	3.9
Ocean 60°N	1–4	3.2	0–2	1.2	1–4	3.1	−1–4	1.2
Land 30°N		3.1		1.9		1.1		2.7
Ocean 30°N		2.3		1.1		1.3		2.0
Ocean 0°N	1–3	1.9	1–2	1.5	1–2	1.2	1–2	1.3

Differences in Averages Between Regions

	GFDL	MPI	NCAR	HC
Arctic Ocean–Land 60°N	1	4	0.6	1
Arctic Ocean–Ocean 60°N	2	5	0	4
Land–Ocean 60°N	0.5	0.7	−0.7	2.7
Land–Ocean 30°N	0.8	0.8	−0.2	0.7

[a] IPCC (1992).

Uncertainties

There are important uncertainties in these results. Major ones include:

(a) Radiative forcing. As discussed earlier, the forcing of the climate system by changes in the environment is imperfectly known. This problem is not confined to predicting future greenhouse gas concentrations; sulphate aerosols are thought to be important in compensating the warming due to greenhouse gases but they are only just being included in the climate models, and their past as well as their future concentrations are still uncertain. Evidence so far indicates that the predictions of warming omitting aerosol effects may be too large, particularly in the regions of large emissions and immediately downwind. Interestingly, just as the effects are being recognized, they are decreasing in two of the most important areas, Europe and North America, as fuel use turns increasingly to less sulphurous forms. If such a decrease is of global scale and the effects have been large, one may expect an acceleration of warming in the immediate future as the braking effect of aerosols is reduced.

(b) The ocean model. The flux corrections applied to keep the ocean temperatures and salinities near to reality are significant fractions of the actual fluxes, suggesting that the modeled changes in temperature and salinity due to the ocean currents are too small.

(c) The representation of cloud–radiation interactions. Sensitivity to this was highlighted by a series of experiments in which global mean warming ranged from 1.9 to 5.2 K for doubled CO_2, depending on which of three cloud representations was used (Senior and Mitchell, 1993).

(d) The representation of the land surface hydrology. One improvement needed is to make the resistances imposed by plant stomata dependent on atmospheric variables such as temperature, vapor pressure deficit below saturation, and solar radiation, as well as on increased CO_2. An increase in evaporation as the climate warms is likely to be reduced by these dependences; they are now being included in the MO model.

Possible Interactions with Vegetation

One of the main climate changes predicted as a consequence of increased greenhouse gases is a warming of high latitude land. This may be expected to favor changes in vegetation, such as a contraction of the tundra region, and a northward shift of boreal forest. There is a possibility of a feedback on such changes due to the lower albedos of forested than deforested regions both in the snow-free season when coniferous forest has typical albedos of 0.12, a few percent below that of tundra and deciduous forest, and more importantly in the presence of snow when the masking of the snowcover by trees can reduce albedos by up to about 60% compared with exposed snow (Bonan et al., 1992; Robinson and Kukla, 1985; Thomas and Rowntree, 1992). Using a slab model of the ocean, Bonan et al. found a large cooling of the climate in middle and high latitudes when boreal forest was removed. The zonal averages of temperature changes over land near 60°N reached 12 K in April and 5 K or more in other midseason months. Thomas and Rowntree (1992) obtained smaller changes, partly because ocean surface temperatures were kept the same, and partly because they deforested a smaller area.

Provided soil conditions are favorable, the forest may spread north faster than it is eroded on the south side, in which case the tree belt would widen and provide a positive feedback to warming. On the other hand, a negative feedback is possible if the forest is eroded on the south side by either onset of a warmer, less favorable, climate or by human activities and does not advance as quickly northward because soils are unsuitable.

Warming of high latitude land is likely to lead to partial melting of permafrost. This will involve an increase in summer thaw depth. Christensen and Cox (1995) studied the response of modeled methane emissions to changes in temperature and precipitation. Broadly, their results indicate that a significant increase in methane release could occur with increases in temperature and small changes or increases in soil moisture; drying of the soil in summer could lead to reduced emissions despite a warming.

Validation

Validation of the model's global mean temperature predictions is difficult because of the wide range of model sensitivities (e.g., the sensitivity to different cloud representations discussed above), the uncertainty in the total changes in greenhouse forcing to date, and the influence of natural variations of the climate. If only the enhancements of the greenhouse effect are considered, the observed warming of about 0.5 K over the last century (Folland et al., 1990) fits best with the least sensitive of Senior and Mitchell's (1993) experiments, which found a 1.9 K response of global mean temperature for a doubling of CO_2. However, if the contributions of decreased ozone and increased sulphate aerosol have been as important as suggested in the 1992 IPCC Supplement (Isaksen et al., 1990), a greater sensitivity is needed to fit the observed record.

Natural variations introduce a similar uncertainty: in their absence, the observed warming might have been either larger or smaller, implying a larger or smaller sensitivity. However, an analysis of trends on decadal and century timescales for a 1000-year run with the GFDL model (Stouffer et al., 1994) revealed that the observed warming trend for the last 100 years was larger than any over a comparable period in the model.

Validation of regional effects is at an early stage. Comparison of the spatial pattern of warming in the 1980s relative to the preceding 30 years with model predictions (Gates et al., 1992, Figure B4; Folland et al., 1992, Figure C5) shows some marked similarities, notably in the strong warming over the northern continents, and the cooling near southern Greenland.

Summary

The conclusions of this chapter may be summarized as follows.

1. The predictions of climate change over Arctic land in the HC model show the following features:

a. Temperatures increase generally, but with a wide range of magnitudes—most over melted snow, least over land with ocean or sea-ice to windward, especially where the ocean warming is small.

b. In all seasons precipitation increases over most of the area; for a given mean rainfall, intensity, at least in summer, tends to increase, and frequency to decrease. Soil moisture increases generally except in summer.

c. Snowfall increases in winter over most of the Arctic; snowmass increases over a smaller area; snowfall and snowmass decrease in summer except over parts of Greenland.

d. Circulation changes forced by spatial variations in warming are responsible for regional features including decreases in precipitation and relatively weak warming, and other exceptions to the above general statements.

2. Comparison of the four available transient response experiments reveals large differences in the gradients between the Arctic Ocean and 60°N and the ocean–land contrast at 60°N. For precipitation and soil moisture, the models' results are similar for winter but less so for summer.

3. There are important uncertainties associated with the radiative forcing, including the effects of sulphate aerosols, with ocean simulation, with cloud representations, and with land surface hydrology. However, because the underlying physical mechanisms are understood, the general conclusions in No. 1 above are not likely to be affected by these uncertainties except possibly the reductions of precipitation and soil moisture in summer. Evidence from observations encourages some credence in the predictions, at least on the larger scales.

4. As vegetation responds to any change in climate, there is a possibility of important feedbacks due to changes in the extent of masking of highly reflective snow.

Acknowledgments. The Hadley Centre 75-year transient response experiment, which supplied many of the results reported here, was supported by the Department of the Environment.

References

Bonan, G.B., Pollard, D., and Thompson, S.L. 1992. Effects of boreal forest vegetation on global climate. *Nature* 359:716–718.

Bretherton, F.P., Bryan, K., and Woods, J.D. 1990. Time-dependent greenhouse-gas-induced climate change. In: Houghton. J.T., Jenkins, G.J., and Ephraums, J.J. (eds.), *Climate Change: The IPCC Scientific Assessment* (pp. 173–193). Cambridge: Cambridge University Press.

Cattle, H., and Thomson, J.F. 1993. The Arctic response to CO_2-induced warming in a coupled atmosphere-ocean general circulation model. In: Peltier, W.R. (ed.), *Ice in the Climate System* (pp. 579–596). Proceedings of a NATO Advanced Research Workshop. New York: Springer-Verlag.

Christensen, T.R., and Cox, P.M. 1995. Response of methane emissions from Arctic Tundra to climatic change: results from a model simulation. *Tellus* 47B:301–309.

Cubasch, U., Hasselmann, K., Hock, H., Maier-Reimer, E., Santer, B.D., and Sausen, R. 1992. Time-dependent greenhouse warming computations with a coupled ocean-atmosphere model. *Climate Dynam.* 8:55–69.

Folland, C.K., Karl, T.R., Nicholls, N., Nyenzi, B.S., Parker, D.E., and Vinnikov, K.Ya. 1992. Observed climate variability and change. In: Houghton, J.T., Callander, B.A., and Varney, S.K. (eds.), *Climate Change 1992: The Supplementary Report to the IPCC Scientific Assessment* (pp. 135–170). Cambridge: Cambridge University Press.

Folland, C.K., Karl, T., and Vinnikov, K.Ya. 1990. Observed climate variations and change. In: Houghton, J.T., Jenkins, G.J., and Ephraums, J.J. (eds.), *Climate Change: The IPCC Scientific Assessment* (pp. 195–238). Cambridge: Cambridge University Press.

Gates, W.L., Mitchell, J.F.B., Boer, G.J., Cubasch, U., and Meleshko, V.P. 1992. Climate modelling, climate prediction and model validation. In: Houghton, J.T., Callander, B.A., and Varney, S.K. (eds.), *Climate Change 1992: The Supplementary Report to the IPCC Scientific Assessment* (pp. 97–134). Cambridge: Cambridge University Press.

Gates, W.L., Rowntree, P.R., and Zeng, Q-C. 1990. Validation of climate models. In: Houghton, J.T., Jenkins, G.J., and Ephraums, J.J. (eds.), *Climate Change: The*

IPCC Scientific Assessment (pp. 93–130). Cambridge: Cambridge University Press.

Gregory, J.M., and Mitchell, J.F.B. 1995. Simulation of daily variability of surface temperature and precipitation over Europe in the current and $2 \times CO_2$ climates using the UKMO Climate model. *Q. J. R. Meteorol. Soc.* 121:1451–1476.

Hall, N.M.J., Hoskins, B.J., Valdes, P.J., and Senior, C.A. 1994. Storm tracks in a high resolution GCM with doubled CO_2. *Q. J. R. Meteorol. Soc.* 120:1209–1230.

Hare, F.K., and Hay, J.E. 1974. The climate of Canada and Alaska. In: Bryson, R.A., and Hare, F.K. (eds.), *Climates of North America. World Survey of Climatology* (Vol. 11). New York: Elsevier Scientific Publishing Company.

Hibler, W.D., Jr. 1979. A dynamic-thermodynamic sea ice model. *J. Phys. Oceanogr.* 9:817–846.

Houghton, J.T., Jenkins, G.J., and Ephraums, J.J. (eds.). 1990. *Climate Change. The IPCC Scientific Assessment.* Cambridge: Cambridge University Press.

Isaksen, I.S.A., Ramaswamy, V., Rodhe, H., and Wigley, T.M.L. 1992. Radiative forcing of climate. In: Houghton, J.T., Callander, B.A., and Varney, S.K. (eds.), *Climate Change 1992: The Supplementary Report to the IPCC Scientific Assessment* (pp. 47–67). Cambridge: Cambridge University Press.

Lydolph, P.E. 1977. *Climates of the Soviet Union. World Survey of Climatology* (Vol. 7). New York: Elsevier Scientific Publishing Company.

Manabe, S., Spelman, M.J., and Stouffer, R.J. 1992. Transient responses of a coupled ocean-atmosphere model to gradual changes of atmospheric CO_2. Part II: Seasonal response. *J. Climate* 5:105–126.

Matson, M., Ropelewski, C.F., and Varnadore, M.S. 1986. An atlas of satellite-derived northern hemisphere snow cover frequency. NOAA Atlas. U.S. Dept. of Commerce.

Meehl, G.A., Washington, W.M., and Karl, T.R. 1993. Low-frequency variability and CO_2 transient climate change. Part 1. Time-averaged differences. *Climate Dynam.* 8:117–133.

Mitchell, J.F.B., Manabe, S., Tokioka, T., and Meleshko, V. 1990. Equilibrium climate change. In: Houghton, J.T., Jenkins, G.J., and Ephraums, J.J. (eds.), *Climate Change: The IPCC Scientific Assessment* (pp. 131–164). Cambridge: Cambridge University Press.

Murphy, J.M. 1995. Transient response of the Hadley Centre coupled ocean-atmosphere model to increasing carbon dioxide. Part I. Control climate and flux correction. *J. Climate* 8:36–56.

Murphy, J.M., and Mitchell, J.F.B. 1995. Transient response of the Hadley Centre coupled ocean-atmosphere model to increasing carbon dioxide. Part II. Spatial and temporal structure of response. *J. Climate* 8:57–80.

Robinson, D.A., and Kukla, G. 1985. Maximum surface albedo of seasonallly snow-covered lands. *J. Climatol. Appl. Meteorol.* 24:402–411.

Rowntree, P.R. 1991. Atmospheric parametrization schemes for evaporation over land: basic concepts and climate modelling aspects. In: Schmugge, T.J., and Andre, J-C. (eds.), *Land Surface Evaporation: Measurement and Parametrization* (pp. 5–29). New York: Springer-Verlag.

Semtner, A.J., Jr. 1976. A model for the thermodynamic growth of sea ice in numerical investigations of climate. *J. Phys. Oceanogr.* 6:379–389.

Senior, C.A., and Mitchell, J.F.B. 1993. CO_2 and climate; the impact of cloud parametrizations. *J. Climate* 6:393–418.

Stouffer, R.J., Manabe, S., and Vinnikov, K.Ya. 1994. Model assessment of the role of natural variability in recent global warming. *Nature* 367:634–636.

Thomas, G., and Rowntree, P.R. 1992. The boreal forests and climate. *Q. J. R. Meteorol. Soc.* 118:469–497.

Washington, W.M., and Meehl, G.A. 1989. Climate sensitivity due to increased CO_2: Experiments with a coupled atmosphere and ocean general circulation model. *Climate Dynam.* 4:1–38.
Watson, R.T., Rodhe, H., Oeschger, H., and Siegenthaler, U. 1990. Greenhouse gases and aerosols. In: Houghton, J.T., Jenkins, G.J., and Ephraums, J.J. (eds.), *Climate Change: The IPCC Scientific Assessment* (pp. 1–40). Cambridge: Cambridge University Press.
Wilson, M.F., and Henderson-Sellers, A. 1985. A global archive of land cover and soil data for use in general circulation climate models. *J. Climatol.* 5:119–143.

2. Effects of Anticipated Global Change on Organisms and Populations

5. Photosynthesis and Respiration in Mosses and Lichens

Bjartmar Sveinbjörnsson and Mats Sonesson

Bryophytes and lichens constitute a large part of the total biomass of present-day tundras (Figure 5.1), and in *Rangifer* distribution areas the latter are an important part of the food web. Both are important in the structure and function of the ecosystems because of their effects as insulators and filters between the above- and belowground parts. Their insulating properties result partly from increased reflectance and partly from the large proportion of air pore space when dry. Thus, soil under bryophyte and lichen mats tends to be cold and moist. In addition to their effect through lowered soil temperature, mosses and lichens also affect nutrient cycling of the ecosystem by intercepting aerial deposition and leaching from dripping aboveground vascular plant parts (Figure 5.2) (Cowles, 1984; Rosswall and Granhall, 1980). These nutrients are subsequently released upon the decomposition of the bryophyte and lichen tissues, an exceedingly slow process, largely due to the chemical composition of these organisms (Greenfield, 1993; Longton, 1992; Moore, 1984; Parinkina and Piin, 1990; Wetmore, 1982). Lichens with cyanobacterial symbionts and bryophytes with cyanobacterial associations provide the main input of nitrogen into the ecosystem (Alexander et al., 1978; Baselier et al., 1978; Kallio, 1975). Lichens, and to lesser extent bryophytes, have also been shown to inhibit growth of neighboring vascular plants and thus further reduce ecosystem productivity (Brown and Mikola, 1974; Cowles, 1984; Fisher, 1979; Mallik and Newton, 1988).

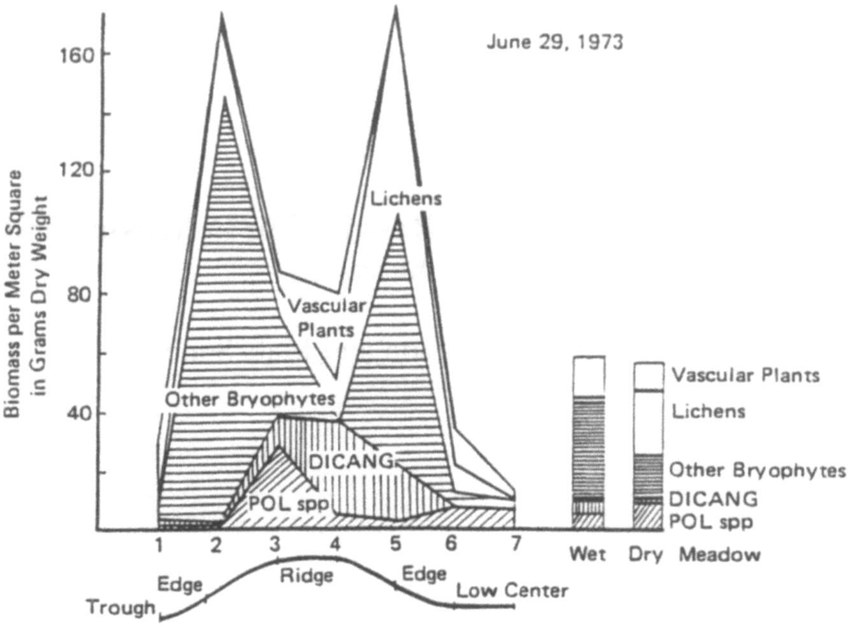

Figure 5.1. Distribution of early season aboveground biomass of lichens, bryophytes, and vascular plants along microtopographic gradients on the tundra near Barrow, Alaska. From Oechel and Sveinbjörnsson (1978), with permission from Springer-Verlag.

As there have recently been published thorough reviews of lichen and bryophyte ecophysiology and climate change (Longton, 1992; Sveinbjörnsson and Oechel, 1992; Tenhunen et al., 1991), we will not attempt a thorough review of the field but will instead focus on the aspects particularly important to tundra communities and in situ conditions. First we will assess the hitherto considered limitations to net photosynthesis and estimate responses based on these. Then we will evaluate the evidence of new information on CO_2 responses, and finally we will assess the adequacy of present information.

Photosynthesis and Respiration

General Pattern

There is a great deal of similarity in the CO_2 exchange patterns of bryophytes and lichens (Longton, 1992). Both have simple structures and little control over water loss. They are metabolically active when wet; short respiratory burst is sometimes experienced when they are demoistened

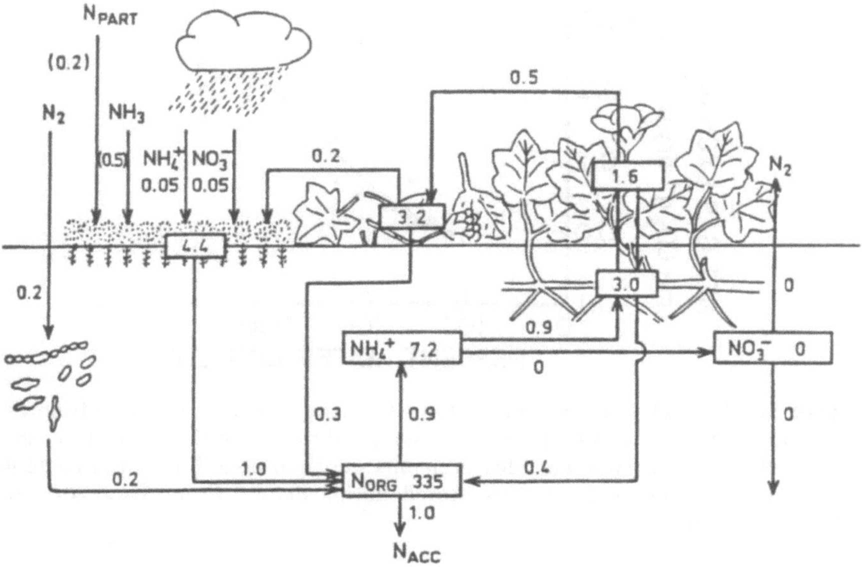

Figure 5.2. Nitrogen budget of a nutrient poor mire in northern Sweden. Note the importance of the moss layer in intercepting and retaining nitrogen. From Rosswall and Granhall (1980), with permission from *Ecological Bulletins* (Stockholm).

after desiccation during which internal activity is quickly resumed. As they dry out, they first increase their net CO_2 exchange rate, reach a maximum net CO_2 exchange rate at intermediate water contents, and then with further drying they decrease their activity, which eventually ceases (Figure 5.3). Any future change in moisture regimes will affect their distribution patterns and hence the cryptogamic species composition of arctic vegetation. Arctic bryophytes show lower rates of photosynthesis under ideal short-term conditions than do their subarctic, temperate, and tropical conspecifics (Oechel and Sveinbjörnsson, 1978; Sveinbjörnsson, 1980), whereas the present authors have found similar rates in congeneric arctic and tropical lichens (unpublished data) and Lechowicz (1982) found no latitudinal differences in maximum photosynthetic rates among lichen species.

Arctic bryophytes and lichens exhibit distinct temperature optima, which are lower than those of temperate conspecifics and congenerics (Lechowicz, 1982; Sveinbjörnsson and Oechel, 1980), and they are able to photosynthesize at lower temperature than can temperate ones. Arctic lichens were previously thought to have a higher rate of respiration than southern ones, but further studies have not supported these early observations. An opposite general trend (Sveinbjörnsson and Oechel, 1983) has been observed in geographically diverse populations of the moss *Polytrichum commune*.

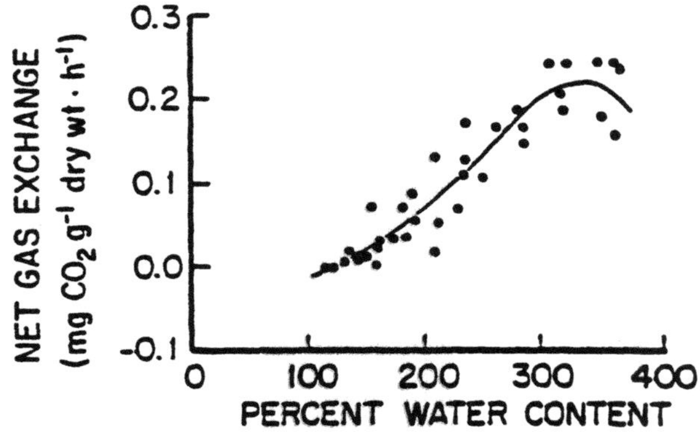

Figure 5.3. The net photosynthetic rate of the lichen *Cetraria cucullata* from the Seward Peninsula, Alaska, at optimum temperature, saturating irradiance, and varying thallus water contents. Measurements were started at high water contents achieved by soaking and were monitored during drying. From Sveinbjörnsson (1990), with permission from *Rangifer.*

Photosynthesis and Irradiance

Where they are intermingled with vascular plants, future irradiance levels reaching the bryophyte and lichen surface can be assumed to decrease as vascular plants may become taller and denser. Irradiance penetration under the vascular plant canopy is uneven during the arctic day (Sveinbjörnsson and Oechel, 1981) and this unevenness can be assumed to continue and perhaps increase. Because understory bryophytes and lichens are adaptable (Oechel and Sveinbjörnsson, 1978), they will survive all but the most deeply shaded situations. The situation will perhaps be ameliorated by increased moss/lichen level CO_2 concentration, which allows equal photosynthesis at lower light levels (Silvola, 1985).

Photosynthesis and Temperature

The temperature of bryophytes and lichens varies considerably from air temperature, particularly that measured at weather stations. Thus, even for the frequently overcast coastal tundra near Barrow, Alaska, maximum daily bryophyte temperature averaged 7°C above the weather station temperature whereas minimum bryophyte temperature was on average slightly lower than that of the weather station. Even with the temperature of bryophytes elevated above the ambient and with seasonally varying temperature relations of photosynthesis (Oechel, 1976; Oechel and Sveinbjörnsson, 1978), temperature is mostly below optimum for Barrow bryophytes (Figure 5.4) (Oechel, 1976). This lends credence to the assumption

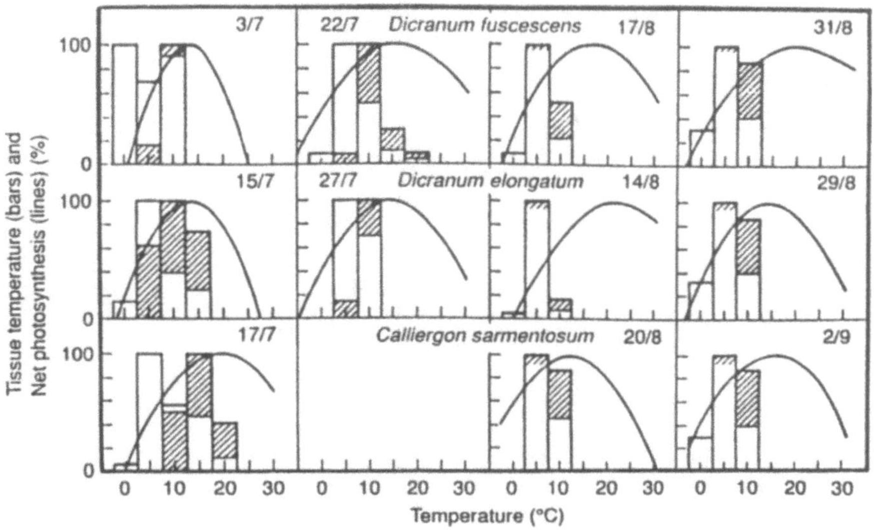

Figure 5.4. Frequency of in situ tissue temperature classes 5 days prior to measured temperature relations of net photosynthesis for three moss species growing on the tundra near Barrow, Alaska. The open bars refer to all temperatures throughout the day, and hatched bars refer to the 4 midday hours (from Oechel, 1976, with permission from *Photosynthetica*). When daily and midday intervals showed the most frequent temperatures in the same temperature range the bars are shaded at the top. From Oechel (1976), with permission from *Photosynthetica.*

that bryophyte photosynthesis will benefit from the anticipated air temperature increase. On clear, calm days and nights, temperatures deviate most whereas during rainy windy days there is the least difference between the macro- and bioclimate (Figure 5.5). Hence, it is on these latter types of days that there is likely to be a significant effect of globally increased temperatures. During these days lichen and bryophyte water content is likely to become near-optimal (see below) for net photosynthesis and growth. In addition, there is evidence, in bryophytes, for increasing temperature optima of net photosynthesis at increasing CO_2 concentrations (Figure 5.6) (Silvola, 1985).

Photosynthesis and CO_2 Concentrations

In order to predict the effects of future CO_2 concentrations, one needs to know what the present-day CO_2 concentrations are, and how these surface organisms respond to them. Field CO_2 concentrations surrounding tundra bryophytes and lichens are unknown, but drawing on inferences from studies in boreal and temperate zone it is reasonable to assume that they are sometimes considerably higher than the atmospheric ambient. Thus,

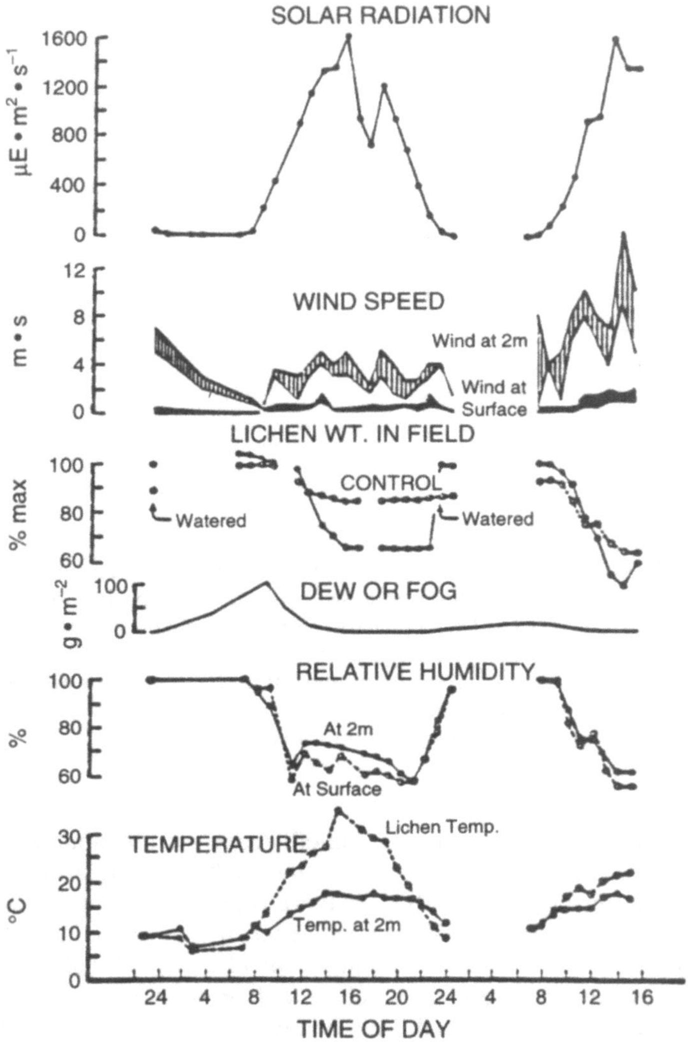

Figure 5.5. Lichen water content and temperature in relation to weather parameters over a palsa mound near the airport on Nunivak Island, western Alaska. From Sveinbjörnsson (1990), with permission from *Rangifer*.

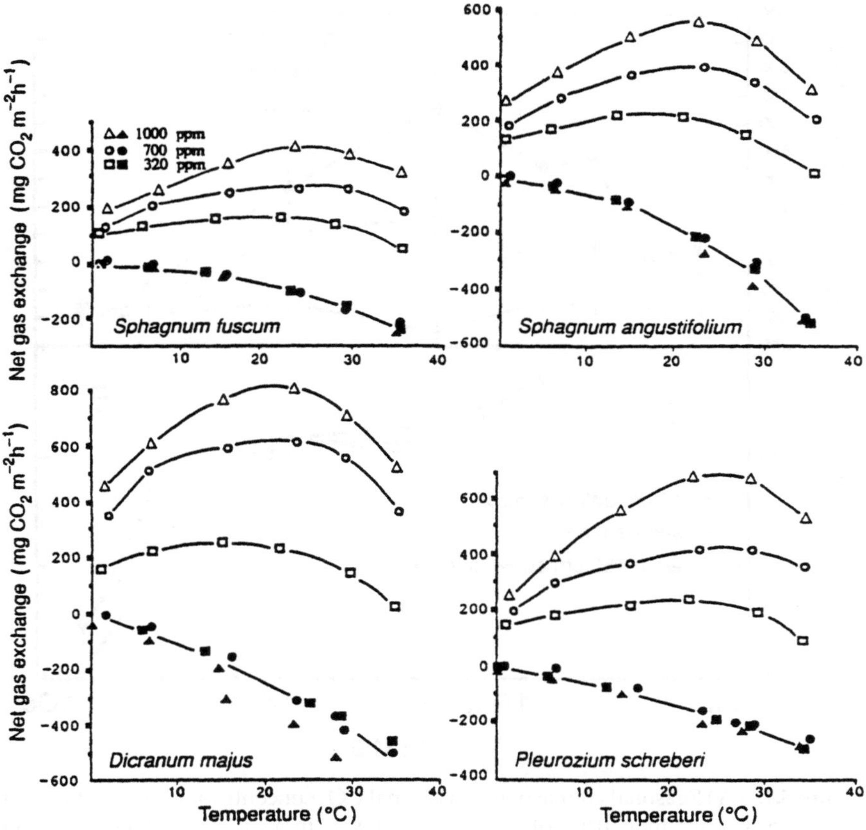

Figure 5.6. Temperature relations of net CO_2 gas exchange in four boreal bryophyte species. The measurements were made at $500 \mu mol\, m^{-2} s^{-1}$ (open symbols) and in darkness (closed symbols) at a constant CO_2 concentration. Note increasing temperature optima at increasing CO_2 concentrations. From Silvola (1985), with permission from *Lindbergia*.

Sonesson et al. (1992) found that the average daily CO_2 concentration for *Hylocomium splendens* in a subarctic birch forest averaged about 400 ppm and reached as high as 700 ppm (Figure 5.7). Similarly, Sveinbjörnsson and Butler (cited in Sveinbjörnsson and Oechel, 1992) found that CO_2 concentration levels around shoots of *Pleurozium schreberii* and *Ptilium crista-castrensis* in a subarctic black spruce woodland were highest in the early morning hours and lowest during midday (Figure 5.8). Soil respiration and nocturnal moss respiration is the most probable source of this elevated CO_2 concentration, which is reduced by wind (Silvola, 1986). It is of interest to note that the lowered rate of photosynthesis with age that has been found in bryophytes (Oechel and Van Cleve, 1986) and lichens (Nash et al., 1980;

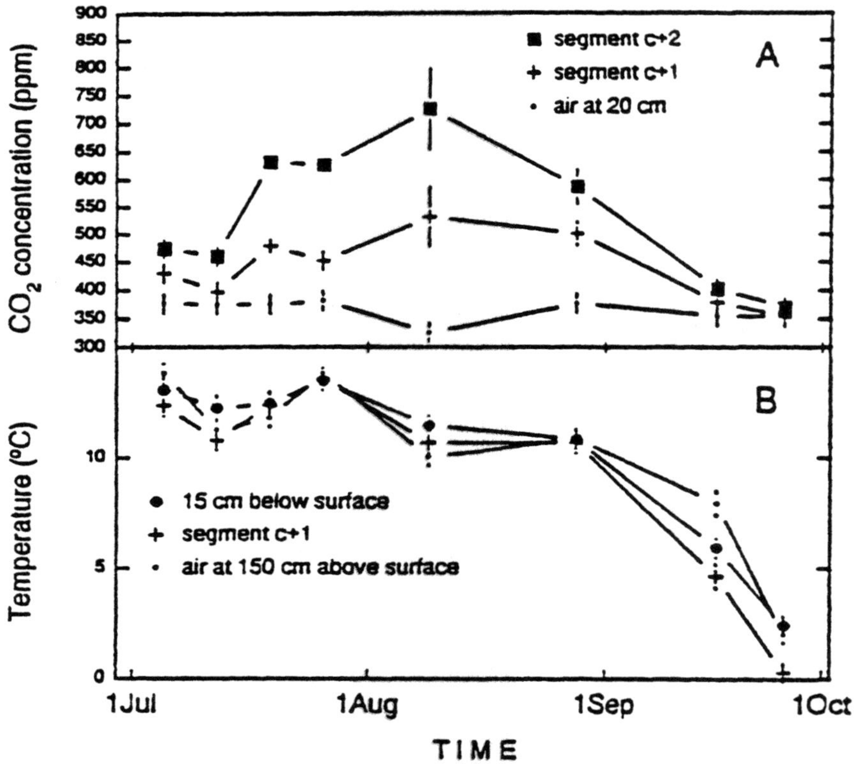

Figure 5.7. (A) Seasonal course of mean diurnal CO_2 concentrations during times of irradiance greater than $10 \mu mol\, m^{-2} s^{-1}$ at two levels in a moss mat (and one level above it) dominated by *Hylocomium splendens* growing at Abisko, Sweden. C + 1 (current plus 1 year) and C + 2 (current plus 2 years) refer to tissue age, C + 1 being last year's frond and C + 2 being 2 years old. The C + 1 frond is about 0.5 to 1.0 cm below the moss surface and the C + 2 frond is 2 to 4 cm below the moss surface. (B) Mean temperatures at the three levels in and above the moss mat. From Sonesson et al. (1992), with permission from *Oecologia.*

Sveinbjörnsson, 1987) may not be expressed as strongly in field conditions because of the gradient in CO_2 concentration through the soil–cryptogram–air continuum (see Figure 5.8).

Net photosynthesis of subarctic and temperate bryophytes and lichens generally increases linearly in response to short-term CO_2 exposure up to a certain saturating CO_2 concentration, the value of which varies between species and time of year (Figures 5.9 and 5.10) (Silvola, 1985; Sonesson et al., 1992). No short-term study of the photosynthetic response to a range of different CO_2 concentrations has been conducted on arctic bryophytes and lichens, but Russian researchers (Gerasimenko and Zalensky, 1973; Shvetsova and Voznesensky, 1970) have studied "potential" photosynthesis

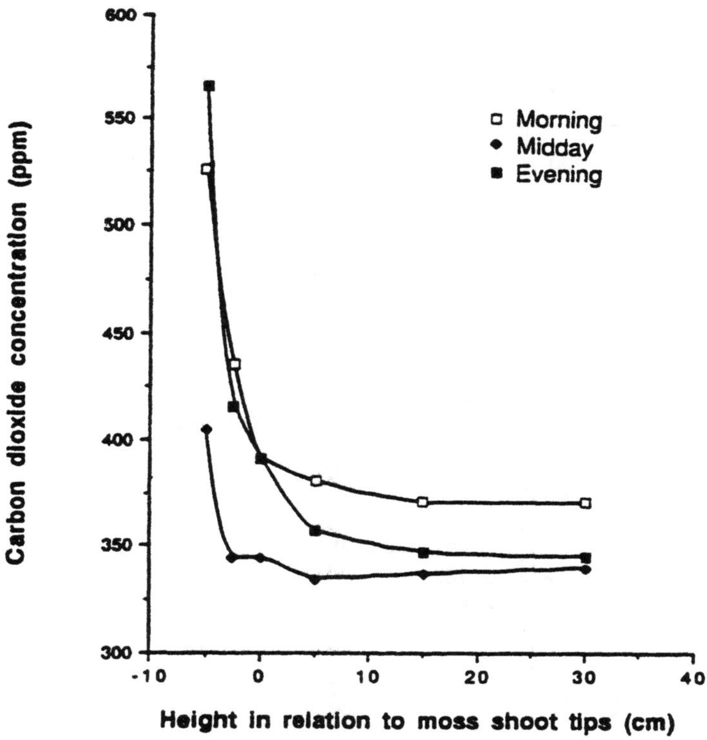

Figure 5.8. Concentration of CO_2 concentrations in air at different levels in and above a moss mat dominated by *Ptilium crista-castrensis* and *Pleurozium schreberii* growing on a black spruce forest floor in south central Alaska. The different lines are for different times during the day. From Sveinbjörnsson and Oechel (1992), with permission from Clarendon Press.

at very high concentrations on plants in situ on the Taimyr Peninsula and on Wrangel Island. Long-term studies have not been conducted on any lichen or bryophyte material from any biome.

Comprehensive studies of bryophytes at Barrow, Alaska (Figure 5.11) and Sweden (Johansson and Linder, 1980; Oechel and Sveinbjörnsson, 1978), and of lichens in Alaska (Tenhunen et al., 1992) were all carried out at stable "normal" CO_2 concentrations. Assuming that bryophytes and lichens express steady-state responses to CO_2 in the field, it becomes obvious that the pattern hitherto found may not reflect the natural one. Several additional factors need to be considered.

The photosynthetic depression found at high tissue water contents, such as occur in the field following rainshowers, is caused by high diffusion resistance in lichens and bryophytes and disappears when CO_2 concentra-

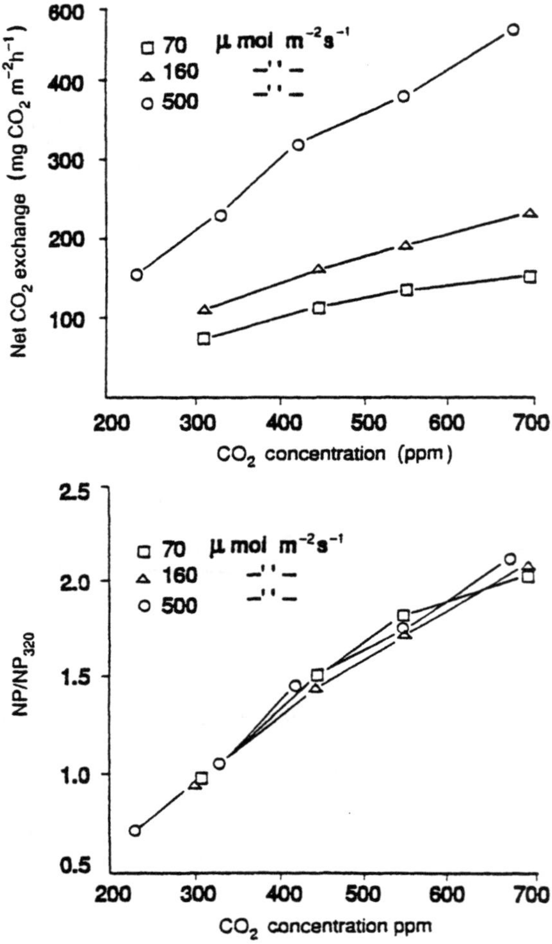

Figure 5.9. Absolute (upper graph) and relative (lower graph) net photosynthetic rates at different irradiance levels as a function of CO_2 concentration in boreal *Dicranum majus* in Finland. From Silvola (1985), with permission from *Lindbergia.*

tion is high (Figure 5.12) (Lange and Tenhunen, 1981; Silvola, 1990). Thus, in a future high CO_2 environment the diurnal pattern would be altered from what has previously been assumed (i.e., a period of lowered photosynthesis following soaking by rainshowers may not be observed). In fact, the measured present-day diurnal pattern (Figure 5.13) (Tenhunen et al., 1992) may differ considerably from the actual present-day pattern during wetting by dew because this early morning dew formation generally coincides with calm weather and low light, when CO_2 concentrations at bryophyte and lichen level are likely to be high (Figure 5.11) (Sveinbjörnsson, 1990).

Figure 5.10. Seasonal variation in net photosynthetic responses of top segments of *Hylocomium splendens* to irradiance at different CO_2 concentrations. The measurements were conducted at 15°C and optimal tissue hydration. From Sonesson et al. (1992), with permission from *Oecologia*.

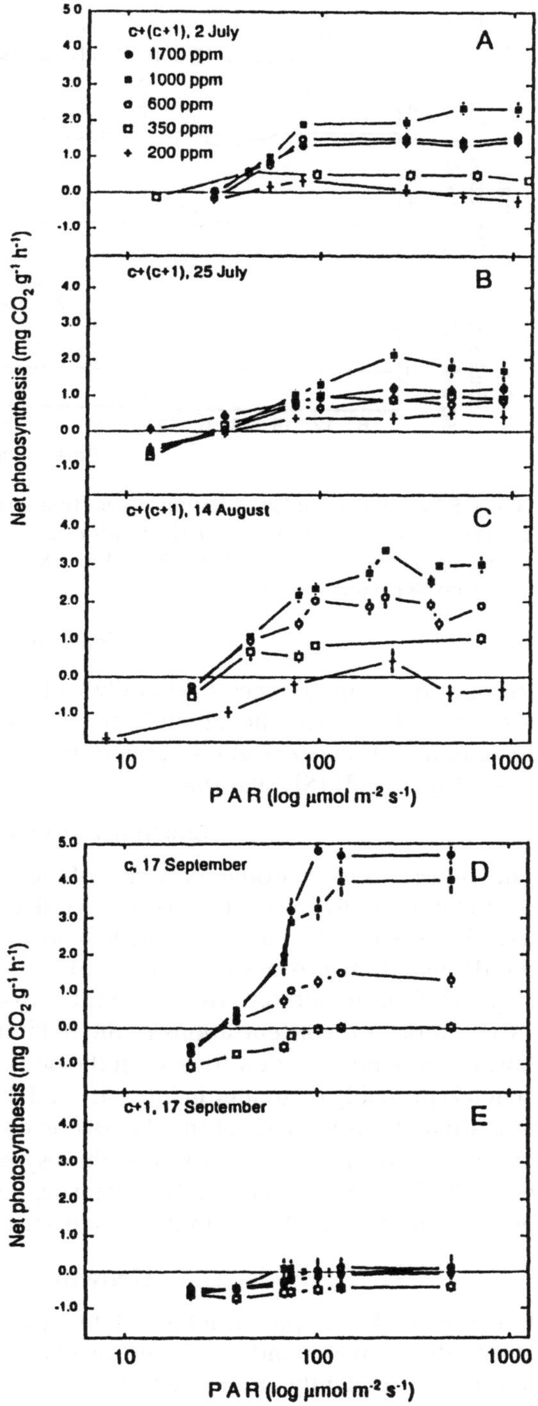

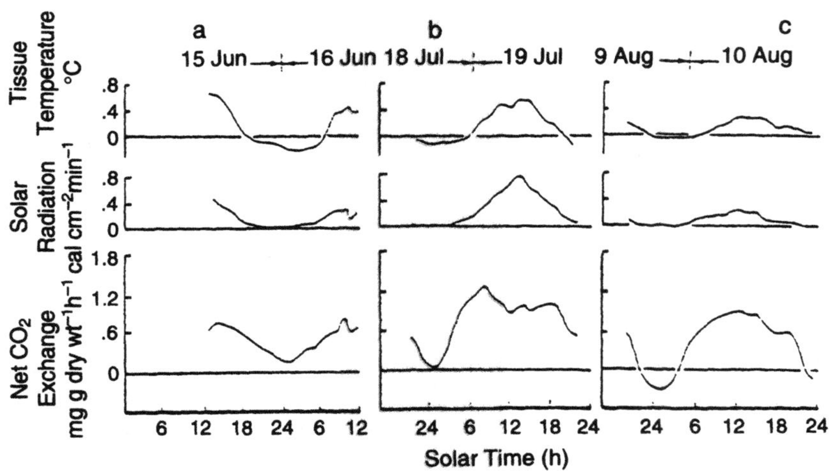

Figure 5.11. Diurnal solar radiation, tissue temperature, and net CO_2 exchange of the moss *Polytrichum alpinum* on the tundra near Barrow, Alaska, at different times during the growing season. From Oechel and Sveinbjörnsson (1978), with permission from Springer-Verlag.

Respiration

Respiration of arctic species generally varies much less than does net photosynthesis (it ranges from about 25 to 50% of the net photosynthetic rate). Increased CO_2 concentration appears to have little effect on respiration rates (Figure 5.8) (Silvola, 1985).

Diurnal Patterns

The diurnal and seasonal patterns of net photosynthesis as observed in arctic bryophytes near Barrow, Alaska (Figure 5.11) (Oechel and Sveinbjörnsson, 1978) and in arctic lichens in the foothills tundra north of the Brooks Range, Alaska (Figure 5.13) (Tenhunen et al., 1992) are probably significantly different from the real ones because of suspected diurnal variation of CO_2 concentrations in the habitat, as stated above. Thus, following early morning dew formation the depicted brief activity burst in the lichen is probably of greater in intensity and duration. The seasonal change in diurnal photosynthesis of the Barrow bryophytes is similarly likely to be different, showing greater morning photosynthesis in the late season particularly, because of seasonally increasing soil respiration, aboveground plant respiration, and low morning wind speeds (Figure 5.7).

Growth

Because of the simple structure of bryophytes and lichens, which exist essentially aboveground, modeling productivity using net CO_2 exchange has been frequently attempted. This approach is, as pointed out by

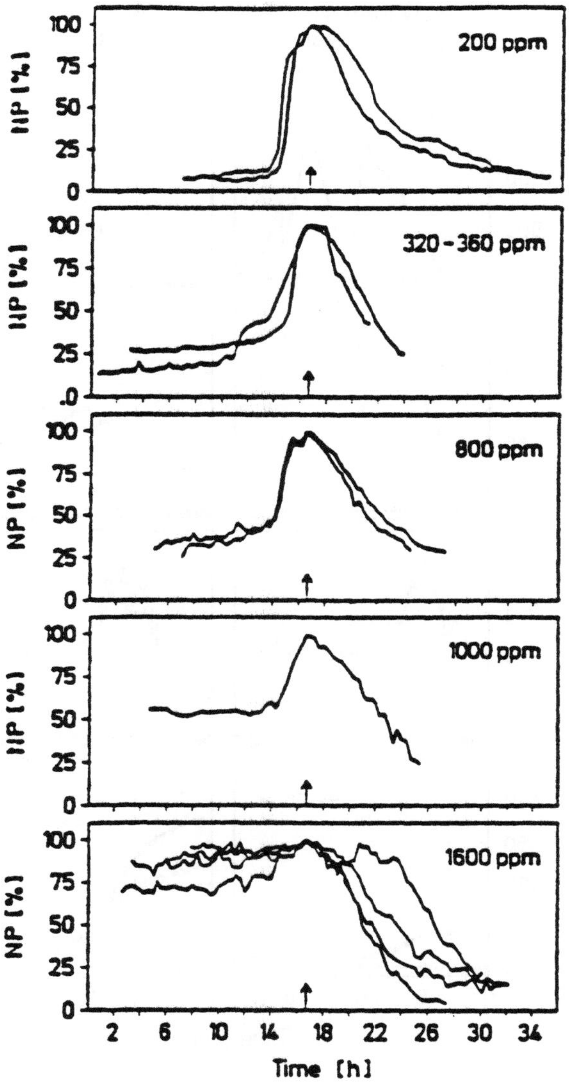

Figure 5.12. The interaction between CO_2 concentration and water content effects on relative net photosynthesis in the temperate lichen *Ramalina maciformis*. The measurements were made at $750 \mu mol\, m^{-2} s^{-1}$ and at 17°C. The graphs depict the time course following moistening to maximal water holding capacity during which the lichen thallus lost water. The response curves have been synchronized at time of maximum net CO_2 uptake rate. From Lange and Tenhunen (1981), with permission from *Oecologia*.

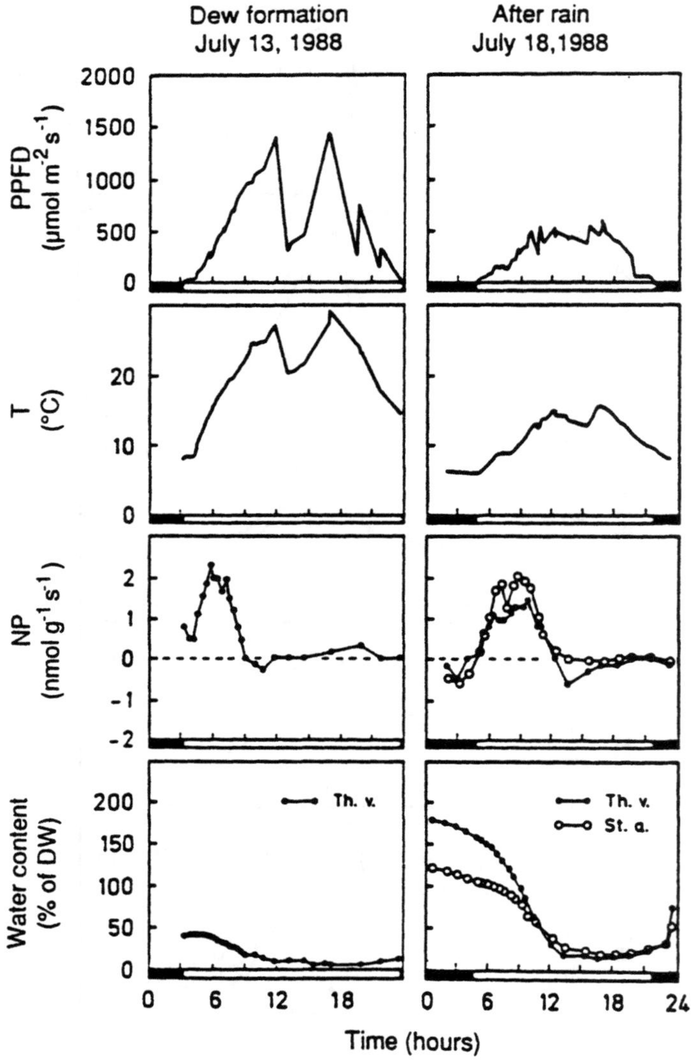

Figure 5.13. Diurnal irradiance, tissue temperature, net CO_2 exchange, and water content of two lichen species in arctic Alaska trough 2 days differing in water availability. On the right the lichens are moistened by dew formation during the low light period; on the left the lichens have been remoistened by night rain. From Tenhunen et al. (1992), with permission from Academic Press.

Sveinbjörnsson (1990), based on the assumption that there is no heterotrophic organic uptake and no leaching of assimilate to the environment. A recent report on cellulase activities in Spanish lichens suggests the possibility of heterotrophic carbon uptake (Guerrero et al., 1992). Sveinbjörnsson cited studies of leaching from epiphytic lichens in rain forests, but failed to find any leaching from subarctic *Cladina stellaris* in Alaska in an experimental study tracing the fate of assimilated ^{14}C.

Bruns-Strenge and Lange (1992), in a study of fruticose lichen on a North Sea island, conclude, based on differences between modeled and observed weight gain, that much of the assimilated carbon is lost through leaching. Whether their interpretation is correct, and if so, applicable to arctic lichens and bryophytes, is not known. The ability to predict growth responses to global changes in atmospheric CO_2 concentrations depends on our understanding of all carbon uptake and loss, not only gas exchange.

References

Alexander, V., Billington, M., and Schell, D.M. 1978. Nitrogen fixation in arctic and alpine tundra. In: Tieszen, L.L. (ed.), *Vegetation and Production Ecology of an Alaskan Arctic Tundra* (pp. 539–558). New York: Springer-Verlag.

Baselier, K., Granhall, U., and Stenström, T.-A. 1978. Nitrogen fixation in wet microtrophic moss communities of a subarctic mire. *Oikos* 31:236–246.

Brown, R.T., and Mikola, P. 1974. The influence of fruticose soil lichens upon mycorrhizae and seedling growth of forest trees. *Acta Forest. Fenn.* 141:1–23.

Bruns-Strenge, S., and Lange, O.L. 1992. Photosynthetische Primärproduktion der Flechte *Cladonia portentosa* auf der Nordseeinsel Baltrum. III. Anwendurng des Photosynthesemodells zur Stimulation von Tagesläufen des CO_2-Gaswechsels und zur Abschätzung der Jahresproduktion. *Flora* 186:127–140.

Cowles, S. 1984. Effects of lichen cover on conifer growth in the northern taiga. Ph.D. dissertation, McGill University, Montreal.

Fisher, R.F. 1979. Possible allopathic effects of reindeer-moss (Cladonia) on jack pine and white spruce. *Forest Sci.* 25:256–260.

Gerasimenko, T.V., and Zalensky, O.V. 1973. Diurnal and seasonal dynamics of photosynthesis in plants of Wrangel Island. *Bot. Zhurn.* 58:1655–1666.

Greenfield, L.G. 1993. Decomposition studies on New Zealand and antarctic lichens. *Lichenologist* 25:73–82.

Guerrero, J., Yague, E., and Esteves, M.P. 1992. Cellulase production by lichens. *J. Plant Physiol.* 140:508–510.

Johansson, L.-G., and Linder, S. 1980. Photosynthesis of Sphagnum in different microhabitats on a subarctic mire. In: Sonesson, M. (ed.), *Ecology of a Subarctic Mire. Ecol. Bull. (Stockh.)* 30:181–190.

Kallio, P. 1975. Kevo, Finland. In: Rosswall, T., and Heal, O.W. (eds.), *Structure and Function of Tundra Ecosystems. Ecol. Bull. (Stockh.)* 20:193–223.

Lange, O.L., and Tenhunen, J.D. 1981. Moisture content and CO_2 exchange of lichens. II. Depression of net photosynthesis in *Ramalina maciformis* at high water content is caused by increased thallus carbon dioxide diffusion resistance. *Oecologia* 51:426–429.

Lechowicz, M. 1982. Ecological trends in lichen photosynthesis. *Oecologia* 53:330–336.

Longton, R.E. 1992. The role of bryophytes and lichens in terrestrial ecosystems. In: Bates, J.W., and Farmer, A.M. (eds.), *Bryophytes and Lichens in a Changing Environment* (pp. 32–76). Oxford: Clarendon Press.

Mallik, A.U., and Newton, P.F. 1988. Inhibition of black spruce seedling growth by forest-floor substrates of central Newfoundland. *Forest Ecol. Manage.* 23:273–283.

Moore, T.R. 1984. Litter decomposition in a subarctic spruce-lichen woodland, eastern Canada. *Ecology* 65:299–308.
Nash, T.H., III., Moser, T.J., and Link, S.O. 1980. Nonradom variation of gas exchange within arctic lichens. *Can. J. Bot.* 58:1181–1186.
Oechel, W.C. 1976. Seasonal patterns of temperature response of CO_2 flux and acclimation in arctic mosses growing *in situ. Photosynthetica* 10:447–456.
Oechel, W.C., and Sveinbjörnsson, B. 1978. Primary production processes in arctic bryophytes at Barrow, Alaska. In: Tieszen, L.L. (ed.), *Vegetation and Production Ecology of an Alaskan Arctic Tundra* (pp. 269–298). New York: Springer-Verlag.
Oechel, W.C., and Van Cleve, K. 1986. The role of bryophytes in nutrient cycling in the taiga. In: Van Cleve, K., Chapin, F.S., III., Flanagan, P.W., Viereck, L.A., and Dyrnes, C.T. (eds.), *Forest Ecosystems in the Alaskan Taiga* (pp. 121–137). New York: Springer-Verlag.
Parinkina, O.M., and Piin, T.Kh. 1990. Decomposition of soil-growing fruticose lichens in boreal zone conditions. *Soviet J. Ecol.* 20:154–160.
Rosswall, T., and Granhall, U. 1980. Nitrogen cycling in a subarctic ombotrophic mire. In: Sonesson, M. (ed.), *Ecology of a Subarctic Mire. Ecol. Bull. (Stockh.)* 30:209–234.
Shvetsova, V.M., and Voznesensky, V.L. 1970. Diurnal and seasonal variations in the rate of photosynthesis in some plants of Western Taimyr. *Bot. Zhurn.* 55:66–76.
Silvola, J. 1985. CO_2 dependence of photosynthesis in certain forest and peat mosses and simulated photosynthesis at various actual and hypothetical CO_2 concentrations. *Lindbergia* 11:86–93.
Silvola, J. 1986. Carbon dioxide dynamics in mires reclaimed for forestry in eastern Finland. *Ann. Bot. Fenn.* 23:59–67.
Silvola, J. 1990. Combined effects of varying water content and CO_2 concentration on photosynthesis in *Sphagnum fuscum. Holarctic Ecology* 13:224–228.
Sonesson, M., Gehrke, C., and Tjus, M. 1992. CO_2 environment, microclimate, and photosynthetic characteristics of the moss *Hylocomium splendens* in a subarctic habitat. *Oecologia* 92:23–29.
Sveinbjörnsson, B. 1987. Reindeer lichen productivity as a function of mat thickness. *Arctic Alpine Res.* 19:437–441.
Sveinbjörnsson, B. 1990. Reindeer lichen productivity: Problems and possibilities. *Rangifer Special Issue.* 3:91–98.
Sveinbjörnsson, B., and Oechel, W.C. 1981. Controls on CO_2 exchange of two *Polytrichum* moss species. 1. Field studies on the tundra near Barrow, Alaska. *Oikos* 36:114–128.
Sveinbjörnsson, B., and Oechel, W.C. 1983. The effect of temperature preconditioning on the temperature sensitivity of net CO_2 flux in geographically diverse populations of the moss *Polytrichum commune. Ecology* 64:1100–1108.
Sveinbjörnsson, B., and Oechel, W.C. 1992. Controls on growth and productivity of bryophytes: Environmental limitations under current and anticipated conditions. In: Bates, J.W., and Farmer, A.M. (eds.), *Bryophytes and Lichens in a Changing Environment* (pp. 77–102). Oxford: Clarendon Press.
Tenhunen, J.D., Lange, O.L., Hahn, S., Siegwolf, R., and Oberbauer, S.F. 1992. The ecosystem role of poikilohydric tundra plants. In: Chapin, F.S., III., Jeffries, R.L., Reynolds, J.F., Shaver, G.R., and Svoboda, J. (eds.), *Arctic Ecosystems in a Changing Climate* (pp. 213–238). San Diego: Academic Press.
Wetmore, C.M. 1982. Lichen decomposition in a black spruce bog. *Lichenologist* 14:267–271.

6. Impacts of Climate Change on Demographic Processes and Population Dynamics in Arctic Plants

Terry V. Callaghan and Bengt Å. Carlsson

Most studies on the impact of climate change on plants have focused on short-term physiological and ecosystem processes such as gas exchange, while a few studies have addressed long-term responses such as biome redistribution (e.g., Emanuel et al., 1985a, 1985b; Prentice, 1993). Studies on an intermediate, population-orientated, level are largely lacking. Yet it is on this level that the physiological responses to the environment are integrated into the major discrete step events of reproduction and death, which are measures of a plant's fitness and lead to subsequent changes in biodiversity. Also, it is at the population level that many intra- and interspecific interactions between populations take place, which ultimately decide future community structure.

Plants of the tundra are usually long-lived and many of them contain morphological records of past events of vegetative "births," flowering, and deaths (Callaghan and Collins, 1976). Such plants represent a chronosequence of demographic events that can be analyzed to present long-term information on how the plants grow, reproduce, and die in relation, for example, to climatic changes. Where these records do not exist, or for validation of predictions based on the retrospective analysis of growth, population monitoring is required. Both monitoring and the retrospective analysis of growth are important tools for interpreting short-term

This chapter was submitted in November 1993.

physiological processes within the context of larger scales of both time and space. They also allow sensitivity analyses to be applied whereby stages of a life cycle particularly sensitive to environmental constraints can be identified.

Despite the importance of population dynamics, few data exist, although many anecdotal records are available, particularly on flowering. The aim of this chapter cannot be to synthesize the available data on arctic plant demography, or generalize, because the data are so few. Therefore, the aim will be to highlight examples of the likely impacts of climate change on a range of demographic parameters that, together, control population dynamics and the spreading and dispersal of a species.

Characteristics of Plant Populations in the Arctic

Arctic plants are generally long-lived, perennial species adapted to harsh climatic conditions and low nutrient availability (Billings and Mooney, 1968; Bliss, 1971; Sørensen, 1941; Warming, 1909). Annuals occur only rarely and include hemiparasites (e.g., *Euphrasia* and *Gentianella* spp.), which can quickly acquire the resources needed to be able to complete their life cycle in the short growing season typical of the Arctic.

The clonal life-form is dominant among vascular plants in the Subarctic and low Arctic, and recruitment of seedlings to populations is usually rare in closed vegetation, even though genetic diversity can be surprisingly high (Jonsdottir, 1995). Seedlings are often restricted to disturbed areas such as polygons (Jonasson, 1992) and frost boils (Gartner et al., 1986), and where seedlings occur, mortality can be high (e.g., Jonsdottir, 1995). However, seed banks can sometimes be extensive (Jonsdottir, 1995; McGraw, 1980) and are even present in the high Arctic (Levesque and Svoboda, 1995). During the meristem–meristem life cycle, connections between ramets in the few arctic clonal plants studied are often long-lived and enable the plants to be very efficient both in locating scarce resources and in translocating them to priority sinks such as growing points (e.g., Allessio and Tieszen, 1985). These connections contribute directly to the fecundity of individuals and to the survival ability of the offspring. They also allow the development of internal controls, such as apical dominance, which affect both survival and fecundity. However, the meristem–meristem cycle leads to stable or decreasing biodiversity at the subspecific taxonomic level as each plant thus formed in the population has the parent's genotype. Therefore, during competitive exclusion of species relying on sexual reproduction by a clonally spreading plant, the number of genotypes will be reduced.

In the more barren habitats of the high Arctic, where expanses of bare ground are frequent and often extensive, plant populations recruit individuals from seed and viviparous propagules more frequently than in lower arctic latitudes (Grulke and Bliss, 1985), and clonally spreading species are relatively less common. Winter seed rain is relatively low but can reach 70 diaspores m^{-2}, which is representative of species found within a distance of 1 km: exotic forms are not reported (Grulke and Bliss, 1983). Characteristic growth forms of higher plants of the high Arctic tend to be cushions, tussocks, and rosettes, which often arise through recruitment of seedlings, and genotypic differentiation may be widespread with high biodiversity at the subspecific level (Crawford et al., 1993).

Flowering throughout the Arctic is generally an extended process. Growth is slow and it usually takes several years before a plant reaches the size or stage necessary for flowering (Billings, 1987). Also, flower initiation often takes place 1 year before an individual shoot flowers (Fries, 1913; Sørensen, 1941). Thus, flowering is a consequence of environmental conditions over an extended time period and this allows short-term adverse conditions to be buffered (Jonasson, Chapter 19). Flowering fluctuates widely between years, and is often reported as being synchronized within species over considerable areas (Shaver et al., 1986). Whereas flowering in many temperate nonclonal plants may be triggered by stress, flowering in arctic plants is often related to maximal growth and favorable climatic conditions (Carlsson and Callaghan, 1994; Havström et al., 1995b).

Population structures of arctic plants often show low frequencies of young age classes and intermittent low frequencies of older age classes associated with sporadic recruitment (Callaghan and Emanuelsson, 1985; Petersen, 1981; Philipp et al., 1990; Wager, 1938). In populations of some clonally proliferating species typical of the Subarctic, young individuals are lacking as the process of clonal fragmentation bypasses the youngest age classes (i.e., new individuals are already relatively old when they are separated from others by the decomposition of old connecting tissues) (Callaghan and Emanuelsson, 1985). In sexually reproducing species, particularly associated with fellfields and the high Arctic, frequencies of the youngest age classes can be very high followed by a steep decline, representing low survival probabilities of the youngest age classes (Figure 6.1; Philipp et al., 1990). Within the high arctic Saxifrages, species reproducing vegetatively had a high ratio of immature to mature individual plants (equivalent to ramets) whereas those species reproducing by seed had a high ratio of mature to immature individual plants/genets (Grulke and Bliss, 1985), illustrating lower survival probabilities of sexually recruited individuals (Callaghan and Emanuelsson, 1985).

Few population growth rates have been calculated for arctic plants, whether ramets or genets. Those that have, show that most population growth rates range from 0.99 (1 = stable population) in *Carex bigelowii* (Carlsson and Callaghan, 1991) to 1.17 for *Dryas octopetala* (McGraw and Antonovics, 1983). However, Fetcher and Shaver (1983) reported a high population growth rate of 1.65 for *Eriophorum vaginatum* growing on a recently fertilized site in Alaskan tussock tundra.

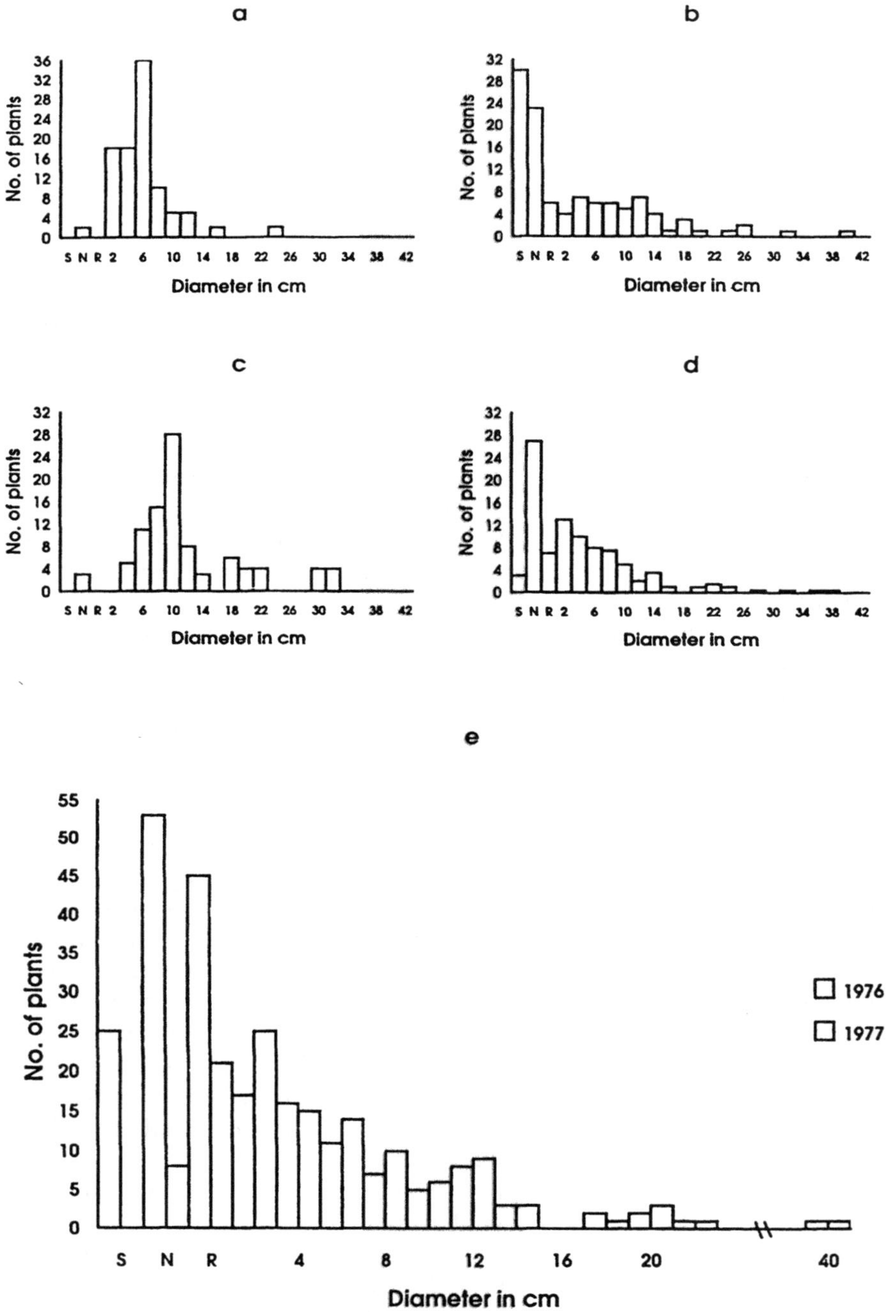

Figure 6.1. Population structure of *Silene acaulis* from Disko Island, West Greenland, illustrating sporadic recruitment from seedlings (a, b, c, and d) in various microsites and low survival probabilities of young individuals (e). From Philipp et al. (1990).

Impact of Environmental Change on Demographic Parameters

The dynamics of a population is controlled basically by the interaction between survival of individuals and the numbers of offspring produced. In long-lived organisms such as arctic plants, the relationships between age and/or stage and survival and fecundity are particularly important (Callaghan, 1976; Fetcher and Shaver, 1983; McGraw and Antonovics, 1983; Svensson and Callaghan, 1988). Also, in stressed environments where resource accumulation is slow, we can expect that the relationships between size (resource pool) and survival and fecundity are also important (Carlsson and Callaghan, 1991a; Fetcher and McGraw, 1992).

Climate change could affect arctic plant population characteristics in numerous ways. It could act directly on fecundity or survival (e.g., temperature effects on seed set, waterlogging on root death) or indirectly by influencing some aspect of growth, which, in turn, might control flowering and reproductive success (seed weight, germinability, etc.) or vegetative expansion rates.

Survival

Survival probabilities can be affected directly by many aspects of climate. A major cause of mortality of young individuals in the Arctic is frost heave (Wager, 1938) whereas physical disturbance, which is often mediated by climate, is a major cause of mortality of all age classes in many arctic plants. Disturbance events occur over a wide range of spatial areas and also at a wide range of frequencies (Walker and Walker, 1991). Slope processes (Heilbron and Walton, 1984), erosion, and movement of soil due to freeze–thaw cycles are particularly important causes of local and frequent mortality. Thawing permafrost could lead to relatively large-scale erosion, slumping, and death of plants through mechanical disturbance or flooding. Changes in snow depth and duration can lead to winter desiccation and death of exposed individuals if winter snowcover is reduced. However, with higher winter temperatures, and particularly if snowcover increases, winter mortality risks through low temperatures and desiccation should be lower.

Laboratory studies suggest that death due to high temperatures may occur in many alpine plants (Gauslaa, 1984), but the significance of this remains to be proven in the field; some of these species can often grow in warmer conditions in the absence of competition, and experimental warming of subarctic ecosystems has not yet resulted in death of individuals (Havström et al., 1993; Wookey et al., 1993).

Changes in soil moisture during the growing season could also directly impact survival, either through drought or lack of oxygen in the rooting zone through flooding. Although many arctic plant species can withstand extremes of soil moisture associated with flooding after spring thaw or late-season drought in areas with low summer precipitation, changes in soil moisture can directly kill individual plants or growing points of plant modules. For example, when the demand for water by aboveground tissues exceeds the supply capacity of roots, water potentials of growing points decrease (Headley et al., 1988) and death can occur. Although changes in soil moisture have the potential for killing plants and plant modules, it is difficult to predict what might happen during a changing climate because predictions of changes in soil moisture by the GCMs are insecure (Mitchell et al., 1990).

The indirect effects of changes in climate on survival are numerous. Environmental changes that increase growth are likely to increase survival as there is usually a strong correlation between resource accumulation and survival, the largest individuals having the greatest survival probabilities (Figure 6.2; Carlsson and Callaghan, 1990; McGraw and Fetcher, 1992). Temperature increases would be expected to result in increased vegetative growth, but comparisons of plant responses to simulations of climate change along a latitudinal gradient show that increased temperature directly affects vegetative growth, mainly in the most extreme environments at high latitudes and high altitudes (Havström et al., 1993; Wookey et al., 1993). Direct effects of temperature on vegetative growth in the Subarctic at low altitude are generally small whereas increased temperature could, in some cases, decrease left growth (Callaghan and Jonasson, 1995). Therefore, the relationship of increased temperature, increased vegetative growth, and increased survival may not always apply.

An indirect effect of increasing temperatures is suggested to be increased decomposition rates and an increase in plant-available nutrients (Melillo et al., 1990). Although the magnitude and generality of this effect have recently been questioned (Callaghan and Jonasson, 1995; Jonasson et al., 1993), the responsiveness of growth in many arctic plant species to increases in nutrient supply (e.g., Havström et al., 1993; Henry et al., 1986; Jonasson, 1992; Shaver and Chapin, 1980; Wookey et al., 1993, 1994) suggests that any increase in plant-available nutrients could, via increased growth, increase survival probabilities. It is unlikely, however, that changes in the nutrient status of soils—either decreases due to microbial immobilization, or increases—will directly kill plants or plant modules as the impact of changes in nutrient availability would be mediated through changes in growth and competitive ability.

In rhizomatous clonal plants, high probabilities of survival of young modules are often achieved by the import of resources—carbon, water, nutrients—from older parts of the plants. Experimental breaking of connections between modules of different ages in the sedge *Carex bigelowii* showed that survival probabilities were reduced most, and for the greatest duration, when young modules were separated from clones growing in infertile conditions (Jonsdottir and Callaghan, 1988). Such mechanisms supporting young offspring by transport of resources from older parts of a clone increase survival probabilities of young offspring in occasional extreme weather conditions, in resource-patchy habitats, and in unsafe

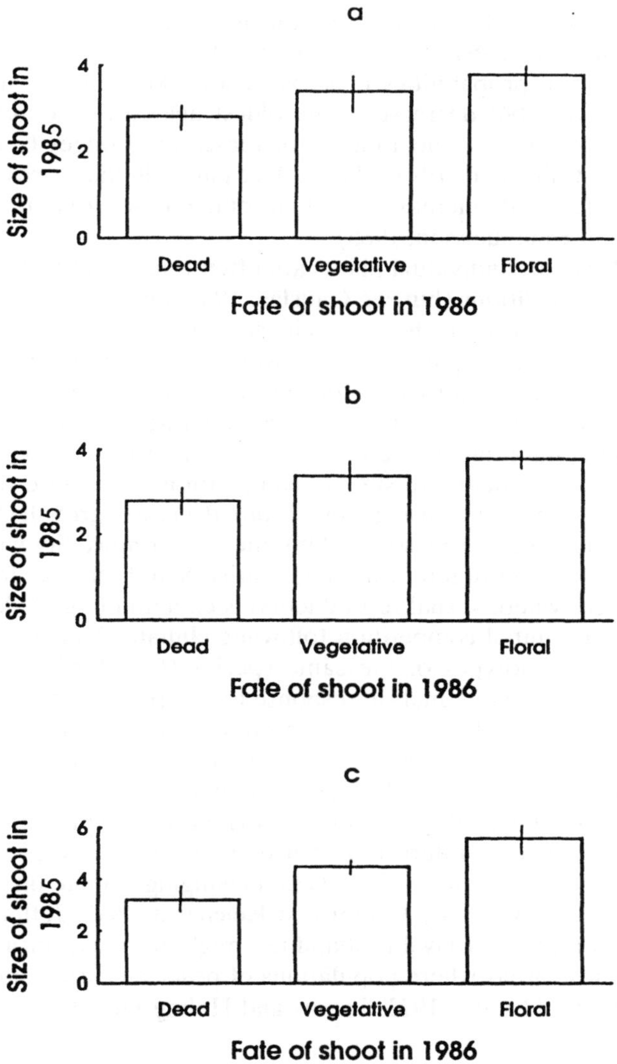

Figure 6.2. The relationship between shoot size (number of green leaves per tiller) in *Carex bigelowii* growing at three sites along an altitudinal gradient (a, high altitude to c, low altitude) in 1 year, and the fate of the tiller in the following year: the smallest tillers have the highest risks of death and the largest tillers have the highest probabilities of flowering. From Carlsson and Callaghan (1990a).

microsites. These mechanisms are also efficient in enhancing competitive abilities. Under climate warming scenarios, clonal growth forms might gain increasing dominance in the high Arctic where their superior competitive abilities would displace the current extreme stress-tolerant species.

Although increases in temperature, nutrients, and CO_2 might increase growth and survival probabilities overall, the effect is likely to be age/stage specific. Thus, the probabilities of survival are likely to increase in the lowest age classes, but decrease in the oldest age classes as turnover of individuals (both genets and ramets) increases and longevity decreases (Carlsson and Callaghan, 1994). This effect parallels the increase in longevity associated with increases in environmental severity in the north (Callaghan and Emanuelsson, 1985).

In general, mature individual plants are often tolerant of widely varying environmental conditions (but see Gauslaa, 1984, for an alternative view): it is often a short phase of the life cycle, such as seed setting or seedling establishment, which is particularly sensitive to environmental factors. However, one of the major causes of mortality may be competition from neighboring plants. Species with conservative strategies such as *Cassiope tetragona* and cushion plants are generally unresponsive to environmental change and may be unable to survive competition from more responsive species. In cases where increasing temperature decreases growth (Welker et al., 1993; Callaghan and Jonasson, 1995), the risk of mortality will be particularly high. Competition may not necessarily be a threat just from other species: in areas where sexual reproduction is common (e.g., fellfields and the high Arctic), initial competition following climate change may be between different genotypes of the same species (Crawford et al., 1993). Mortality of mature individuals in response to competition from neighboring species is likely to be particularly pronounced at the boundaries between species and communities (e.g., tree line, snowbed margins, latitudinal limits of distribution, etc.) (Wijk, 1986a, 1986b; Havström et al., 1993).

Another cause of mortality of mature individuals is the pressures exerted by herbivores and pests. Interactions between grazers, browsers, and pests on the one hand, and their host plants in a changing climate on the other, are complex (see reviews by Bryant and Reichardt, 1992; Jefferies et al., 1992). However, plant survival probabilities might decrease during climate warming in those areas where populations of pests, etc., are controlled by low temperatures (Tenow, 1972; Tenow and Holmgren, 1987).

Reproduction: Flowering

We would expect that flowering should be particularly sensitive to increases in temperature because there has been an evolution of mechanisms within arctic flowers to increase their temperature (e.g., solar tracking flowers of *Dryas octopetala* increase the weight of seed) (Kjellberg et al., 1982), parabolic flowers increase stigma temperatures, and dark colored *Salix catkins* increase temperatures of reproductive organs (Krog, 1955), particularly in female catkins (Kevan, 1972). Temperature also controls insect pollinator activity, which generally increases with increasing temperature, although some insect species are more commonly found in flowers when temperatures are low during the diurnal cycle (Philipp et al., 1990). The sensitivity of flowering development to small differences in temperature is also evident between microsites and even between sides of individual cushion plants such as the "compass plant" *Silene acaulis*, with different aspects. Surprisingly, however, flowering frequency of *Cassiope tetragona* was not increased in temperature enhancement experiments (Havström et al., 1995b), although flowering in the current warm period has been shown to be more frequent than in the cooler period preceding the Little Ice Age on Ellesmere Island, as would be expected (Havström et al., 1996). In field manipulation experiments, shading dramatically reduced flowering frequency, suggesting great sensitivity to competition from taller species.

The initiation of flowering in *Carex bigelowii* is highly correlated with summer temperature (Figure 6.3; Carlsson and Callaghan, 1994) and flowering initiation also depends on resource accumulation (Figure 6.2). During the year after initiation of flowering, seed set is controlled by the climatic conditions in that year but it cannot usually exceed the constraints imposed by the previous year's weather on number of flowering meristems available to set seed (Philipp et al., 1990). During seed setting, both higher temperatures during the season and the duration of the growing season can be critical in controlling seed set. In the high Arctic, a mean temperature increase of 3.5°C over the growing season had an immediate effect on the flowering phenology of *Dryas octopetala* and allowed set to occur, although this was almost absent in control plots during the cool season of 1991 (Figure 6.4(a); Wookey et al., 1993). Attempts to alleviate other environmental constraints on growth and reproduction in this polar semidesert

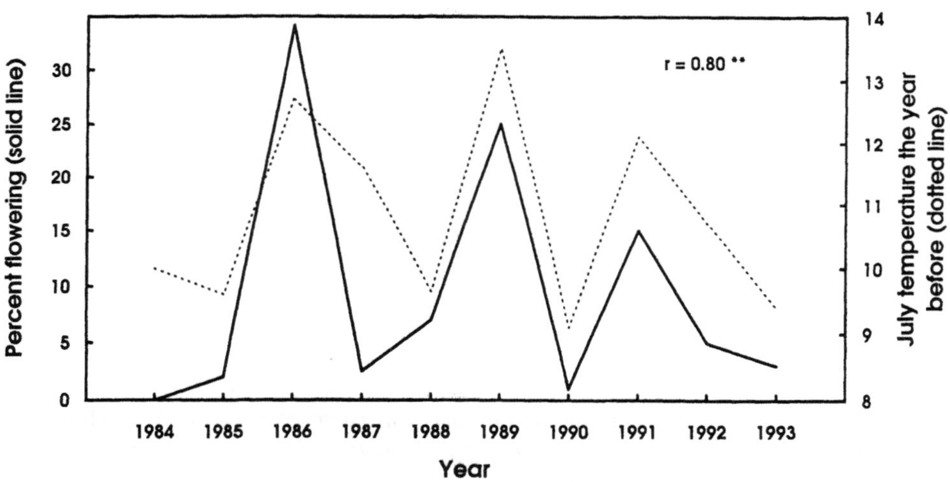

Figure 6.3. Relationship between percentage of tillers of *Carex bigelowii* flowering, and July temperatures in the preceding year. From Carlsson and Callaghan (1994).

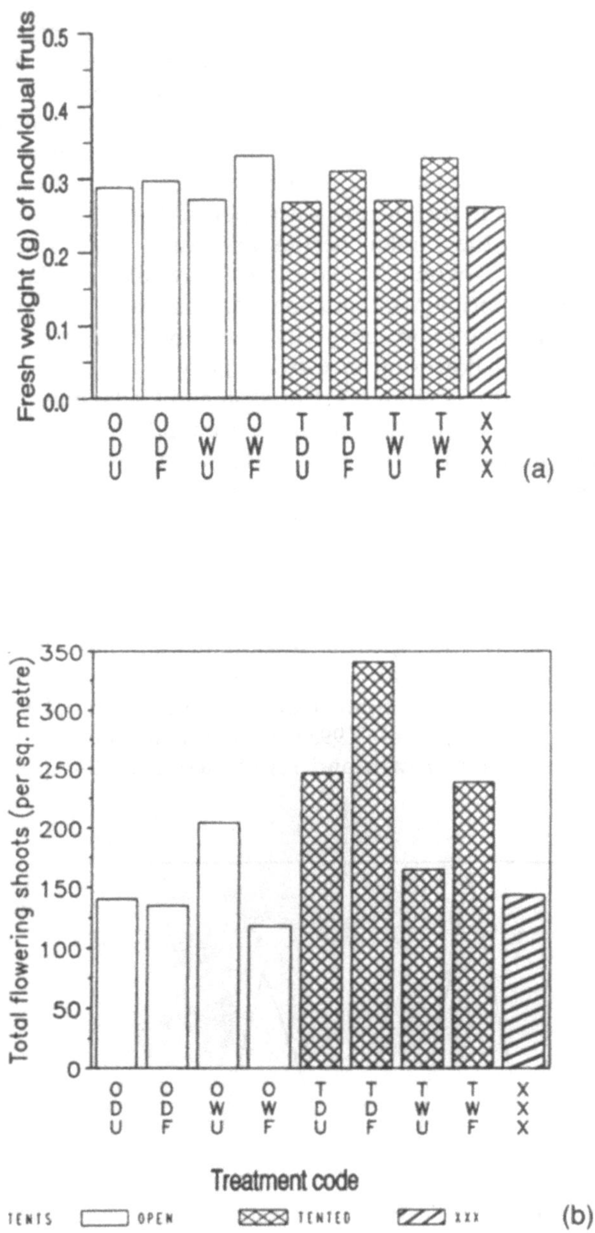

Figure 6.4. Response of reproductive output to environmental perturbations simulating aspects of climate change. (a) *Empetrum hermaphroditum* from the Subarctic, (b) seed setting in *Dryas octopetala* from the High Arctic. T = temperature enhancement (3.5°C for *Dryas*, 2.7°C for *Empetrum*), N = nutrient addition (see sources for quantities and frequency of application), W = addition of 50% of summer precipitation. From Wookey et al. (1993).

(i.e., adding nutrients or extra precipitation) had no initial effect on flowering. In contrast, in a subarctic dwarf shrub heath, temperature exerted little effect on flowering but nutrients were important in determining reproductive output [Figure 6.4(b); Wookey et al., 1993].

Environmental conditions control seed "quality" (i.e., germinability) in addition to number of seeds produced. In a 9-year study of seed germinability of 10 arctic-alpine species growing along an altitudinal gradient in the northern Scandes, Laine et al. (1995) found that germinability was high in the upper part of the low alpine zone. During their study period, the June to August mean temperature exceeded the long term mean by 0.5 to 1.5°C. When temperatures dropped below this, seed germinability was poor. Laine et al. (1995) concluded that global warming would increase the germinability of arctic and alpine species.

Changes in the length or timing of the growing season, and changes in the photoperiod associated with this, are uncertain in a warming Arctic. Snow-free periods may be extended by up to a month (Maxwell, 1992), but in some areas, late-season drought might leave growing season length the same. In other areas, increased snowfall in responses to warming winters might delay the onset of the growing season, thereby decreasing the input of solar radiation and altering daylength. As the seed set of some species (e.g., *Phleum alpinum*) (Callaghan, 1974) is prevented in short growing seasons of less than 100 days, and this limits their distribution, any increase in the effective length of the growing season might increase seed set and allow a northward spread of the species to occur. On the other hand, species with photoperiodic constraints (Heide, 1992) such as those that cease growth under short-day conditions at the end of season, may be unable to utilize the end of extended or delayed snow-free periods.

Many biological phenomena in the Arctic are cyclical, perhaps because of the period of resource accumulation which is required for flowering. A model of cyclicity in the flowering of *Carex bigelowii*, related to climate, population growth rate, and a vegetative phase of resource accumulation, predicted that cyclicity would break down under warmer conditions as these would increase the frequency of flowering, reduce the period of resource accumulation, and increase population growth rate (Figure 6.5; Carlsson and Callaghan, 1994). In the field, cyclicity in the flowering of *Carex bigelowii* populations from a subarctic fellfield in Sweden (Carlsson and Callaghan, 1994) contrasted strongly with the lack of cyclicity found in Icelandic populations that experience longer, more benign growing seasons (Jonsdottir, 1995). The Icelandic situation may, to some extent, be regarded as an analogue of a warmer climate for the Swedish population and agrees with the model's output.

In graminoids, such as *Carex bigelowii*, flowering occurs by the conversion of apical meristems and may remove apical dominance, thereby stimulating the development of new lateral meristems. These relationships between flowering and meristem dynamics result in the control of the

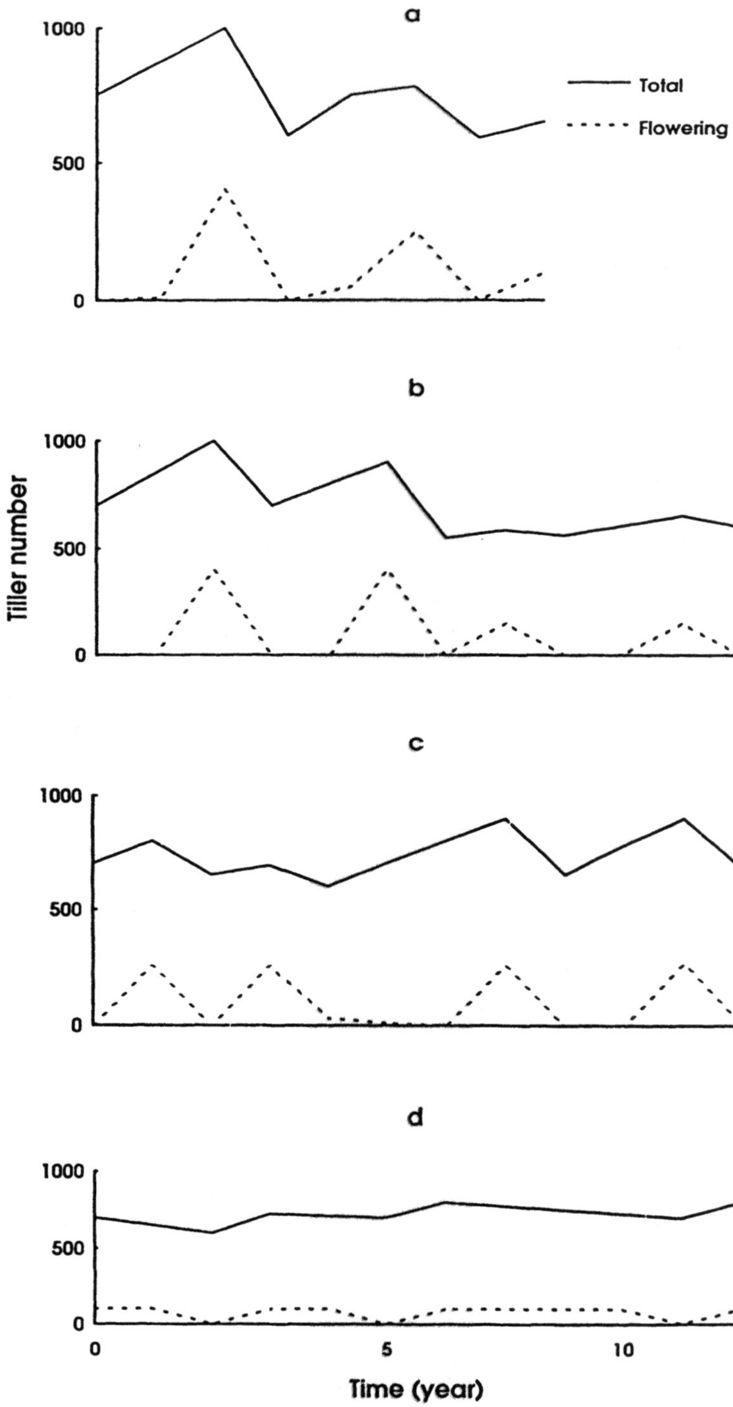

a
b
c
d
Total
Flowering
Tiller number
Time (year)
1000
500
0
5
10

population dynamics of *Carex* by flowering—even in the absence of seedling recruitment (Carlsson and Callaghan, 1990a). Thus, during climate warming, with a decreased frequency of cool summers that limit flowering, flowering cyclicity will be dampened and fluctuations in total population numbers (tillers) will be reduced. Initially, this will lead to a reduced frequency of high-risk small populations of tillers occurring.

The sporadic recruitment of seedlings combined with low population numbers in some years is a mechanism that must limit the occurrence of annual species, particularly those without persistent seed banks. Traditionally, their rarity in the Arctic has been associated with lack of time for resource accumulation. However, the patterns of population dynamics typical of the Arctic (i.e., years in which flowering is very low or absent) might contribute to their lack of representation.

Other environmental variables also affect flowering, of course, but mainly by contributing to resource accumulation as noted above. An exception is the increase in concentrations of atmospheric CO_2. Although these should lead to an increase in resource accumulation (e.g., in *Carex bigelowii* over a 3-year period; Oechel and Billings, 1992), it has been shown that they may decrease the time threshold for flowering to occur (references in Tissue and Oechel, 1987). If this applies also to long-lived perennials such as those of the Arctic, generation times will be significantly decreased and the rate of spread in a changing environment could increase considerably, for example, by more than double in *Carex bigelowii* (Carlsson and Callaghan, 1994).

Reproduction: Meristem–Meristem Cycle

The number of meristems produced (e.g., tillers in graminoids) results from complex interactions between environmental factors and internal control mechanisms.

The reproductive output in terms of meristems and their modules is related directly to climatic factors predicted to change by GCMs. Vegetative reproductive output has been shown to increase:

(a) as temperature increases (e.g., in *Carex bigelowii*; Carlsson and Callaghan, 1991b),
(b) as nutrients increase (e.g., in *Carex bigelowii*; Carlsson and Callaghan, 1990b),
(c) as concentrations of atmospheric CO_2 increase (e.g., in *Eriophorum vaginatum*; Tissue and Oechel, 1987).

Figure 6.5. Simulations of changes in cyclicity of tiller populations of *Carex bigelowii* during climate warming. Solid line: total numbers of tillers, dotted line: numbers of flowering tillers. (a) Observed fluctuations from 1984 to 1991. (b) Simulated fluctuations during this period. (c) and (d) Simulations at two levels of climate change. From Carlsson and Callaghan (1994).

However, there may be complex interactions between such increased vegetative proliferation and other developmental changes such as the threshold age/stage for meristem initiation (i.e., generation time), age/stage-specific fecundity rates, and parental life span—all controlled by changing environmental conditions. Such complex effects and interactions make the computing of overall population growth rate changes difficult without recourse to modeling (Carlsson and Callaghan, 1994).

In clonally proliferating species common in the Subarctic, apical dominance is a major internal control on meristem production and subsequent development (Jonsdotti and Callaghan, 1988; Svensson and Callaghan, 1988). The production of meristems may be discontinued in older generations, but renewed by death of the dominant meristem. In *Carex bigelowii*, death of young tillers leads to the proliferation of meristems in old, formerly moribund, generations (Jonsdottir and Callaghan, 1988). Such mechanisms provide a buffer against disturbance and herbivory that might increase in a changing climate.

In such vegetatively proliferating species, under typical environmental conditions, a plant's strategy might be to invest all of its resources in a single line of generations to increase the success of that line (i.e., the "guerrilla strategy"). This strategy is found in *Lycopodium annotinum* (Callaghan et al., 1990). However, the part of the plant often most sensitive to environmental conditions is the apical meristem, which suppresses the growth of subdominant meristems (Svensson and Callaghan, 1988). Should this meristem be destroyed, other meristems will proliferate (Figure 6.6). This leads to an inverse correlation between the survival of apical meristems and the formation of new meristems (Callaghan et al., 1986), which acts as a mechanism to buffer interannual variations in weather conditions (Figure 6.6). As the apical meristems are particularly sensitive to water stress, increases in temperatures without equivalent increases in summer precipitation might lead to an initial change from current large, relatively unbranched clones, to complex clones with many ramifications. It is not known for how long this strategy can survive—currently, when adversity is overcome, the typical equilibrium strategy returns. In a warmer climate, insufficient resources in each of many offspring might result in death of whole clones.

Reproduction: Vivipary

Vivipary is a reproductive mechanism that is intermediate between sexual reproduction and vegetative proliferation in terms of survival probabilities (Callaghan and Emanuelsson, 1985). Propagules are physically and physiologically independent units, but those studied contain more resources than seed and can become established quickly. In the high Arctic, viviparous species are particularly abundant (Grulke and Bliss, 1985) and may reproduce in poor weather conditions when sexually reproducing species cannot (Bell and Bliss, 1980). Factorial experiments in the high Arctic simulating

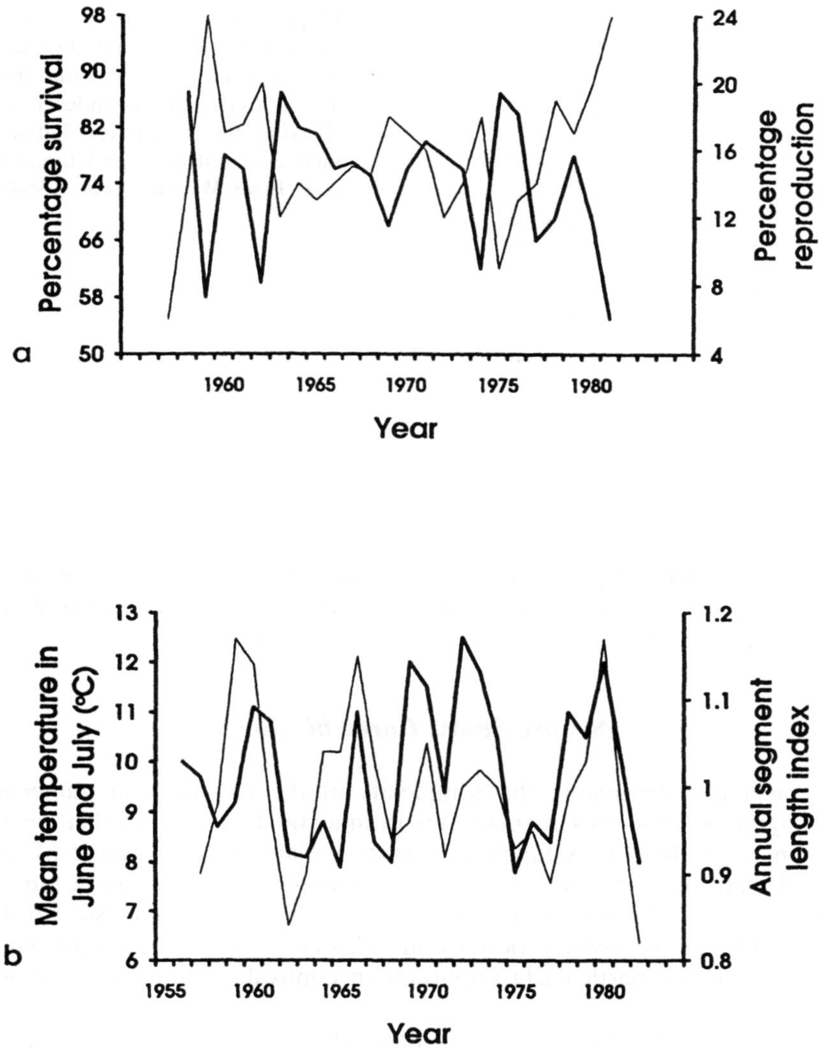

Figure 6.6. Relationship between (a) meristem death (thin line), release of apical dominance, and meristem proliferation (thick line), and (b) annual growth (thin line) and climate (thick line) in *Lycopodium annotinum*. From Callaghan et al. (1986). Climate can be seen to control annual growth, and the response of meristem initiation to the death of the apical meristem is also illustrated.

various aspects of climate change showed that both increases in temperature and nutrients were important in increasing the reproductive output of *Polygonum viviparum* (Figure 6.7; Wookey et al., 1994). All aspects of reproductive development investigated were dependent upon both nutrient

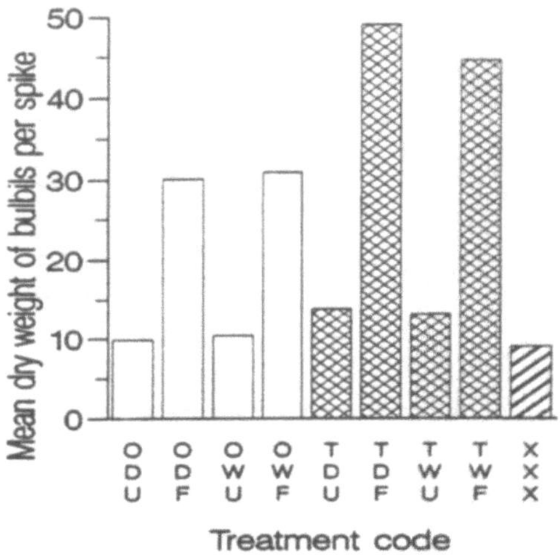

Figure 6.7. Effect of environmental perturbations on bulbil production in *Polygonum viviparum* from a high arctic polar semidesert site showing the interaction between increased nutrients and temperature. From Wookey et al. (1994).

status and temperature. This contrasted with *Dryas octopetala*, growing in the same experimental plots, which are affected only by temperature increases (see above).

A Case Study: *Carex bigelowii*

In the above discussion, the complexity of the responses in numerous demographic processes to many environmental factors, and their interactions is evident. It is far from easy to collate these (Figure 6.8) and to calculate the net outcome for a change in climate. However, a modeling exercise has been performed in an attempt to do this, and also in order to identify demographic processes, life history stages, and environmental factors that are particularly important in controlling population growth rate.

In Table 6.1 we have assembled data on the responses of the clonally growing sedge *Carex bigelowii* to the environmental factors that are likely to alter under current climate change scenarios, and the relationships between environment and various aspects of population dynamics are summarized in Figure 6.8. These data have been used for making assumptions on how survival and reproductive probabilities would be altered following climate change (Carlsson and Callaghan, 1994).

The data suggest that, under current climate change scenarios, increases in CO_2, temperature, and nutrients would promote growth in *Carex bigelowii* in a qualitatively similar way. The evidence suggests that tiller size and daughter tiller production may increase, and be shifted towards production of a type of tiller—phalanx tillers—which have a greater propensity for flowering. Furthermore, age at tillering as well as tiller life span may decrease, whereas survival of younger age classes might be higher (Table 6.1).

Mathematical models incorporating these hypotheses suggest that the expected increases in population growth rate following climate change would lie within the range of population growth rates measured in the field. A sensitivity analysis shows that young, large, spreading "guerrilla tillers" are the most important tiller category contributing to population growth rate, even though their proportion was initially decreased in response to increased nutrients. The rate of vegetative spread was calculated to more than double—from 1.3 to 3.0 cm per year as a result of interactions between increases in tiller rhizome lengths and decreases in generation time (Carlsson and Callaghan, 1994). Surprisingly, flowering intensity was reduced under simulated climate warming as the changes between tiller morphological forms, which are associated with different probabilities of flowering, outweighed the effects of increased temperature and nutrients on flowering probabilities through increasing tiller size.

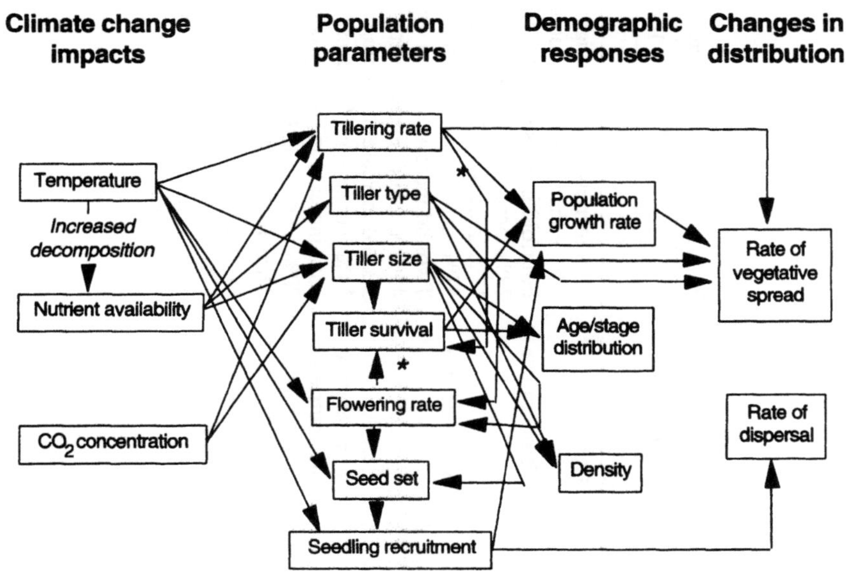

Figure 6.8. Diagrammatic representation of direct and indirect controls of climate change variables on demography and distribution using the sedge *Carex bigelowii* as an example. Asterisks denote negative responses. Based on Carlsson and Callaghan (1994).

Table 6.1. Responses of Life History Traits and Population Dynamics of *Carex bigelowii* to Components of Climate Change

Plant parameter	Factor	Response	Reference
Primary environmental effects			
Tiller size	CO_2	11% increase in total weight (under high nutrients)	Oberbauer et al., 1986
		60% enhancement of photosynthesis (under high nutrients)	Oechel and Billings, 1992
	Temperature	Fourfold increase in total dry weight in 12 vs. 2°C	Kummerow and Ellis, 1984
		29% increase in leaf length in warm microsite	Carlsson and Callaghan, 1991b
		17% increase in leaf length in warm microsite	Carlsson and Callaghan, unpubl.
	Soil nutrients	57% increase in aboveground phytomass	Chapin and Shaver, 1985
		Twofold increase in total dry weight	Oberbauer et al., 1986
		49% increase in leaf length	Carlsson and Callaghan, 1990b
Tiller initiation	CO_2	Sixfold increase in tillering in *Eriophorum vaginatum*	Tissue and Oechel, 1987
	Temperature	13% increase in tiller density in warm microclimate	Carlsson and Callaghan, 1991b
	Soil nutrients	56% increase in shoot production in *Eriophorum vaginatum*	Chapin and Shaver, 1985
		70% increase in tiller density after 3 years	Carlsson and Callaghan, 1990b
		Threefold increase in the proportion of new tillers	Carlsson and Callaghan, unpubl.
Tiller type	Soil nutrients	Prop. phalanx tillers increased from 20 to 36%	Carlsson and Callaghan, 1990b
Age at tillering	Soil nutrients	Decrease from 2.2 years to ca. 1 year	Carlsson and Callaghan, unpubl.
Autumn senescence	CO_2	Prolonged photosynthetic activity in autumn in *Eriophorum vaginatum*	Tissue and Oechel, 1987

Flowering	Temperature	From 0.2 to 0.7 flowering shoots per plant in 9 vs. 12°C	Heide, 1992
		17-fold increase in warm microsite	Carlsson and Callaghan, unpubl.
Seed set	Temperature	Increase from 0 to ca. 10 mature seeds/shoot in ca. 1°C higher mean summer temperature	Carlsson and Callaghan, 1990a
Seedling recruitment	Temperature	Thermally induced disturbances, e.g., patterned ground, thaw slumps, and detachment of the active layer above permafrost favor seedling recruitment	Lawson, 1986; French, 1987; Edlund, 1989; Jonasson, 1992
Secondary, indirect effects			
Tiller survival	Tiller size	Positive correlation between tiller size and survival	Carlsson and Callaghan, 1990a
Tiller life span	Tillering and flowering	Initiates senescence	Carlsson and Callaghan, 1990b
Flowering	Tiller size	Positive correlation between tiller size and flowering probability	Carlsson and Callaghan, 1990b
	Tiller type	Phalanx tillers have twice the flowering probability of guerrilla tillers	Carlsson and Callaghan, 1990b
Seed set	Tiller size and flowering	Positive correlation between tiller size and seed set	Carlsson and Callaghan, unpubl.

This is a compilation of data from the literature and recent environmental manipulation experiments on *C. bigelowii* unless otherwise stated.

Summary

It is important to realize that even if we are able to predict how a demographic process, such as flowering or age-specific survival, will respond to changes in various environmental factors associated with climate change, it is the integrated responses of all demographic parameters to a range of interacting environmental factors together with species-specific internal control mechanisms that determine population growth rate and structure. Although changes in population growth rate, particularly decreases, might lead to inferences about changes in biodiversity and community structure, it is often the interactions between populations, rather than death triggered directly by the environment, that finally determine how community structure and biodiversity will change. Determining these interactions is, perhaps, the greatest challenge for population biologists.

Acknowledgments. Many colleagues were involved in the research summarized here. We are particularly grateful to: M. Havström, Prof. S. Jonasson, Prof. J.A. Lee, Dr. A. Parsons, Dr. M.C. Press, Dr. J.M. Welker, and Dr. P.A. Wookey. For hospitality and logistic support, we thank Prof. M. Sonesson and Nils Åke Andersson (Abisko) and Nick Cox and the KBKC company, Ny Ålesund, Svalbard. We are grateful to J. Kirkham for preparing the figures. T.V.C. wishes to acknowledge support from the UK NERC Arctic Terrestrial Ecology Special Topic Programme.

References

Allessio, M.L., and Tieszen, L.L. 1975. Patterns of carbon allocation in an arctic tundra grass, *Dupontia fischeri* (Gramineae), at Barrow, Alaska. *Am. J. Bot.* 62(8):797–807.

Bell, K.L., and Bliss, L.C. 1980. Plant reproduction in a high arctic environment. *Arctic Alpine Res.* 12:1–10.

Billings, W.D. 1987. Constraints on plant growth, reproduction, and establishment in arctic environments. *Arctic Alpine Res.* 19:357–365.

Billings, W.D., and Mooney, H.A. 1968. The ecology of arctic and alpine plants. *Biol. Rev.* 43:481–529.

Bliss, L.C. 1971. Arctic and alpine life cycles. *Annu. Rev. Ecol. Syst.* 2:405–438.

Bryant, J.P., and Reichardt, P.B. 1992. Controls over secondary metabolite production by arctic woody plants. In: Chapin, F.S., III, Jefferies, R.L., Reynolds, J.F., Shaver, G.R., Svoboda, J., and Chu, E.W. (eds.), *Arctic Ecosystems in a Changing Climate: An Ecophysiological Perspective* (pp. 377–390). San Diego: Academic Press.

Callaghan, T.V. 1974. Intraspecific variation in *Phleum alpinum* L. with special reference to polar populations. *Arctic Alpine Res.* 6:361–401.

Callaghan, T.V. 1976. Strategies of growth and population dynamics of tundra plants. 3. Growth and population dynamics of *Carex bigelowii* in an alpine environment. *Oikos* 27:402–413.

Callaghan, T.V., and Collins, N.J. 1976. Strategies of growth and population dynamics of tundra plants. 1. Introduction. *Oikos* 27:383–388.

Callaghan, T.V., and Emanuelsson, U. 1985. Population structure and processes of tundra plants and vegetation. In: White, J. (ed.), *The Population Structure of Vegetation* (pp. 399–439). Dordrecht: Junk.
Callaghan, T.V., and Jonasson, S. 1995. Implications for changes in arctic plant biodiversity from environmental manipulation experiments. In: Chapin, F.S., III, and Körner, Ch. (eds.), *Arctic and Alpine Biodiversity: Patterns, Causes and Ecosystem Consequences* (pp. 151–164). Berlin: Springer-Verlag.
Callaghan, T.V., Svensson, B.M., Bowman, H., Lindley, D.K., and Carlsson, B.Å. 1990. Models of clonal plant growth based on population dynamics and architecture. *Oikos* 57:257–269.
Callaghan, T.V., Svensson, B.M., and Headley, A.D. 1986. The modular growth of *Lycopodium annotinum*. *Fern Gazette* 13(2):65–76.
Carlsson, B.Å., and Callaghan, T.V. 1990a. Effects of flowering on the shoot dynamics of *Carex bigelowii* along an altitudinal gradient in Swedish Lapland. *J. Ecol.* 78:152–165.
Carlsson, B.Å., and Callaghan, T.V. 1990b. Programmed tiller differentiation, intraclonal density regulation and nutrient dynamics in *Carex bigelowii. Oikos* 58:219–230.
Carlsson, B.Å., and Callaghan, T.V. 1991a. Simulation of fluctuating populations of *Carex bigelowii* tillers classified by type, age and size. *Oikos* 60:231–240.
Carlsson, B.Å., and Callaghan, T.V. 1991b. Positive interactions in tundra vegetation and the importance of shelter. *J. Ecol.* 79:973–983.
Carlsson, B.Å., and Callaghan, T.V. 1994. Simulated effect of climate change on population dynamics of *Carex bigelowii. Ecography* 17:321–330.
Chapin, F.S., III, and Shaver, G.R. 1985. Individualistic growth response of tundra plant species to environmental manipulations in the field. *Ecology* 66:654–576.
Crawford, R.M.M., Chapman, H.M., Abbott, R.J., and Balfour, J. in press. Potential impact of climatic warming on Arctic vegetation. *Flora* 188:367–381.
Emanuel, W.H., Shugart, H.H., and Stevenson, M.P. 1985a. Climate change and the broad-scale distribution of terrestrial ecosystem complexes. *Climate Change* 7:29–43.
Emanuel, W.H., Shugart, H.H., and Stevenson, M.P. 1985b. Response to comment: climate change and the broad-scale distribution of terrestrial ecosystem complexes. *Climate Change* 7:457–460.
Fetcher, N., and Shaver, C.R. 1983. Life histories of tillers of *Eriophorum vaginatum* in relation to tundra disturbance. *J. Ecol.* 71:131–147.
Fries, T.C. 1931. Botanische Untersuchungen im nördlichsten Schweden. Vetensk. och prakt. unders. i Lappland, *Flora Fauna* 2:361.
Gartner, B.L., Chapin, F.S., III, and Shaver, G.R. 1986. Reproduction of *Eriophorum vaginatum* by seed in Alaskan Tussock Tundra. *J. Ecol.* 74:1–18.
Gauslaa, Y. 1984. Heat resistance and energy budget in different Scandinavian plants. *Holarctic Ecol.* 7:1–78.
Grulke, N.E., and Bliss, L.C. 1983. A note on winter seed rain in the High Arctic. *Arctic Alpine Res.* 15:261–265.
Grulke, N.E., and Bliss, L.C. 1985. Growth forms, carbon allocation and reproductive patterns of high artic Saxifrages. *Arctic Alpine Res.* 17:241–250.
Havström, M., Callaghan, T.V., and Jonasson, S. 1993. Differential growth responses of *Cassiope tetragona*, an arctic dwarf shrub, to environmental perturbations among three contrasting high- and subarctic sites. *Oikos* 66:389–402.
Havström, M., Callaghan, T.V., and Jonasson, S. 1995a. Effects of simulated climate change on the sexual reproductive effort of *Cassiope tetragona*. In: Callaghan, T.V., Oechel, W.C., Gilmanov, T., Holten, J.I., Maxwell, B., Molau, U., Sveinbjörnsson, B., and Tyson, M. (eds.), *Global Change and Arctic Terrestrial*

Ecosystems. Proceedings of contributed and poster papers from the international conference, 21–26 August 1993, Oppdal, Norway. Brussels: Commission of the European Communities Ecosystems Research Report (pp. 109–114).

Havström, M., Callaghan, T.V., Jonasson, S., and Svoboda, J. 1995b. Little Ice Age temperature estimated by growth and flowering differences between subfossil and extant shoots of *Cassiope tetragona*, an arctic heather. *Funct. Ecol.* 9:650–654.

Headley, A.D., Callaghan, T.V., and Lee, J.A. 1988. Water uptake and movement in the evergreen clonal plants *Lycopodium annotinum* L. and *Diphasiastrum complanatum* (L.) Holmb. *New Phytol.* 110:497–502.

Heide, O.M. 1992. Experimental control of flowering in *Carex bigelowii*. *Oikos* 65:371–376.

Henry, G.H.R., Freedman, B., and Svoboda, J. 1986. Effects of fertilization on three tundra plant communities of a polar desert oasis. *Can. J. Bot.* 64:2502–2507.

Heilbron, T.D., and Walton, D.W.H. 1984. Plant colonization of actively sorted stone stripes in the Subantarctic. *Arctic Alpine Res.* 16:161–172.

Jefferies, R.L., Svoboda, J., Henry, G.H.R., Raillard, M., and Ruess, R. 1992. Tundra grazing systems and climatic change. In: Chapin, F.S., III, Jefferies, R.L., Reynolds, J.F., Shaver, G.R., Svoboda, J., and Chu, E.W. (eds.), *Arctic Ecosystems in a Changing Climate: An Ecophysiological Perspective* (pp. 391–412). San Diego: Academic Press.

Jonasson, S. 1992. Growth responses to fertilisation and species removal in tundra related to community structure and clonality. *Oikos* 63:420–429.

Jonasson, S., Havström, M., Jensen, M., and Callaghan, T.V. 1993. *In situ* mineralisation of nitrogen and phosphorus of arctic soils after perturbations simulating climate change. *Oecologia* 95:179–186.

Jonsdottir, I.S. 1995. The importance of sexual reproduction in arctic clonal plants and their evolutionary potential. In: Callaghan, T.V., Oechel, W.C., Gilmanov, T., Holten, J.I., Maxwell, B., Molau, U., Sveinbjörnsson, B., and Tyson, M. (eds.), *Global Change and Arctic Terrestrial Ecosystems*. Proceedings of contributed and poster papers from the international conference, 21–26 August 1993, Oppdal, Norway. Brussels: Commission of the European Communities Ecosystems Research Report (pp. 81–88).

Jonsdottir, I.S., and Callaghan, T.V. 1988. Interrelationships between different generations of interconnected tillers of *Carex bigelowii*. *Oikos* 52:120–128.

Kjellberg, B., Karlsson, P.S., and Kerstensson, I. 1982. Effects of heliotropic movements of flowers of *Dryas octopetala* L and gynoecium temperature and seed development. *Oecologia* 54:10–13.

Kevan, P.G. 1972. Insect pollination of high Arctic flowers. *J. Ecol.* 60:831–847.

Krog, J. 1955. Notes on temperature measurements indicative of special organization in arctic and sub-arctic plants for utilization of radiated heat from the sun. *Physiol. Plantar.* 8:836–839.

Kummerow, J., and Ellis, B.A. 1984. Temperature effect on biomass production and root/shoot biomass ratios in two arctic sedges under controlled environmental conditions. *Can. J. Bot.* 62:2150–2153.

Laine, K., Malila, E., and Siuruainen, M. 1995. How is annual climatic variation reflected in the production of germinable seeds of arctic and alpine plants in the northern Scandes? In: Callaghan, T.V., Oechel, W.C., Gilmanov, T., Holten, J.I., Maxwell, B., Molau, U., Sveinbjörnsson, B., and Tyson, M. (eds.), *Global Change and Arctic Terrestrial Ecosystems*. Proceedings of contributed and poster papers from the international conference, 21–26 August 1993, Oppdal, Norway. Brussels: Commission of the European Communities Ecosystems Research Report (pp. 89–95).

Levesque, E., and Svoboda, J. 1995. Germinable seed bank from polar desert stands Central Ellesmere Island, Canada. In: Callaghan, T.V., Oechel, W.C., Gilmanov, T., Holten, J.I., Maxwell, B., Molau, U., Sveinbjörnsson, B., and Tyson, M. (eds.), *Global Change and Arctic Terrestrial Ecosystems*. Proceedings of contributed and poster papers from the international conference, 21–26 August 1993, Oppdal, Norway. Brussels: Commission of the European Communities Ecosystems Research Report (pp. 97–107).

Maxwell, B. 1992. Arctic climate: Potential for change under global warming. In: Chapin, F.S., III, Jefferies, R.L., Reynolds, J.F., Shaver, G.R., Svoboda, J., and Chu, E.W. (eds.), *Arctic Ecosystems in a Changing Climate: An Ecophysiological Perspective* (pp. 11–34). San Diego: Academic Press.

McGraw, J.B. 1980. Seed bank size and distribution of seeds in cottongrass tussock tundra. *Can. J. Bot.* 58:1607–1611.

McGraw, J.B., and Antonovics, J. 1983. Experimental ecology of *Dryas octopetala* ecotypes. II. A demographic model of growth, branching, and fecundity. *J. Ecol.* 71:899–912.

McGraw, J.B., and Fetcher, N. 1992. Response of tundra plant populations to climatic change. In: Chapin, F.S., III, Jefferies, R.L., Reynolds, J.F., Shaver, G.R., Svoboda, J., and Chu, E.W. (eds.), *Arctic Ecosystems in a Changing Climate: An Ecophysiological Perspective* (pp. 359–376). San Diego: Academic Press.

Melillo, J.M., Callaghan, T.V., Woodward, F.I., Salati, E., and Sinha, S.K. 1990. Effects on ecosystems. In: Houghton, J.T., Jenkins, G.J., and Ephraums, J.J. (eds.), *Climate Change, the IPCC Scientific Assessment* (pp. 282–310). Cambridge: Cambridge University Press.

Mitchell, J.F.B., Manabe, S., Tokioka, T., and Meleshko, V. 1990. Equilibrium climate change. In: Houghton, J.T., Jenkins, G.J., and Ephraums, J.J. (eds.), *Climate Change, the IPCC Scientific Assessment* (pp. 131–172). Cambridge: Cambridge University Press.

Oberauer, S.F., Sionit, N., Hastings, S.J., and Oechel, W.C. 1986. Effects of CO_2 enrichment and nutrition on growth, photosynthesis, and nutrient concentration of Alaskan tundra plant species. *Can. J. Bot.* 64:2993–2998.

Oechel, W.C., and Billings, W.D. 1992. Effects of global change on the carbon balance of arctic plants and ecosystems. In: Chapin, F.S., III, Jefferies, R.L., Reynolds, J.F., Shaver, G.R., Svoboda, J., and Chu, E.W. (eds.), *Arctic Ecosystems in a Changing Climate: An Ecophysiological Perspective* (pp. 139–168). San Diego: Academic Press.

Petersen, P.M. 1981. Variation of the population structure of *Polygonum viviparum* L. in relation to certain environmental conditions. *Medd. Grφnl. (Bioscience)* 4.

Philipp, M., Böcher, J., Mattsson, O., and Woodell, S.R.J. 1990. A quantitative approach to the sexual reproductive biology and population structure in some arctic flowering plants: *Dryas integrifolia*, *Silene acaulis*, and *Ranunculus nivalis*. *Medd. Grφnl. (Bioscience)* 34.

Prentice, C. 1993. Process and Production. *Nature* 363:209–210.

Shaver, G.R., and Chapin, F.S., III. 1980. Response to fertilisation by various plant growth forms in an Alaskan tundra: Nutrient accumulation and growth. *Ecology* 61:662–675.

Shaver, G.R., Fetcher, N., and Chapin, F.S., III. 1986. Growth and flowering in *Eriophorum vaginatum*: Annual and latitudinal variation. *Ecology* 67:1524–1525.

Sørensen, T. 1941. Temperature relations and phenology of the north-east Greenland flowering plants. *Medd. Grφnl. Kφbenhavn* 125(9):1–305.

Svensson, B.M., and Callaghan, T.V. (1988) Apical dominance and the simulation of metapopulation dynamics in *Lycopodium annotinum*. *Oikos* 51:331–342.

Tenow, O. 1972. The outbreaks of *Oporinia autumnata* Bkh. and *Operophthera* ssp. (Lep. Geometridae) in the Scandinavian mountain chain and northern Finland 1862–1968. *Zool. Bidrag Uppsala Suppl.* 2.

Tenow, O., and Holmgren, B. 1987. Low winter temperatures and an outbreak of *Epirrita autumnata* along a valley of Finnmarksvidda, the "cold-pole" of northern Fennoscandia. In: Axelsson, H., and Holmgren, B. (eds.), Climatological extremes in the mountain, physical background, geomorphological and ecological consequences. Dept. of Physical Geography of the University of Uppsala, UNGI Rapport 15:203–216.

Tissue, D.T., and Oechel, W.C. 1987. Responses of *Eriophorum vaginatum* to elevated CO_2 and temperature in the Alaskan arctic tundra. *Ecology* 68:401–410.

Wager, H.G. 1938. Growth and survival of plants in the Arctic. *J. Ecol.* 26:390–410.

Walker, D.A., and Walker, M.D. 1991. History and pattern of disturbance in Alaskan arctic terrestrial ecosystems: A hierarchical approach to analyzing landscape change. *J. Appl. Ecol.* 28:244–276.

Warming, Eug 1909. *Oecology of Plants, An Introduction to the Study of Plant-Communities* (pp. 163–322). Oxford: Clarendon Press.

Welker, J.M., Wookey, P.A., Parsons, A.N., Press, M.C., Callaghan, T.V., and Lee, J.A. 1993. Leaf carbon isotope discrimination and vegetative responses of *Dryas octopetala* to temperature and water manipulations in a high arctic polar semi desert, Svalbard. *Oecologia* 95:463–469.

Wijk, S. 1986a. Influence of climate and age on annual shoot increment in *Salix herbacea*. *J. Ecol.* 74:685–692.

Wijk, S. 1986b. Performances of *Salix herbacea* in an alpine snow-bed gradient. *J. Ecol.* 74:675–684.

Wookey, P.A., Parsons, A.N., Welker, J.M., Potter, J., Callaghan, T.V., Lee, J.A., and Press, M.C. 1993. Comparative responses of phenology and reproductive development to simulated environmental change in sub-arctic and high arctic plants. *Oikos* 67:490–502.

Wookey, P.A., Welker, J.M., Parsons, A.N., Press, M.C., Callaghan, T.V., and Lee, J.A. 1994. Differential growth, allocation and photosynthetic responses of *Polygonum viviparum* to simulated environmental change at a high arctic polar semi-desert. *Oikos* 70:131–139.

7. Phenology and Reproductive Success in Arctic Plants: Susceptibility to Climate Change

Ulf Molau

Reproductive success in tundra plants is highly influenced by phenology and microclimatic variations in time and space (Molau, 1993a; Walker et al., 1995). Climate change and associated alteration of the growing season could therefore have a strong impact on the performance of arctic and alpine plants (Bazzaz, 1990; Galen and Stanton, 1995). Recent autecological studies (e.g., Eriksen et al., 1993; Molau, 1993b; Stenström and Molau, 1992) have shown that there are strong correlations between reproductive strategies and flowering phenology in tundra plant species (see Molau, 1993a, for review). In this context, Bliss's (1956) categorization of the species in phenoclasses (vernal, early aestival, and late aestival) reflecting their flowering time is most useful. Early flowering ("vernal") species tend to be predominantly outcrossing, and they spend considerable time on selective seed abortion prior to final maturation of the fruits. At the other end of the phenological gradient, "late aestival" species, such as many snowbed species, show the opposite strategy, being predominantly inbreeding and with low levels of seed and fruit abortion (Eriksen et al., 1993; Molau, 1993a, 1993b). Sexual reproduction in vernal species is mainly pollen-limited, since pollination is hazardous in early summer in the Arctic, but there is usually some seed output every year. Late-flowering species have a reliable, although mainly autogamous, pollination, but successful seed maturation and, thereby, reproductive success does not take place all years due to variable onset of winter (Molau, 1993a).

Even though there is a continuum of phenological strategies in tundra plants from vernal to late aestival, the variation among species in reproductive strategy shows a certain bimodality, the species being either pollen- or seed-risk strategists, with few species showing intermediate strategies (Molau, 1993a). Reproductive systems requiring substantial interplant pollen transfer, such as dioecy, gynodioecy, and reduced self-fertility in hermaphrodites, are found mainly in the vernal phenoclass. Late aestival species frequently exhibit means of reproduction in addition to the more or less infrequent outcrossing (e.g., asexual formation of vegetative diaspores, high levels of selfing, and apomixis), traits that are rare or absent in vernal species. Furthermore, the variation in ploidy levels increase from vernal to late aestival phenoclasses, and the presence of high ploidy levels among snowbed species is interpreted as a means for increasing self-fertility, which is often low in diploids (Molau, 1993a).

The impact of phenology is easily understood in the light of available time for growth, flowering, fertilization, and subsequent maturation of fruits, when the length of the growing season itself is perhaps the most important limiting factor for performance in tundra plants. The depth of the winter snowpack varies with topography on a local scale (within landscapes), and the sites for individual plants will successively become snow free during the arctic summer as thawing proceeds. Snow accumulation patterns during winter are rather constant among years, resulting in spatial thawing sequences that are highly predictable. On the contrary, onset of winter is sharp and equal for all sites in an area, and will affect all plants simultaneously. The time for final onset of winter, terminating the growing season, however, is highly stochastic in most areas and may vary up to a month between consecutive years. Generally, thawing time (onset of growing season) in a particular site is less variable among years than is onset of winter, and this is especially true for early thawing, windblown habitats.

Within the hilly subarctic–alpine tundra site of Latnjajaure in northernmost Swedish Lapland, vernal species inhabiting hilltops, rock ledges, and ridges experience an average thaw period of 120 days and a total annual of some 600 to 800 thawing degree days (TDD) at the ground surface. For the snowbed species, the available growing season is only 60 days or less, and the total thawing degree days is below 300. In years with very early onset of winter (as in 1991, when the final onset occurred already in the first days of September, about three weeks earlier than normal) there is a complete loss of all sexually formed diaspores in late aestival species whereas vernal species remain unaffected. However, vernal species may also be at risk. Late May and early June is a hazardous time period with regard to climate, and sudden periods of days with drifting snow can cause physical damage to vegetative as well as reproductive tissues and suppress pollinator activity.

Climatic impact on plant reproduction is of particular interest in the Arctic, where changes in duration of the growing season may entail dramatic changes to the organisms. According to all General Circulation Models (GCMs), the rate of the greenhouse gas-induced climate change is predicted to be much greater than that in the past, even though the range of warming within foreseeable time is not unique (Callaghan et al., 1992, and references therein; Root and Schneider, 1993).

There are several recent studies on the impact of climate change on growth responses in tundra plants (e.g., Havström et al., 1993; Wookey et al., 1993), but few studies deal with the impact on reproduction more specifically. Our lack of insight is elegantly phrased by Root and Schneider (1993): "The only forecast that seems certain is that the more rapidly the climate changes, the higher the probability of substantial disruption and surprises within natural systems." Here I present an overview of the interaction between climatic conditions and reproductive ecology in tundra plants, with special emphasis on phenology, pollination, and reproductive success. I will focus on the following issues: (1) climatic factors controlling the length of the growing season in the tundra in the ground–air interface where plants reproduce, (2) plasticity in plant responses to climatic fluctuations within and among years, and (3) gradation of adaptive specializations in different phenoclasses. The main question addressed is: Are some plants more susceptible to global warming than others?

Materials and Methods

Climate monitoring and most investigations of plant reproductive ecology were performed at the Latnjajaure Field Station (68°22′N, 18°13′E) in northern Swedish Lapland during 1990 to 1993. The station is the main Swedish site of the International Tundra Experiment (ITEM), and is situated at ca. 1000 m a.s.l. in a subarctic–alpine tundra valley. Permafrost is widespread, and the depth of the active layer in most situations ranges between 0.5 and 1.5 m (J. Kling, unpubl. data). The topography is hilly with some windblown ridge tops and rock ledges snow free almost year-round, and with depressions with an early May snowpack of 1.5 to 2.5 m; in extreme snowbeds that melt out completely during favorable years (e.g., 1990; see Table 7.1), the snowpack is 8 to 12 m deep in early May. Flowering of chasmophytic (i.e., rock-inhabiting) plants of *Saxifraga oppositifolia* in the dark cliffs above the station commences between 10 and 20 May, whereas the first soil-inhabiting plants come into flower 1 to 2 weeks later. Onset of winter, defined as continuous snowcover and below-freezing ground surface temperatures, may occur from the first week of September (1991) to the second week of October (1992). Mean July and August temperatures vary among years in the range of 5.4 to 8.7°C. The highest daily maximum temperature recorded was 18.4°C (June 1992). The annual mean temperature (August 1992 to July 1993) was –2.49°C.

Table 7.1 Climate Data from the Latnjajaure Field Station, North Swedish Lapland, for the Growing Seasons of 1990–93

Period	Mean Temperature (C)	Thawing Degree Days	Growing Degree Days	Integrated Global Radiation (MJ/m^2)	Precipitation (mm)
1990					
May	+0.39	35.4	3.6		72.1
June	+6.09	171.6	67.1		27.5
July	+9.10	282.3	135.1		55.0
August	+7.88	232.5	96.8		70.7
September	+2.54	87.9	20.1		22.3
Total		809.7	322.7		247.6
1991					
May	−1.82	31.6	0.0	587.9	50.0
June	+4.94	157.9	39.1	549.3	72.0
July	+8.15	247.7	96.6	407.8	44.2
August	+8.50	258.1	110.5	351.1	46.3
September	−0.47	43.0	5.8	208.0	104.7
Total		738.3	252.0	2104.1	317.2
1992					
May	+1.91	96.0	24.5	570.4	66.7
June	+5.69	175.6	70.2	709.6	27.8
July	+5.43	169.3	48.2	454.6	133.0
August	+5.60	173.9	40.0	307.3	52.0
September	+3.82	118.6	20.5	177.7	20.6
Total		733.4	203.4	2219.6	300.1
1993					
May	−0.06	52.4	10.6	611.6	80.0
June	+1.60	65.3	7.9	572.3	60.1
July	+8.46	262.4	115.7	509.8	63.6
August	+6.57	203.6	67.0	277.3	63.3
September	+0.91	39.6	0.6	214.9	30.7
Total		623.3	201.8	2185.9	297.7

The Latnjajaure Field Station was staffed from mid-May to early September in 1992 and 1993; somewhat shorter periods in 1990 and 1991. During the field season there is a standard manual weather station where temperatures and relative humidity are measured in a Stevenson Screen at 2.0m aboveground at 0700 and 1900 hours normal time; cloudiness is estimated and precipitation are measured as well (see Molau, 1993c, for detailed description). In addition, there is an automatic station equipped with a Delta T™ datalogger and sensors for maximum, minimum, and mean temperatures, soil temperature at 0, 5, 10, and 20cm depths, relative humidity, global solar radiation, PAR (photosynthetically active radiation), wind velocity, and wind direction, values being recorded every 30s (for wind velocity counted during 10-min periods) and stored as hourly means (except for wind direction). Onset of winter is defined as a stable snow depth of at least

10 cm; this is recorded by an unsheltered thermocouple mounted at 10 cm aboveground, where temperature variation ceases almost completely when the probe becomes snow-covered. The automatic station has been operating year-round since 1992.

Additional data were collected in West Greenland in 1991 and in the Chukotka Peninsula, Russian Beringia, in 1993. Literature data on reproductive ecology in tundra plants are more extensively reviewed elsewhere (Molau, 1993a).

Methods for investigation of reproductive systems follow Eriksen et al. (1993) and Molau (1993a, 1993b). Statistical analysis and graphical illustration of plant and climate data were performed using the StatView™ 4.0 and SuperANOVA™ 1.2 software packages (Abacus Concepts, Inc.).

Results and Discussion

Thawing, Flowering, and Growing Season

Snowmelt at Latnjajaure is clearly a function of ambient air temperature as measured in shade and shelter in a Stevenson Screen (Figure 7.1; simple regression, $R^2 = 0.283$, $n = 293$, $p < 0.0001$). The average daily snowmelt during the summer of 1992 was 3.88 ± 1.90 cm (mean ± SD; $n = 293$), and melting of 1 cm of the snowpack thus requires on average 1.48 TDD heat units. The total heat accumulation during the cold summer of 1992 (Table 7.1) was not sufficient for melting of more than 4 m snowpack, and snow accumulation occurred in the extremely late-melting snowbeds.

Due to the high albedo of snow, variation in global solar radiation energy influx showed little correlation with thawing. As soon as a site is snow-free, however, global radiation instantly becomes the major determinant of soil surface temperature. Ambient temperature (standard measurements in Stevenson Screen) varies a lot in early summer, but the difference between ambient and surface temperatures is significantly dependent on global radiation influx (Figure 7.2; simple polynomial regression, $R^2 = 0.655$, $n = 1665$, $p < 0.0001$). Global solar radiation turned out to be the most important environmental parameter governing the length of prefloration time among individuals of a number of tundra species (Marsden, 1992). However, the intensity of global radiation declines as the summer proceeds (Figure 7.3), and from late July ambient air temperature becomes the main determinant of field surface temperature and tuning of temperature-related plant processes.

The four seasons (1990 to 1993) of investigations at Latnjajaure were quite different with regard to climate (see Table 7.1). Comparisons with the long-term climatic record at the nearby Abisko Scientific Research Station (from 1913 until present) show that all winters, 1990 to 1993, were much milder than normal and with a much deeper snowpack than normal. The growing season of 1990 was much warmer than normal in general, and seed

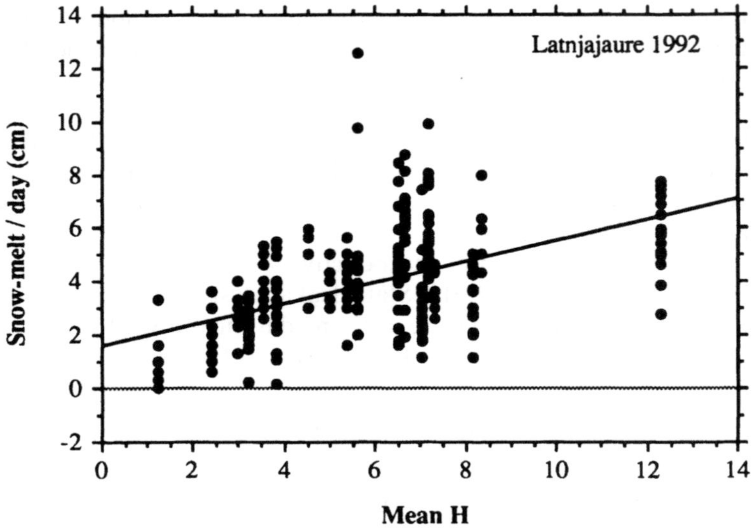

Figure 7.1. Scattergram showing the significant effect of ambient temperature on rate of snowmelt at Latnjajaure, North Sweden. Data from 1992. Snowmelt was measured at every 5 m every 3 days along two parallel 100 m transects down a north-facing slope at ca. 1000 m altitude. Each dot thus represents a daily average, and is the mean of five repeated probings at each point. Values of 10 cm/d and higher occur only during very rainy conditions. Temperature is expressed as mean daily heat accumulation (H) (i.e., the temperature sum over 0°C) and is calculated from logger data (see text). For regression data, see text.

set was good in all plant phenoclasses. In 1991, summer temperatures were close to normal, but July was windy and rainy; a very sudden onset of winter occurred September 3 to 4 with a harsh snowstorm and wind velocities up to 50 m/s. and the ground was continuously snow-covered from this occasion until next summer. In 1992, the growing season started early and proceeded rapidly in very good weather until mid-June, when the climate deteriorated, and the rest of the summer was very poor and far below normal temperatures; the July mean was 3.0° below the average. Snowstorms occurred several times in late June and lasted several days each, causing lots of physical damage to the species in flower at that time (see below). September was warmer than normal, and some more late-flowering species together with vernal species had a decent seed set. The winter 1992 to 1993 was characterized by frequent storms (even hurricanes), relatively high temperature, and lots of snow. The growing season of 1993 started, despite the deep snowpack, extremely early, and *Saxifraga oppositifolia* started flowering already in the first half of May (see Figures 7.4 and 7.5). A very cold period in late May, and lasting almost throughout June, with frequent hard winds and snowfall, normalized the situation. July and August were somewhat warmer than normal, and there was a good seed set in all phenoclasses. The cold summer climate in 1992 and (partly) 1993 all over the Arctic and Subarctic was mainly induced by the Mt. Pinatubo eruption in the Philippines in 1991, stratospheric dusts reducing the influx of global solar radiation (Minnis et al., 1993). Also the trend towards earlier thawing and onset of the growing season observed at Latnjajaure during 1990 to 1993 follow a general trend for the Northern Hemisphere (Groisman et al., 1994).

In late spring (May) global solar radiation is very high at the same time as ambient air temperature is mostly below freezing (Figure 7.3). Chasmophytic plants thriving in snow-free dark rock cliffs may, however, commence their reproductive cycle at this stage. In 1992, a chasmophytic clone of *Saxifraga oppositifolia* opened its first flowers on May 18, when ambient air temperature was +0.3°C and with some drifting snow in the air; the temperature inside the tuft was at this time as high as +7.8°C. Bumblebee queens (particularly *Bombus alpinus* and *B. hyperboreus*) started to be active after hibernation at this time, and 2 days later all open flowers of this

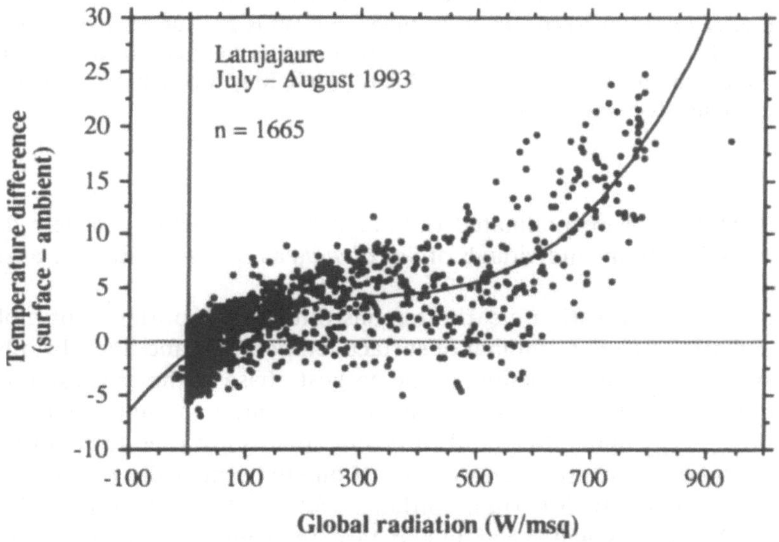

Figure 7.2. Scattergram showing the significant effect of global solar radiation (in W $m^{-2} s^{-1}$) on ground surface temperature in a mesic site at Latnjajaure, North Sweden. Data from July and August 1993. The dots represent hour means of logger data. The *y* axis represents the difference between temperature of ambient air (at 2.0 m in shade and shelter) and ground surface. Negative differences (ambient higher than surface temperature) occur during clear nights, often resulting in ground frost. Msq = meter square. For regression data, see text.

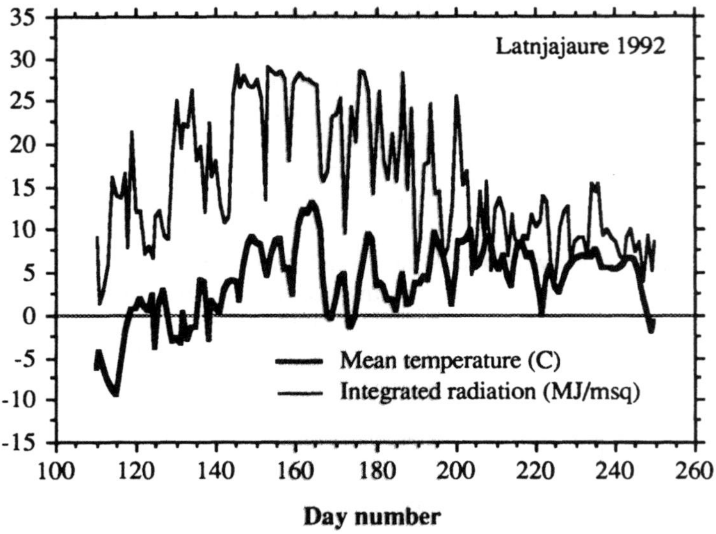

Figure 7.3. Scattergram showing daily mean temperature and integrated daily global radiation (in MJ/m^2) during the growing season at Latnjajaure, North Sweden, in 1992. Daily means are calculated from hour means of logger data. Day number (Julian day) 120 is 29 April, 180 is 28 June, and 240 is 27 August. Note that the annual optimum of the radiation energy curve occurs long before the potential temperature optimum. The radiation influx is high in May, when ambient temperature is still mostly below freezing.

particular plant were pollinated (although probably selfed). A few days later, rock-inhabiting individuals of *Salix lanata* as *Draba lactea* were also in flower.

The above observation clearly demonstrates the importance of global solar radiation as a determinant of surface temperature and, thereby, plant development and reproduction, in the earliest phase of the arctic summer. From Figure 7.3 it is easily seen that there is a large amount of incoming radiation energy in late May and early June that is not used for photosynthesis by plants because of snowcover. If one hypothetically raises the air temperature curve by 3°C (in accordance with most predictions; see Maxwell, 1992), thawing will speed up and early thawing sites will permit plant growth 2 to 3 weeks earlier than at present, at the same time as the radiation climate is close to the annual maximum. A similar temperature increase in late season (September) would have little effect on plant growth, because the available radiation influx is getting very low at this time. Increased winter precipitation expected from the GCMs will be spatially unevenly distributed through the normal redistribution by wind, and windblown ridge tops will not experience an increase in snowcover and duration. The extremely high winter precipitation at Latnjajaure, 1992 to 1993 (on the average almost 3 m of snow) just accentuated the usual distribution pattern with the summits snow-free in early May and up to 8 m accumulation in the extreme snowbeds. Summarizing these results, a 3°C increase in average ambient temperature would prolong the growing season at least 2 weeks, and (taking the magnitude of radition energy influx into account) increase the potential annual plant productivity per unit area in the order of 20 to 30%, provided that there are no other factors limiting plant growth (e.g., soil moisture, nutrient availability).

Phenology and Floral Traits

The species' annual sequence of onset of flowering varies little between consecutive years, even though climatic fluctuations will displace parts of the sequences but leave the species' order unchanged (Figures 7.4 and 7.5).

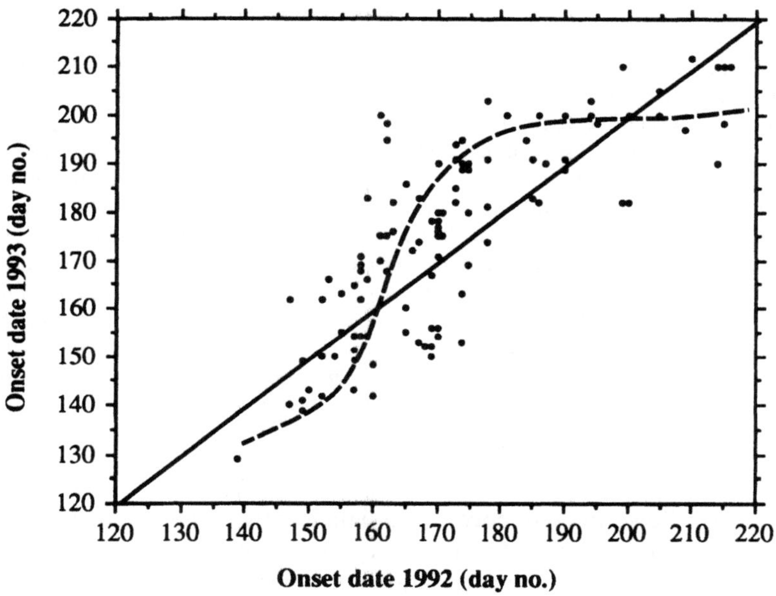

Figure 7.4. Congruence of phenological sequences of first flowering of 124 different tundra species at Latnjajaure, North Sweden, during two consecutive years, 1992 and 1993. The order of the phenological sequences between years is highly constant, but actual dates are displaced due to climatic between-year variation (see GDD curves in Figure 7.5). The straight diagonal line represents a hypothetical scenario of equal onset dates for all species in both years. The eye-fitted curve (dashed line) shows that flowering phenology was delayed in June 1993 relative to the same period in 1992, whereas the early and late-flowering species were ahead of the schedule. For day number decoding, see Figure 7.3.

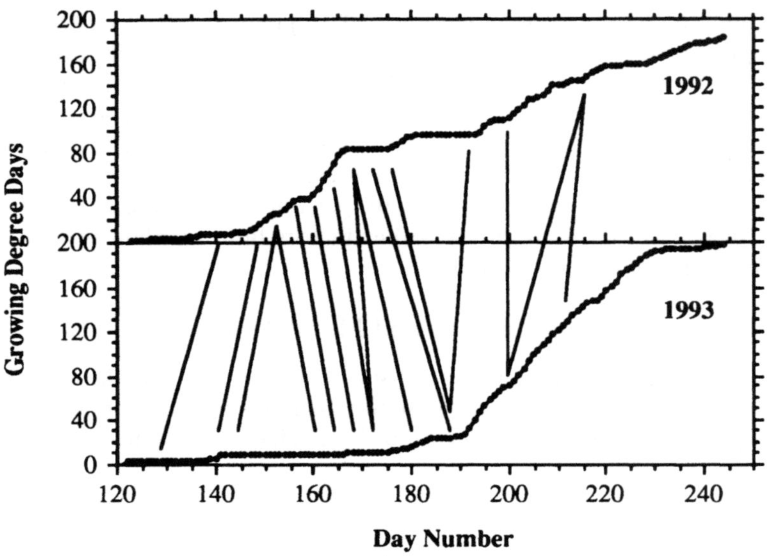

Figure 7.5. Congruence of phenological sequences of first flowering of particular individuals or clones of 15 different tundra species at Latnjajaure, North Sweden, during two consecutive years, 1992 and 1993. The curves show growing degree days (GDD), the cumulative sum of daily integrated temperature above 5°C, from 1 May to 1 September during the two seasons. Each line connecting the two curves represents first flowering of an individual plant. The order of the phenological sequences between years is highly constant, but actual dates are displaced due to climatic between-year variation (GDD curves). For day number decoding, see Figure 7.3. The plant species are, from left to right, *Saxifraga oppositifolia*, *Salix lanata*, *Draba lactea*, *Rhododendron lapponicum*, *Cassiope tetragona*, *Diapensia lapponica*, *Dryas octopetala*, *Bartsia alpina*, *Hierochloë alpina*, *Campanula uniflora*, *Saxifraga aizodes*, *Chamorchis alpina*, *Viscaria alpina*, *Phippsia algida*, and *Saussurea alpina*. Compare the inclination of the lines connecting the curves with the displacement of the curve in Figure 7.4.

The species' flowering sequences were significantly correlated over the 1992 and 1993 seasons using the Spearman rank correlation test (rho = 0.814, $p <$ 0.0001, $n = 124$ species); the correlation coefficient obtained by ranking was higher than that obtained by Pearson product-moment correlation ($r =$ 0.784) in the same material, the latter being influenced by the above-mentioned climatic fluctuations. Prefloration time (i.e., time lapse from thawing to first flower) is basically a function of accumulated global solar radiation (Marsden, 1992), but there is also a stable among-individual variance that is probably genetic in nature (as demonstrated in *Bartsia alpina*; Molau, 1991).

Most entomophilous vernal plants have colored corollas (68% possess blue, purple, red flowers), a means of lowering flowering albedo and increasing the absorption of solar radiation energy. The significance of higher temperatures in dark-colored flowers has been pointed out previously (see Savile, 1972, and references therein). Typical representatives of the vernal phenoclass are purple-flowered *Saxifraga oppositifolia* and *Rhododendron lapponicum*, and the deep yellow *Ranunculus nivalis*. The proportion of white- or pale-flowered species increases significantly in the early and late aestival phenoclasses (50 and 64%, respectively). The frequencies of dark and light flower colors were significantly different among the phenoclasses in a contingency table analysis of data from 100 species of entomophilous species ($DF = 2$, $G^2 = 15.429$, $p = 0.0004$; data from Molau, 1993a). These observations from the species-rich flora of Latnjajaure were confirmed during the shorter visits in West Greenland and Chukotka, despite deviating composition of the flora.

Furthermore, plant stature changes gradually though the growing season. Vernal plants have their flowers 1 to 5 cm from the ground, but as ambient temperature increases and radiation influx reaches its maximum and starts to decrease, inflorescences become gradually taller, and late aestival plants generally have their flowers at 10 to 25 cm aboveground (Savile, 1972; this chapter). Vernal species flowering at low height aboveground (e.g., *Ranunculus nivalis*) often prolong their peduncles during ripening of the fruits later in the summer. In late aestival plants there is no such large postfloral elongation of the inflorescence. The increase in stature along the growing season appeared to be similar in the West Greenland and Chukotka sites. Again, the importance of solar radiation energy early in the season in manifested. A good example is provided by the woolly *Pedicularis* species, *P. lanata* and *P. hirsuta*, often growing side by side in West Greenland. *Pedicularis lanata* is a vernal species with deep red flowers densely arranged in a short subglobose spike that later elongates four to six times before seed dispersal; *P. hirsuta* flowers later in the season, has pinkish flowers on a taller inflorescence, and shows less postfloral elongation, ending up in infructescences of about the same height as in *P. lanata*.

Pollinator Activity

Early flowering tundra plants are mainly insect pollinated (Molau, 1993a, and references therein), and anticipated depletion of pollination in several vernal and early aestival species at Latnjajaure during periods of cold weather could not be verified in most cases. Bumblebee queens forage even during light snowfall, provided that radiation influx is high (i.e., from mid-May to early July; Bergman et al., 1996), and pollination of vernal species is reliable in quality though variable in quantity. In the case of early aestival species, pollination by flies (and to some extent *Colias* and *Erebia* butterflies) is more important, and those insect groups are more weather dependent, making pollination less reliable. Late aestival plants, on the other hand, always achieve stable and high levels of fertilization, even in the absence of pollinators (Molau, 1993a), but their reproductive success in terms of seed output is stochastic and may be low or absent (cf. below).

Effects of Intermittent Freezing

Periods with cold weather and drifting snow are not uncommon in the arctic summer, but their occurrence is highly stochastic. A 5-day period of blizzards with drifting snow and temperatures below freezing in late May 1993 caused little damage (<10% of the flowers) to the flowering population of *Saxifraga oppositifolia* (vernal; purple flowers, ca. 1 cm aboveground), whereas a similar event in late June 1992 caused severe damage to *Cassiope tetrugona* and *Diapensia lapponica* (white flowers, 3 to 10 cm aboveground; more than 90% destroyed). In all these evergreen species, however, we have noted accumulation of frost damage in vegetative tissue, in northern Scandinavia as well as in northern Alaska, and this impact particularly pronounced in the case of *Diapensia*. Over the 5-year period 1990 to 1994, the decrease in live tissue in this extremely long-lived cushion plant was 22% in exposed sites at Latnjajaure, and the decrease in flower number per unit area as much as 55% (Molau, 1996). Many of the medium-sized (50 to 100 years old) cushions died off completely during the study period, and similar episodes of "sudden death" have been observed in evergreen *Saxifraga oppositifolia* in sites that normally have a more long-lasting snowcover than during the last few years (M. Stenström, unpubl. data). Such damages to foliage and apical meristems occur when they are triggered to "premature" vernalization by an earlier onset of the growing season when the annual frost hardening/dehardening cycle is at its most sensitive phase (Pihakaski, 1988; Pihakaski and Junnila, 1988). Thus, evergreen and relatively early flowering species, such as the *Diapensia*, *Dryas*, and *Cassiope* species, seem to be particularly vulnerable, whereas the effects on deciduous perennial species are less severe and do not accumulate in the same way. This risk of population decrease in evergreen species is accentuated by their long life spans and slow growth; for example, the annual radial growth rate of *Diapensia lapponica* cushions at Latnjajaure is on average only 0.6 mm, and decreases at higher age (Molau, 1996). Cushions with a 30-cm radius and one single taproot (i.e., representing a single genet) are quite common, and must have a minimum age of 300 years or more. High mean genet ages has also been documented in other tundra plants [e.g., *Bartsia alpina* (194 years; Molau, 1990) and *Eriophorum vaginatum* (158 years; Mark et al., 1985)].

Effects of Early Onset of Winter

Onset of winter varies in time from early September to mid-October at the Latnjajaure site. Early onset, as in 1991, resulted in a complete loss of the entire seed crop in all late aestival species, whereas vernal and most early aestival species had already successfully completed reproduction and dispersed their seed. As seen from Figure 7.3, a displacement of the entire growing season towards the early end by 1 to 2 weeks would reduce the risk for late aestival plants of being caught by winter.

Phenotypic Plasticity

Vernal tundra species show highly stable flowering progress curves with regard to the shape of the curve, expressed as open flowers per unit area (Eriksen et al., 1993; Molau, 1993a). However, they show a high plasticity in that the entire flowering curve may be displaced among years according to climatic regime. The flower buds of vernal species are highly differentiated already in the preceding season (Sørensen, 1941), and their final development into open flowers is easily adjusted by climatic factors. Late aestival plants, on the other hand, over-winter with poorly developed floral meristems, and the floral response is highly determined by the accumulated heating during the summer. All in all, vernal species are able to modify the timing but not the number of flowers (absolutely determined during the preceding season), whereas late aestival species can vary their reproductive effort in terms of numbers of flowers according to the contemporary climatic regime (Molau, 1993a; Sørensen, 1941).

At the level of individual plants, phenological diversity seems to be highest among individuals of vernal and early aestival species, whereas late aestival species appear more homogeneous genetically as well as with regard to phenotypical plasticity (Eriksen et al., 1993; Molau, 1991; Molau, 1993a, and references therein; Stenström and Molau, 1992).

Temperature enhancement experimentation during ITEX (International Tundra Experiment) field work at Latnjajaure has also shown that vernal and early aestival evergreen species may suffer increased competition from co-occurring deciduous species (Molau et al., submitted; Welker et al., submitted). These findings are corroborated by recent data from Alaskan North Slope, where Chapin et al. (1995) found a significant decrease in biomass of evergreen *Eriophorum vaginatum* and a concurrent increase in deciduous *Betula nana* in unmanipulated control plots over a 10-year period, associated with the strongest ongoing warming trend anywhere in the Arctic (Chapman and Walsh, 1993).

During cold summers, individuals of vernal and early aestival plants in the most late-thawing sites within their habitat range may postpone flower bud opening, and these buds then will—if differentiation has not proceeded too far—open up very rapidly after thawing the next season without any signs of reduced performance. This phenomenon was observed at Latnjajaure in *Cassiope tetragona* and *Diapensia lapponica* during the extremely cold summer of 1992. In the most late-thawing (late June) of my permanent $10\,m^2$ plots for monitoring of *D. lapponica* (implemented in 1990), 22 flower buds were conspicuous by late July, but only 10 came into flower in 1992, the remainder were postponed; 2 of these flowered in early August 1993, the other 10 were still visible but dormant, postponed even further (they later turned out to flower in 1995, after having been postponed for 3 years). Similar abilities to postpone flower buds has been seen in *Eriophorum vaginatum* (Shaver and Molau, unpubl. data).

Implications for Over-Wintering Theory

The lengthy debate on over-wintering theory for arctic and alpine plants and plant communities during Pleistocene glaciations (see Dahl, 1992, and references therein) has hitherto very rigidly focused on two opposing views, the "nunatak" and "tabula rasa" hypotheses. The former states that arctic and alpine plants survived glaciations in refugia, such as nunataks and coastal ice-free areas, and the latter states that all species of the present-day tundra have been immigrating from over-wintering areas at lower latitudes and altitudes. With the differentiation in reproductive strategies between early and late-flowering species, both hypotheses may very well be correct—in part. Vernal and most early aestival species at Latnjajaure show normal levels of reproductive success when average summer temperature is several degrees below normal, and can probably survive hundreds of seasons at 5°C or more below normal, although in smaller populations in snow-free or early thawing habitats (nunatak hypothesis). The plant record from the recently investigated Jensen's Nunataks in Greenland is mainly composed of vernal and early aestival species (data from Gjærevoll and Ryvarden, 1977), and the same view is supported by regional morphological differentiation in Fennoscandian populations of species of those phenoclasses (e.g., *Bartsia alpina*) (Molau, 1990).

Late aestival plants, on the other hand, have no means of survival in situ during such adverse conditions (tabula rasa). These species often have adaptations for long-distance dispersal, and are better preadapted for more rapid migration than early flowering species. Seeds of vernal and early aestival tundra plant species are mostly larger and with no adaptations for wind dispersal over longer distances. For example, if the tabula rasa hypothesis should apply in the case of *Bartsia alpina* in the absence of long-distance dispersal, my data (Molau, 1990) indicate that this species should by now have reached the German–Danish border on its northward migration after the last ice age. Thus, the distance from present-day arctic habitats to the closest site for glacial over-wintering of a species is probably positively correlated with its phenoclass, and many early flowering species have the capability of in situ glacial survival. This more differentiated view was also supported in the summary of a recent Norwegian workshop on the subject (Fægri, 1993), taking quaternary geology and plant geography into account.

Conclusions

Both climatic warming and cooling will have substantial effects on reproductive success in tundra plants, thereby affecting diaspore output and recruitment. During long or warmer-than-normal growing seasons, vernal and early aestival species may suffer from temperature-induced reflowering in late summer, resulting in a depletion of overall reproductive success if the phenomenon becomes regular. Species belonging to these phenoclasses are much better adapted to survive periods of climate colder than today, because they are able to postpone flower buds for one or several consecutive seasons (Savile, 1972; Sørensen, 1941; this chapter). Late aestival plants, on the other hand, respond more positively to temperature amelioration, and an earlier thawing of the snowbed habitats in combination with a general speeding-up of the reproductive cycle results in more stable reproductive success (in terms of seed dispersal and subsequent recruitment potential of seedlings).

In conclusion, a climatic warming of the rate indicated by GCMs (3°C on average for the Arctic during a 50-year period from now in the "business-as-usual" scenario; see Maxwell, 1992) will increase reproductive success in most species, but in the long run turn into a fitness depression in vernal and early aestival species. Evergreen vernal species are particularly at risk due to their extreme life spans of individual plants and the competition for the light and nutrient resources imposed by co-occurring deciduous species increasing their leaf surface areas. Clones of vernal evergreen species have very low growth rates, and it is mainly the plant individuals present in the tundra today that will be affected by climatic change.

Furthermore, early aestival species starting to flower earlier than today in the season face another risk, namely that of short periods of cold weather and damage on reproductive tissues by freezing and drifting snow. The proportion of late aestival species' representation in the seed pool of arctic habitats will increase steadily due to prolongation of the growing season and, thereby, stabilization of reproductive success among years.

Climatic cooling will have the opposite effects, imposing increasing adversity selection pressures on late aestival plants because of lower and more stochastic reproductive success and decreased thawing of snowbeds. Vernal and—to some extent—early aestival species, showing some reproductive success most years at present, will remain relatively unaffected. Especially important features are the long life span of individuals, short prefloration time, the possibility for postponing of flower bud dehiscence to following seasons, high phenotypic plasticity, and high genetic variation within populations.

For modeling purposes it is often assumed that arctic plant communities will migrate as entities in response to climate change. However, all tundra species have unique species-specific properties (or unique combinations of traits) in vegetative and reproductive strategies, and the plant communities familiar in the present-day tundra are constellations that, at least partly, are stabilized by relatively low climatic variation during the present postglacial period. A rapid climatic change, as the warming predicted by GCMs, will most certainly cause a disintegration of communities due to differential species-specific responses in combination with interspecific variation in seed dispersal ability, rather than an en bloc movement of entire communities.

With regard to biodiversity and nature conservation of the tundra plants in an anticipated warmer climate in a near future, I conclude that vernal and early aestival species are the most vulnerable. Evergreen dwarf shrubs and cushion plants are highly susceptible to physical damage by too early an onset of the growing season, but total extinction within an area is unlikely for this reason only, because there always will be some "safe" sites. However, evergreen plant species show little or no responses to temperature enhancement, and will also suffer competitive disadvantage when interacting with rapidly responding deciduous dwarf shrubs, herbs, and graminoids. Hardy, long-lived vernal species with outbreeding mating systems are particularly endangered, especially when occurring in small populations. An adequate assessment of their susceptibility requires detailed analyses of mating systems, seed dispersal and dormancy, population genetics, and demography.

Acknowledgments. I thank all the field assistants working for long periods of time gathering data at the remote Latnjajaure Field Station: Juha Alatalo, Pia Emanuelsson, Markus Engelhart, Anna Lindskog, Eva Nilsson, Urban Nordenhäll, Martin Sommerkorn, Mikael Stenström, and Michael Svensson. I also thank the Abisko Scientific Research Station and its staff for support and maintenance of the field station. Furthermore, I thank Björn Holmberg for helping in installing the weather station, supplying part of the equipment, and for training the weather observers, Kjell Ericsson, Anders Eriksson, and Bengt Wahnatalo for construction and maintenance of equipment, and Roger Barry, Jerry Brown, and Barrie Maxwell for additional methodological advice. Helicopter transportation of equipment, personnel, and supplies was carried out by skillful pilots Bertil "Berra" Björk and Lars "Lidas" Lidström. Bente Eriksen and three anonymous reviewers critically read the manuscript. Financal support was received from the Swedish Natural Sciences Research Council (NFR), the Kempe Foundation, and the Swedish Society for Nature Conservation.

References

Bazzaz, F.A. 1990. The response of natural ecosystems to the rising global CO_2 levels. *Annu. Rev. Ecol. System.* 21:167–196.

Bergman, P., Molau, U., and Holmgren, B. 1996. Micrometeorological impacts on pollinator activity and plant reproductive success in an arctic environment. *Arctic Alpine Res.* 28:196–202.

Bliss, L.C. 1956. A comparison of plant development in microenvironments of arctic and alpine tundras. *Ecol. Monogr.* 26:303–337.

Callaghan, T.V., Sonesson, M., and Sømme, L. 1992. Responses of terrestrial plants and invertebrates to environmental change at high latitudes. *Philos. Trans. R. Soc. Lond.* [B] 338:279–288.

Chapin, F.S., III, Shaver, G.R., Giblin, A.E., Nadelhoffer, K.G., and Laundre, J.A. 1995. Responses of arctic tundra to experimental and observed changes in climate. *Ecology* 76:694–711.

Chapman, W.C., and Walsh, J.E. 1993. Recent variations of sea ice and air temperature in high latitudes. *Bull. Am. Meteorol. Soc.* 74:33–47.

Dahl, E. 1992. Nunatakkteori. IV. Hvor fantes isfrie områder og hva slags planter kunne Ieve på dem? *Blyttia* 50:23–35 [in Norwegian with English summary].

Eriksen, B., Molau, U., and Svensson, M. 1993. Reproductive strategies in two arctic *Pedicularis* species. *Ecography* 16:154–166.

Fægri, K. 1993. Overvintringssymposium i Bergen. *Blyttia* 51:70–72 (in Norwegian).

Galen, C., and Stanton, M.L. 1995. Responses of plant snowbed species to changes in growing-season length. *Ecology* 76:1546–1557.

Gjærevoll, O., and Ryvarden, L. 1977. Botanical investigations on J.A.D. Jensens Nunatakker in Greenland. Det Konglige Norske Videnskablige Selskabs Skrifter 1977(4):1–40.

Groisman, P.Y., Karl, T.R., and Knight, R.W. 1994. Observed impact of snow cover on the heat balance and the rise of continental spring temperatures. *Science* 263:198–200.

Havström, M., Callaghan, T.V., and Jonasson, S. 1993. Differential growth responses of Cassiope tetragona, an arctic dwarf-shrub, to environmental perturbations among three contrasting high- and subarctic sites. *Oikos* 66:389–402.

Mark, A.F., Fetcher, N., Shaver, G.R., and Chapin, F.S., III. 1985. Estimated ages of mature tussocks of *Eriophorum vaginatum* along a latitudinal gradient in central Alaska, U.S.A. *Arctic Alpine Res.* 17:1–5.

Marsden, R. 1992. Snow melt, flowering phenology and reproductive success in tundra plants. M.Sc. thesis, University of North Wales, Bangor.

Maxwell, B. 1992. Arctic climate: potential for change under global warming. In: Chapin, F.S., et al. (eds.), *Arctic Ecosystems in a Changing Climate* (pp. 11–34). San Diego, CA: Academic Press, Inc.

Minnis, P., Harrison, E.F., Stowe, L.L., Gibson, G.G., Denn, F.M., Doelling, D.R., and Smith, W.R., Jr. 1993. Radiative climate forcing by the Mount Pinatubo eruption. *Science* 259:1411–1414.

Molau, U. 1990. The genus *Bartsia* (Scrophulariaceae-Rhinanthoideae). *Opera Bot.* 102:1–99.

Molau, U. 1991. Gender variation in *Bartsia alpina*, a subarctic perennial hermaphrodite. *Am. J. Bot.* 78:326–339.

Molau, U. 1993a. Relationships between flowering phenology and life history strategies in tundra plants. *Arctic Alpine Res.* 25:391–402.

Molau, U. 1993b. Reproductive ecology of the three Nordic *Pinguicula* species (Lentibulariaceae). *Nord. J. Bot.* 13:149–157.

Molau, U. 1993c. ITEX climate stations. In: Molau, U. (ed.), *ITEX Manual* (pp. 6–10). Copenhagen: Danish Polar Center.

Molau, U. 1996. Climatic impacts on flowering, growth, and vigour in an arctic cushion plant, *Diapensia lapponica*, under different snow cover regimes. *Ecol. Bull.* 45 (in press).

Pihakaska, K. 1988. Seasonal changes in the chloroplast ultrastructure of *Diapensia lapponica. Nord. J. Bot.* 8:361–367.

Pihakaski, K., and Junnila, S. 1988. Cold acclimation of subarctic *Diapensia lapponica. Funct. Ecol.* 2:211–228.

Root, T.L., and Schneider, S.H. 1993. Can large-scale climatic models be linked with multiscale ecological studies? *Conserv. Biol.* 7:256–270.

Savile, D.B.O. 1972. Arctic adaptations in plants. Canada Department of Agriculture, Research Branch, Monograph No. 6.

Sørensen, T. 1941. Temperature relations and phenology of the northeast Greenland flowering plants. *Medd. Grønl.* 125(9):1–305 + Pl. 1–15.

Stenström, M., and Molau, U., 1992. Reproductive biology in *Saxifraga oppositifolia*: Phenology, mating system and reproductive success. *Arctic Alpine* Res. 24:337–343.

Walker, M.D., Ingersoll, R.C., and Webber, P.J. 1995. Effects of interannual climate variation on phenology and growth of two alpine forbs. *Ecology* 76:1067–1083.

Wookey, P.A., Parsons, A.P., Welker, J.M., Potter, J.A., Callaghan, T.V., Lee, J.A., and Press, M.C. 1993. Comparative responses of phenology and reproductive development to stimulated environmental change in sub-arctic and high arctic plants. *Oikos* 67:490–502.

8. A Kinetic Approach to Microbial Ecology in Arctic and Boreal Ecosystems in Relation to Global Change

Nikolai S. Panikov

Kinetics is a branch of natural science that deals with the rates and mechanisms of dynamic processes. The kinetic method is based on combinations of dynamic experimental studies or observations with mathematical modeling. The model describes postulated mechanisms of studied reactions, so comparisons of observations and predictions allow for discarding wrong assumptions. Experimentally tested models are analyzed by mathematical methods for better understanding of studied natural phenomena. Particularly important targets of such studies are predictions and developments of strategies for management of high latitude ecosystems in relation to anticipated climate changes.

The terms "kinetic method" and "systems approach" are often treated as synonyms, but there is a subtle difference. The former generally requires detailed experimental data obtained for well-shaped objects and focuses on the problems related to mechanistic considerations, whereas the latter is mainly used for developing and better understanding of highly complicated and poorly studied natural systems.

In this chapter I will demonstrate the applicability of a kinetic approach to IGBP's studies of global change. The main emphasis is given to effects of elevated temperature on microbial activity in arctic and boreal ecosystems.

There is a great body of experimental evidence that elevated temperature dramatically accelerates decomposition and mineralization of organic matter in many soils. For arctic ecosystems, temperature is considered to be the most important environmental factor and best predictor of decomposition rates (Flanagan and Veum, 1974). On the other hand, there is an opinion, that "microbial activity is relatively temperature-insensitive across much of the 2–10°C range of growing-season temperatures typical of most arctic soils" (Nadelhoffer et al., 1992). This statement is supported by numerous facts of weak or negative correlations between temperature and rates of such microbial processes as organic matter decomposition, microbial respiration (CO_2 evolution and O_2 uptake), and N mineralization in arctic soils, where microbial activity was frequently somewhat higher at 2 to 4°C than at 8 to 10°C (Flanagan and Scarborough, 1974; Flanagan and Veum, 1974; Nadelhoffer et al., 1991). To resolve this contradiction I will start from mechanistic considerations, which stem from kinetic analysis of temperature effects on the rates of chemical reactions.

Short-Term Effects of Temperature

Kinetic Analysis of Temperature Effects on Chemical Reactions

In chemical kinetics the dependence of reaction rate on temperature is explained by transition-state theory developed by Eyring as early as 1930 to 1935. It is based on the use of thermodynamics and principles of quantum mechanics. The reaction proceeds through a continuum of energy states and must surpass the state of maximum energy, when transient activated complex is formed. Then the dependence of reaction rate constant, k, on absolute temperature, T, is expressed as follows:

$$\frac{d \ln k}{dT} = \frac{\Delta H^* + RT}{rT^2} \tag{1}$$

where R is the gas constant and ΔH^* is the enthalpy of activated complex formation. The classic Arrhenius equation may be derived from Eq. (1) under a simplified condition $\Delta H^* + RT \approx \Delta H = E_a$ (where E_a is the so-called activation energy). Most often the Arrhenius equation is used in its integrated form

$$\ln k = \ln A - E_a / RT \quad \text{or} \quad k = A \exp(-E_a / RT) \tag{2}$$

where A is the integration constant, interpreted as the frequency of collisions of reacting molecules.

Apart from mechanistic derivations there are a number of empirical expressions relating k and Celsius temperature T_c. The most popular is the exponential formula:

$$\ln k = \ln A + \alpha T_c \quad \text{or} \quad k = A \exp(\alpha T_c) \tag{3}$$

where α is an empiric constant related to widely used temperature coefficient $Q_{10} = \exp(10 * \alpha)$.

The Specificity of Temperature Effects on Biological Processes

All presented mathematical expressions predict exponential or almost exponential increases of chemical reaction rates with temperature. However, biological reactions deviate from this relationship at high temperature because of thermal denaturation of enzymes. To describe this phenomenon we propose the following alternative:

(i) denaturation of active enzyme molecules E to inactive E' is reversible with equilibrium constant $K_T = [E']/[E]$;
(ii) denaturation is irreversible and obeys first-order kinetics with specific rate, which depends on T according to Eqs. (1) to (3).

In the first case we combine Eq. (3) with the van't Hoff relationship for K_T ($-RT \ln K_T = \Delta G° = \Delta H° - T\Delta S°$)

$$v = A\exp(\alpha T_c)/\left[1 + \exp(\Delta S°/R - G°/RT)\right] \qquad (4)$$

where $\Delta G°$, $\Delta S°$, and $\Delta H°$ are the standard Gibbs free energy, enthalpy, and entropy of denaturation reaction, respectively, and v is the observed rate of biological process.

In the second case there are no possibilities for simple mathematical expressions, because temperature effects depends on time t. If denaturation rates for free and substrate-bound enzymes are the same, we have:

$$v(t) = A\exp(\alpha T_c)\exp\left[-A'\left(\exp(-E_a'/RT)\right)t\right] \qquad (5)$$

where A and E_a are the parameters of studied biological process, whereas A' and E_a' refer to denaturation reaction.

To discriminate between the two assumed mechanisms, we have to assess experimental data on T-activity dependence in long-term dynamics. Figure 8.1 demonstrates the results of one such experiment: soil respiration was recorded across a wide temperature range during two cycles of heating and cooling. Under the first assumption [Eq. (4)] both curves should be identical, whereas the second assumption implies that the longer the incubation time, the lower should be microbial activity at given T. Moreover, optimum temperature at prolonged incubation should be shifted towards low T-values as compared with short incubation. In fact, we did not observe mentioned shift-down phenomena: both experimental curves in Figure 8.1 retained the same shape with two local maxima (inflection points) around 15°C. Some decrease of respiration activity did take place during the second cycle, but it could be explained by depletion of oxidizable soil substrates

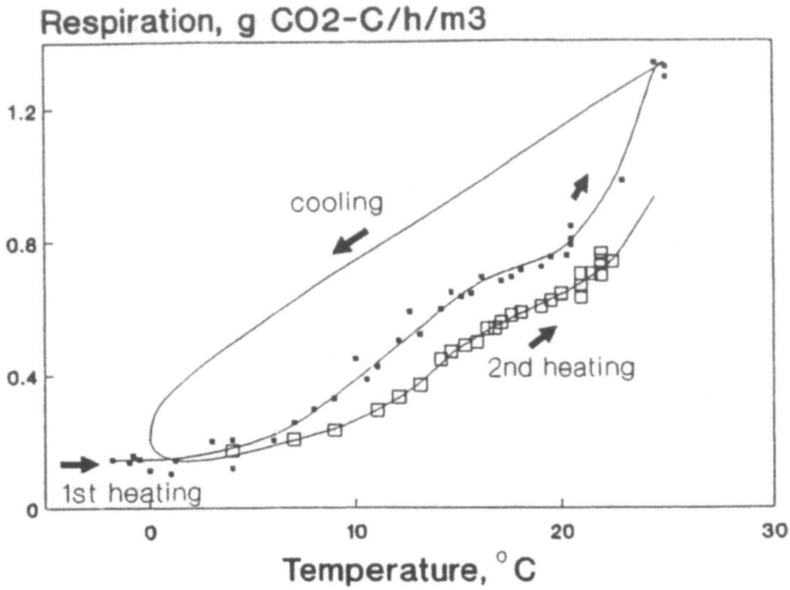

Figure 8.1. The effect of temperature on soil microbial respiration. CO_2 evolution rate was followed in fresh samples of tussock peat bog (Vasyugan wetland, West Siberia) during two consecutive heatings. Temperature gradient is 1°C/h. Automatic recording of temperature and CO_2 with IRGA.

rather than by thermoinactivation. Therefore, Eq. (5) should be discarded as incorrect, whereas Eq. (4) may be regarded as suitable for simulations. In particular, it describes the influence of temperature on soil cellulase activity (Figure 8.2).

The Specificity of Temperature Effects on Microbial Activity in Tundra Soils

However, an account of reversible denaturation (Figure 8.1) is not sufficient for some tundra soils. They often display deviation from simple Arrhenius kinetics not only at high but also at low T: as a temperature decreases from 20 to 0°C microbial activity generally declines somewhat slower than would be predicted by Eq. (4) (Figure 8.3, bottom), or even could be observed "temperature-insensitivity of microbial activity across 2–10°C range" (Nadelhoffer et al., 1992). The most reasonable explanation of this deviation is heterogeneity of soil microbial community in respect to temperature response (i.e., simultaneous action of mesophilic and psychrophilic populations). Psychrophilic and thermophilic microorganisms, capable of functioning at temperature extremes, are widely distributed in nature. There is no distinct boundary between mesophilic microorganisms and either cold-

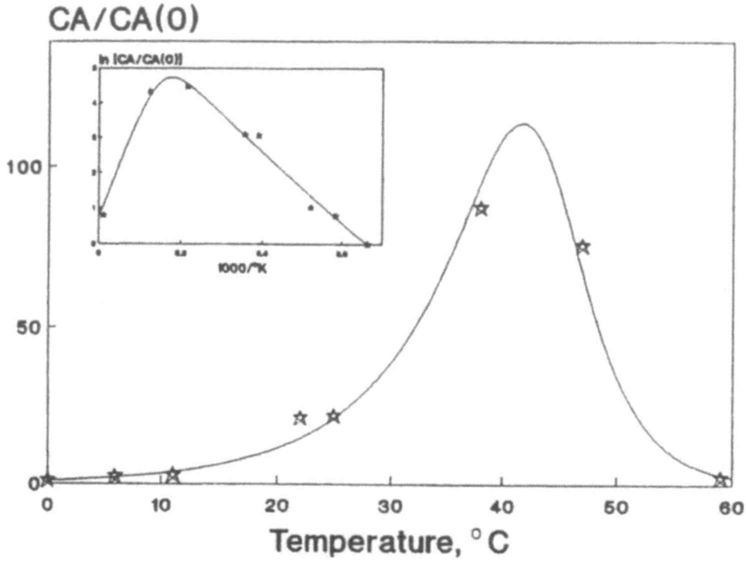

Figure 8.2. Cellulase activity as a function of incubation temperature. Soil samples were taken from the layer 10 to 20 cm of ombrotrophic bog under dwarf pine forest (Tver region). Cellulase activity (CA) is expressed as mg of reducing sugars released/g soil/h. CA(0) is the CA at 0°C (4 mg/g/h).

or warm-adapted microbes. Rather, there is a continuum of temperature characteristics of individual populations. To express this continuum, Eq. (4) should be modified in the following way:

$$\sum_{i=1}^{n} \frac{A_i e^{\alpha_i T_c}}{1+e^{\beta_i(1-\Theta_i/T)}}, \quad i=1, 2, \ldots, n \tag{6}$$

where $\beta_i = \Delta G^\circ_i/R$ and $\Theta_i = \Delta H^\circ_i/\Delta S^\circ_i$. Kinetic properties of ith microbial group are explicitly defined by three parameters: α_i, β_i, and θ_i. Parameter β sets the steepness of activity decline at elevated temperature, whereas θ is numerically equal to that temperature (in K) at which metabolic activity is suppressed by 50% (formally half of enzyme molecules are reversibly denatured). As a rule, θ is close to optimum temperature corresponding to the highest observed activity.

It is surprising that Eq. (6), even at $n = 2$, was able to provide fairly good fit to experimental points (Figure 8.3, see continuous simulation lines on the bottom panel). It may be explained by the fact that distribution of soil microorganisms with respect to parameter θ is bimodal. Therefore, an identification procedure may be used for assessment of the relative contribution of psychrophiles and mesophiles to mineralization of organic matter in soils. In the samples taken from the bog of West Siberian wetland, the

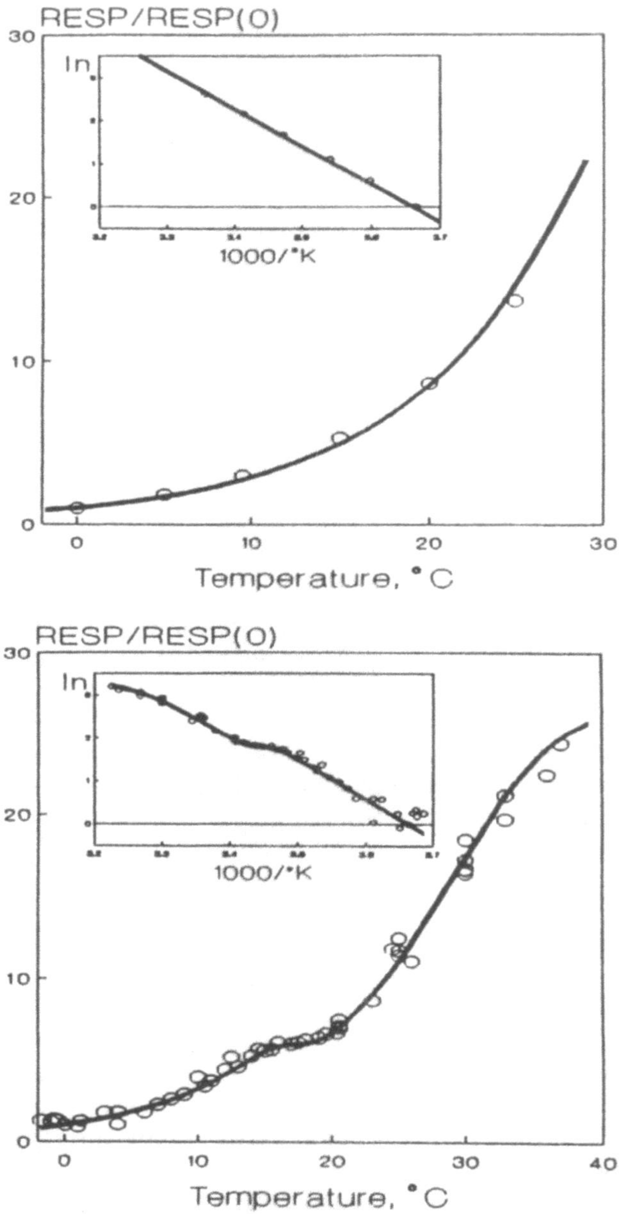

Figure 8.3. The effects of incubation temperature on soil respiration. Top: samples from 0 to 10 cm layer of temperate ombrotrophic bog (Tver region). Bottom: samples from 0 to 10 cm of northern tussock bog (Vasyugan wetland). Continuous lines represent best fit of experimental data to Eq. (6) with $n = 1$ (top panel) and $n = 2$ (bottom panel).

contribution of microorganisms with optimum temperature 13.7°C was as high as 33% (Figure 8.3). In a temperate bog of Tver region the psychrophilic component was too small to be revealed by the described kinetic approach.

Thus, we have found an adequate mathematical expression to characterize short-term responses of soil microbial populations to temperature variation. However, global change predictions should also take into account long-term responses resulting from alterations in the structure of soil community.

Long-Term Effects of Temperature: Simulation of Microbial Community Dynamics Caused by Climate Change

Climate change, fire, soil damage, and other perturbations cause shifts in microbial community structures of soils. Intense perturbations may lead to partial soil sterilization, followed by microbial succession. Understanding of succession mechanisms and prediction of their dynamics has been traditionally one of the most important problems in soil microbiology. This problem was highlighted during the last decade in connection with new IGBP targets (calls for mechanistic understanding of biospheric process, predictions of the consequences of anticipated global changes, etc.). Here I will try to relate the dynamics of soil community with kinetic properties of individual microbial populations.

Growth Kinetics and Microbial Life Strategies

Extensive kinetic studies of the growth of diverse microbial isolates under different laboratory conditions and of indigenous soil populations in situ indicate the existence of several primary types of microbial behavior or life strategy (Panikov, 1991a). Life strategy characterizes the so-called "survival triad": (1) the ability to compete with other populations, (2) to recover after perturbations, and (3) to survive stresses. Respectively one can distinguish three types of natural selection.

(1) *K*-Selection operates in climax or stable ecosystems in favor of efficient competitors with high substrate use efficiency and "economic" metabolism. *K*-Selected species are also referred to as autochthonous microflora (Winogradsky, 1949), oligotrophic microorganisms (Kuznetsov et al., 1979; Poindexter, 1979), and "microflora of dispersion" (Zavarsin, 1970). In tundra soils they are represented mainly by *Arthrobacter*, with similar behavior displayed by other corynebacteria (*Rhodococcus*), and some fungi (*Mortierella ramanniana*) (Parinkina, 1989).

(2) *r*-Selection operates on the pioneer stages of succession after perturbations of climax ecosystems. *r*-Selected species are successful colonizers, adapted to survive in ephemeral, unpredictable habitats. Characteristically, they have high growth rate and react rapidly to environmental changes. But

they have lower substrate use efficiency than *K*-selected species and are characterized by wasteful catabolic machinery. Synonyms applied to this ecological group are: opportunistic populations, zymogenic microflora (Winogradsky, 1949), copiotrophic (Poindexter, 1979), and eutrophic microorganisms (Kuznetsov et al., 1979). In tundra soils *r*-selection favors *Pseudomonas* as a dominant bacterial species, as well as other non-spore-forming bacteria (*Achromobacter*), and rapidly growing fungi (*Mucor, Trichoderma*).

(3) *L*-Selection operates under adverse environmental conditions caused by various stresses (extremely high or low pH and salinity, frost, drought, starvation, etc.). *L*-Selected microorganisms are able to form dormant structures such as endospores of *Bacillus* and *Clostridium*. For unknown reasons these bacteria are not abundant in arctic soils.

Mathematical Model

To simulate the population dynamics of three organisms, *Arthrobacter*, *Pseudomonas*, and *Bacillus*, representing the three growth strategies discussed above, I used the simplified version of so-called "synthetic chemostat model" (SCM). SCM was specially designed (Panikov, 1991b) for description and mechanistic explanation of complicated dynamic behavior of microorganisms: growth, multiplication, survival under starvation, steady-state and transient processes, adaptive responses to alteration of growth conditions, modification of cell composition, etc.

The main feature of the SCM is the use of physiological state variable *r*. It is interpreted in the following way. There are two types of cell constituents: P and U. P-components are absolutely necessary for growth and their relative amount increases with increasing of growth rate (RNA, ribosomal proteins, enzymes of the central metabolic pathways). U-components are responsible for survival of the cells under any kind of growth limitation; their contribution to total cell mass decreases with growth acceleration (reserved substances, enzymes of secondary metabolism, protective pigments, CN-resistant oxidases, transport systems of high affinity, etc.). The contents of P- and U-components vary within the upper (P^{max}, U^{max}) and lower (P^{min}, U^{min}) limits in response to changes of limiting substrate concentration. State variable *r* is defined as a ratio for one particular P-component, the intracellular content of RNA:

$$r = \left(P - P^{min}\right) / \left(P^{max} - P^{min}\right), \quad 0 \le r < 1.0. \qquad (7)$$

As there is an intercorrelation between different P-components, then the single scalar variable *r* can be used to assess the content of any other P-constituent:

$$P_i = P^{min}{}_i + r\left(P^{max}{}_i - P^{min}{}_i\right). \qquad (8)$$

Moreover, it follows from the mass balance of cell composition that there is a reverse relationship between P and U, hence,

$$U_j = U^{\min}{}_j + (1-r)\left(U^{max}{}_j - U^{min}{}_j\right). \tag{9}$$

Equations (7) to (9) significantly simplify the mathematical model. Thus, many features of complicated dynamics behavior of *Pseudomonas* may be realistically approximated by SCM, which contains no more than three state variables: limiting substrate concentration s (common for all competing species), cell biomass x, and physiological state variable r.

To simulate the behavior of *Arthrobacter* (K-selected bacteria), I used additionally the forth variable, the intracellular concentration of a metabolic by-product, h, which is responsible for growth inhibition at extremely high s (characteristic feature of majority of oligotrophic microorganisms).

To describe dynamics of L-strategists exemplified by *Bacillus*, the basic SCM was augmented by the variable x_4, spore biomass. There is exchange between vegetative cells and spores: under high s this flux is directed to vegetative cells, and under low s it is in the opposite direction.

It was assumed that the growth of all selected heterotrophic bacteria is limited by the source of carbon and energy. Maintenance requirements include direct substrate consumption for osmoregulation, turnover of cell components, and wasteful catabolic activity (contributes significantly to cell energetic budget only after sudden transition from famine to oversupply of nutrients).

The final set of ordinary differential equations used for simulation (Table 8.1) includes, apart from s, three other common variables: the biomass of Protozoa grazing on microbes, y, dead organic matter, w, and temperature, T. The microbial predators are the most important sink for bacterial biomass. Grazing was assumed to be suspended if prey biomass declines below some threshold level x^*. Elimination of Protozoa, represented in tundra soils mainly by amoebas, was presumed to be caused by lysing of their cells.

Dead organic matter w was presented by a stable component of plant litter; hence, it was not immediately available for microorganisms. Decomposition of w into s was assumed to follow first-order kinetics; the value of decomposition constant DECOMP and its dependence on temperature was calculated from the data on the Barrow site (Flanagan and Veum, 1974).

Net primary production of the tundra ecosystem was assumed to be in equilibrium with temperature conditions, plant growth being adjusted to new steady-state rate as early as the first year after temperature reset. The phytomass as a separate state variable was omitted. The input of organic carbon to soil was postulated to be performed as (i) soluble and readily available root exudates (40% of net photosynthesis), and (ii) insoluble necromass of plants (60% of net photosynthesis). The first flux was directed

Table 8.1. Mathematical Model Used for Simulation of Population Dynamics of Three Common Bacteria in Tundra Soil

State Variable	Mathematical Expression
Air temperature	$T = 30 * \sin 4[(t-10) * \pi/360)] - A$, $A = 14 \cdots 24°C$
Temperature factor for bacteria	$\text{TBACT} = 0.10736 * \exp(0.166 * T)/(1 + \exp(67.5 * (1 - 290/(t + 273))))$
Temperature factor for plant community	$\text{TPLANT} = (\text{arctg}(1000 * \pi * T) + \pi/2) * \exp(0.03 * T)$
Available substrate	$ds/dt = 0.4 * \text{TPLANT} + \text{DECOMP} * w + a_1 r_1 x_1 + a_2 r_2 x_2 + a_3 r_3 x_3 + a_y y - (q_{s1} + q_{s1}')x_1 - (Q_{s2} + Q_{s2}')x_2 - (Q_{s3} + Q_{s3}')x_3$
Arthrobacter (*K*-selected)	
Biomass	$dx_1/dt = (\mu_1 - \text{GRA1}) * x_1$
Sp. substrate uptake rate	$q_{s1} = r_1 * Q_1 * s/(K_1 + s)/(1 + b * h)$
Waste catabolic activity	$q_{s1}' = 0.025 * Q_1 * s/(K_1 + s)$
Sp. growth rate	$\mu_1 = Y_1 * (q_{s1} - m_1 * r_1) - a_1 * r_1$
Grazing rate	$\text{GRA1} = k_y * y,\ k_y = 0.009$ if $x_1 > x_1 *$ else $k_y = 0$
Physiol. state variable	$dr_1/dt = (\mu_1 + a_1 * r_1) * [s/(K_{r1} + s) - r_1]$
Self-inhibitory by-product	$dh/dt = k_h q_{s1} - \mu_1 * h$
Pseudomonas (*r*-selected)	
Biomass	$dx_2/dt = (\mu_2 - \text{GRA2}) * x_2$
Sp. substrate uptake rate	$q_{s2} = r_2 * Q_2 * s/(K_2 + s)$
Waste catabolic activity	$q_{s2}' = 0.1 * Q_2 * s/(K_2 + s)$
Sp. growth rate	$\mu_2 = Y_2 * (q_{s2} - m_2 * r_2) - a_2 * r_2$
Grazing rate	$\text{GRA2} = k_y * y,\ k_y = 0.009$ if $x_2 > x_2 *$ else $k_y = 0$
Physiol. state variable	$dr_2/dt = (\mu_2 + a_2 * r_2) * [s/(K_{r2} + s) - r_2]$
Bacillus (*L*-selected)	
Biomass of active cells	$dx_3/dt = (\mu_3 - \text{GRA3}) * x_3 TR_1 + TR_2$
Sp. substrate uptake rate	$q_{s3} = r_3 * Q_3 * s/(K_3 + s)$
Waste catabolic activity	$q_{s3}' = 0.1 * Q_3 * s/(K_3 + s)$
Sp. growth rate	$\mu_3 = Y_3 * (q_{s3} - m_3 * r_3) - a_3 * r_3$

Table 8.1. *Continued*

State Variable	Mathematical Expression
Grazing rate	$GRA3 = k_y * y$, $k_y = 0.009$ if $x_3 \geq x_3*$ else $k_y = 0$
Physiol. state variable	$dr_3/dt = (\mu_3 + a_3 * r_3) * [s/(K_{r3} + s) - r_3]$
Bacillus spores	
Flux from active cells to spore compartment	$TR_1 = 100 * x_3/(x_3 + x_4)$ if $s \geq s*$ else $TR_1 = 0$
Flux from spores to active cells	$TR_2 = 100 * x_4/(x_3 + x_4)$ if $s \leq s*$ else $TR_2 = 0$
Cell turnover rate	$a_3 = 0.1$ if $s \leq s*$ else $a_3 = 0$
Biomass of spores	$dx_4/dt = TR_1 + TR_2 - GRA4 * x_4$
Grazing rate	$GRA4 = k_y * y$, $k_y = 0.009$ if $x_4 > x_4*$ else $k_y = 0$
Biomass of *Protozoa*	$dy/dt = Y_y * (GRA_i * x_i) - a_y * y^2$
Soil organic matter	$dw/dt = 0.6 * TPLANT - DECOMP * *TBACT * w$
Soil respiration	$dp/dt = [(1 - Y_i) * q_{si} + q_{si}' + Y_i * m_i * r_i] * x_i + (i - Y_y) * (GRA_i * x_i)$

Dimensions: gC/m² (biomass), TBACT*gC/day/m² (fluxes), and TBACT*day⁻¹ (specific rates).

to the s-pool, whereas the second was directed to the w-reservoir of soil organic matter.

Annual dynamics of temperature was accepted as an average one for the Barrow IBP site in 1930 to 1960 (French, 1974). Temperature response of plants was approximated from the data for the same site (Tiezen, 1974).

The parameters of the model were identified for three bacterial strains: *Arthrobacter globiformis* VKM-658, *Pseudomonas fluorescens* VKM-1472, and *Bacillus subtilis* VKM-428. They were grown in chemostat, feed-batch, dialysis, and simple-batch cultures; details of experimental techniques and results of kinetic studies are published elsewhere (Panikov, 1991a). The reliability of obtained parameters may be questioned by ecologists, as many properties of "domesticated" microbes are not the same as their indigenous counterparts. However, we have no alternatives to such an approach. To minimize the possible "laboratory artifacts," we used exceptionally wide ranges of laboratory conditions (temperature, substrate concentration, composition of nutrient medium, etc.) to comprise all possible in situ situations. As a scaling-up criterion an index of microbial specific growth rate was used. It is generally accepted that natural microbial populations multiply in situ rather slowly, producing no more than 10 generations per season (doubling time varies from 1 to 2 days to 30 to 100 days). In our experiments microorganisms were grown with doubling times varying from 2 to 3h to 3 months (Panikov, 1991a).

Numeric Methods

Curve fitting at the stage of parameter identification was carried out by nonlinear regression with the computer routine POISK-4 (Krut'ko et al., 1988). For a number of relatively simple kinetic equations it was possible to apply a simple linear or general linear model, which allowed obtaining true global minimization of square residuals (Cornish-Bouden, 1976).

Numeric integration of the set of ordinary differential equations was made by a method specially designed for solving of so-called stiff equations (Novikov et al., 1987). All computations were made with IBM PC/AT.

Simulations Results: The Outcome of Competition Between *r*-, *K*-, and *L*-Strategists

As a starting point I have chosen the temperature condition of cold continental climate characteristic for such IBP sites as Barrow, Devon Island, and Taimyr. Parameters of the model were adjusted to obtain stable (steady-state or sustained oscillations) running of all variables. Then the temperature regime was changed by shifting the entire annual temperature curve. For this purpose, parameter *A* (see equation for "Air temperature" in Table 8.1) was changed stepwise from –24 to –14°C. Results of these simulations are shown in Figure 8.4. Under normal cold climate conditions for Barrow and Taimyr bacterial communities were dominated by *Arthrobacter* and *Pseudomonas*, whereas bacilli displayed only sporadic small rises of activity during springtime. These results are highly consistent with experimental data for Canadian and Russian arctic tundra soils (Nelson and Visser, 1978; Parinkina, 1989). The most striking result is the agreement between observed and simulated *Bacillus* population dynamics. Almost all authors reported the abundance in arctic soils of non-spore-forming bacteria, and higher probabilities of detecting bacilli at spring thaw time. It is important that we did not introduce any special temperature requirements for these bacteria. Their competitive exclusion in arctic soils was provided by a single factor—the trophic status of environment. Sensitivity analysis of the mathematical model revealed that only strong eutrophication (net primary production above $1000 g/m^2$!) would lead to competitive advantage of *Bacillus* over other species. In accord with this conclusion are the empirical data that bacilli are abundant in very rich and highly productive soils (alluvial, meadow, chernozemic, and manured soils).

The mechanistic explanation of the simulated *Bacillus* behavior is as follows. These bacteria are characterized by a well-known ability to form

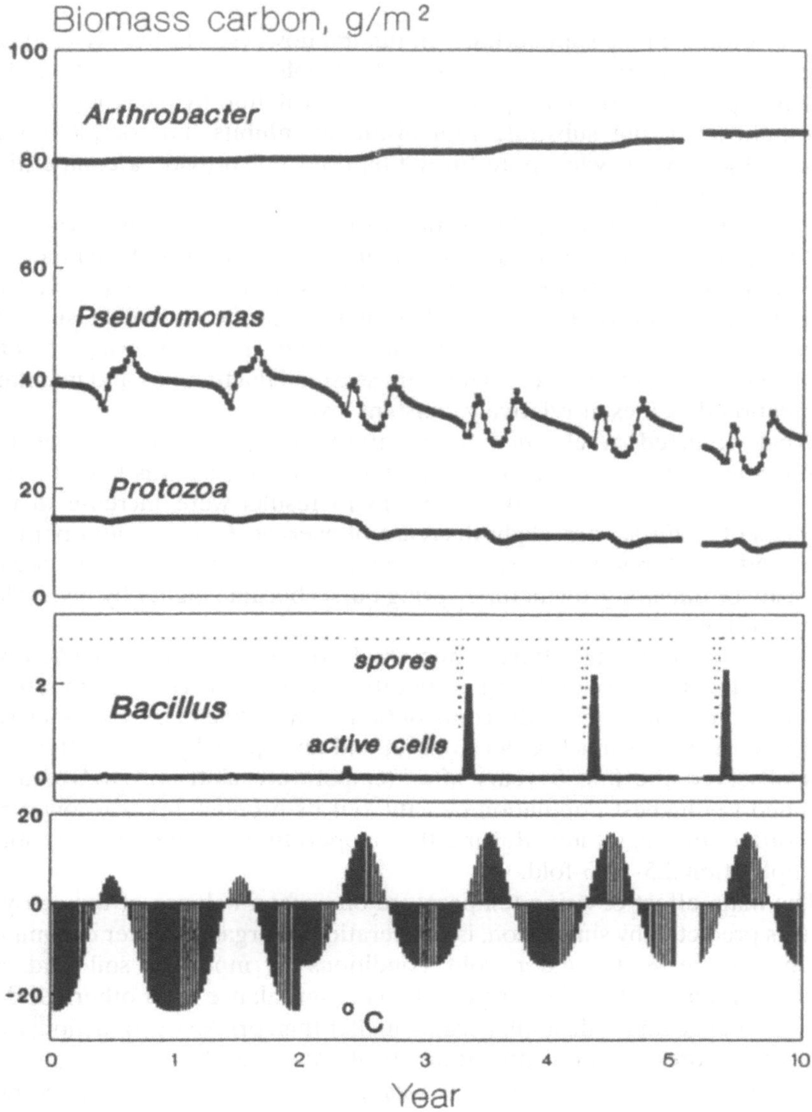

Figure 8.4. Simulated soil biota responses to elevated temperature: dynamics of bacterial biomass. Assumed changes in the climatic and ecosystem parameters: increase of the January temperature from –24 to –14°C, the number of frost-free days from 81 to 135, and index of degree-days from 293 to 1365.

endospores in response to starvation and to recover from dormant state after nutrient addition. So they adapted to transition from feast to famine. These transition events do occur in natural temperate habitats where high microbial activity leads to practically complete removal of easily degradable substrates in the summertime. In tundra, however, the supply of available substrates is slow but continuous. This trophic regime sets a "trap" for bacilli populations: they are provoked for sporulation by decreases in s to threshold level, but substrate replenishment inhibits the completion of sporulation. That is why spore formation does not provide a competitive advantage in tundra.

Pseudomonas and *Arthrobacter* do not form resting dormant structures but they survive starvation due to adaptive r variation and maintaining action of low temperatures. *Pseudomonas* grow faster; these bacteria provide the second maximum of microbial biomass at the end of the summer due to increased supply of organic matter from growing plants. *Arthrobacter* do not show any significant seasonal fluctuations and maintain population densities at relatively constant levels.

What occurred at elevated temperature? It was surprising that the changes in community structure at temperature shift-up was low at ΔA = 5°C and modest at ΔA = 10°C. The main results were increase in the biomass of *Arthrobacter*, slight decrease of average *Pseudomonas* populations under the pressure of predation, and (the most significant) increase of intensity of *Bacillus* growth: their spring bursts became higher by one order of magnitude.

The contrast among different microbial taxa is more obvious when we compare the dynamics of their growth rates (Figure 8.5). The growth rate of *Pseudomonas* was highest; the contribution of these bacteria to total microbial activity was as much as 90 to 98%, and an especially large difference was observed the first 3 years after temperature shift-up. *Arthrobacter* retained the highest population density, but its productivity was lower by two orders of magnitude. Raising the temperature accelerated microbial multiplication 2.5- to 3-fold.

The main effect of rising temperature on carbon balance of the ecosystem, as predicted by simulation, is acceleration of organic matter decomposition (Figure 8.6). Under cold conditions C input to soil and its mineralization (CO_2 evolution) exactly counterbalance each other. Under climate change mineralization became higher than primary production; as a result the stores of plant litter dramatically decreased (15% for the first year). New steady-state value of w was attained after 20 to 30 years (not shown) at the level about 740 g C/m^2 (50% of the initial value).

Competition Between Psychrophiles and Mesophiles

As soon as *Pseudomonas* played a predominant role in organic matter transformation I decided to omit other bacteria and simulate the competition between two subpopulations of *Pseudomonas* that have different temperature–response characteristics. The psychrophilic subpopulation (optimum temperature about 10°C) absolutely dominated in the bacterial community under normal cold conditions (Figure 8.7). Raising the soil

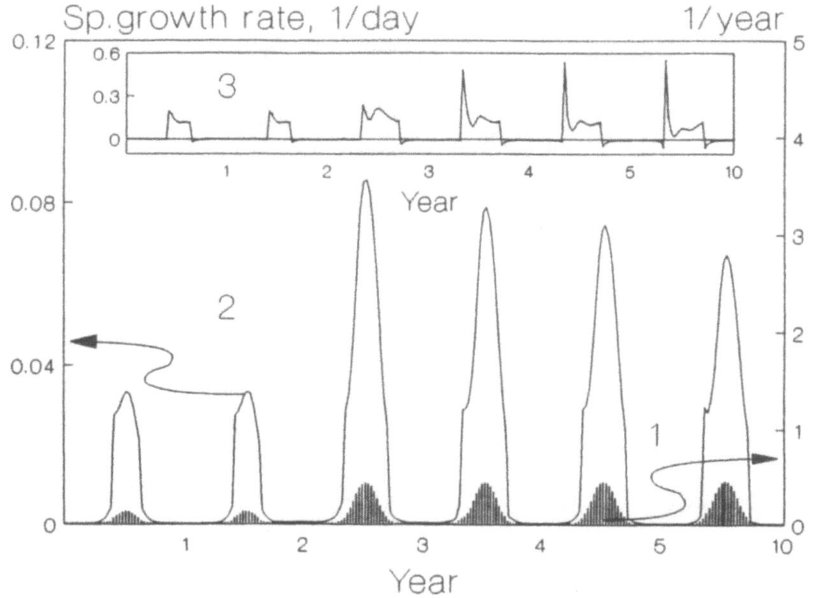

Figure 8.5. Simulated soil biota response to elevated temperature: dynamics of bacterial growth rates. 1: *Arthrobacter*, 2: *Pseudomonas*, 3: *Bacillus*. Changes in the climatic parameters are shown in Figure 8.4.

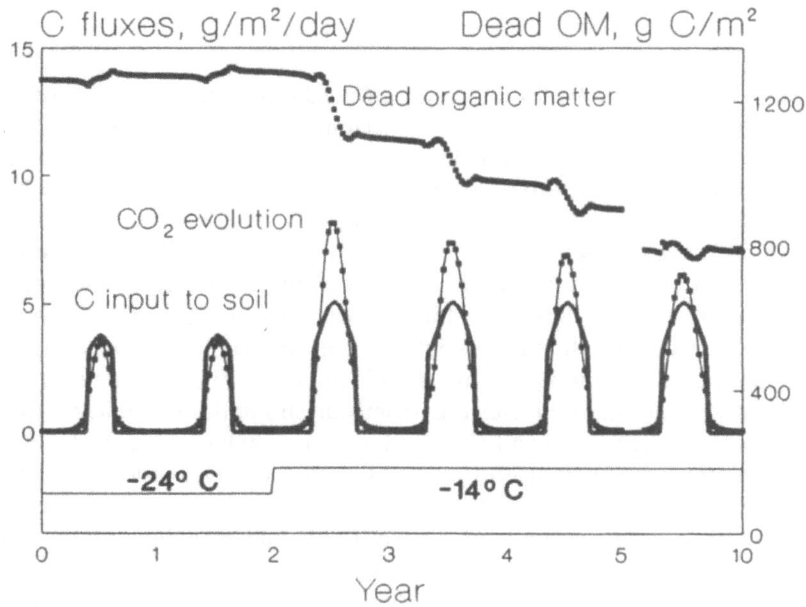

Figure 8.6. Simulated response of tundra ecosystem to elevated temperature: dynamics of the main sources and sinks of organic carbon. Changes in the climatic parameters are shown in Figure 8.4.

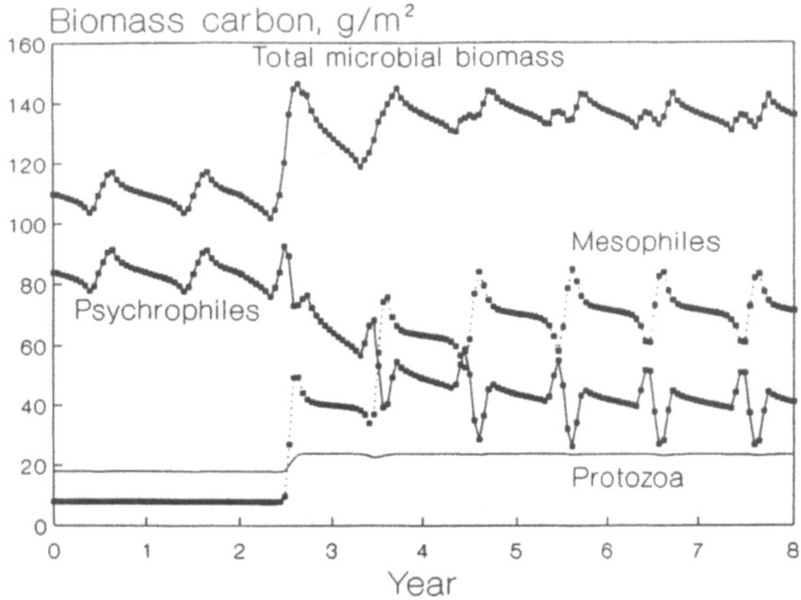

Figure 8.7. Simulation of competition between psychrophilic and mesophilic subpopulations of *Pseudomonas* in tundra soil after temperature shift-up. Assumed changes in the climatic and ecosystem parameters are the same as in the previous simulation (see Figures 8.4 to 8.6).

surface temperature by 5°C (increase of the January temperature from –24 to –19°C, the number of frost-free days from 81 to 108, and index of degree-days from 293 to 746) did not have any suppressive effects on psychrophiles. Further warming (shift-up of January temperature to –14°C, the number of frost-free days to 135, and index of degree-days to 1365) resulted in rather quick 50% substitution of psychrophiles by mesophiles. As compared with initial growth conditions under warmer climate, bacteria revealed two maxima in seasonal dynamics: the first one (in spring) was due to rapid growth of psychrophilic subpopulations, and the second one was caused by an increase of the number of mesophilic moiety in the late summer.

What are the main conclusions from our simulations? First, the agreement between predicted and observed microbial behavior suggests that our mathematical model is useful as a research tool. The model adequately describes seasonal dynamics with two characteristic maxima at spring and late summer, explains the competitive disadvantage of *Bacillus* compared with *Arthrobacter* and *Pseudomonas*, demonstrates the ability of soil Protozoa to control the bacterial populations via grazing, and reveals the role of opportunistic populations in the transformation of organic matter in tundra soils.

Under simulated shift in temperature regime the microbial community revealed a rather high degree of stability. Significant changes in the composition of microbial community took place only after an increase of mean annual temperature by 10°C. As a result of activation of mineralization activity and (to a lesser degree) due to elevated primary production, the pressure of *K*-selection was partially relieved (the growth limitation became less severe). It resulted in an increase in population densities of weaker competitors (*L*-selected species) without displacement of the former dominants. In other words, climate warming caused some stabilization of tundra microbial community.

The simulation results should be interpreted with required care. It is obvious that any modification of the temperature will affect the moisture content in the air and soil, the supply of nutrients other than available organic compounds, and will modify soil structure and soil aeration, etc. To answer all these questions we need to develop more comprehensive simulation models and use them in combination with efficient experimental approaches.

Summary

We have shown that northern soil is characterized by an unusual pattern of temperature–activity response. The deviation from an Arrhenius-type relationship was explained by the heterogeneity of microbial communities and significant contribution to total activity of psychrophilic microorganisms. A relatively simple and robust mathematical model was developed, which was able to simulate seasonal dynamics of dominant microbial populations. Simulations displayed a rather high degree of stability of microbial community towards anticipated warming of the climate. Significant changes in the microbial composition took place only after a rise of mean annual temperature by 10°C. As a result of increased mineralization activity and (to a lesser degree) due to elevated primary production, the pressure of *K*-selection was relieved. It resulted in stabilization of tundra microbial community.

Acknowledgment. The research described in this chapter was made possible in part by Grant No. MIQ 000 from the International Science Foundation, and Russian Fund of Fundamental Research, Grant No. 96-04-49321.

References

Cornish-Bowden, A. 1976. *Principles of Enzyme Kinetics*. London: Butterworths.

Flanagan, P.W., and Scarborough, A.M. 1974. Physiological groups of decomposer fungi on tundra plants remains. In: Holding, A.J., Heal, O.W., MacLean, S.F., Jr., and Flanagan, P.W. (eds.), *Soil Organisms and Decomposition in Tundra* (pp. 159–181). Stockholm: Tundra Biome Steering Committee.

Flanagan, P.W., and Veum, A.K. 1974. Relationship between respiration, weight loss, temperature, and moisture in organic residues on tundra. In: Holding, A.J.,

Heal, O.W., MacLean, S.F., Jr., and Flanagan, P.W. (eds.), *Soil Organisms and Decomposition in Tundra* (pp. 249–277). Stockholm: Tundra Biome Steering Committee.

French, D.D. 1974. Classification of IBP tundra biome sites based on climate and soil properties. In: Holding, A.J., Heal, O.W., MacLean, S.F., Jr., and Flanagan, P.W. (eds.), *Soil Organisms and Decomposition in Tundra* (pp. 3–25). Stockholm: Tundra Biome Steering Committee.

Krut'ko, P.D., Maximov, A.I., and Skwortzov, L.M. 1988. *Algorithms and Computer Programs in Designing of Automatic Systems*. Moscow: Radyo y Svyaz (In Russian).

Kuznetsov, S.I., Dubinina, G.A., and Lapteva, N.A. 1979. Biology of oligotrophic bacteria. *Annu. Rev. Microbiol.* 33:377–387.

Nadelhoffer, K.J., Giblin, A.E., Shaver, G.R., and Laundre, J.L. 1991. Effects of temperature and substrate quality on element mineralization in six arctic soils. *Ecology* 72:242–253.

Nadelhoffer, K.J., Giblin, A.E., Shaver, G.R., and Linkins, A.E. 1992. Microbial processes and plant nutrient availability in arctic soils. In: *Arctic Ecosystems in a Changing Climate* (pp. 281–300). New York: Academic Press.

Nelson, L.M., and Visser, A.S. 1978. Effects of spring thaw on microorganisms in the arctic meadow site. *Arctic Alpine Res.* 10:679–688.

Novikov, V.A., Novikov, E.A., and Jumatova, L.A. 1987. Resolution of Jacobian matrix in Rosenbroke type method of second order of precision. *J. Comput. Math. Phys.* [Zurnal vychislitelnoy matematiki i matematicheskoy physiki] 27(3):385–390 (in Russian).

Oechel, W.C., and Billings, W.D. 1992. Effects of global change on the carbon balance of arctic plants and ecosystems. In: *Arctic Ecosystems in a Changing Climate* (pp. 139–167). New York: Academic Press.

Panikov, N.S. 1991a. *Kinetics of Microbial Growth: General Principles and Ecological Applications*. Moscow: Nauka Publ. (in Russian).

Panikov, N.S. 1991b. Synthetic chemostat model as a tool for description of complicated dynamic behavior of microorganisms. *Microbiologia* 60:343–357 (English translation).

Parinkina, O.M. 1989. *Microflora of Tundra Soils*. Leningrad: Nauka Publ. (in Russian).

Poindexter, J.S. 1979. Oligotrophy: Feast and famine existence. *Adv. Microbial Ecol.* 5:63–89.

Tiezen, L.L. 1974. Photosynthesis in relation to primary production. In: Wiegolaski, F.E., and Rosswall, Th. (eds.), *Tundra Biome* (pp. 52–62). Stockholm: Tundra Biome Steering Committee.

Winogradsky, S.N. 1949. *Microbiologie du Sol*. Paris: Masson.

Zavarzin, G.A. 1970. On the role of dispersion microflora in carbon cycle. *Z. Obtschey Biologii* 31:386–393 (in Russian).

9. Responses of Caribou and Reindeer to Global Warming

Anne Gunn and Terje Skogland

Global warming will initiate a cascade of changes throughout the arctic ecosystems; predicting their magnitude and direction becomes more complex at the higher trophic levels. For top herbivores such as reindeer and caribou *Rangifer tarandus*, we have to consider the effects of climate change on lower trophic levels (forage), as well as the more direct effects on the animals themselves.

In arctic Europe, Asia, and North America, caribou and reindeer number in the millions and are still the staple diet of many indigenous people. Caribou and reindeer are widely distributed in the circumpolar regions and are found in polar deserts, arctic tundra, coniferous forests, and from sea level to mountains. That wide circumpolar distribution and diverse habitats means that predicting responses to global warming limits the value of generalizations.

Caribou and reindeer widely fluctuate in their numbers, even within the scale of the climatic variation of this century. We predict that global warming will add variability to fluctuations in population size and that the effect will vary regionally and according to a population's recent history. Those changes in caribou population dynamics will have far-reaching consequences for arctic ecology, including effects on humans.

We have taken Maxwell's (1992, and Chapter 1) predictions for global warming in the arctic. Regional differences are a certainty, but for the Canadian, arctic mainland winters will be at least a month shorter and warmer with as much as a third more snowfall. The warmer temperatures will likely increase the frequency of freeze–thaw cycles. Summers will see an increase in temperature and precipitation comparable to the increases projected for the winters. A similar pattern of warmer temperatures and more precipitation is forecast for northern Europe (Holten, 1993).

Before we can predict the effects of global warming we need to understand the role of weather in caribou ecology. To this end we briefly review the caribou's annual cycle of physical condition as it relates to fecundity and survival for an individual. Then we describe how weather affects forage growth and availability, and its more direct effects on caribou seasonal migration and insect harassment. Whether those effects are additive, compensatory, or synergistic will determine their cumulative effect on the caribou's physical condition, and thus on fecundity and calf survival. Our understanding of caribou population dynamics is incomplete, but we describe the possible role of weather at the level of the population.

We give our predictions for the global warming first for individual ecology and second for population dynamics. Then we close the chapter with what those changes could mean to the people who depend on caribou and reindeer.

Caribou Ecology and Weather

Caribou Forage

Weather figures largely in the relationship between caribou and their forage. Forage availability in summer through plant growth (absolute availability) is modified in winter by snow and ice (relative availability). Absolute forage availability has two components: the amount of growth (productivity) and its timing (plant phenology).

Absolute forage availability largely determines the physical condition of caribou in summer and calf survival (Elorante and Nieminen, 1986; White et al., 1981). Snowmelt triggers green-up—the burst of plant growth (Woodley and Svoboda, 1994)—and new growth has the highest nutrient value (Chapin et al., 1980; Kuropat and Bryant, 1983). Caribou calve either just before or during the spring greenup, and they depend on foraging more than their body reserves to provide milk for their calves. They selectively feed, switching between plant species to track the flowering and leafing sequence. The timing of the availability of high-quality forage is then a key element in calf growth. Spring weather largely determines the timing of the flush of new growth. But the amount of flowering lags 1 year as it depends more on the previous summer's temperatures. The disadvantage of late springs is obvious, but earlier springs are not advantageous to the lactating cows if peak to forage nutrition precedes the cow's needs.

Relative forage availability depends on the amount of the forage (plant biomass) and snow conditions. Weather during the previous plant growing seasons influences the forage available in winter. For example, lichens tend to grow on drier sites and their productivity is reduced during dry summers. Dry sites often are relatively exposed and have shallow snowcover, where caribou will preferentially feed. Caribou are energetically efficient at cratering through the snow to reach forage. Nevertheless, those costs are some 30% higher than just walking, and costs increase with depth. It is not just the depth of snow but its density and whether crusted or not that exerts a controlling effect over the energetic costs of cratering to reach the forage and to move though the snow.

Insect Harassment

Mosquitos (*Culicidae*), nasal bot flies (*Cephenemyia trompe*), and warble flies (*Hypoderma tarandi*) in summer drive caribou to stop feeding and try to escape the insects either by running from mosquitos or watching for and evading warble and nose bot flies. The combined effects can be severe enough to reduce body weight: for example, in Finland warm summers are correlated with a linear decline in the dressed weight of calves in the fall (Helle and Tarvainen, 1984; White et al., 1981).

Physical Condition, Fecundity, and Survival

A cow's physical condition, pregnancy, and calf survival are tightly related (Cameron et al., 1993; Eloranta and Nieminen, 1986; Skogland, 1984; Thomas, 1982). A cow's physical condition is usually at its annual lowest point at calving and at that time she has to depend more on foraging than her body reserves to provide milk for her calf (Parker et al., 1990). During the summer a cow rebuilds her body reserves of muscle and fat and, particularly for caribou not exposed to insect harassment, those summer weight gains are substantial. The reindeer on Svalbard in the fall can add as much fat as a third of their body weight, and this can be readily visualized from the thickness of their rump fat, which can be 50 mm (Tyler, 1987).

Conception in October depends on the cow reaching a threshold level of condition. Then her physical condition during winter, especially in the last third of pregnancy, largely determines her fetus's birth weight: light calves are less likely to live (Eloranta and Nieminen, 1986). Caribou usually draw on their body reserves during winter when snow reduces forage availability and movements and digging for forage are energetically costly (e.g., Adamczewski et al., 1988). In Norway, for example, calves born after winters marked by shortfalls in forage availability were underweight and fewer survived (Skogland, 1984).

Seasonal distribution is the cumulative response to range conditions, predation, and traditional behavior. For caribou the annual return to their calving grounds is the strongest example of traditional fidelity to a specific site. On the other hand, caribou rotate their use of winter ranges depending on snow and forage conditions (e.g., Russell et al., 1993).

Caribou Population Dynamics

Caribou populations characteristically fluctuate in size over decades. Debate about causes of the fluctuations in numbers of barren-ground caribou have focused on predation or forage as regulating factors (e.g., Bergerud, 1983; Messier et al., 1988; Skogland, 1985, 1991). Previously the prevailing opinion about weather and caribou population dynamics was weather was only a limiting factor for caribou on islands, the northern edge of their tundra ranges, or in maritime areas (Bergerud, 1983). Attention focused on winter weather when snow and ice reduced forage availability to the extent that caribou deaths caused populations to decline. This is almost certainly a simplistic view. Spring (timing of snowmelt) and summer weather modify the timing and amount of plant available as forage. Those effects become more significant toward the edge of the ranges for plant growth, as on the arctic islands.

Support for this point of view comes from Svalbard when caribou (*R. t. platyrhynchus*) had decreased during three winters (Tyler, 1987). More deaths coincided with below-average summer rainfall preceding a cold winter and high caribou densities. Winter ranges on Svalbard are heath and scree communities whose exposures ensure that their snow is shallow. However, the productivity of such dry sites is likely limited by summer moisture, and thus it was unclear as to what extent annual variations in winter forage availability were due to snow and ice conditions compared to the amount of plant growth (Tyler, 1987).

Tyler's (1987) analysis of reindeer dynamics on Svalbard introduces a cautionary note in assuming that limitations in forage availability from winter severity drives the dynamics of the caribou on the Canadian Arctic Islands. Peary caribou in winter also feed on xeric-exposed sites dominated by dwarf cushion plants whose primary productivity is relatively low and moisture limited (Svoboda, 1977). The influences of summer rainfall and caribou density on absolute forage availability, as well as the effect of winter conditions on relative forage availability, have to be figured into determining the effect of severe winters on the population dynamics of Peary caribou.

For continental populations weather is assumed to have either a transitory effect—one severe winter—or to have an effect on population dynamics only when densities are high. For example, in Norway, more reindeer calves died when high densities reduced forage availability during severe winters (Skogland, 1985). The difficulties of retroactively trying to partition the causes of changes in population sizes with weather, hunting, and predation have hampered interpretations of the effect of weather on continental populations (Crête and Payette, 1990).

When we turn to other herbivores we find that population fluctuations are common and weather is often a factor in their population dynamics. Even relatively minor and short-lived weather events may have prolonged and cascading population effects. For example, Albon et al. (1987) showed for red deer (*Cervus elaphus*) that temperature during the last 2 months of gestation affects the entire cohort's birth weight, age at first reproduction, survivorship during adulthood, and the birth weights of the next generation. The result was a fivefold difference in number of surviving young produced by cohorts born after differing temperature regimes.

Fluctuations in the size of caribou populations have been documented in most circumpolar caribou herds whether or not there are predators. This suggests that there is an underlying cause, which we argue is the dynamic relationship between caribou, weather, and forage (Caughley and Gunn, 1993). The pattern of recovering and declining numbers is understandable in terms of changeable weather, its effects on forage supplies, and those effects of forage on the herbivore's rate of increase (Caughley and Gunn, 1993). Arctic weather is annually unpredictable and variable. For example, the coefficient of variation for snowfall at the end of May is 87% and for the length of the growing season is 47% on Banks Island, one of the western Arctic Islands; serial correlations between years are insignificant.

And again we find help in understanding the effects of weather when we look at the dynamics of other herbivores. A similar pattern of unpredictable variability in weather was found to be the key feature in modeling the fluctuations in numbers of kangaroos in Australia. The system components were rainfall and forage with kangaroos as the herbivores. Forage biomass as mediated by the unpredictable and variable rainfall was found to drive the rate of change of the kangaroo populations. But the kangaroos are not completely at the weather's mercy. Their grazing acts through their density as a feedback loop to dampen the fluctuations: the higher the kangaroo density, the greater the grazing reduction in the forage.

A significant finding from the model of kangaroos–plants–rainfall was that the short-term swings of weather and plant biomass explained the longer-term (decades) fluctuations in the numbers of kangaroos. There are marked similarities between the system dynamics of caribou and kangaroos (Caughley and Gunn, 1993), which suggests that weather and its effect on absolute and relative forage availability drives swings in caribou numbers. Predation and hunting may accentuate or dampen those swings.

Climate Change Effects

Caribou Ecology

It is a relatively secure prediction that global warming's increased variability in weather will change forage availability, migration conditions, and insect harassment, which in turn will translate into variations in physical condition and then survival and fecundity. It is less secure as to what those changes will mean at the population level and then how their consequences will ripple through the caribou and reindeer's ecosystem.

Predicting changes for the absolute availability of forage is complex given the numbers of plant species. Warmer temperatures and increased soil fertility (from more rapid nutrient cycling) will increase the plant's net productivity but individual species will differ (Chapin et al., Chapter 11). Increased evapotranspiration and lowering the water table may decrease productivity on drier sites. Those sites are also less likely to see an accentuated nutrient cycle. But those drier sites are also those that usually have shallow snow in winter and are preferred for feeding, and that preference is likely to be accentuated as snowfall is heavier as a consequence of global warming.

Even if the amount of forage available in summer through plant growth increases, the relative amount available in winter as mediated by snow will become energetically more costly to obtain. Snowfall is predicted to increase, and with warmer temperatures, snow will be denser and freeze–thaw cycles could cause ice layering, all of which will impede caribou foraging. Faced with deep snow, caribou initially save energy by reducing their efforts to feed and spend more time lying down (Russell et al., 1993), but such tactics have limitations. An earlier snowmelt would compensate energetically for some reduction in relative availability of forage caused by snow conditions.

A warmer summer would increase the caribou's burden of insect harassment. Mosquito activity correlates with temperature and wind speed, and peaks with winds <6 m/s and temperature 18°C (Russell et al., 1993). A trend toward warmer summers (2 to 4°C) could result in a linear increase in mosquito activity that, based on Russell et al.'s (1993) energetic model, could cause a 7% decrease in foraging (Russell, 1993).

Global warming will change the timing and duration of spring melt. An earlier springtime melt (2 to 3 weeks) may increase annual energetic costs. The predicted heavier snowfall and increased spring cloud cover would add to migration's energetic cost if caribou had to trudge through deep and wet or crusted snow. Costs of moving in snow increase exponentially with sinking depth, and dense or crusted snow drains even more energy (Fancy and White, 1985). A crust on the snow that would almost support the caribou's weight before it collapsed raised the cost of walking by about 570%. Deep snow and crusted snow also favor lighter-bodied wolves, which would give them the advantage in chasing caribou.

In years when deep snow delays spring migration, caribou cows do not always reach the traditional calving grounds. When deep and late snow forces cows of the Porcupine herd to calve in the foothills rather than on the coastal plain, calf survival is reduced as predators are more abundant (Fancy and Whitten, 1991).

The timing of freeze-up can have consequences for caribou during fall migration. A later freeze-up when caribou are below the tree line can cause deaths as the caribou try to travel on across lakes rather than detour through the bush. Caribou will attempt to cross on thin ice and cannot always extract themselves if they break through the ice (Miller and Gunn, 1986). Similar mortality has occurred when caribou are trying to cross between the arctic islands. In treed areas, caribou loaf on the open expanses of frozen lakes as an antipredator strategy—a later freeze-up in the fall will decrease the availability of frozen lakes, which conjecturally may be one of those subtle factors influencing caribou–wolf relationships.

Changes in the timing of break-up and freeze-up of rivers, lakes, and marine waters are another factor influencing movements and thus seasonal distribution. The predicted changes are not so much energetic costs but more the consequences of physical barriers. Migration may be blocked when ice fails to form and caribou are trapped on islands where forage is insufficient, as happened in the 1930s off Russia's coast (Klein and Kuzyakin, 1982).

Caribou Population Dynamics

The net effect of global warming on caribou ecology will be seen as annual variations in survival and fecundity, whose magnitude will be depend on whether the effects of the climate change are subtractive, additive, or compensatory. Predicting the scale of those effects requires translating energetic gains and costs into changes in fecundity and survival. Russell et al. (1993) modeled caribou energetics using data collected on diet, forage composition, and seasonal activity budgets of the Porcupine caribou herd—a large herd of Grant's caribou (*R. t. granti*) that migrates between the arctic coastal plain and forested interior of Alaska and the Yukon. Their model allows estimation of the energetic costs if, for example, snowfall increased by 30 to 50% and insect harassment increased. The model's output is projected changes in body weight (Russell, 1993); fall body weight influences birth rate (Cameron et al., 1993).

The fluctuations in population size introduce another aspect to predicting the effects of global warming. A population's recent history and size—whether it is increasing, stable, or decreasing will mediate weather effects on forage. Food limitation in high-density reindeer populations in Norway reduced the condition of pregnant cows in winter: calves were born later and apparently incurred higher mortality (Skogland, 1985). In those herds with newborn calves dying, calves also died in the summer and winter: lack of forage left those herds unbuffered against the effects of weather (Skogland, 1985).

Global warming will shift system dynamics and increase amplitude of population fluctuations, especially where the margins between survival and mortality are the slimmest, such as those populations near the limits of their distribution on the high arctic islands or in regions where snowfall is already substantial. And those margins may become slimmer for populations that have increased in size and have high densities relative to forage supplies. In the 1990s, many of North America's large continental herds are at a high after a decade of mostly increases in numbers.

Ecological Effects

Caribou directly or indirectly affect their forage. A plant's compensatory growth after being grazed or browsed depends on the plant species and is in turn affected by weather (Noy-Meir, 1981). Herbivores can also boost plant growth when the rapid breakdown of their dung accelerates the nutrient cycle, although for caribou this is probably not a strongly coupled feedback relationship (Jefferies et al., 1992) as caribou usually do not feed intensively in one spot. However, we lack many details about how caribou affect their forage and how plants may respond to interactions between the direct effects of global warming on the plants and on the caribou utilizing the plants.

The caribou's principal predator is the wolf, but grizzly bears (*Ursus arctos*) and, less commonly, wolverine (*Gulo gulo*) hunt caribou. How bears and wolverine hunt caribou and reindeer is less known than that for wolves, but deep snow may favor their hunting success. In Europe, wolverine hunt reindeer in the forests and have the advantage in deep snow (Bevanger, 1992). At least initially, more severe winters for caribou may benefit predators and scavengers—both wolverine and grizzly bears are among the suite of scavengers supported by caribou. However, if increased predation coincided with increased caribou deaths or decreased fecundity and accelerated a population decline, then in the absence of alternate prey, a decline in predators and scavengers would follow.

Parasite ecology is an oft neglected subject and this is true for caribou parasites. We have already predicted that warmer summers will increase the number of days over which insect parasites will be active, but we have no comparable information of how weather affects the free-living life history phases of intestinal or muscular parasites. Predictions for the effects of global warming on caribou and reindeer include poor physical condition, either from increased insect harassment or severe winter weather, and that could in turn change levels of parasitism. There is, however, uncertainty about whether poor physical condition increases vulnerability to parasites or the parasites reduce physical condition. We are not, then, in much of a position to speculate about what global warming may mean directly or indirectly to caribou parasites.

Caribou and People

There are about 4 million caribou and wild reindeer and about 3 million domesticated reindeer in North America, Europe, and Asia (Skjenneberg, 1984; Williams and Heard, 1986). Large as those numbers may seem, they are spread among many populations and our experience has been that even large populations of tens of thousands can collapse in just a few years. Both the wild caribou and their domesticated counterparts figure largely in the culture and economy of many northern people. The replacement costs of caribou meat runs into millions of dollars, but any such estimates ignore the high intangible values. Indigenous people, such as the Inuit and Dene in Canada's Northwest Territories, are largely dependent on hunting caribou and have strong cultural ties with the herds. The same people harvest furbearers such as wolves, wolverine, and foxes, whose populations fall and rise to varying degrees with the caribou populations.

The risk is that some caribou populations will decline following global warming and that unless those declines are detected, they could be accelerated by hunting. Many caribou herds are currently at a high, which has encouraged commercial harvesting (large-scale harvests of hundreds of caribou for meat sales). And meat sales remains the impetus for the domestic reindeer industry. But particularly for caribou or reindeer not under private ownership, there is inherently, in common with most resource harvesting, an economic impetus toward overharvesting (Clark, 1976). This gives force to the argument that harvesting will have to be monitored with finer resolution as levels increase. And there is another suite of longer-term problems that follow from global warming. An ameliorated arctic and subarctic climate could increase the numbers of people and industrial development, which will have the potential to impinge upon caribou and their habitats.

Summary

Our account of effects of climate change on caribou is sprinkled with equivocations. Some uncertainties are the result of the precision of the predictions of the global climate change models. The climatic predictions are those based on a doubling of carbon dioxide. There are other scenarios, such as those that incorporate cooling forced by sulphate aerosols (Taylor and Penner, 1994). In their model, the warming in the Arctic will be on the order of 0 to 1°C rather than 2 to 3°C in the doubled carbon dioxide model.

The uncertainties are also engendered by our slow progress in unravelling climate's role in arctic ecology. Without understanding how weather limits current numbers and distributions, we are hard put to predict the magnitude and direction of global warming's effects. In particular, we lack understanding about causes of changes at the level of population sizes. Our thumbnail sketch of the dynamics between caribou, their forage, and weather suggests a role for weather to drive fluctuations in numbers of caribou. Substantiation of those predictions is possible through recasting the predictions as testable hypotheses rather than relying only on correlative reasoning.

Within decades, if the degree of warming is as predicted for doubling the amount of atmospheric carbon dioxide, we could expect to see greater variability in the rates of births and deaths for the caribou. These are likely to translate into population declines and even loss of some populations of arctic island caribou. For the mainland populations, we can expect changes in the timing of annual migrations and shifts in distribution. It is less certain as to whether populations would decline or increase.

Longer-term responses of caribou populations are less predictable and will be in part tied to changes in plant communities. Chapin et al. (Chapter 11) emphasize difficulties in predicting community changes but on the tundra, deciduous shrubs may increase dominance over other plant groups. Caribou are flexible in their feeding habits and adapt to availability (Klein, 1986; Staaland et al., 1993). But it is too soon to predict what shifts in plant communities, such as northward migration of the tree line, may mean to caribou distribution and population dynamics.

Changes in caribou populations will be echoed in a host of other ecological changes. The ramifications for predators and parasites are that the fate of their populations will probably follow those of the caribou. And this in turn will have consequences for the people who also harvest the caribou and reindeer.

References

Adamczewski, J.Z., Gates, C.C., Soutar, B.M., and Hudson, R.J. 1988. Limiting effects of snow on seasonal habitat use and diets of caribou (*Rangifer tarandus groenlandicus*) on Coats Island, Northwest Territories, Canada. *Can. J. Zool.* 66:1986–1996.

Albon, S.D., Clutton-Brock, T.H., and Guinness, F.E. 1987. Early development and population dynamics in red deer. II. Density-independent effects and cohort variation. *J. Anim. Ecol.* 56:69–81.

Bergerud, A.T. 1983. The natural population control of caribou. In: Bunnell, F.L., Eastman, D.S., and Peek, J.M. (eds.), Symposium on natural regulation of wildlife populations. Forest Wildlife and Range Experimental Station, University of Idaho, Moscow, Idaho, Proceedings 14:14–61.

Bevanger, K. 1992. Report on the Norwegian wolverine (*Gulo gulo L.*). IUCN/SSC Mustelid, Viverrid and Procyonid Specialist Group Newsletter 6:8–10.

Cameron, R.D., Smith, W.A., Fancy, S.G., Gerhart, K.L., and White, R.G. 1993. Calving success in relation to body weight. *Can. J. Zool.* 71:480–486.

Caughley, G., and Gunn, A. 1993. Dynamics of large herbivores in deserts: Kangaroos and caribou. *Oikos* 67:47–55.

Chapin, F.S., III, Johnson, D.A., and Mckendrick, J.D. 1980. Seasonal movement of nutrients in plants of different growth form in an Alaskan tundra ecosystem: Implications for herbivory. *J. Anim. Ecol.* 68:189–209.

Clark, C.W. 1976. *Mathematical Bioeconomics: The Optimum Management of Renewable Resources*. New York: Wiley Interscience.

Crête, M., and Payette, S. 1990. Climate changes and caribou abundance in northern Québec over the last century. *Rangifer* 3:159–165.

Eloranta, E., and Nieminen, M. 1986. Calving of the experimental reindeer herd in Kaamanen during 1970–1985. *Rangifer Special Issue* 1:115–122.

Fancy, S., and White, R.G. 1985. Energy expenditure by caribou while cratering in the snow. *J. Wildl. Manag.* 49:987–993.

Fancy, S., and Whitten, K.R. 1991. Selection of calving sites by the Porcupine herd caribou. *Can. J. Zool.* 69:1736–1743.

Helle, T., and Tarvainen, L. 1984. Effects of insect harassment on weight gain and survival in reindeer calves. *Rangifer* 4(1):28–34.

Holten, J.I. 1993. Potential effects of climate change on distribution of plant species with emphasis on Norway. In: Holten, J.I., Paulsen, G., and Oechel, W.C. (eds.), *Impacts of Climate Change on Natural Ecosystems with Emphasis on Boreal and Arctic/Alpine Areas* (pp. 84–104). Trondheim, Norway: Norwegian Institute for Nature Research (NINA) and the Directorate for Nature Management (DN).

Jefferies, R.L., Svoboda, J., Henry, G., Raillard, M., and Ruess, R. 1992. Tundra grazing systems and climatic change. In: Chapin, F.S., Jefferies, R.L., Reynolds, J.F., Shaver, G.R., Svoboda, J., and Chu, E.W. (eds.), *Arctic Ecosystems in a Changing Climate an Ecophysiological Perspective* (pp. 391–412). New York: Academic Press Inc.

Klein, D.R. 1986. Latitudinal variation in foraging strategies. In: Gudmundsson, O. (ed.), *Grazing Research at Northern Latitudes* (pp. 237–246). New York: Plenum Publishing Corporation.

Klein, D.R., and Kuzyakin, V. 1982. Distribution and status of wild reindeer in the Soviet Union. *J. Wildl. Manag.* 46:728–733.

Kuropat, P., and Bryant, J.P. 1983. Digestability of caribou summer forage in arctic Alaska in relation to nutrient, fiber, and phenolic constituents. *Acta Zool. Fenn.* 175:51–52.

Maxwell, B. 1992. Arctic climate: Potential for change under global warming. In: Chapin, F.S., Jefferies, R.L., Reynolds, J.F., Shaver, G.R., Svoboda, J., and Chu, E.W. (eds.), *Arctic Ecosystems in a Changing Climate an Ecophysiological Perspective* (pp. 11–34). New York: Academic Press Inc.

Messier, F., Huot, J., le Henaff, D., and Luttich, S. 1988. Demography of the George River Herd: Evidence of population regulation by forage exploitation and range expansion. *Arctic* 41:279–287.

Miller, F.L., and Gunn, A. 1986. Observations of barren-ground caribou travelling on thin ice during fall migration. *Arctic* 39:85–88.

Noy-Meir, I. 1981. Responses of vegetation to the abundance of mammalian herbivores. In: Holt, S. (ed.), *Problems in Management of Locally Abundant Wild Mammals.* New York: Academic Press.

Parker, K., White, R.G., Gillingham, M.P., and Holleman, D.F. 1990. Comparison of energy metabolism in relation to daily activity and milk consumption by caribou and muskox neonates. *Can. J. Zool.* 68:104–114.

Russell, D.R. 1993. Effects of global warming on the biology and management of the Porcupine caribou herd. In: Wall, G. (ed.), Impacts of climate change on resource management in the North (pp. 91–99). University of Waterloo, Department of Geography Publication Series, Occasional Paper No. 16.

Russell, D.R., Martell, A.M., and Nixon, W. 1993. Range ecology of the Porcupine Caribou Herd in Canada. *Rangifer Special Issue* 8.

Sjenneberg, S. 1984. Reindeer. In: Mason, I.L. (ed.), *Evolution of Domesticated Animals* (pp. 128–138). London: Longman.

Skogland, T. 1984. Wild reindeer foraging-niche organization. *Holarctic Ecol.* 7:345–379.

Skogland, T. 1985. The effects of density-dependent resource limitation on the demography of wild reindeer. *J. Anim. Ecol.* 54:359–374.

Skogland, T. 1991. What are the effects of predators on large ungulate populations? *Oikos* 61:401–411.

Staaland, H., Scheie, J.O., Grøndahl, F.A., Persen, E., Leifseth, A.B., and Holand, Ø. 1993. The introduction of reindeer to Brøggerhalvøya, Svalbard: Grazing preference and effect on vegetation. *Rangifer* 13:15–20.

Svoboda, J. 1977. Ecology and primary production of raised beach communities. Truelove Lowland. In: Bliss, L.C. (ed.), *Truelove Lowland, Devon Island, Canada—a High Arctic Ecosystem* (pp. 185–216). Edmonton, Alberta: University of Alberta Press.

Taylor, K.E., and Penner, J.E. 1994. Response of the climate system to atmospheric aerosols and greenhouse gases. *Nature* 369(6483):734–737.

Thomas, D.C. 1982. The relationship between fertility and fat reserves of Peary caribou. *Can. J. Zool.* 60:597–602.

Tyler, N.J.C. 1987. Body composition and energy balance of pregnant and non-pregnant Svalbard reindeer during winter. *Symp. Zool. Soc. Lond.* 57:203–229.

White, R.G., Bunnell, F.L., Gaare, E., Skogland, T., and Hubert, B. 1981. Ungulates on arctic ranges. In: Bliss, L.C., Heal, W., and Moore, J.J. (eds.), *Tundra Ecosystems; a Comparative Analysis* (pp. 397–483). Cambridge: Cambridge University Press.

Williams, T.M., and Heard, D.C. 1986. World status of wild *Rangifer tarandus* populations. *Rangifer Special Issue* 1:19–28.

Woodley, E.J., and Svoboda, J. 1994. Effects of habitat on variations of phenology and nutrient concentration among four common plant species of Alexandra Fiord lowland. In: Svoboda, J., and Freedman, B. (eds.), *Ecology of a polar desert. Alexandra Fiord, Ellesmere Island, Canada* (pp. 157–176). Toronto, Canada: Captus University Publications.

10. Impacts of Global Change on Arctic-Breeding Bird Populations and Migration

Hugh Boyd and Jesper Madsen

Few birds breeding in the Arctic remain there for more than 3 months in the year. This complicates any assessment of how they might be affected by climatic or other environmental changes in the Arctic, because they are also subject to what happens elsewhere. This chapter is restricted to the likely effects of possible changes to bird populations either in the Arctic or in subarctic and north-temperate sites where Arctic-breeding migrants stop on migration.

All published 2 × CO_2 scenarios suggest that in most parts of the Arctic there are likely to be substantial increases in temperature and, less certainly, precipitation in winter, with relatively small changes at other seasons. Were the suggested global changes to occur, their greatest effects on Arctic-nesting birds would probably be experienced in their wintering, rather than their breeding, areas.

Before anything useful can be said about how Arctic birds might respond to future global change, it is necessary to know how they have been affected by, and dealt with, climatic variations in the recent past. In many parts of the Arctic, notably in northern Canada, the network of meteorological stations was very sparse until about 40 years ago. Quantitative records of the numbers and breeding success of Arctic birds are scarce, and are available for even shorter periods. In many cases the only relevant observations have been made in staging or wintering areas, rather than in the Arctic. Consequently, the runs of data that can be used to relate population changes to climatic changes are few, brief, and mostly imprecise.

What we can say about the responses of birds to climate is also determined by choices of scale, both geographically and in time. Birds depend on vegetation, and plants on soils. The responses of soils and vegetation to climatic variation show many interactions and lags. Jefferies et al. (1995) provided examples of the complex interactions between geese and their food-plants in the wet meadows along the west coast of Hudson Bay. Many more studies at those micro- and mesolevels are needed. Here, we concentrate on direct large-scale influences of climate and weather on birds, affecting their distribution, nesting effort, and success over large parts of their breeding ranges.

Weather conditions in spring staging areas are also very important because nesting effort is partly determined by the birds' physiological state on arrival in the Arctic (e.g., Ebbinge, 1989). That depends not only on the stresses of the last migratory flight, but on their condition before departure. Recent studies have been concerned with the physiological limits to the performance of migrating individuals and with the thermoregulatory demands on their remaining fat and protein reserves that are imposed by the near-ground weather they have to live in before and while nesting (Piersma, 1994).

We consider three general questions: (1) how far are the present breeding distributions of birds in the Arctic determined by the present climates? (2) how might the distribution, abundance and breeding success of Arctic-nesting birds be changed should rapid warming and changes in seasonal precipitation occur? and (3) how might global warming and changes in weather conditions affect migratory flights?

We have to be concerned with events at regional and local scales; and our bird data are drawn from the recent past. For both those reasons, it seems appropriate to use scenarios of future change derived from analogues from the instrumental records of the 20th century, rather than from global general circulation models (GCMs). We are concerned with what may happen in the next few decades, when birds are likely to encounter environmental changes not greatly different from those they have met during the last half-century. The climate of the 22nd century may be very different, but the birds have to get there from here, so that a modest approach seems likely to be more instructive than a dramatic one.

The most useful analogue scenarios known to us are those of Palutikof et al. (1984). We have used only one of their four sets of scenarios—Scenario C—based on comparisons between the warmest and coldest sets of 20 consecutive years' data from the average annual Northern Hemisphere temperatures for the period 1901 to 1980 inclusive. The warmest 20-year period was 1934 to 1953, the coldest 1901 to 1920. Scenario C has the advantage, not shared by others based on shorter periods or on groups of the warmest or coldest individual years, of enabling the interannual variability of temperature and precipitation to be examined at the regional level. That kind of variability is of crucial importance to birds, and hence to this study. One limitation of the study by Palutikof et al. (1984) is that, though it dealt with the whole of North America, in the Old World it was limited to Europe, west of 50°E. Thus, we are unable to use it to deal with possible changes in central and eastern Siberia.

Effects on Distribution of Species Breeding in the Arctic

About 115 species of birds breed in the Arctic. In Table 10.1 they are assigned to climatic regions, on the basis of recently published maps of their breeding distributions. Though auks (*Alcidae*) are the most numerous Arctic birds we do not discuss them here, because their responses to global change will be driven by changes in the marine environment and because it would be impracticable to deal with both terrestrial and marine changes in the space available. The families of terrestrial birds best represented, by both numbers of species and of individuals, are those associated with wetlands: the waders (shorebirds) and waterfowl (ducks, geese, and swans). Though the Order Passeriformes (perching birds) includes by far the most species worldwide, there are few in the Arctic.

During the 20th century there have been no known extinctions of species breeding in the Arctic. There have been some extensions of range north

Table 10.1. Distribution of Arctic-Breeding Birds Across the Major Climatic Regions of the Palearctic and Nearctic

Family	Arctic Only	A + S	A + S + B	A + S + B + T	Total
Gaviidae—loons		2	2	1	5
Procellaridae—fulmar				1	1
Anatidae—geese, swans	3	7	3		13
—ducks	3	1	3	10	17
Accipiteridae—hawks		1			1
Falconidae—falcons			1	2	3
Tetraonidae—grouse				3	3
Charadriidae—plovers	1		3	2	6
Scolopacidae—sandpipers	8	19	4	5	36
Laridae—skuas, gulls	6		4		10
Alcidae—auks, puffins	1	1	3		5
Strigidae—owls	1				1
Alaudidae—larks		1			1
Corvidae—crows				1	1
Turdidae—thrushes			2	2	4
Motacillidae—pipits		1	2	1	4
Emberizidae—finches	2	1	1		4
Total	25	34	28	28	115

Arctic (A), Subarctic (S), Boreal (B), and Temperate (T). Species extending into Alpine areas at lower latitudes are listed only under Arctic. Numbers of species are listed by Family and by the regions used in addition to the Arctic.

from, and south to, the Subarctic, but few north of 70°N. This relative stability may well be due to the comparative lack of habitat destruction by human activities at high latitudes, unlike the massive and still-increasing havoc at lower latitudes, matched in the Arctic only in parts of Siberia (Dewdney, 1982; Rogacheva, 1992).

Several biologists have looked at why Arctic-nesting birds are found where they are, for different reasons and in different ways. Johansen (1956) reviewed their circumpolar distribution on a broad scale, paying particular attention to the influence of climate change. Johansen noted that, "We witness to-day how the arctic birds spread farther into the polar region, and that even many typical boreal species penetrate into the Arctic." That comment, from nearly 40 years ago, makes three important points about changes in the distribution of birds in the Arctic: (1) they are continual; (2) they include long-term, as well as short-term, responses to climatic variation; and (3) some species have been moving north for much of the 20th century. Thus, it will never be easy to show that changes in the distribution and abundance of Arctic-breeding birds in the 21st century are due to climate changes resulting from current levels of human production of "greenhouse gases," rather than being lagged responses to earlier events, or due to causes that are not related to climate (e.g., human persecution).

Tables 10.2 and 10.3, which deal respectively with waders breeding in northern Canada and in central Siberia, show that most Arctic nesters also breed in the Subarctic, with many extending into the Boreal and Temperate Regions. This breadth of range implies that the linkage of most species to summer conditions in the Arctic is not close. The published scenarios for a $2 \times CO_2$ world, based on GCMs, all show the Subarctic extending much farther north than at present, greatly reducing the area defined as Arctic. Tables 10.1 to 10.3 suggest that further summer warming would pose few threats to the continued presence of most species now breeding in the Arctic, though there would probably be many changes in their relative abundance. For example, Meltofte (1985) found that in northeast Greenland, highest nesting densities of waders were found in the middle region; to the south, high winter precipitation and late snowmelt prevents birds from nesting, and to the north, low precipitation gives arctic desert-like conditions with scarce food supplies. In case of global warming, the optimal zone is likely to move northwards, with a consequent compression of the available breeding range.

If primarily subarctic species increase and move northwards, this may lead to increased interspecific competition in the Low and Mid-Arctic. It seems unlikely that High Arctic species would be affected by such competition, though, as is the suggestion for Northeast Greenland nesting waders, their breeding ranges may be compressed.

The advantages of wide north–south spread of breeding ranges in offsetting climatic risks are complemented for most High Arctic species by wide longitudinal spread. For example, the range of the Knot *Calidris canutus*

Table 10.2. Numbers of Species of Waders (*Charadriidae* and *Scolopacidae*) Breeding in Arctic Canada and in Additional Wetland Regions

Breeding in	Arctic	(and)	Subarctic	(and)	Boreal	Total
Arctic	10	+	5	+	5	20
Subarctic			2	+	7	9
Boreal					4	4
Regional total	20		19		16	33

islandica, a medium-sized wader breeding wholly north of 70°N, extends from 20°W in northeast Greenland to 120°W on Prince Patrick Island, Northwest Territories, Canada. As it is rare for snowmelt to be delayed across the entire breeding range, total nesting failure is rendered improbable, and seems not to have occurred in the last 20 years (Boyd, 1992).

Rogacheva (1992) has shown that Central Siberia (Table 10.3) is occupied by a mixture of species breeding principally in western Siberia with others centered much further east. The zoogeographic boundary runs to the east of the Yenisey River. The higher, drier, and rocky northern Taimyr Peninsula forms an ecological barrier that seems to have prevented many species breeding in the boggy lowland tundras of West Siberia from spreading eastwards. Even if that region becomes wetter, its elevation and lack of fertile soils will presumably continue to deter immigration from the west. This is important, because another reason for interest in Central Siberia is that it has been identified both as an area where future summer warming is likely to be great (Palutikof et al., 1984) and where warming has already been occurring (Chapman and Walsh, 1993).

Another means of reducing the risk of being prevented from nesting operates at the local level. Though some form of philopatry by birds is usual, in the High Arctic it seems to be area rather than site specific. The

Table 10.3. Distribution of Arctic-Nesting Waders (Shorebirds) Across Climatic Regions of Central Siberia, as Described by Rogacheva (1992)

	Arctic Tundra		Typical Tundra		Shrub Tundra		Forest Tundra		Taiga	Total
Arctic Tundra	3	+	4	+	1	+	2	+	1	11
Typical Tundra			0	+	1	+	2	+	2	5
Shrub Tundra					0	+	0	+	3	3
Forest Tundra							0	+	2	2
Taiga									5	5
Total	3		4		2		4		13	26

south plain of Bylot Island, Northwest Territories, Canada (73°N, 80°W) is the home of nearly one-fifth of the population of Greater Snow Geese *Anser caerulescens atlantica*. In a river valley at the northwestern end of the plain, the largest concentrations of nests have been in different areas each year from 1987 to 1993. In 1992, when snowmelt was late over most of the Canadian Arctic, the geese were unable to find cleared nest sites in that valley before mid-June. Most of them abandoned any attempt to nest. Some moved about 35 km to the southeast, where they nested successfully. Having done so, many of the parents led their broods back to the valley in which they themselves had been reared and with which they were familiar, as shown by sightings and recaptures of marked birds (G. Gauthier and A. Reed, pers. comm.). Such a combination of flexibility and tenacity is highly effective in exploiting food resources. In many cases, such as small islands, nest site opportunities are restricted, so that there is little scope for flexibility.

Masses of the white seed-heads of *Eriophorum* spp. were visible on many parts of the south plain in early August 1992. In other years hardly any were to be seen, thanks to heavier grazing by geese. A "bad year" for geese gave plants the chance to grow and set seed. Boyd and Maltby (1979) suggested that the Brent Geese *Branta bernicla* breeding in the Queen Elizabeth Islands, Northwest Territories, Canada, north of 75°N, are obliged to adopt a nomadic strategy because in cool summers their food plants can make little or no growth, so that they cannot provide grazing year after year without interruption. By shifting their nesting and brood-rearing places frequently, the geese can continue to occupy sparsely vegetated areas, while giving the plants recovery time. For those interested in wildlife habitat conservation, this nomadic use requires the recognition that the total feeding area these geese require for secure existence is much larger than the area they happen to occupy in any one year. If summer warming were to encourage regular annual plant growth, the geese might become more sedentary, perhaps leading to overgrazing of the kind reported from La Perouse Bay, Manitoba, Canada (Jefferies et al., 1995) and other breeding places at lower latitudes. (Analogies with the conversion of the Inuit people from nomads to hamlet-dwellers suggest themselves.)

The summer warming that has occurred in north-central Siberia (Chapman and Walsh, 1993) seems to have rendered the Dark-bellied Brent Geese *Branta b. bernicla* breeding on the coast of the Taimyr Peninsula much less prone to breeding failures due to late snowmelt. Failures were frequent in the 1950s and 1960s (Boyd, 1987), when the population fell to 16,000 to 25,000 birds. By the mid-1980s the population had increased to about 250,000 (Madsen, 1987), principally as a result of cessation of hunting in its wintering areas in northwest Europe and the growing readiness of the geese to feed on improved grasslands and winter wheat, instead of being restricted to *Zostera* spp. and other plants of the intertidal zone, their traditional foods. Climatic change on the Taimyr may have led to increased frequency of years when the snowcover opens relatively early, but this has not simply increased nesting success. The nests of Brent Geese and waders continue to be at risk from Arctic foxes *Alopex lagopus* and other predators that are primarily dependent on lemmings (*Dicrostonyx* spp. and *Lemmus* spp.), but that switch to taking birds' eggs when lemming numbers crash after peaking in a cyclical way (Summers and Underhill, 1987; Underhill et al., 1993).

The most important lesson to be learned from that example, and one with general implications for the study of faunal responses to climatic change, is that explanations of population regulation that seem satisfactory at one period and at one level of numbers are unlikely to remain valid at other times and population levels.

An extreme example of unexpected behavior by a population is the recent successful establishment of a nesting population of the Barnacle Goose *Branta leucopsis* on islets in the Baltic. The colonies were primarily founded by North Russian High Arctic breeding birds, which used the islands on spring passage (Larsson et al., 1988). Not only did the new colonizers advance nesting by more than 1 month, they also adapted to conditions where hatching occurs after the quality of vegetation has started to deteriorate (Larsson and Forslund, 1991), in strong contrast to the High Arctic, where hatching generally takes place at the time of spring flush of vegetation.

Until recently, much more attention has been paid to the effects of "bad" conditions (late snowmelt, cold Junes, wet Julys) than to survival and success in "good" years (with early snowmelt, warm and dry summers). Boyd (in prep.) found that the breeding success of the five species of geese breeding in Arctic Canada was more highly correlated with the mean daily temperature in June than with measures of snow depth and rate of snowmelt, or with the date on which the daily minimum temperature first exceeded 0°C. Presumably, the monthly mean serves as an integrator of measures of factors affecting the ability of geese to nest at all, or to lay large clutches: these include the resumption, and rate, of growth by food plants. Fewer young geese were reared in years when June temperatures were below, and more when they were above, the local means. The temperature in early July, when hatching occurs, seemed unimportant, though cold, wet weather at, and shortly after, hatching can kill goslings, at a stage when their thermoregulation is poor.

In the context of possible summer warming, the response of breeding birds to unusually warm Junes is of special interest. Over the range of values observed since the 1950s, Boyd (in prep.) found linear relationships between June temperatures and the numbers of geese reared, although apparent output was also affected by the numbers of yearlings and other prebreeders present in each population. A crucial question for the future breeding success of birds in the High and Mid-Arctic zones is whether Junes will become milder, or remain cold enough to reduce or prevent nesting.

Experience in the latest decade, in which the period mean summer temperature has increased in several parts of the Arctic, suggests that cold Junes may not become less frequent with further summer warming. Scenario C of Palutikof et al. (1984) suggested that the amplitude of year-to-year changes in summer temperatures is likely to increase in Alaska, northwestern Canada, and the White Sea region, to be reduced in the eastern Canadian Arctic and west Greenland, and to change very little in Iceland or northwest Europe. A possible implication of increasing amplitude is that there will be higher variation in breeding output.

Effects on Timing of Breeding and Predation Pressure

Longer and warmer summers at high latitudes will only benefit birds if the timing of their nesting remains as closely linked as it now is to the timing of plant growth and the emergence of insects used as food during their entire stay in the Arctic. Phenological adjustments may well not occur in optimum ways for the birds, because the processes driving plant and insect development will not necessarily follow the same timetables as those determining when birds will nest. Predicted Global Climate Change and Migratory Bird Management, a mimeographed note prepared by the U.S. Fish & Wildlife Service, based on an unpublished report by The Climate Institute to the Environmental Protection Agency in May 1988, refers to "potential calamitous loss among Arctic nesting shorebirds, where hatching is acutely timed to the availability of flightless insects on which the newly hatched shorebirds feed; but an increase in Arctic nesting geese, whose young will have warmer temperatures, richer vegetation as forage, and reduced predation." That assessment seems to assume no problems with changing phenology of plant growth, while denying the likelihood that insects will adjust their times of emergence in response to a longer summer. Nettleship (1974), who found that at Lake Hazen, Northwest Territories, Canada (81.50°N, 70.25°W), the hatching of waders was timed to coincide with the peak emergence of insects, drew attention to the risk for birds of a loss of coincidence between the time of their greatest need for food and of its greatest supply.

However, he noted that other species of insect were present, which hatched either earlier or later than the peak emergence. This seems to make it possible for total insect biomass to increase in a longer, warmer summer, by changes in the relative abundance of species, even if the timing of emergence of the species now dominant were to prove inflexible. This is the case in Eqalungmiut Nunaat, Low Arctic West Greenland (67°N, 50°W), where a prolonged insect season permits larger variation in nesting phenology of Lapland Bunting *Calcarius lapponicus* (Fox et al., 1987).

The short arctic summer constrains the breeding season of geese, the time being compressed between spring thaw and die-back of the vegetation.

Arctic-nesting geese generally synchronize egg-laying with snowmelt (MacInnes et al., 1974; Prop and de Vries, 1993; Raveling, 1978). Prop and de Vries (1993) argue that the egg-laying date is a trade-off between (1) an advantage of early settling to enhance survival rate of goslings, hatching being timed to the spring growth of vegetation, the quality of which quickly deteriorates (Sedinger and Raveling, 1986), and (2) an advantage of postponing settling to increase hatching success of the eggs (settling later the incubating geese can more easily find food on the tundra, exposing the nest for a shorter time to avian predators).

Longer and warmer breeding seasons will probably result in earlier snowmelt and growth of food plants. To synchronize with the primary growth and food quality, the geese will probably have to advance egg-laying to the time of snowmelt, in which case mammalian predators will still have easy access to the nests, thus opposite to the expectations of the U.S. Fish & Wildlife Service. However, if primary productivity in the High Arctic were to increase the incubating females should be able to minimize feeding bouts away from the nests, and hence additional exposure of the nest to predation. Furthermore, if longer and warmer summers lead to more plant biomass in late summer, this could be beneficial to the geese during the premigratory period, when they have to put on weight. This period is regarded as one of the "bottlenecks" limiting the Svalbard nesting Barnacle Goose population (Owen and Black, 1989) and possibly also other arctic-nesting goose populations. Thus, global warming may offset some of the existing limitations to population growth of Arctic-nesting geese.

The Arctic fox is a major predator of eggs and young of birds in arctic regions. In many goose species, fox predation limits the available nesting space, and in years with high densities of foxes, they can cause a complete breeding failure, not only of colonial birds but also dispersed species (Underhill et al., 1993). To assess the effect of global warming on birds, it is thus also very important to evaluate the likely effects on the fox populations and their fluctuations. As far as we know, no such analysis has been carried out.

How often heavy predation occurs in the same year over large parts of the breeding ranges of birds is not clear. Arctic foxes are known to be able to make long-distance movements (Shilyaeva, 1967). In 2 out of 7 years of study in Tusenöyane, eastern Svalbard (77°N, 22°E) Arctic foxes seriously hampered breeding attempts of Light-bellied Brent Geese *Branta bernicla hrota*. The fox years coincided with fox peaks in Taimyr. This suggests that, because the foxes can cross the sea-ice during winter, fox abundance in Taimyr, driven by lemming cycles, may affect the breeding success of geese nesting on islets more than 1000 km away and in lemming-free conditions (Madsen et al., 1992, and unpubl.).

Boyd (in prep.), using exports of fox pelts as a proxy for the abundance of Arctic foxes in different regions of Arctic Canada, found that most foxes were trapped in the winters following summers when geese had bred well, both in the northwest, where lemmings seem not to be cyclic, and in the east, where they show a 6-year cycle (C.J. Krebs, in litt.). Further analyses are needed before the relative importance of climate (in winter, as well as in summer) and of food supplies in determining fox survival becomes clear.

Effects of Changes in Sea Level

Sea level rise as a consequence of global warming is expected to have dramatic effects on birds breeding and staging in the temperate coastal zone (Goss-Custard et al., 1990), although the effects may be moderated by human protective measures. The tidal areas of the Wadden Sea and the British estuaries on the east Atlantic, as well as the Bay of Fundy, Chesapeake Bay on the North American Atlantic coast, are crucial staging grounds for many Arctic waders and waterfowl, where they build up body reserves necessary for migration and breeding. Reductions in carrying capacity of these areas could make it difficult for many birds to complete their migrations.

In Arctic Canada, particularly around Foxe Basin and Hudson Bay, most birds are breeding on areas that have emerged from the sea during the last 3000 years. This uplift is continuing at a rapid rate. Presumably, rising sea level will alter the rate of change, or even reverse it. In western North America and parts of Eurasia there is already an eustatic rise in sea level. Further rising could result in submergence of substantial areas of the coastal wetland that are of high value to birds.

In other parts of the Arctic, where the ground is more elevated and few new areas are emerging, the effects of rising sea level on birds may be small. The consequences for the larger river deltas, where many birds breed, stage, and molt, appear to be complex and difficult to judge.

Climate and the Metabolic Costs of Living in the Arctic

Current research on the physiology and energetics of Arctic-breeding migrants is providing new information and ideas on the metabolic costs of living in the Arctic, as well as the costs of the migratory flights from and to staging areas. One important component of summer maintenance costs is the variability in the amounts of solar radiation directly available to the birds. In the High Arctic the period of effective insolation is only about 3 months. It peaks while most birds are still incubating, or in the early stages of brood rearing. The rapid decline in solar radiation in August may be an important reason for the early departure of most adult female waders, well before their young are capable of flight, parental care being left to males, and for a general exodus well before freeze-up.

The annual cycle of incoming solar radiation shares with changes in photoperiod the characteristics of being fixed for any particular latitude and independent of annual fluctuations in climate, so that they are not subject to direct anthropogenic interference. However, what happens to incoming solar radiation is greatly affected by cloud cover, so that one of the most influential consequences of global change in the Arctic seems likely to be changes in the formation, character, and persistence of clouds, perhaps as much for their effects on the radiation balance as because of changes in resulting precipitation.

The historical scarcity of instrumental records of precipitation at high latitudes, aggravated by the difficulties of measurement of blowing snow, has resulted in inability to model changes in the Arctic with the confidence possible in dealing with events at lower latitudes. The historical records of cloud cover are even sketchier and more subjective. Very few Arctic stations record solar radiation. Only in the last 25 years has satellite imagery begun to provide objective, large-scale records of day-by-day changes in ice, snow, and cloud cover. Thus, it seems unlikely that enough is yet known about year-to-year variations in the net amount of radiation available to birds in different regions to enable physiologists to test their estimates of the energetic costs of living in the Arctic over the full range of conditions now being encountered.

Maps of the global distribution of cloud cover over land in summer (Warren et al., 1986) show that the frequency of total cover increases fairly steadily northward from 50°N: it averages 68% between 60°N and 70°N and 76% between 70°N and 80°N. In the vicinity of Bering Strait it averages over 80%. Elsewhere north of 60°N there is remarkably little longitudinal variation. The mean frequency of total cloud cover in each month increases from 66% in May to 70% in June and July and 74% in August, again with little longitudinal variation. With such widespread persistent cover in summer prevalent in the immediate past, any major changes in its frequency as a consequence of global change could be of considerable importance to the well-being of Arctic birds.

The Scenario C maps of likely changes in sea level pressure (Palutikof et al., 1984, Figures 4 and 16) suggest that pressures may become a little higher in summer over Fennoscandia and northwest Russia, and lower over Iceland and the whole of Arctic North America, especially north of 70°N. If those changes were to happen, summer cloud cover might decrease over northern Europe, but would be more likely to increase than decrease over the Canadian Arctic and Alaska. The changes in cloud cover seem unlikely to be beneficial to Arctic birds; and the suggested pattern of pressure reduction in the Canadian Arctic might also result in increases in mean wind speeds. As mean summer wind speeds there are already rather high (18 to 22 km/h), except in the extreme north, the metabolic costs of living for birds seem more likely to increase than to decrease.

Weather Variability and the Migrations of Arctic-Breeding Birds

Food availability, temperature, wind speed and direction, precipitation, and cloudiness are often cited among the potential causes of variation in the phenology of bird migration. For long-distance migrants, "food availability" includes important geographical (and climatic) elements: the positions of, and distances between, their wintering, staging, and breeding areas, and whether alternative sites exist. Noting that, "Temperature, wind direction and speed, barometric pressure, and precipitation patterns are often highly correlated in a larger meteorological context," Smith and Hayden (1984) used principal component analysis to show that the arrivals and departures of Greater Snow Geese wintering at a wildlife refuge in North Carolina, U.S.A., in the years 1957 to 1977 were correlated with variations in the frequency of extratropical storms moving across North America: "Years in which autumn migration is delayed occur when the Atlantic coast storm track is displaced northward.... Increased cyclone activity through the Ohio River valley and over the Great Lakes tends to advance the onset of spring migration." This synoptic approach to the timing of migration has yet to be adopted widely. That example would be worth reexamining using data for recent years, because the timing of the migrations of the same geese between their High Arctic breeding areas and their principal staging area on the St. Lawrence estuary, Quebec, Canada, has also been varying (A. Reed, unpubl.).

Changes in the numbers and breeding success of Greater Snow Geese have been tracked with unusual precision and accuracy during the last 25 years. Widespread use of the "synoptic method" is unlikely because of the great scarcity of long runs of counts of birds at major staging or wintering areas, or of direct observations (visually or with radar) of migratory movements (Gudmundsson, 1992).

In his book *Bird Migration*, Alerstam (1993) gives detailed accounts of long-distance migration by Arctic-breeding waders and geese. The most spectacular accomplishments are by those waders that fly south in autumn from eastern Canada and the northeastern U.S.A. over the western Atlantic Ocean to staging areas in South America, traveling nonstop for more than 4000 km in some cases. Some of the waders wintering in western Europe make northward flights of comparable length in spring, to breeding places in Siberia. Those spring journeys seem not to be dangerous (though much of the distance is over sparsely peopled lands from where dropouts are unlikely to be reported). The journeys made by Knots, Turnstones *Arenaria interpres* and Brent Geese from Iceland over the Greenland icecap are shorter (2000 to 2400 km) but present the special problems of crossing the high inland ice, which is often shrouded in cloud and where the wind direction and strength may be very different from the conditions in which the birds left Iceland. Occasionally, these spring migrants suffer catastrophic losses in west Greenland or the Queen Elizabeth Islands, when they arrive with unusually depleted fat and protein reserves, to find the terrain still snow-covered, so that they starve to death. This is known to have happened in 1974 (Morrison, 1975) and 1979 (Boyd, unpubl.), but not since.

In the context of global change, the general question that these icecap crossings illustrate in a dramatic way is: will the conditions for spring migration become less rigorous, or will they become more hazardous, as a result of stronger adverse winds, heavier persistent cloud cover, and more precipitation? The Scenario C illustrations of Palutikof et al. (1984) suggest that in Iceland springs and summers are likely to be warmer, with spring temperatures more variable. Iceland is omitted from their maps of precipitation, though those showing sea level pressure patterns suggest increased frequency of low pressure systems there in spring. East Greenland is not shown. The Baffin Bay region (west Greenland and the northeastern islands of Canada) is expected to have higher pressures in spring, though lower in summer, with warmer temperatures in both seasons, with less and less variable, precipitation. Should these expectations be fulfilled, leaving Iceland in spring may be more difficult, but conditions in the "receiving area" should tend to be better, reducing the risk of catastrophic losses.

The smaller the species the higher the expected risk of long-distance sea crossings. Salomonsen (1967) suggested that Wheatear *Oenanthe oenanthe*, crossing the Atlantic from breeding grounds in west Greenland to wintering grounds in southwest Europe, take advantage of low pressure systems, moving on the north side of the systems on their way to Greenland in spring, and on the south side in autumn to benefit from tailwinds. Changes in the position, timing and speed of movements of the low pressure systems could threaten this adaptation.

Most scenarios of global warming suggest increases in frequency and severity of droughts in the prairie regions of north-central North America and the steppes of Eurasia. Birds migrating through these areas are likely to suffer from the drying up of lakes and damage to their food supplies. Davies and Cooke (1983) found that in years of drought Lesser Snow Geese *Anser c. caerulescens* staging in spring in southern Manitoba, Canada, failed to build up sufficient body reserves to reproduce successfully in their breeding colonies on the shores of Hudson Bay.

Because there are many differences in the suggested patterns of change in other regions of northern North America and northern Europe, no broad generalizations about likely changes in the spring and fall migration can yet be made. It would certainly be unwise to assert that a general reduction of risks is likely.

Climatic changes are likely to be reflected in the migration schedules of arctic-nesting birds. We are aware of few examples of such, but the Svalbard population of Pink-footed Goose *Anser brachyrhynchus* spring staging in western Jutland, Denmark (56°N, 08°E), and stopping-over in Vesterålen

(69°N, 15°E), northern Norway (a distance of ca. 1500km) prior to departure to the breeding grounds on Svalbard provides such an example. Migrating from Denmark to northern Norway the geese take advantage of an early growth of pasture grasses with a high nutritive value and almost 24h of daylight (Madsen, in prep.). From the early 1980s to the early 1990s, the arrival in Vesterålen has advanced by more than a week (Figure 10.1). It is tempting to ascribe the development to warmer springs; on the west coast of Norway there has been an increasing trend in spring temperatures during the interval (Prop et al., in prep.), even though there is only a nonsignificant tendency in Vesterålen (Madsen, unpubl., based on data from the Norwegian Meteorological Institute). However, at least two alternative explanations for the advanced arrival cannot be ruled out. First, since the early 1980s the geese staging in northern Norway have increasingly shifted feeding from saltmarshes to fertilized pastures, which provide an earlier spring growth. Second, on the Danish staging grounds, the geese have increasingly fed on newly sown grain, enabling them to achieve an earlier premigratory fattening (Madsen, 1985, and unpubl.).

The example shows that the geese can rapidly respond to changing environmental conditions, but the trigger may be difficult to unravel.

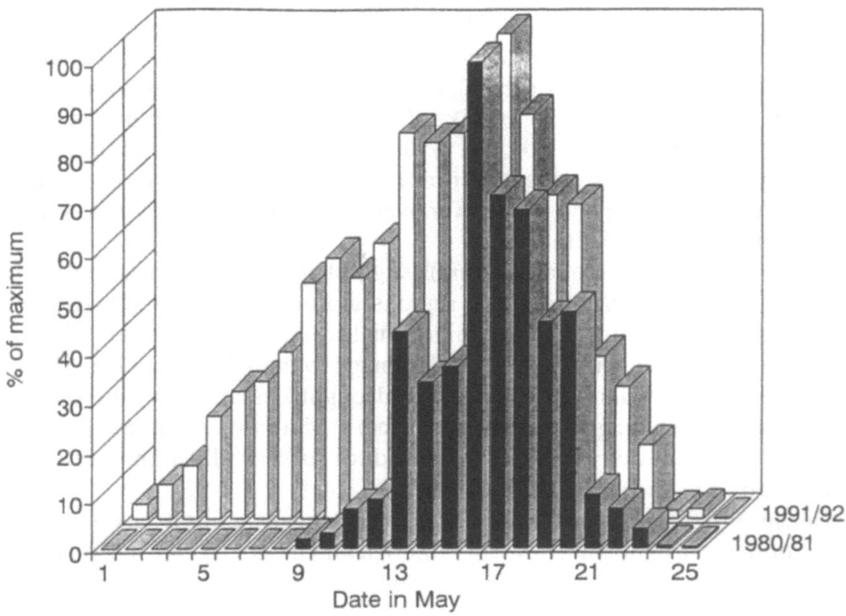

Figure 10.1. Development in numbers of Pink-footed Geese *Anser brachyrhynchus* spring staging in Vesterålen, northern Norway 1980–81 to 1991–92, expressed as the proportion of peak numbers recorded. In 1980–81 average peak number was 9485 individuals, and in 1991–92 it was 10,890 individuals. Sources: 1980–81: Rikardsen (1982) and 1991–92: B. Røsshag (unpubl.).

Summary

In summary, changes in the Arctic are likely to be within the adaptive capacity of the birds. Earlier snowmelt and increasing summer temperatures would be largely beneficial to their breeding success. Further increases in cloud cover and precipitation may, on the other hand, be harmful.

Probably the most serious threats from global change to Arctic breeding birds will occur at other times of their life cycle, particularly for those birds which make long-distance flights.

Loss of crucial coastal staging areas due to sea level rise is the most obvious point of concern. The effects of increasing frequency of drought may be serious for birds using inland staging areas. Long-distance flights push the birds to their physiological limits so that they are very sensitive to changes in weather on the journey.

In order to be able to better understand the mechanisms causing changes in the bird fauna and their interaction with the environment, we call for the continuation of intensive and long-term studies at the meso- and microscales, in line with the Lesser Snow Goose project and its spin-offs at the Hudson Bay colonies and work on geese in the Yukon-Kuskokwim Delta, Alaska, in Svalbard and, more recently, in Taimyr. We are not aware of ongoing long-term studies of waders in the Arctic, and urge a revival of former studies such as the Point Barrow, Alaska, project.

In order to monitor the large-scale effects of global change on the Arctic bird fauna there is an urgent need to create a circumpolar sampling network to record changes in distribution, abundances, and phenologies. Good reasons for using birds for monitoring are that they respond quickly to environmental change, they are easy to observe, and it is possible to enlist the help of large numbers of skilled enthusiasts, both to count them and to spread knowledge of the results. Such an international venture would require a secretariat and agreement on standards, almost as difficult to obtain as long-term funding.

References

Alerstam, T. 1993. *Bird Migration*. Cambridge, UK: University Press.

Boyd, H. 1987. Do June temperatures affect the breeding success of Dark-bellied Brent Geese? *Bird Study* 34:155–159.

Boyd, H. 1992. Arctic summer conditions and British Knot numbers: An exploratory analysis. In: Piersma, T., and Davidson, N. (eds.), *The Migration of Knots* (pp. 144–152). Wader Study Group Bulletin 64, Supplement.

Boyd, H., and Maltby, L.S. 1979. The Brant of the western Queen Elizabeth Islands, N.W.T. In: Jarvis, R.L., and Bartonek, J.C. (eds.), *Management and biology of Pacific Flyway geese, a symposium*, (pp. 5–21). Corvallis, OR: OSU.

Chapman W.L., and Walsh, J.E. 1993. Recent variations of sea ice and air temperature in high latitudes. *Bull. Am. Meteorol Soc.* 74(1):33–41.

Davies, J.C., and Cooke, F. 1983. Annual nesting productivity in Snow Geese: Prairie droughts and arctic springs. *J. Wildl. Manage.* 47:291–296.

Dewdney, J.C. 1982. *USSR in Maps*. London, UK: Hodder & Stoughton.
Ebbinge, B. 1989. A multifactorial explanation for variation in breeding performance of Brent Geese *Branta bernicla*. *Ibis* 131:196–204.
Fox, A.D., Francis, I.S., Madsen, J., and Stroud, J.M. 1987. The breeding biology of the Lapland Bunting *Calcarius lapponicus* in West Greenland during two contrasting years. *Ibis* 129:541–552.
Goss-Custard, J.D., McGrorty, S., and Kirby, R. 1990. Inshore birds of the soft coasts and sea-level rise. In: Beukema, J.J., Wolff, W.J., and Brouns, J.J.W.M. (eds.), *Expected Effects of Climatic Change on Marine Coastal Ecosystems. Developments in Hydrobiology* (Vol. 57, pp. 189–193). Dordrecht: Kluwer Acad. Publ.
Gudmundsson, G.A. 1992. Flight and migration strategies of birds at polar latitudes. Doctoral thesis, Department of Ecology, Lund University, Sweden.
Jeffries, R.L., Gadallah, F.L., Srivastova, D.S., and Wilson, D.J. 1995. Desertification and trophic cascades in arctic coastal ecosystems: A potential climate change scenario. In: *Global Change and Arctic Terrestrial Ecosystems* (pp. 201–256). Ecosystems Research Report No. 10. EUR 15519EN. Luxembourg.
Johansen, H. 1956. Revision und Entstehung der Arktischen Vogelfauna. Acta Arctica, Fasc. VIII. Munksgaard, Köbenhavn, Denmark.
Larsson, K., and Forslund, P. 1991. Environmentally induced morphological variation in the Barnacle Goose, *Branta leucopsis*. *J. Evol. Biol.* 4:619–636.
Larsson, K., Forslund, P., Gustafsson, L., and Ebbinge, B.S. 1988. From the high Arctic to the Baltic: The successful establishment of a Barnacle Goose *Branta leuçopsis* population on Gotland, Sweden. *Ornis Scand.* 19:182–189.
MacInnes, C.D., Davies, R.A., Jones, R.N., Lieff, B.C., and Pakulak, A.J. 1974. Reproductive efficiency of McConnell River Small Canada Geese. *J. Wildl. Manage.* 38:686–707.
Madsen, J. 1985. Relations between change in spring habitat selection and daily energetics of Pink-footed Geese *Anser brachyrhynchus*. *Ornis Scand.* 16:222–228.
Madsen, J. 1987. Status and management of goose populations in Europe, with special reference to populations resting and breeding in Denmark. *Danish Rev. Game Biol.* 12(4):1–76.
Madsen, J., Bregnballe, T., and Hastrup, A. 1992. Impact of the Arctic Fox *Alopex lagopus* on nesting success of geese in southeast Svalbard, 1989. *Polar Res.* 11:35–39.
Meltofte, H. 1985. Populations and breeding schedules of waders, *Charadrii*, in high arctic Greenland. *Medd. Grønl. Biosci.* 16:1–43.
Morrison, R.F.G. 1975. Migration and morphomatrics of European Konts and turnstones on Ellesmere Island, Canada. *Bird-Bonding* 46:790–301.
Nettleship, D.N. 1974. The Breeding of the Knot *Calidris canutus* at Hazen Camp, Ellesmere Island, N.W.T. *Polarforschung* 44:8–26.
Owen, M., and Black, J.M. 1989. Factors affecting the survival of Barnacle Geese on migration from the breeding grounds. *J. Anim. Ecol.* 58:603–617.
Palutikof, J.P., Wigley, T.M.L., and Lough, J.M. 1984. Seasonal climate scenarios for Europe and North America in a high-CO_2, warmer world. DOE/EV/10098-5. Office of Energy Research, Washington, DC.
Piersma, T. 1994. Close to the edge: Energetic bottlenecks and migratory pathways in Knots. PhD thesis, The Netherlands: University of Groningen.
Prop, J., and de Vries, J. 1993. Impact of snow and food conditions on the reproductive performance of Barnacle Geese *Branta leucopsis*. *Ornis Scand.* 24:110–121.
Raveling, D.G. 1978. The timing of egg-laying by nothern geese. *Auk* 95:294–303.

Rikardsen, F. 1982. Migration studies of Pink-footed Geese *Anser brachyrhynchus* on Andøya, Nordland, *Vår Fuglefauna* 5:163–168 (In Norwegian, English summary).
Rogacheva, E.V. 1992. *The Birds of Central Siberia.* Germany: Husum.
Salomonsen, F. 1967. *The Birds of Greenland.* Copenhagen: Ejnar Munksgaord.
Sedinger, J.S., and Raveling, D.G. 1986. Timing of nesting by Canada Geese in relation to the phenology and availability of their food plants. *J. Anim. Ecol.* 55:1083–1102.
Shilyaeva, L.M. 1967. Studying the migration of the Arctic Fox. *Probl. Severa* 11:91–98.
Smith, T.J., III, and Hayden, B.P. 1984. Snow Goose phenology is related to extra-tropical storm climate. *Int. J. Biometeorol.* 28(3):225–233.
Summers, R.W., and Underhill, L.G. 1987. Factors relating to breeding populations of Brent Geese (*Branta b. bernicla*) and waders (*Charadrii*) on the Taimyr Peninsula. *Bird Study* 34:161–171.
Underhill, L.G., Prys-Jones, R.P., Syroechkovski, E.E., Jr., Groen, N.M., Karpov, V., Lappo, H.G., van Roomen, M.W.J., Rybkin, A., Schekkerman, H., Spiekman, H., and Summers, R.W. 1993. Breeding of waders (*Charadrii*) and Brent Geese *Branta bernicla bernicla* at Pronchishcheva Lake, northeastern Taimyr, Russia, in a peak and a decreasing lemming year. *Ibis* 135:277–292.
Warren, S.G., Hahn, C.J., London, J., Chervin, R.M., and Jenne, R.L. 1986. Global distribution of total cloud cover and cloud type amounts over land. DOE/ER/60085-Hi. Office of Energy Research, Washington, DC.

3. Effects of Anticipated Global Change on Ecosystem Structure and Ecological Complexity

11. Impacts of Global Change on Composition of Arctic Communities: Implications for Ecosystem Functioning

F. Stuart Chapin III, Sarah E. Hobbie, and Gaius R. Shaver

As a result of the increasing atmospheric concentrations of radiatively active greenhouse gases, global temperatures are expected to rise (Houghton et al., 1990), particularly at high latitudes. Current global temperature trends are consistent with these predictions (Maxwell, 1992). Accompanying this climatic warming will be changes in growing season length, irradiance, permafrost depth, soil moisture, and nutrient availability (Chapin et al., 1992; Kane et al., 1992; Maxwell, 1992), although the exact nature and magnitude of these changes are less certain than those of temperature change. Other, more direct, human impacts on arctic ecosystems may be even more important than climatic change in the next few decades, including air pollution, disturbance associated with resource exploitation, and altered grazing regime due to changing patterns of reindeer husbandry and hunting. Together these global changes in environment are certain to affect arctic communities. The challenge is to predict how community composition will respond to these environmental changes and what will be the consequences for arctic ecosystems. In this chapter we present a framework for predicting how arctic plant communities will respond to global change and discuss the implications for ecosystem functioning. Detailed justification for this approach is provided elsewhere in general terms (Chapin, 1993; Hobbie et al., 1993; Shaver et al., 1992) and for tundra (Chapin et al., 1995).

Effects of Global Change on Communities

All species differ to some degree in physiological, morphological, and developmental traits and, therefore, in their response to the environment. Thus, species differ in their distribution along environmental gradients (Webber, 1978; Whittaker, 1953), their response to environmental manipulations (Chapin and Shaver, 1985b), and their rates and patterns of migration in response to climatic change (Davis, 1981). To the extent that these individualistic responses of species to environment cause "equivalent species" to replace one another in activity or abundance, any changes in community composition will have minimal impact on ecosystem processes such as carbon gain or herbivory. Alternatively, if the types of species that increase in activity or abundance in response to global change differ functionally from those species that are negatively affected by the change, then ecosystem processes will also be affected.

Most tundra plant species show an initial positive response to increasing nutrient supply (Shaver and Chapin, 1980) because of the slow rates of nutrient mineralization under tundra conditions of waterlogging and low temperature (Gersper et al., 1980; Nadelhoffer et al., 1991). The nutrient response is most pronounced in forbs, graminoids, and deciduous shrubs and least pronounced in evergreen shrubs, mosses, and lichens. In the long term this would lead to dominance by deciduous shrubs and graminoids and decreased abundance of evergreen shrubs, mosses, and lichens (Shaver and Chapin, 1986; Shaver et al., 1995). Thus, enhanced soil fertility predictably increases biomass and production of most species in the short term but changes species composition in the long term (Figure 11.1).

Graminoids and mosses respond positively to increased soil moisture in the short term, although presumably for different reasons. Graminoids respond more to the effects of water movement on lateral nutrient transport (Chapin et al., 1988) and tolerance of low soil aeration than to the direct effects of moisture (Oberbauer and Miller, 1982). As with nutrients, long-term increases in soil moisture change community composition, leading to dominance by aerenchymatous graminoids and decreases in shrubs and lichens (Webber, 1978). Both the short-term and long-term community responses to changes in soil moisture lead us to expect changes in functioning of ecosystems.

Decreases in light intensity below current ambient levels generally have small, compensatory effects on different species (Chapin and Shaver, 1985b), suggesting little predictable ecosystem impact. However, over the long term there is a decline in production of all but the most shade-tolerant species (an overall decline of 50%; Figure 11.1) (Chapin et al., 1995). This is particularly pronounced in the High Arctic (Havström et al., 1993). The high carbohydrate concentrations of tundra plants (Billings and Mooney, 1968) indicates a lack of carbon limitation to growth under current environmental conditions (Chapin and Shaver, 1985a), suggesting that tundra

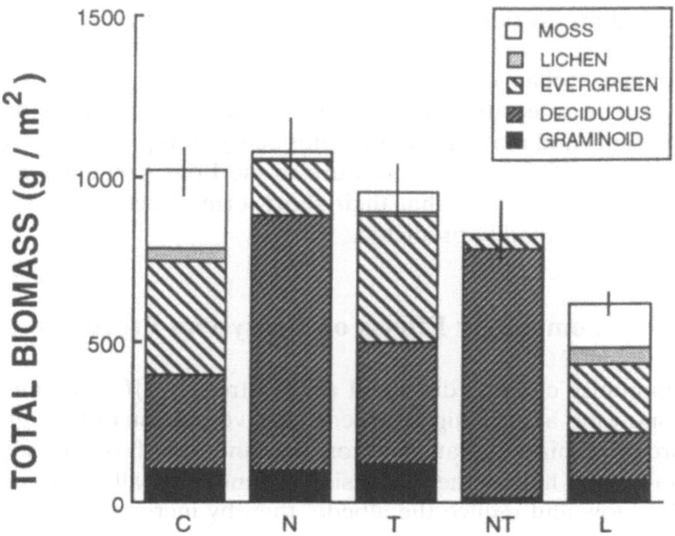

Figure 11.1. Total biomass (excluding roots) of plant functional groups in arctic tussock tundra after 9 years of different environmental manipulations: control (C), nutrient addition (N), greenhouse that raised summer air temperature by 3°C (T), fertilized greenhouse (NT), and shading to reduce light by 50% (L) (Chapin et al., 1995).

plants would show little response to increased light availability. In summary, tundra ecosystems should not be strongly affected by any initial changes in light intensity, but a long-term reduction in irradiance associated with increased cloudiness could reduce rates of most ecosystem processes.

Elevated CO_2 has little effect on photosynthesis by dominant tundra species (Tissue and Oechel, 1987) or on net ecosystem CO_2 flux (Oechel and Billings, 1992), so we assume that the trend of increasing atmospheric CO_2 will have minor direct effects on arctic ecosystems, at least in the short term.

Tundra plants are well adapted to low temperature and generally show relatively small direct growth responses to increased temperature in the Low Arctic (Chapin and Shaver, 1985b; Havström et al., 1993; Wookey et al., 1993). Here, some species respond positively to temperature, whereas others respond negatively, in ways that currently are difficult to predict a priori (Chapin and Shaver, 1985b; Chapin et al., 1995; Figure 11.1). However, in the High Arctic, growth of all plants increases substantially in response to increased air temperature caused by microtopography (Warren Wilson, 1966) or temperature manipulation (Havström et al., 1993; Romer et al., 1983; Wookey et al., 1993), so we expect a stimulation of community biomass and production in response to warmer temperatures, primarily in the High Arctic.

In summary, changes in soil resources (nutrients and water) and decreases in light intensity have predictable, contrasting effects on different types of plants and may, therefore, alter ecosystem processes, whereas the direct effects of CO_2 and temperature in the Low Arctic are more subtle and less predictable, such that their short-term effects on ecosystem processes cannot be readily predicted.

Community Effects on Ecosystem Processes

The community changes discussed above strongly affect biophysical processes. Snow has a much higher albedo than vegetation or bare ground and, therefore, minimizes radiation absorption and heat flux into soil. Any increase in canopy height due to invasion of tundra by tall shrubs or trees will mask the snow and reduce the albedo, thereby increasing canopy and soil temperatures (Bonan et al., 1992). These effects on ecosystem energy budget are large enough that, if extensive, they could permanently alter regional and global temperature (Bonan et al., 1992). Mosses also have strong biophysical effects because they are excellent thermal insulators when dry and reduce heat flux to soils. Thus, a warmer and drier arctic climate will not necessarily increase soil temperature and nutrient availability (Jonasson et al., 1993).

Biogeochemical processes are determined by (1) plant uptake, internal recycling, and loss of nutrients (Chapin, 1980; Shaver, 1981; Shaver and Melillo, 1984), and (2) litter quantity, quality, and location (Hobbie, 1992). Any change in nutrient, water, or light availability that alters community biomass and production will cause corresponding changes in nutrient uptake and loss by vegetation (Chapin et al., 1988; Shaver and Chapin, 1991), even in the absence of qualitative changes in community composition. However, changes in species composition also have important implications. Deciduous shrubs and graminoids may have higher rates of growth and nutrient cycling per unit biomass than do evergreen shrubs or lichens (Chapin, 1980). However, these generalizations rely heavily on studies of leaf turnover, and differences in uptake, loss, and turnover may be much less pronounced when considered on a whole-plant or community level (Shaver and Chapin, 1991; Shaver et al., 1995). From these observations, we predict that increases in nutrient supply and subsurface water flow will enhance rates of nutrient cycling, and that these would be counteracted by any decrease in light intensity or soil drying.

Changes in quality and location of litter inputs differ strongly among different plant growth forms and could profoundly affect nutrient cycling. Mosses have low concentrations of nitrogen and high concentrations of lignin-like polymers (Chapin et al., 1986), explaining their generally low rates of decomposition (Johnson and Damman, 1991) and their important role in peat formation (Clymo and Hayward, 1982). Deciduous shrubs have leaves with high nitrogen and low lignin concentrations (Chapin and Shaver, 1988) but a large stem allocation (Shaver, 1986) and, therefore, a high overall lignin content (Chapin et al., 1986). Evergreen shrubs have low leaf litter quality and a substantial stem allocation, suggesting low rates of decomposition and mineralization. Graminoids have a low stem allocation but a high allocation to roots (Shaver and Billings, 1975), where low temperature, low oxygen, and high lignin content retard root decomposition in situ (Chapin et al., 1988). The net consequence of these multiple species effects on decomposition requires further study. However, we predict that, under field conditions, mosses most strongly reduce decomposition and nitrogen mineralization and promote carbon storage through their effective thermal insulation and low litter quality, followed by evergreen shrubs with their low leaf and stem litter quality, followed by deciduous shrubs with their high leaf but low stem litter quality, and that graminoids enhance mineralization.

Future Projections

Global changes in climate and land use will have profound effects on arctic ecosystems. Many of the most important of these impacts will be mediated by changes in community composition. The short-term, direct effects of climatic warming are difficult to predict in the Low Arctic, because increases in growth of some species will be balanced by reductions in others. In contrast, any increased nutrient supply or turnover resulting from atmospheric deposition, enhanced mineralization in soils, or changes in species composition (e.g., decline in moss biomass) could enhance production and nutrient cycling, particularly of deciduous shrubs, which are favored by these conditions. Because the major stimulatory effect would be on production rather than decomposition, ecosystem carbon storage would increase (Shaver et al., 1992). Any soil drying associated with climatic warming would reduce water flow and nutrient transport to root surfaces (reducing rates of nutrient uptake) but could increase abundance of deciduous shrubs relative to graminoids and mosses. We anticipate that drier conditions would initially inhibit production more than decomposition, thus reducing carbon storage. However, over the long term, the nutrients released by decomposition would stimulate shrub production and minimize further declines in carbon storage. Declines in light intensity associated with increased cloudiness could reduce productivity and associated demand for nutrients and reduce nutrient cycling rates. This would reduce productivity more strongly than decomposition and reduce carbon storage. Obviously, if global change involves increased temperature (promoting production), reduced soil moisture (promoting decomposition), increased cloudiness (reducing production), and nitrogen deposition (promoting production), many outcomes are possible, depending on the relative magnitude and timing of these effects. However, an understanding of the response of plant growth forms to the components of global environmental change is critical to the development of a mechanistic framework for predicting the future role of arctic ecosystems in the global system.

References

Billings, W.D., and Mooney, H.A. 1968. The ecology of arctic and alpine plants. *Biol. Rev.* 43:481–529.

Bonan, G.B., Pollard, D., and Thompson, S.L. 1992. Effects of boreal forest vegetation on global climate. *Nature* 359:716–718.

Chapin, F.S., III. 1980. The mineral nutrition of wild plants. *Annu. Rev. Ecol. System.* 11:233–260.

Chapin, F.S., III. 1993. Functional role of growth forms in ecosystem and global processes. In: Ehleringer, J.R., and Field, C.B. (eds.), *Scaling Physiological Processes: Leaf to Globe* (pp. 287–312). San Diego: Academic Press.

Chapin, F.S., III, and Shaver, G.R. 1985a. Arctic. In: Chabot, B.F., and Mooney, H.A. (eds.), *Physiological Ecology of North America* (pp. 16–40). New York: Chapman and Hall.

Chapin, F.S., III, and Shaver, G.R. 1985b. Individualistic growth response of tundra plant species to environmental manipulations in the field. *Ecology* 66:564–576.

Chapin, F.S., III, and Shaver, G.R. 1988. Differences in carbon and nutrient fractions among arctic growth forms. *Oecologia* 77:506–514.

Chapin, F.S., III, McKendrick, J.D., and Johnson, D.A. 1986. Seasonal changes in carbon fractions in Alaskan tundra plants of differing growth form: Implications for herbivores. *J. Ecol.* 74:707–731.

Chapin, F.S., III, Fetcher, N., Kielland, K., Everett, K.R., and Linkins, A.E. 1988. Productivity and nutrient cycling of Alaskan tundra: Enhancement by flowing soil water. *Ecology* 69:693–702.

Chapin, F.S., III, Jefferies, R.L., Reynolds, J.F., Shaver, G.R., and Svoboda, J. (eds.) 1992. *Arctic Ecosystems in a Changing Climate: An Ecophysiological Perspective.* San Diego: Academic Press.

Chapin, F.S., III, Hobbie, S.E., Bret-Harte, M.S., and Bonan, G. 1995. Causes and consequences of plant functional diversity in arctic ecosystems. In: Chapin, F.S., III, and Körner, C. (eds.), *Arctic and Alpine Biodiversity: Patterns, Causes and Ecosystem Consequences* (pp. 225–237). Berlin: Springer-Verlag.

Chapin, F.S., III, Shaver, G.R., Giblin, A.E., Nadelhoffer, K.G., and Laundre, J.A. 1995. Response of arctic tundra to experimental and observed changes in climate. *Ecology* 76:694–711.

Clymo, R.S., and Hayward, P.M. 1982. The ecology of *Sphagnum*. In: Smith, A.J.E. (eds.), *Bryophyte Ecology* (pp. 229–289). London: Chapman and Hall.

Davis, M.B. 1981. Quaternary history and the stability of forest communities. In: West, D.C., Shugart, H.H., and Botkin, D.B. (eds.), *Forest Succession: Concepts and Applications* (pp. 132–153). New York: Springer-Verlag.

Gersper, P.L., Alexander, V., Barkley, S.A., Barsdate, R.J., and Flint, P.S. 1980. The soils and their nutrients. In: Brown, J., Miller, P.C., Tieszen, L.L., and Bunnell, F.L. (eds.), *An Arctic Ecosystem: The Coastal Tundra at Barrow, Alaska* (pp. 219–254). Stroudsburg: Dowden, Hutchinson & Ross.

Havström, M., Callaghan, T.V., and Jonasson, S. 1993. Differential growth responses of *Cassiope tetragona*, an arctic dwarf-shrub, to environmental perturbations among three contrasting high- and sub-arctic sites. *Oikos* 66:389–402.

Hobbie, S.E. 1992. Effects of plant species on nutrient cycling. *Trends Ecol. Evol.* 7:336–339.
Hobbie, S.E., Jensen, D.B., and Chapin, F.S., III. 1993. Resource supply and disturbance as controls over present and future plant diversity. In: Schulze, E.-D., and Mooney, H.A. (eds.), *Ecosystem Function of Biodiversity* (pp. 385–407). Berlin: Springer-Verlag.
Houghton, J.T., Jenkins, G.J., and Ephraums, J.J. (eds.) 1990. *Climate change: The IPCC Scientific Assessment*. Cambridge: Cambridge University Press.
Johnson, L.C., and Damman, A.W.H. 1991. Species controlled *Sphagnum* decay on a South Swedish raised bog. *Oikos* 61:234–242.
Jonasson, S., Havström, M., Jensen, M., and Callaghan, T.V. 1993. *In situ* mineralization of nitrogen and phosphorus of arctic soils after perturbations simulating climate change. *Oecologia* 95:179–186.
Kane, D.L., Hinzman, L.D., Woo, M., and Everett, K.R. 1992. Arctic hydrology and climate change. In: Chapin, F.S., III, Jefferies, R.L., Reynolds, J.F., Shaver, G.R., and Svoboda, J. (eds.), *Arctic Ecosystems in a Changing Climate: An Ecophysiological Perspective* (pp. 35–57). San Diego: Academic Press.
Maxwell, B. 1992. *Arctic Climate: Potential for Change Under Global Warming*. In: Chapin, F.S., III, Jefferies, R.L., Reynolds, J.F., Shaver, G.R., and Svoboda, J. (eds.), *Arctic Ecosystems in a Changing Climate: An Ecophysiological Perspective* (pp. 11–34). San Diego: Academic Press.
Nadelhoffer, K.J., Giblin, A.E., Shaver, G.R., and Laundre, J.A. 1991. Effects of temperature and substrate quality on element mineralization in six arctic soils. *Ecology* 72:242–253.
Oberbauer, S., and Miller, P.C. 1982. Growth of Alaskan tundra plants in relation to water potential. *Holarctic Ecol.* 5:194–199.
Oechel, W.C., and Billings, W.D. 1992. Effects of global change on the carbon balance of arctic plants and ecosystems. In: Chapin, F.S., III, Jefferies, R.L., Reynolds, J.F., Shaver, G.R., and Svoboda, J. (eds.), *Arctic Ecosystems in a Changing Climate: An Ecophysiological Perspective* (pp. 139–168). San Diego: Academic Press.
Romer, M.J., Cummins, W.R., and Svoboda, J. 1983. Productivity of native and temperate "crop" plants in the Keewatin District, N.W.T. *Nat. Can.* 110:85–93.
Shaver, G.R. 1981. Mineral nutrition and leaf longevity in an evergreen shrub, *Ledum palustre* ssp. *decumbens*. *Oecologia* 49:362–365.
Shaver, G.R. 1986. Woody stem production in Alaskan tundra shrubs. *Ecology* 56:401–410.
Shaver, G.R., and Billings, W.D. 1975. Root production and root turnover in a wet tundra ecosystem, Barrow, Alaska. *Ecology* 56:401–410.
Shaver, G.R., and Chapin, F.S., III. 1980. Response to fertilization by various plant growth forms in an Alaskan tundra: Nutrient accumulation and growth. *Ecology* 61:662–675.
Shaver, G.R., and Chapin, F.S., III. 1986. Effect of fertilizer on production and biomass of tussock tundra, Alaska, U.S.A. *Arctic Alpine Res.* 18:261–268.
Shaver, G.R., and Chapin, F.S., III. 1991. Production: biomass relationships and element cycling in contrasting arctic vegetation types. *Ecol. Monogr.* 61:1–31.
Shaver, G.R., and Melillo, J.M. 1984. Nutrient budgets of marsh plants: Efficiency concepts and relation to availability. *Ecology* 65:1491–1510.
Shaver, G.R., Billings, W.D., Chapin, F.S., III, Giblin, A.E., Nadelhoffer, K.J., Oechel, W.C., and Rastetter, E.B. 1992. Global change and the carbon balance of arctic ecosystems. *BioScience* 61:415–435.
Shaver, G.R., Giblin, A.E., Nadelhoffer, K.J., and Rastetter, E.B. 1995. Plant functional types and ecosystem change in arctic tundras. In: Smith, T., Shugart, H.H.,

and Woodward, F.I. (eds.), *Plant Functional Types*. Cambridge: Cambridge University Press.

Tissue, D.T., and Oechel, W.C. 1987. Response of *Eriophorum vaginatum* to elevated CO_2 and temperature in the Alaskan tussock tundra. *Ecology* 68:401–410.

Warren Wilson, J. 1966. An analysis of plant growth and its control in arctic environments. *Ann. Bot.* 30:383–482.

Webber, P.J. 1978. Spatial and temporal variation of the vegetation and its productivity, Barrow, Alaska. In: Tieszen, L.L (ed.), *Vegetation and Production Ecology of an Alaskan Arctic Tundra* (pp. 37–112). New York: Springer-Verlag.

Whittaker, R.H. 1953. A consideration of climax theory: The climax as a population and pattern. *Ecol. Monogr.* 23:41–78.

Wookey, P.A., Parsons, A.N., Welker, J.M., Potter, J.A., Callaghan, T.V., Lee, J.A., and Press, M.C. 1993. Comparative responses of phenology and reproductive development to simulated environmental change in subarctic and high arctic plants. *Oikos* 67:490–502.

12. Effect of Climate Change on Biodiversity of Arctic Plants

Boris A. Yurtsev

In this chapter I shall try, starting from an analysis of the modern spatial–climatic trends of biodiversity of arctic flora, to predict major temporal changes in connection with the expected global changes, caused by the greenhouse effect. The ground for the prognosis is provided by the floristic subdivision of the circumpolar Arctic (Yurtsev, 1992, 1994; Yurtsev et al., 1978, 1985), by the subdivision of the Arctic into phytogeographic zones, and the interpretation of both in terms of their dependence upon the climatic differentiation of the Arctic territory and its history. The nature of the data at my disposal does not permit any precise quantitative estimations or calculations. But this phytogeographic approach (in contrast, for example, to the ecophysiological one) permits us to trace the results of climatically induced spatial changes of plant cover as a whole, with all its multidimensional complexity.

This chapter deals with the taxonomical diversity of arctic flora (primarily species diversity) and with different levels of biochorological diversity of the Arctic, that is, the diversity of natural assemblages of plants (such as plant communities, elementary or local floras, and floras of different phytochoria) within some larger area.

The Arctic itself represents a phytochorion of the rank of a floristic region (Yurtsev, 1992, 1994; Yurtsev et al., 1978, 1985). Its flora includes four to five endemic and subendemic genera of vascular plants, most of which are common throughout its territory, and several intergeneric hybridogenous taxa, a number of endemic sections, subsections or series, and at least 10% endemic species and subspecies (about 200). It is not so few, taking into account the youth of the arctic landscapes: deforestation about 2.5 million years B.P. (Akhmetiev et al., 1978; Bennike, 1990; Bennike and Böcher, 1990; Matthews and Ovenden, 1990) and the forming of permanent ice cover over the Arctic Ocean ca. 1 million years P.B. (Herman and Hopkins, 1980).

The Arctic has a peculiar major type of environment and set of habitats, in contrast to that in the Boreal region, which suggests different directions and modes of evolution. There is a different set of growth forms and a distinct lack of trees in the true Arctic, in addition to the absence of gymnosperms in the arctic flora (beyond some ecotonal areas), which so dramatically contrasts with the dominance of coniferous trees throughout the Boreal region.

The integrity of arctic flora is very significant, which can be illustrated by the high proportion of circumpolar species in local floras of the Arctic (from one third to four fifths and still higher in some high-Arctic floras).

The following concept of the Arctic seems to be the simplest and most natural: northern, cold, treeless region. Certainly, treelessness is an indication of an extreme environment, because in any other thermal zone, forest or woodlands exist and usually dominate the landscape at some favorable balance of summer warmth and precipitation.

One can distinguish three subregions within the Arctic from the above wide interpretation. The central subregion (Arctic proper) is generally restricted to the coast of polar seas with almost permanent ice cover, frosty winters, deep permafrost, and the dominance of tundra vegetation. The continental species predominate in its flora along with those more or less indifferent, mostly circumpolar, ones. The other subregions, namely North Atlantic (South Greenland, Iceland, forestless northernmost Fennoscandia) and North Pacific (Alaska Peninsula, Aleutian, and Commander Islands, Pribilof Islands), embrace the areas having typically oceanic climate with almost permanently open sea, mild winters, absence of permafrost, dominance of mesic meadows in association with heaths, very high proportion of boreal oceanic nonarboreal species, and oceanic arctic–alpine species, which correlates with a low proportion of the true arctic plants. These phytochoria, and especially the Commander–Aleutian chain, situated at lower latitudes are very often (and not unreasonably) included in the Boreal region. In this chapter I will speak mostly about the Arctic proper.

How great is the taxonomical diversity of vascular plants in the whole Arctic? The data at my disposal are not quite accurate because different taxonomical schools and traditions exist in different major sectors of the Arctic, and the circumpolar revision, started by "Panarctic flora" project, is just at its starting point. According to the cytotaxonomical atlas by Love and Love (1975), 1629 vascular plant species (with subspecies 1899 taxa) of

404 genera and 106 families (among them Pteridophyta 3.2%, gymnosperms 0.5%, all being conifers of the ecotonal zone, monocots about 27%, and dicots about 70%) grow in the Arctic, including Iceland and south Greenland.

But, according to our "Arctic Flora of the USSR" (1960 to 1987), in the Russian Arctic itself there are 1640 species of vascular plants (with subspecies, 1779 taxa). Consequently, the total number of species and subspecies of vascular plants in the Arctic cannot be less than 2000 (Yurtsev, 1991b).

The species diversity of bryophytes and lichens, according to draft estimation by my colleagues (Olga M. Afonina—mosses, A.D. Potemkin—hepatics, and M.P. Zhurbenko and M.P. Andreyev—lichens), includes 750 species of bryophytes (700 species in the Russian Arctic) and 1500 species of lichens (1000 species in the Russian Sector).

Thus, the contribution of arctic flora to global diversity of plants seems to be modest (because of low endemism). Nevertheless, due to the same reason (poverty of species), the value of each unit here drastically increases as does the functional load on it in the northern ecosystems. So, arctic species (and populations) are irreplaceable components of global biodiversity.

Any major climatic change inevitably creates new ecological niches that could be filled either by some supplementary species (owing to their immigration or speciation in situ) or by some native species, which results in the increase of their population diversity. In the Arctic, the latter is the first in importance, and the long-distance dispersal is the second in importance. The speciation in situ plays an insignificant role.

Biochorological diversity of the arctic flora could be characterized in the threefold system of biogeographic coordinates. These coordinates are the latitudinal and altitudinal zonations and the longitudinal (sectoral) differentiation. The altitudinal zonation is closely interconnected with the latitudinal one in terms of its dependence on thermic parameters of macroclimate.

The shortage of summer warmth increases in the northward direction, which is responsible for the treelessness of the Arctic (tundra zone in its broader interpretation) and the zonation of the arctic vegetation (Chernov, 1989). Within the Arctic there occurs a range of mean July temperatures from almost zero or +2°C at the northern limit of the arctic tundra subzone up to 10 to 12°C at the polar limit of a more or less continuous distribution of woodlands.

Our scheme of subdivision of the Arctic into phytogeographic subzones is presented in Figure 12.1. It is a modified version of the scheme presented by Yurtsev et al. (1978, 1985), according to some new material available (e.g., Edlund and Alt, 1989). The most important difference between our scheme and the widely known zonation scheme by Aleksandrova (1977, 1980) is that we include what is called "the region of polar deserts" (Aleksandrova, 1980) or "the polar desert zone" (Chernov and Matveyeva,

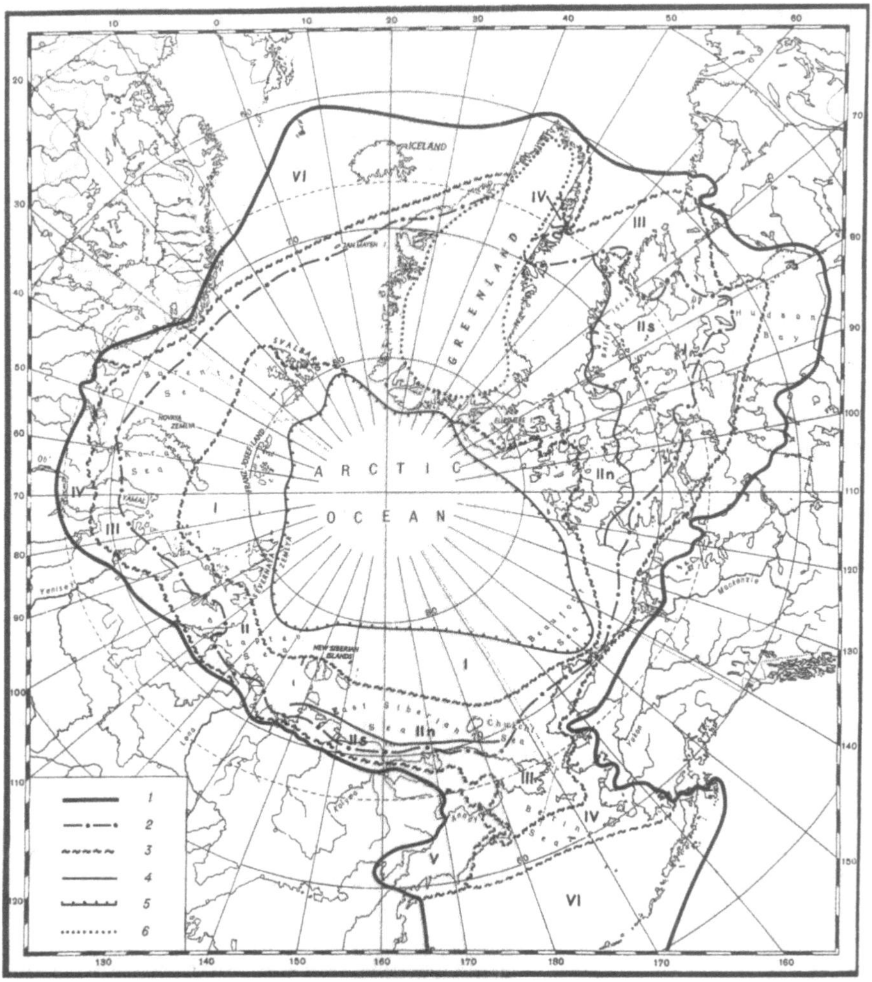

Figure 12.1. Phytogeographic zonation of the Arctic. The names and characteristics of the subzones of the tundra zone (I–VI) are given in the text. The boundaries: 1—of the Arctic (=tundra zone with oceanic treeless areas); 2—between the arctic and hypoarctic groups of subzones; 3—between subzones; 4—between northern and southern variants of a subzone; 5—of the abyssal of Arctic Ocean; 6—of the Greenland Ice Sheet.

1979; Gorodkov, 1956) into the tundra zone as a separate high-Arctic tundra subzone. The mature climax forms of the "polar desert" vegetation (termed by Bliss, 1981, as "polar semi-desert"; see also: Bliss and Svoboda, 1984; Bliss et al., 1984), with a continuous network of closed vegetation in troughs of polygons or sometimes with a very high projective coverage of it (on the habitats with a soil rich in fine-grained material and moisture), fit the general definition of tundra vegetation (Yurtsev, 1991a).

The most universal, really circumpolar units, shown in Figure 12.1, could be naturally united into the following two groups of subzones: the arctic group and the hypoarctic group. The arctic group includes the subzone I (high-Arctic, or polar desert zone in other interpretations) and subzone II (arctic tundras proper, with two zonal variants). The hypoarctic group which roughly corresponds to low-Arctic of Western authors (Bliss and Matveyeva, 1992) includes the units III, IV, and V of our scheme. The unit III, previously referred to as the "northern hypoarctic tundras," actually includes two subzones: the northern hypoarctic tundra proper representing the ecotone with arctic tundras, and the middle hypoarctic tundras, or "typical tundras," as is widely used by many Russian botanists (e.g., Chernov and Matveyeva, 1979; Gorodkov, 1935). The unit IV is southern hypoarctic tundras. The unit V (large stlaniks or *Pinus pumila* subzone) occurs only in northeastern Asia in Anadyr and Penzhina drainages and Koryak Mountains, where landscapes are dominated by thickets of stlaniks (i.e., the creeping tall shrubs or even trees, such as stone pine, *Pinus pumila*, shrubby alder, *Alnus fruticosa*, and birch, *Betula middendorffii*. Groves of arboreal Salicaceae (*Chosenia arbutifolia*, *Populus suaveolens*, *Salix spp.*) are not uncommon in floodplains. Silty plains and terraces, bearing only *Eriophorum vaginatum/Carex lugens* tussock, are very poor in species abundance. The level of summer warmth in the unit V area is the same as that in the northernmost (subarctic) taiga of more continental neighboring areas, such as the Kolyma River drainage. Winters in the unit V are windy, with periodical thaws. The unit VI is the oceanic analogue of the units IV and V outside the Arctic proper.

Here I can only list the major features of hypoarctic versus arctic subzones of the tundra zone (low- vs. high-Arctic). Typical of the hypoarctic tundras is the prevalence of closed vegetation, dominated by a limited number of low and dwarf arboreal, mostly oligotrophic and mycotrophic plants, also some tussock-forming herbs, as well as mosses and fruticulose lichens, growing on and forming acid organic beds over a variety of rocks. Different assemblages of eutrophic to mesotrophic herbs are forced out to some extra- and intrazonal positions, including earlier stages of successions.

The main latitudinal (northward) spatial trend of low-Arctic vegetation is the decrease in area occupied by low to mid-tall shrub vegetation, correlating with the increase in species diversity of arctic and arctic–alpine plants, which primarily consist of herbs. The first major longitudinal trend concerns the constriction of the landscape positions of shrub vegetation in the sectors with more continental climate and heavier permafrost (e.g., the East Siberian Arctic or in the continental parts of the Beringian) in favor of the sedge cottongrass tussock tundra. The second longitudinal trend is the increased importance of mesic meadow (and meadow shrub) vegetation, dominated by boreal and hypoarctic herb species, in the sectors with maritime climate. This trend can be readily traced (e.g., within Chukotka Tundra) when one moves from the Kolyma delta toward the Bering Strait.

Characteristic of both the arctic and high-arctic tundras is the decreased proportion of continuous vegetation (especially on silt or clay) along with much decreased thickness of organic layers in the soil (which are, in fact, organic mineral with almost neutral to basic properties) under fragments of the closed vegetation in patterned-ground tundras. The vegetation, including that on plakors, is dominated by a rather diverse set of herbs, mostly eutrophic and arctic–alpine, and in the unit II also by a limited set of prostrate summer-green arctic–alpine shrubs (*Dryas* and *Salix spp.*). The only semiprostrate evergreen dwarf shrub, *Cassiope tetragona*, marks the southern zonal variant within the unit II. The enlarged ecological ranges of many arctic species within the above two subzones are correlated with much decreased diversity of plant associations (β-diversity).

The dramatic fall in plant biodiversity within the unit I is underlain not only by the extreme lowering of summer warmth critical for most lineages, but also by certain edaphic conditions on exposed bedrock areas. These edaphic conditions include limitation of moisture and nitrogen, which has been recently discovered by Gold and Bliss (1992) and which retain the successions at their initial stage. At least, in the northernmost periglacial areas of Ellesmere Island, where moisture and fine-grained materials are present in the soil, one can observe the replacement of arctic hummock tundras (dominated by *Dryas integrifolia* and *Salix arctica*) by similarly looking high-arctic ones (with cushions of *Saxifraga oppositifolia* on top of a hummock). The set of vascular plants are the same as those in polar deserts or "semideserts." However, in general, the layer of vascular plants, mostly cushion-like ones, within the unit I is sparse and remotely scattered among more continuous ground layers of mosses, crustaceous, and other lichens as well as cyanobacteria.

Here I can only mention an interesting scheme of the floristic zonation of the Arctic presented by Young (1971), which is not entirely congruent with ours due to the following reasons: (1) we take into account not only the presence/absence of a species in a certain zonal unit, but also its role in the plant cover. (2) Young (1971) considers only the process of reduction of the number of species in the vascular flora northward, but the latter in reality results from balance of two mutually corresponding trends: (a) the attenuation of the set of boreal and then hypoarctic species, and (b) the enrichment of the set of arctic and arctic–alpine species in the same direction within the hypoarctic group of subzones.

The general trend of the decrease in biodiversity (in particular, taxonomical diversity) with the decrease of summer warmth is evident and looks trivial. The above dependence is especially great in the Canadian Arctic, where, for example, Rannie (1986) calculated almost 100% correlation between the number of species of vascular plants in a local flora and the mean July air temperature. I can quote the analogous data from some sectors of the Russian Arctic, where a floristic research was conducted using Tolmachev's concrete flora method (Tolmachev, 1931; Yurtsev, 1987b).

According to Rebristaya (1987), in the plains of Yamal Peninsula (West Siberia) local floras of the southern hypoarctic tundra subzone list 181 to 196 species of vascular plants, whereas those of the northern hypoarctic tundra subzone list 136 to 152 species, and those of the arctic tundras list 113 to 120 species. In the neigboring mountainous Ural/Novaya Zemlya sector (Rebristaya, 1977), the species diversity records for the same subzones are higher: there are (180) 201 to 244 (283) species in the southern, 154 to 198 in the northern hypoarctic tundras, 105 to 116 (160) in the arctic tundras, and only 30 to 60 in the high arctic ones. A similar magnitude of floristic richness of local floras in the same subzones is recorded for Taimyr Peninsula by Matveyeva (Chernov, 1989; Matveyeva, 1993) and by Sokolova (1984). These investigators list 238 to 258 species in the southern, 171 to 239 in the middle-northern hypoarctic tundras, 96 to 142 in the arctic tundras, and 59 in the high-arctic ones.

In the Beringian sector (in particular in Chukotka) the situation is more complicated due to diverse topography and the mosaic nature of climates. The species diversity of local floras is, on the average, remarkably higher. Thus, local floras of more than 350 to 400 species of vascular plants are not uncommon in the easternmost Chukotka Peninsula in localities with enclaves of the southern tundra, and more than 300 species in a local flora are recorded in the middle to northern hypoarctic tundras. There are more than 250 species (up to 325) in the arctic tundras of Wrangel Island. The flora of the island lists approximately 400 species and subspecies of vascular plants—much more than that in huge Canadian Arctic Archipelago (Porsild, 1964). The difference is especially remarkable at the level of elementary floras, which in Wrangel Island are at least two to three times richer. The principal reasons for the differences seem to be in the diverse history of the floras (Yurtsev, 1987a).

On the other hand, if we consider the latitudinal trend in biochorological diversity of zonal vegetation at the plant community level, the picture could be just the reverse. There exists very low species diversity in southern hypoarctic tundras (e.g., tussock tundras) and much higher species diversity in arctic tundras—at the expense of much higher diversity of herbaceous plants whereas the diversity of low arboreal ones drastically reduces. For example, in Wrangel Island, the plakor plant communities could list from 40 to more than 50 species of vascular plants. The diversity of communities (β-diversity), however, decreases.

The longitudinal–sectoral biochorological diversity of arctic flora is reflected on the map of floristic subdivision of the Arctic (the Central Subregion) down to the subprovince level (Figure 12.2). The 20 subprovinces are grouped into six provinces marked with Roman numerals as the following: I—East-Siberian (with four subprovinces), II—Chukotkan (with four

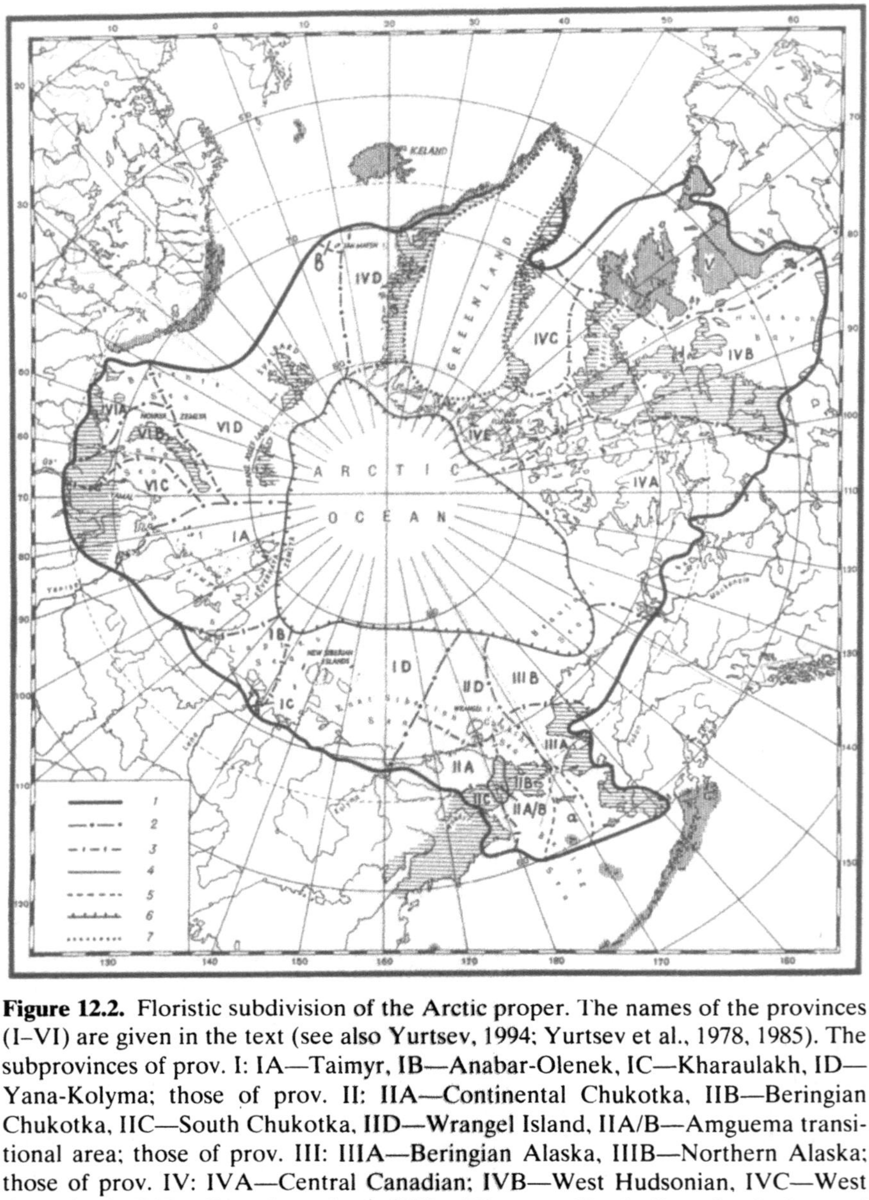

Figure 12.2. Floristic subdivision of the Arctic proper. The names of the provinces (I–VI) are given in the text (see also Yurtsev, 1994; Yurtsev et al., 1978, 1985). The subprovinces of prov. I: IA—Taimyr, IB—Anabar-Olenek, IC—Kharaulakh, ID—Yana-Kolyma; those of prov. II: IIA—Continental Chukotka, IIB—Beringian Chukotka, IIC—South Chukotka, IID—Wrangel Island, IIA/B—Amguema transitional area; those of prov. III: IIIA—Beringian Alaska, IIIB—Northern Alaska; those of prov. IV: IVA—Central Canadian; IVB—West Hudsonian, IVC—West Greenland, IVD—East Greenland, IVE—Ellesmere-North Greenland; those of prov. VI: VIA—Kanin-Pechora, VIB—Ural-Novaya Zemlya, VIC—Yamal-Gydan, VID—Svalbard. The autonomous oceanic areas: a—North Beringian insular, b—Jan Mayen. Shaded vertically are treeless areas (mainly outside the Arctic proper) dominated by oceanic complexes; shaded horizontally are those with stable coexistence of oceanic and continental complexes; not shaded (within the Arctic) is the phytochoria dominated by continental complexes. The boundaries: 1—of the Central subregion of the Arctic floristic region; 2—between provinces; 3—between subprovinces; 4—of a transitional area; 5—of autonomous areas; 6—of the abyssal of Arctic Ocean; 7—of the Greenland Ice Sheet.

subprovinces), III—Alaskan (with two subprovinces), IV—Canadian-Greenland (with five subprovinces), V—Baffin–Labrador (monotypic), and VI—West-Eurasian (with four subprovinces). The detailed description of the above units was published previously (Yurtsev et al., 1978), and updated in 1994 (Yurtsev, 1994).

In Figure 12.2 the unshaded areas (within the Arctic proper) are the territories with the continental type of flora. The horizontally shaded areas are phytochoria with codominance of both continental and oceanic to suboceanic flora elements. The latter phytochoria form the very large Atlantic (essentially Gulf Stream) sector and the rather narrow Central Beringian one, which reflects the difference in the extent of conjunctions of Arctic Ocean with Atlantic and Pacific Oceans, respectively. The vertically shaded areas are oceanic treeless phytochoria, lying mostly beyond the boundaries of the Arctic proper, but protruding into it as two small insular areas: Yan-Mayen in the Atlantic sector, several islands in the Beringian Sector, and Baffin-Labrador province. One can infer that the integrity of the arctic flora formed and then it increased at cold/dry epochs of shelf exposing, and it drastically reduced at those of global marine transgressions and the global warming of climate (including the Holocene one). So, the Arctic in the Quaternary time had two rather contrasting faces, and that of the global sea regression epoch (cryo-arid, or tundra-steppe) had little in common with the traditional concept of tundra. In the epoch of the Holocene sea transgression, the relics of the Pleistocene tundra-steppe (plant populations and communities) persisted in the Arctic in some isolated areas with drier climate (parts of the past Megaberingia, some periglacial areas of Greenland and Canadian Arctic Archipelago) (Yurtsev, 1986).

Analysis of the present scheme permits the conclusion that endemism cannot serve the principal basis for the floristic subdivision of the Arctic, though it is appreciable in some sectors, especially the Beringian one (there are 25 endemics on Wrangel Island and a good set of endemic plants on either coast of Bering Strait). Of importance is also the differential penetration of nonarctic species into the Arctic [e.g., along valleys (mostly boreal) or following mountain ranges (alpine and other montane species)]. Some plain areas (e.g., West Siberian subprovince) are remarkable by their negative floristic features (i.e., the absence of many species common in the neighboring sectors). One of the most universal criteria is the changing proportion of continental and oceanic flora elements in different sectors. The Beringian sector combines most of the above elements of floristic peculiarity, which testifies to its important florogenetic role in the history of the Arctic flora (Hultén, 1973; Yurtsev, 1986). The level of taxonomical and biochorological diversity is the highest here, too. So, only in Chukotka province are there recorded more than 1000 species and subspecies of vascular plants (655 only in the easternmost Chukotka Peninsula). There are about 800 species and subspecies in the arctic Alaska and Yukon (Hultén, 1968), about 600 in the whole Greenland (Böcher et al., 1968), and

360 in Canadian Arctic Archipelago (Porsild, 1964) as in Yamal Peninsula (Rebristaya, pers. commun.), 176 in Svalbard (R. Elvin and A. Elvebakk, pers. commun.).

Let us consider conclusions concerning the possible effect of the expected global climate changes on biodiversity of arctic plants. Because much controversy exists in climatic prognoses themselves, I adopt here only the most common and obvious suppositions, such as "greenhouse effect," the disproportionate warming (seasonally uneven, i.e., winter warming higher than summer warming), and the humidization of climate higher latitudes.

In short, the expected effect could be described as follows.

1. Suboceanic and oceanic elements (including the plants of mesic meadows, heaths, snowbeds) will strengthen in the "mixed" floras (shaded horizontally in Figure 12.2), which results in the expanding of the oceanic situations (e.g., that in the northernmost Fennoscandia). It will stimulate the migrations not only latitudinally (northward), but also longitudinally. Thus, one can expect the expansion of, for example, *Nardus stricta*, and *Calluna vulgaris* in the European Arctic, and *Pinus pumila*, alder, and shrub willow species in northeastern Asia. The processes may be most dramatic in the Atlantic sector (the "gates of the Gulf Stream").

2. Tundra landscapes as a zonal phenomenon are expected to remain in the arctic territories of East Siberia, central to northeastern Canada, and probably in some continental parts of northern Beringia. One can assume that the great Asian anticyclone, interacting with the winter anticyclone over the Arctic Ocean, will promote the frosty and not too snowy character of winters, as well as the permafrost, though the depth of snow in East Siberia will increase. The existence of large massifs of high-Arctic land in Canada is also a stabilizing factor. The American–Asian subbasin is of Polar Basin more or less isolated from the Gulfstream warm water and could keep its permanent more or less stable icecover.

3. We should expect a general constriction of the Arctic as cold, treeless region owing to the expanding of woodlands and a dramatic decrease of the area of the high-Arctic tundra subzone (first) and that of the arctic tundra subzone (second). A greater northward advance of the tree line (that of larch in Asia, and spruce in America) should be expected in the continental sectors. As to the Atlantic or central Beringian sectors, we may prognosticate further activization of winter cyclones with thaws and stormy winds, snowdrift, etc., and the spreading of the Iceland, Aleutian, or Anadyr-Koryak (essentially treeless) situations.

4. The increase of the pressure by zonal hypoarctic oligotrophic acidophilous plant complexes, especially on interfluves, is highly probable, at least for the area of low-Arctic and will expand to mid-Arctic. A part of the area will be occupied by the northern taiga and woodlands where the leaching of soil is still stronger, and the peat accumulation is more intensive. All this will result in the extinction of many components of some intra- and extrazonal plant complexes, the further splitting of populations, and the

gene drifts, etc. The effect of the hypoarctic pressing could be in part weakened, however, due to the increased soil erosion and alluvium accumulation (the factor promoting seral plant complexes), as well as owing to calcium accumulation in the transaccumulative geochemical positions.

5. The expected very high rate of climatic warming and humidization will result in increased floristic undersaturation of plant cover, especially where physical or biotic barriers exist (e.g., high mountain ranges, such as the Brooks Range in Alaska, with respect to the dispersal of tree species) (Billings, 1991). In arctic islands due to a probable delay with diaspore transport, the flora will keep a more northern aspect whereas the vegetation will get a more southern appearance. This is now the case, for example, in the inner Spitzbergen in the "*Cassiope tetragona* zone." So, one can expect first the redistribution of the landscape positions of different ecological–geographic and biological plant groups, already present in the flora. These are: the activization of oceanic to suboceanic and various "southern" elements, present in the flora, with the respective enlarging of their associating ability, at the expense of cryophytes and continental species. One must note that even now many hypoarctic and arctic–boreal species are not uncommon in the arctic tundra subzone (e.g., *Salix pulchra* in Wrangel Island), whereas many arctic–alpine plants (including *Dryas spp*., *Cassiope tetragona*, and *Salix reticulata*) grow on basic rocks under the shelter of subarctic woodlands. Of some importance will be such factors as rapid speciation as hybridization, gene drift in small populations, and apomixis, etc.

6. It is the continental xero- and cryoxerophytes (including tundra-steppe relicts), with the precocious (vernal) seasonal development, that would get into the worst conditions. The plants are usually close to having completed their vegetating by early August or even earlier, despite the fact that August could be warmer than July on some arctic coastal areas. So, many steppe sedges and grasses in Wrangel Island start to turn yellow by mid-July, and arctic Dryas and Salix species in early August. Many steppe species show secondary dysjunction in the vicinities of Bering Strait with maritime climate. Thus, the increased depth of snow and the humidization of climate with a cool June may cause the extinction of relic populations of many such species, though some of them could survive on tors, screes, gravel floodplains, around rodents' burrows, in relic condition.

7. Some other plant complexes would benefit from the expected global changes. These may include plant complexes such as hypoarctic oligotrophic with dwarf birch, willows, ericaceous dwarf shrubs, mire and seepages plants, and also some snowbed plants (in particular, hemichionophytes), which at present are most diverse in more humid sectors of the Arctic.

8. The aggressiveness of zonal hypoarctic plant complexes, supported by the development of acid organic beds over a variety of rocks on plainy

interfluves and in part in valleys (on terraces), will create or strengthen biotic barriers for migrations of many other intrazonal and extrazonal plant complexes. However, it will hardly prevent the spreading of typical taiga plants, which are also organophilous. Carbonate landscapes will remain the oases against the background of floristically poor northern taiga, shrub, and tussock tundra areas. Some expansion of hypoarctic shrublands at the expense of tussocks could also be expected. Under these conditions, long-distance transport of plant diaspores by birds and other animals, man, wind, and water will be of special importance, as will be the migrations along linear geomorphic structures such as river valleys, sea coasts, and mountain ranges, etc.

9. Anthropogenic component of florogenesis will inevitably increase, hopefully not only in its destructive forms. The natural trend to the replacement of low-productive, but stable, "economical" tundra communities by mesic meadows due to grazing, fires, etc., will be supported by the humidization of climate. The unintentional or intentional transport of plant diaspores by man could to some extent compensate for the unprecedented high rate of climatically induced successions of zonal vegetation, both latitudinal and longitudinal.

Summary

As soon as the very high rate of climatic change in the Arctic is provided, one can expect a certain decrease of biodiversity of the arctic flora at the community, landscape (or local flora), and regional levels, because the plant immigration and, even more so, speciation won't have enough time to compensate for the extinctions. The first response of the arctic plant cover could be the redistribution of positions and functions in ecosystems of the available components of the flora, with the subsequent diversifying of populations, involving ecophysiological parameters. The process would hardly be catastrophic, because the modern biochorological diversity of plant complexes in the Arctic guarantees their ability to adapt to the expected global change, and analogous changes have occurred in the Quaternary history of the arctic biota, though possibly at a slower rate. Man could also play some positive role by controlling and regulating natural processes.

What is needed first is to increase and improve our knowledge of the structure, life, dynamics, and evolution of arctic ecosystems. This should include not only elementary ecosystems, but also more extensive and complex ones of a landscape level and still higher levels up to the whole Arctic ecosystem, which possesses certain mechanisms of buffering to climatic changes, originating from some qualities of biota. We also need to establish the monitoring of biodiversity of arctic flora and fauna at different levels. One of the key levels, I am convinced, is that of landscape, of elementary

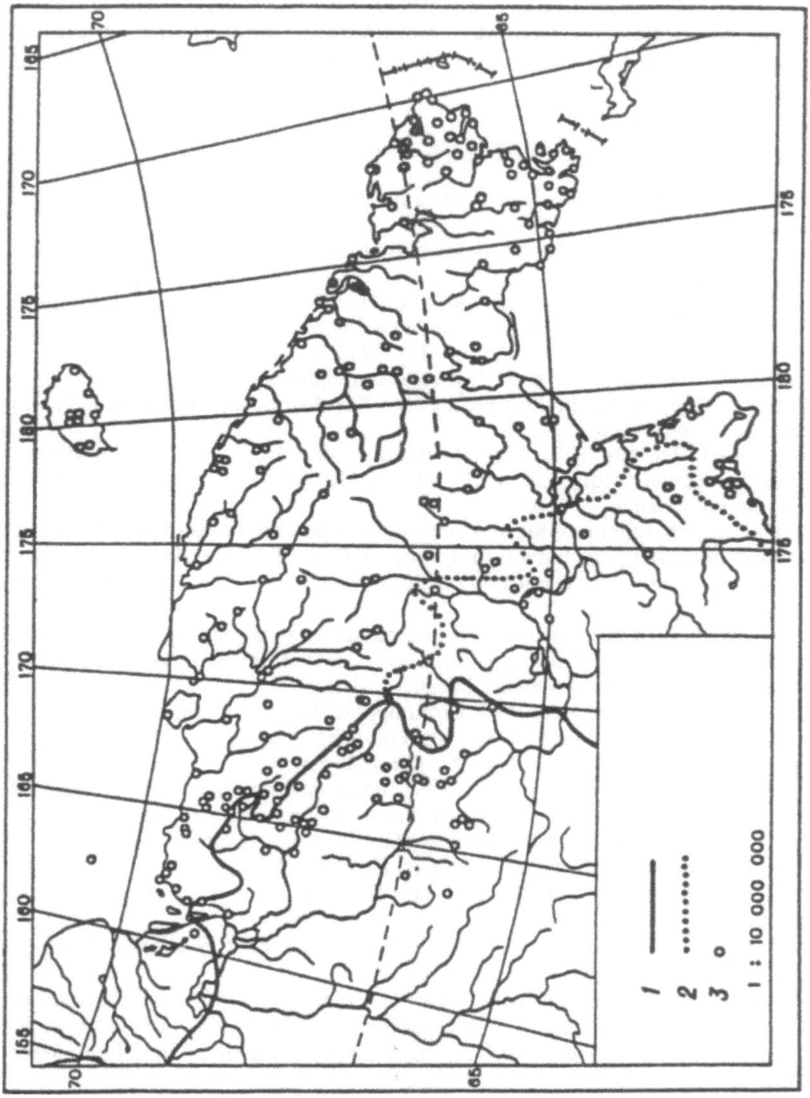

Figure 12.3. The local floras, studied in Chukotka in years 1964–1992. 1—the northeastern boundary of the forest (taiga) zone; 2—the northern and eastern boundary of the large stlaniks area; 3—the location of the floras studied.

(or local) flora, fauna, and biota. The only possible way to the realization of the idea is to establish the international network of well and thoroughly studied basic local floras (faunas, biotas) representative of biochorological diversity of the Arctic, with priority given to protected areas and intense research stations. In this activity the scientific community of nordic countries could use the rich experience, methodology, and concrete methods developed in Russia by A.I. Tolmachev's school (Yurtsev, 1987b) (Figure 12.3).

References

Akhmetyev, N.A. et al. 1978. Late Cenozoic stratigraphy and flora of Iceland. Nauka, Moscow.

Aleksandrova, V.D. 1977. Geobotanical subdivision of the Arctic and Antarctic. Nauka, Moscow.

Aleksandrova, V.D. 1980. The Arctic and Antarctic: their division into geobotanical areas. (Transl. by Doris Löve) (p. 247). Cambridge: Cambridge University Press.

Arctic Flora of the USSR. 1960–1987. Iss. I-X. Nauka, Leningrad.

Bennike, O., and Böcher, J. 1990. Forest-tundra neighboring the North Pole: plant and insect remains from the Plio-Pleistocene Kap Kobenhavn Formation, North Greenland. *Arctic* 43(4):331–338.

Bennike, O. 1990. The Kap Kobenhavn Formation: stratigraphy and paleobotany of a Plio-Pleistocene sequence in Peary Land, North Greenland. Meddel. Grønland, *Geoscience*. 23:1–85.

Billings, D. 1992. Phytogeographic and evolutionary potential of the arctic flora and vegetation in a changing climate. In: Chapin, F.C. et al. (eds.), *Arctic ecosystems in a changing climate: an ecophysiological perspective* (pp. 91–109). San Diego: Academic Press.

Bliss, L.C. 1981. North American and Scandinavian tundras and polar deserts. In: Bliss, L.C. et al. (eds.), *Tundra ecosystems: a comparative analysis* (pp. 8–24). Cambridge: Cambridge University Press.

Bliss, L.C., and Matveyeva, N.V. 1992. Circumpolar Arctic Vegetation. In: Chapin, F.C. et al. (eds.), *Arctic ecosystems in a changing climate: an ecophysiological perspective* (pp. 59–89). San Diego: Academic Press.

Bliss, L.C., Svoboda, J., and Bliss, D.L. 1984. Polar deserts, their plant cover, and plant production in the Canadian High Arctic. *Holarct. Ecol.* 7:304–324.

Bliss, L.C., and Svoboda, J. 1984. Plant communities and plant production in the western Queen Elisabeth Islands. Holarct. Ecol. 7:324–344.

Böcher, T.W., Holmen, K., and Jacobsen, K. 1968. *The flora of Greenland.* Copenhagen: P. Haase & Son.

Chernov, Yu.I. 1989. Warmth conditions and biota of the Arctic. *Ecology (Sverdlovsk)* 2:49–57.

Chernov, Yu.I, and Matveyeva, N.V. 1979. The regularities of zonal distribution of communities on Taimyr. In: Aleksandrova, V.D., and Matveyeva, N.V. (eds.), *Arctic tundras and polar deserts of Taimyr* (pp. 166–200). Leningrad: Nauka.

Edlund, S.A., and Alt, B.T. 1989. Regional congruence of vegetation and summer climate patterns in the Queen Elisabeth Islands, Northwest Territories, Canada. *Arctic* 42:3–23.

* Pulications in Russian are marked by an asterisk.

Gold, W.J., and Bliss, L.C. 1992. Does water availability constrain polar desert plant life? In: *International workshop on classification of circumpolar arctic vegetation* (pp. 53–54). Boulder: University of Colorado at Boulder.

Gorodkov, B.N. 1935. *Vegetation of the tundra zone of the USSR*. Academy of Sciences of the USSR, Moscow-Leningrad.

Gorodkov, B.N. 1956. Vegetation and soils of Kotelnyi Island (Novosibirskyi Archipelago). In: Tikhomirov, B.A. (ed.), *Vegetation of the Far North of the USSR and its utilization*. 2 (pp. 7–132). Academy of Sciences of the USSR, Moscow-Leningrad.

Herman, Y., and Hopkins, D.M. 1980. Arctic oceanic climate in Late Cenozoic time. *Science* 209:557–562.

Hultén, E. 1968. Flora of Alaska and neighboring territories. Stanford: Stanford University Press.

Hultén, E. 1973. Supplements to the Flora of Alaska and neighboring territories. A study in the flora of Alaska and the transberingian connection. *Botaniska Notiser* 126:459–512.

Löve, A., and Löve, D. 1975. *Cytotaxonomical Atlas of the Arctic Flora*. Vaduz: J. Cramer.

Matthews, J.V., Jr., and Ovenden, L.E. 1990. Late Tertiary plant macrofossils from localities in arctic/subarctic North America: A review of the data. *Arctic* 43:364–392.

Matveyeva, N.V. 1993. Climatic gradient and taxonomical structure of the flora of Taymyr Peninsula. In: Gilmanov, T. et al. (eds.), Global change and Arctic terrestrial ecosystems; an International Conference 21–26 August 1993, Oppdal, Norway. Trondheim: NINA.

Porsild, A.E. 1964. Illustrated flora of the Canadian Arctic Archipelago (2nd Ed., revised). Ottawa.

Rannie, W.F. 1986. Summer air temperature and number of vascular species in Arctic Canada. *Arctic* 39(2):133–137.

Rebristaya, O.V. 1977. Flora of the eastern Bolshezemelskaya Tundra. Nauka, Leningrad.

Rebristaya, O.V. 1987. The use of the concrete flora method in the West Siberian Arctic (Yamal Peninsula). In: Yurtsev, B.A. (ed.), *Theoretical and methodical problems in comparative floristics* (pp. 67–90). Leningrad: Nauka.

Sokolova, M.V. 1984. An attempt of quantitative comparison of eight local floras of Taimyr. *Botanicheskyi Zhurnal* 69(2):211–217.

Tolmachev, A.I. 1931. On the methods of comparative floristic investigations. 1. Concept of a flora in comparative floristics. *Journ. of Russian Botanical Society 16(1):171–124.*

Young, S.B. 1971. The vascular flora of St. Lawrence Island with special reference to floristic zonation in the Arctic regions. *Contrib. Gray Herb. Harvard University* 210:11–115.

Yurtsev, B.A. 1986. Megaberingia and cryo-xeric stages in the history of its plant cover. *Komarovian Lectures (Vladivostok)* XXXIII:3–53.

Yurtsev, B.A. 1987a. The significance of historical factor for the adaptations of plants to extreme environments of the arctic tundra subzone (Wrangel Island taken as example). *Botanicheskyi Zhurnal* 72(11):1436–1447.

Yurtsev, B.A., (ed.) 1987b. *Theoretical and methodical problems in comparative floristics* 1987. Leningrad: Nauka.

Yurtsev, B.A. 1991a. The problems of identification of the tundra type of vegetation. *Botanicheskyi Zhurnal* 76(1):30–41.

Yurtsev, B.A. 1991b. The completion of the publication of "The Arctic Flora of the USSR" as the key event on the way to publication of "The Panarctic Flora". *Botanicheskyi Zhurnal* 76(4):489–498.

Yurtsev, B.A. 1992. Floristic subdivision of the Circumpolar Arctic. In: *International workshop on classification of circumpolar tundra vegetation* (pp. 193–194). Boulder: University of Colorado at Boulder.

Yurtsev, B.A., Tolmachev, A.I., and Rebristaya, O.V. 1978. The floristic delimitation and subdivision of the Arctic. In: Yurtsev, B.A. (ed.), *The Arctic floristic region* (pp. 9–104). Nauka, Leningrad.

Yurtsev, B.A., Tolmachev, A.I., and Rebristaya, O.V. 1985. Maps in: Atlas of the Arctic (pp. 125–129). GUGK, Leningrad.

Yurtsev, B.A. 1994. Floristic division of the Arctic. *Journ. Veget.* Sci. 5:765–776.

13. Climate Change and Future Position of Arctic Tree Line

Peter A. Scott, Claude Lavoie, Glen M. MacDonald, Bjartmar Sveinbjörnsson, and Ross W. Wein

Studies of northern circumpolar tree line have related tree distribution to numerous environmental controls, such as permafrost, moisture, heat, radiation, and frontal circulation patterns (Barry, 1967; Bryson, 1966; Hare and Ritchie, 1972; Hustich, 1966; Larsen, 1989). Recent modeling experiments indicate that the circumpolar tree line could be shifted north tens to hundreds of kilometers if anticipated magnitudes of global warming due to increases in atmospheric greenhouse gasses are realized (e.g., Emmanuel et al., 1985; Rizzo and Wiken, 1992). Detecting and migitating the negative impacts of such vegetation shifts requires an understanding of the dynamics of tree line–climate interrelations. In this chapter we review some of the data and current concepts regarding climate–vegetation dynamics at the circumpolar tree line. In an effort to evaluate the mechanisms responsible for forest development we review selected northern European and North American studies of northern forests, the tree line, and forest–tundra, and evaluate the dominant influences and processes of forest development at the forest edge. We synthesize this information to suggest how the tree line might react in real terms to future climate warming.

Defining Northern Tree Lines

Delineation of tree line usually involves some kind of morphological criteria applied to tree line species (Stevens and Fox, 1991), and numerous definitions of tree lines have been applied, particularly along elevational

(Kullman, 1990; Tranquillini, 1979) and north–south gradients (Hustich, 1979; Payette, 1983; Sirois, 1992). The variability in the perception of tree line among researchers corresponds to differences in the species composition and ground cover (Viereck, 1979), but also landforms, climatic characteristics, and history of each tree line region, often results in unique gradations from forest to tundra.

Initially, most definitions of tree line are formed in reference to the physiognomy of single outlying species. For example, in Sweden, *Betula pubescens* extends beyond the range of *Picea abies* and *Pinus sylvestris* and is considered the tree line species (Kullman and Engelmark, 1991). However, as most tree line species can occur in shrub form (krummholz), species presence or absence is generally not used to define tree line. Once a tree species that delineates tree line is identified, some researchers use cover values to quantify stem or crown density and thereby define tree line as the edge of forest cover (e.g., Larsen, 1989). Many tree line classifications also refer to tree heights; for example, areas where trees do not exceed 5 m would be considered the "tree zone" within the forest–tundra of Québec (Payette, 1983). This tree zone extends well north of the zone of continuous forest. Within the forest–tundra another subdivision includes areas where trees are less than 2 m. This zone is considered the "shrub zone" or beyond the limit of (erect) trees (Kullman, 1983; Payette, 1983). Moving from the forest edge across the forest–tundra, the criteria for delineation of discrete tree line shifts from stem density (grouping) to morphological characteristics (size and shape of each individual). In this chapter we will define tree line as the edge of continuous forest cover. We refer to the region between tree line and tree species range limits as the forest–tundra.

Hypotheses for Forest Development

There are at least two processes that can lead to development of forest at the tree line. First, a forest may begin from the initial establishment of a relatively dense group of seedlings near the edge of continuous forest. As seedlings grow they extend more and more influence over the micrometeorological conditions by decreasing wind speeds, changing snowdrift patterns, etc. This in turn could enhance further development. The "cybernetic" (feedback) hypothesis infers an increasingly stable forest edge because of the declining effectiveness of weather to strongly influence within-stand conditions over time. Second, a forest could begin over a long period of time as individual trees scattered beyond the forest edge, which then increase in size and density over time, gradually improving the local environment and allowing the development of offspring under more favorable conditions. The "gradual change" hypothesis infers an initially unstable tree line because individual trees would be highly vulnerable to weather events and changes in climate patterns until sufficient crown density is reached.

Comparative rates of tree establishment in forest stands and forest–tundra trees in close proximity were examined by Scott et al. (1987b) at Churchill, Manitoba (Figure 13.1). Establishment of forest–tundra white spruce (*Picea glauca*) since the 1880s appears to be in proportion to the post-Little Ice Age (LIA) warming and the subsequent cooling from 1940 to 1980. Similar aged populations of forest trees appear to have begun establishing at the same time after the LIA but then establishment accelerated through much of this century. Northern tree line stands developed from an initial establishment phase of 30 to 50 years and then an infilling phase some 30 to 50 years later (Lavoie and Payette, 1992; Morneau and Payette, 1989; Scott et al., 1987a). Scott et al. (1987b) have shown that increases in summer heat availability corresponded to dramatic increases in the number of seeds germinating successfully in the establishment phase of the young forests, corresponded to a moderate increase in the number of seedlings during the infilling phase, and had little effect on older white spruce forests (Figure 13.2). As the trees in the forest become older they are less responsive to variations in summer heat availability. This may indicate that, once triggered, establishment in the forest may moderate the within-stand environment and this could feed back positively by encouraging further establishment.

The relationship between early establishment of seedlings and summer heat availability in white spruce is twofold. First, warmer summers strongly influence development of reproductive buds (Zasada et al., 1991), which leads to greater seed availability and more productive seed. Second, warmer summers promote greater rates of germination and seedling survival. Experiments by Scott (1994) demonstrate that seed germination and survivorship to first fall frost is significantly correlated to summer heat availability 3 years earlier. The typical cycle in tree line white spruce is reproductive bud formation (year 0) followed by flowering, cone production, and seed release (year 1), seed maturation on the ground (year 2), and germination (year 3) (Scott, 1994). The number of establishing seedlings relates to conditions during the year of bud formation. Seeds are often not viable at seed shed (e.g., Elliot, 1979) and after 2 or more years on the ground they have very low viability (Black and Bliss, 1980; Johnson, 1975; Payette et al., 1982). Consequently, in addition to healthy bud development, there must be sufficient heat available to germinate the seeds in the third year. Once germinated, almost all of the seedling mortality occurs before the end of the second growing season (Scott, 1994; Zasada et al., 1978). Thus, the entire stand initiation process occurs in a very short period of perhaps 3 to 5 years when greater heat availability effectively enables a forest to begin developing. Landhäusser and Wein (1993), studying an area that burned in 1968 near Inuvik, N.W.T., showed that most *Picea mariana* seedlings established between 6 and 14 years after the burn. *Picea glauca*,

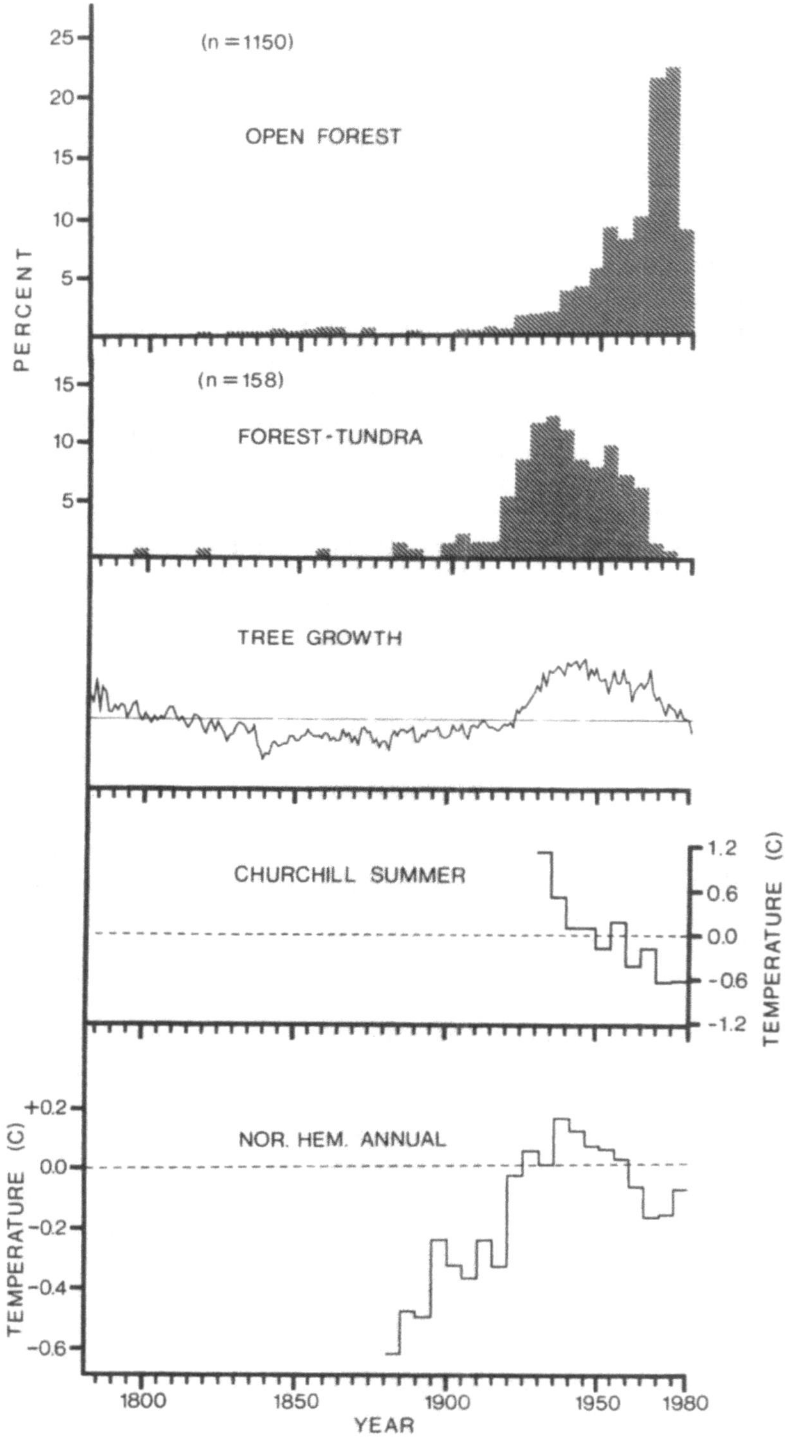

(n = 1150)
OPEN FOREST
PERCENT
25
20
15
10
5
(n = 158)
FOREST-TUNDRA
15
10
5
TREE GROWTH
CHURCHILL SUMMER
1.2
0.6
0.0
-0.6
-1.2
TEMPERATURE (C)
NOR. HEM. ANNUAL
+0.2
0.0
-0.2
-0.4
-0.6
TEMPERATURE (C)
1800
1850
1900
1950
1980
YEAR

Betula papyrifera, and *Populus balsamifera* continued to establish to the end of the 22-year postfire period.

Dynamics of Tree Line Forests

Fire is the major agent of destruction within the northern forests south of the tree line (Timoney and Wein, 1991) and less commonly at the circumpolar tree line and beyond (Sirois, 1992). At the tree line near Inuvik, N.W.T., limited sampling of fire scars on black spruce suggest that fires have burned most treed areas within the past 100 to 150 years. Scott (1994) has found that once established and then disturbed, the rate of white spruce forest regeneration within a patch is proportional to initial establishment rate of the parent population (Figure 13.3). Consequently, it appears that northern forests can accommodate frequent disturbances by fire. The initial opening of the canopy often results in dieback of the moss floor, which can release relatively large amounts of nitrogen held there (Oechel and Van Cleve, 1986) and enhance reproduction. With local sheltering (e.g., Odin and Degermark, 1990) and nearness to seed source (Zasada, 1986), there will be considerable momentum for that patch to infill as a forest.

The North American literature tends to focus on evergreens such as black and white spruce as tree line species; postfire forest development in areas dominated by these species may be slow or may fail (Landhäusser and Wein, 1993; Sirois and Payette, 1991). Often there are deciduous tree species at the tree line especially along north-flowing rivers. *Betula papyrifera* and *Populus balsamifera*, with their wind-dispersed seed, have invaded formerly treeless areas shortly after fires (e.g., in the Mackenzie Region) (Landhäusser and Wein, 1993). These authors suggest that northward forest advance under climate warming may occur in a stepwise fashion with each fire over time and similarly over space on the landscape, with each new burned patch within seed dispersal distance. Patch dynamics are a character of most northern forests as is disturbance, predominantly by fire but also flood, human activity, insect outbreak, and other events (Johnson, 1992; Payette et al., 1989; Pearce et al., 1988; Sveinbjörnsson, 1992). The momen-

Figure 13.1. The age structure of open forest and forest–tundra white spruce populations in primary succession that have developed simultaneously on either side of the tree line (from Scott et al., 1987b). The post-Little Ice Age summer warming (from Jones et al., 1986, bottom) appears to have triggered tree growth (Scott et al., 1988) and a major establishment on either side of the tree line. Subsequent summer cooling, documented from Churchill weather records, corresponds to a decline in the tree growth rate, and establishment/survivorship on the forest–tundra. However, in the developing forest, seedlings continue to establish.

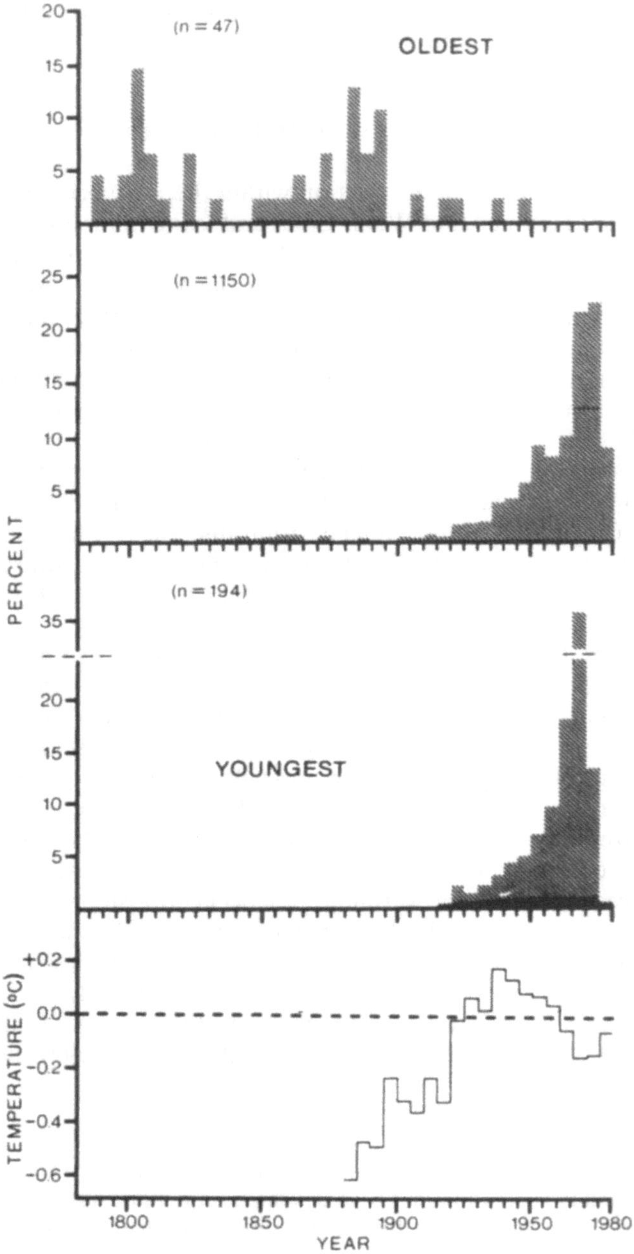
(n = 47)
OLDEST
(n = 1150)
(n = 194)
YOUNGEST
PERCENT
TEMPERATURE (°C)
YEAR
1800
1850
1900
1950
1980

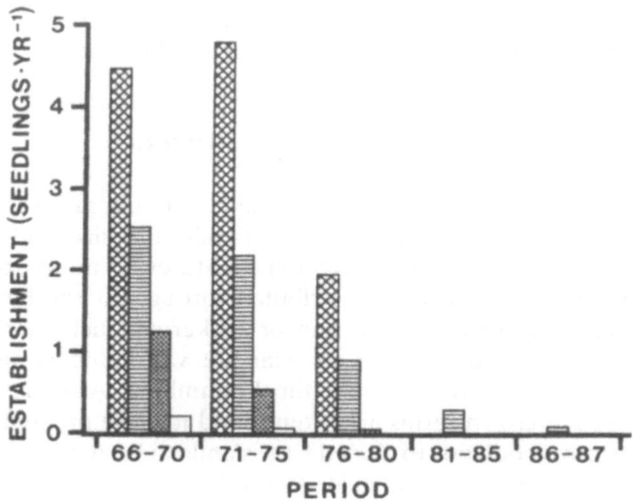

Figure 13.3. Regeneration of tree populations varies with changes in annual heat availability and the history of the parent population. In this example, the current rate of seedling establishment is greatest in a forest that originally established during a warm period (cross-hatching) and much less in forests that originally established during intermediate (lines) and cold (stippled) climatic periods. Seedling establishment on forest–tundra (clear) is minimal (after Scott, 1994).

tum developed in the early phase of tree establishment and the ability to infill a patch accommodates the disturbance mode and frequency.

The strongest establishment of boreal forest tree populations usually occurs after deep burning fires that have removed the surface cover and the organic matter (Johnson, 1992). Olsen (1993) has confirmed that this occurs under ambient conditions. Further, even under simulated climatic change conditions in the field at the arctic tree line, the establishment and subsequent growth of both black and white spruce is more responsive to exposure to mineral soil than either temperature or moisture changes. This has also been shown at Churchill, Manitoba, with white spruce (Scott, 1994). Subsequent studies (Landhäusser, Halliwell, and Wein, unpublished) suggest that there is sufficient energy available for tree growth at, and north of,

Figure 13.2. Three ages of open forest population on primary succession. Invasion and infilling phases in the oldest forest occurred during the LIA. The subsequent warming had no effect. In the middle-aged forest, the accelerated infilling corresponds to a climatic warming and the invasion of the youngest forest appears accelerated during this time as well. The younger forests appear most sensitive to climatic variability.

the present tree line but soil organic matter and vegetation cover insulating the mineral layer prevent successful seedling establishment.

Dynamics of Forest–Tundra

Beyond the forest edge the situation on the forest–tundra is very different. Trees form sparse aggregations of woodlands; the size and density of woodlands is often relative to topographic features. In more exposed situations krummholz are sparsely distributed throughout the landscape. In trees that can reproduce by layering or suckering, such as *Picea abies*, *P. glauca*, *P. mariana*, or *Populus*, for example, various dense aggregations of stems can develop from within the krummholz. Although all wind-abraded types of crown forms are often found in close proximity, there is a gradient from woodlands to cloned tree islands to krummholz or stemless mats (e.g., Kullman, 1990; Scott et al., 1993a, 1993b; Tranquillini, 1979). The dynamics of forest–tundra response to climate change are characterized by rapid changes from one growth type to another, which should not be confused with forest establishment. For example, this could occur when asexually produced mats of spruce produce upright stems in response to changes in ground temperature (Sveinbjörnsson et al., 1993) or snowcover (Filion et al., 1985; Lavoie and Payette, 1992). In this case the shrub zone would become a forest zone very quickly. Conversely, stem loss could reverse the process under severe conditions, as occurred in *Larix laricina* and *Picea glauca* at Churchill, Manitoba, after the severe summer of 1992 following the eruption of Pinatubo (pers. obs.). In areas where the trees do not clone and there are only two zones across the forest–tundra, severe thinning of the forest zone has been noted as well (e.g., Kullman and Högberg, 1989).

A woodland in the forest–tundra zone can appear very similar to the forest. However, the trees reach about 50% of the height of forest trees, most trees are wind altered, indicating very little shelter has developed within a woodland (Scott et al., 1993c), and in winter woodlands fill up with snow and become a snowdrift (Filion and Payette, 1983; Scott et al., 1993a), resulting in late snowmelt (Rouse, 1982, 1984) that enhances the establishment of snowbed and late-flowering plant species. Although undisturbed white spruce forests appear to establish with very little mortality, Lavoie and Payette (1992) show considerable stem mortality in forest–tundra black spruce and the dieback is in synchrony with climate trends. Woodlands also have considerably more asexual reproduction than is found in the forest and, therefore, can continue to develop in the presence of low summer heat availability. This evidence indicates that whereas woodlands appear as aggregations, the trees still function largely as individuals, responding to local conditions and remaining sensitive to environmental change.

Krummholz and stemless mats that dominate at the extreme limit of trees are in fact snow covered very early in the winter and remain protected from wind abrasion throughout the winter. Paradoxically, the smallest and slowest growing of the wind-abraded tree forms are often the least abraded (Scott et al., 1987a, 1993b). This indicates that there are secondary factors that limit the quality and quantity of tree growth, which ultimately determines the vulnerability of a tree to wind effects and other stresses.

Compared with forest trees, vertical shoot elongation is much less in forest–tundra trees on average (Abadie, 1991; Scott et al., 1987a, 1993a; Sveinbjörnsson et al., 1992, 1993). Low air temperatures and reduced sunlight may moderate tree growth but are rarely limiting factors (Day et al., 1991; Goldstein, 1981). Whereas the aboveground summer climate has been implicated in reducing growth, neither the mechanisms nor the controlling factors have been clearly established. Thus, a model based on seasonal net branch photosynthesis and respiration indicated a minimum net CO_2 uptake at northern tree line (Vowinckel, 1975), whereas another model based on dark respiration equivalents at the alpine tree line in Norway indicated insufficient respiration (Skre, 1991). However, both hypotheses remain unproven as recent measurements of CO_2 exchange and carbohydrate levels have failed to substantiate either (Abadie, 1991; Sveinbjörnsson et al., 1992).

Soil conditions have often been shown to relate to tree growth and tree existence. Lack of soil has been shown to affect tree growth at both northern and elevational tree lines (Holtmeier and Broll, 1992; Marr, 1948; Masuzawa, 1985). Soil temperatures have been shown to correlate better with bud burst and seasonal growth than air temperatures (Hansen-Bristow, 1986; Scott et al., 1987a); low soil temperatures have been shown to reduce net photosynthesis at alpine tree line (Delucia, 1986; DeLucia and Smith, 1987). Soil temperatures correlate strongly with subarctic soil decomposition rates and mineralization of carbon and nitrogen (e.g., Flanagan and Van Cleve, 1983; Van Cleve et al., 1981; Van Cleve and Yarie, 1986) and soil nitrogen was significantly lower beyond the mountain birch forest limit in northern Sweden (Davis et al., 1991).

Low nutrient mineralization rates presumably affect available soil nutrients for tree uptake, although the uptake process itself may be limited by low temperature (Chapin, 1986; Van Cleve et al., 1990). Addition of nutrients around mountain birch trees at different elevation resulted in increasing height growth stimulation with increasing elevation (Sveinbjörnsson et al., 1992).

The overall status of the aboveground portions of the tree also influences the rate of uptake of nutrients and water. For example, poorly developed evergreen needles may receive considerable cuticular abrasion during the winter (Hadley and Smith, 1983, 1989). Stomata closure, which usually regulates waterflow in the tree, will be ineffective against water loss without the waxy cuticle on the needles. Consequently, isolated trees may

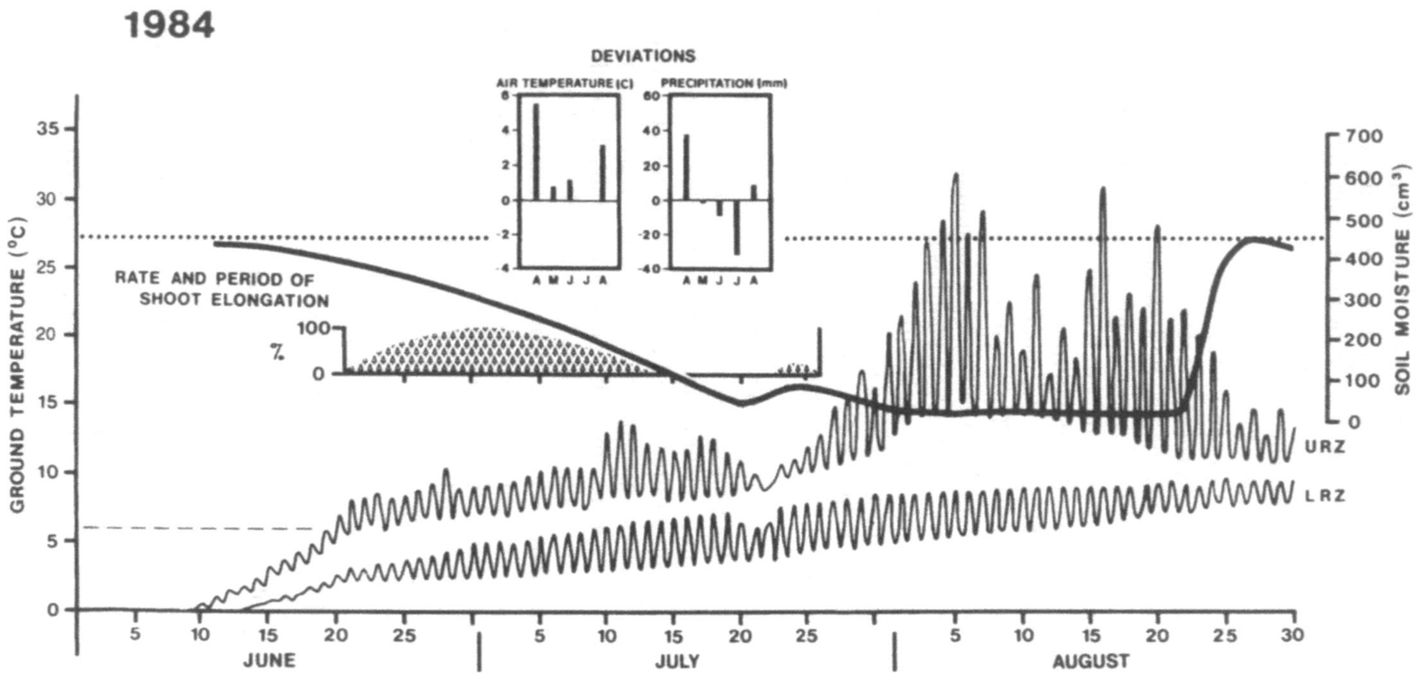

Figure 13.4. A composite model of tree line white spruce growth and environmental conditions based on 1984 data and subsequent studies (after Scott et al., 1987a, 1993b). The upper and lower root zone (URZ & LRZ) begin to thaw in mid-June and shoot elongation begins shortly thereafter before snowmelt has finished. Consequently, the ground is moist (heavy line) and shoot elongation is mostly over before the heat of the summer has arrived. After a rainfall on the 21st of July elongation resumes again for a short period. The trees are snowbed species adapted to growing at low temperatures, largely with meltwater from their own snowdrift. Although moisture stress is not evident, adaptations to prevent moisture stress are.

Table 13.1. The Percentage of Adult White Spruce Trees That Produced Cones at the Tree Line near Churchill, Manitoba, at Random Locations Within the Open Forest, Forest–Tundra Woodlands, Forest–Tundra Islands, and Stemless Krummholz

Site Type	1987	1988	1989	1990	1991	Avg.
Open forest	50.0	100.0	64.0	93.0	3.8	62.1
Woodland	23.0	100.0	54.0	88.8	0.2	53.2
Tree islands	7.0	60.0	27.0	43.8	0.1	27.5
Stemless mats	<0.1	7.0	<0.1	<0.1	<0.1	1.4
Regional avg.[a]	26.6	86.6	48.3	75.2	1.4	

From Staniforth et al. (1993).
[a] Krummholz not included in regional average.

reduce waterflow and other growth-related activities to prevent moisture deficit conditions from developing (Scott et al., 1987a, 1993b) (Figure 13.4). Compared to forest white spruce, the slower growing forest–tundra trees produce needles that have a poorer cuticle and are therefore more vulnerable to abrasion (Scott et al., 1993a) and have poorer height growth. This often means that a stem cannot grow up above the snow level and that sexual reproduction is unlikely to occur (Table 13.1). Loss of evergreen needles increases towards the tree line in northern Québec (Werren, 1979) and with it are lost considerable storage pools of carbohydrates and nutrients in addition to future whole-tree reduction in net photosynthesis. Thus, material acquisition and especially net material balance is reduced towards the tree line.

When a white spruce seedling becomes established in the forest–tundra, the establishment of moss under the crown initially coincides with greater moisture retention and warmer ground conditions (Gill, 1975; Scott et al., 1993b). The corresponding surge in growth often results in a shoot developing from the island. Peat accumulation beneath the moss leads to drying, permafrost intrusion into the roots, frost heaving, and eventual loss of the stem from the inability to repair the increasing damage from snow abrasion (Scott et al., 1993b). Consequently, species that form krummholz can be reservoirs that reproduce asexually and then under more favorable conditions become individuals with erect stems or islands that can reproduce sexually (Filion et al., 1985; Scott et al., 1993c) (Figure 13.5). Black spruce, for example, are better adapted to growing in low nutrient conditions than white spruce. Consequently, black spruce may be more widespread on the forest–tundra than white spruce, but the loss of the stem terminal will be more frequent because of the slow growth.

Changes in the local environment of the tree as well as regional weather variation are all factors that will influence the production and loss of stem terminals of forest–tundra trees (e.g., Arseneault and Payette, 1992). In particular, the acting mechanism on a stressed tree is snow abrasion in

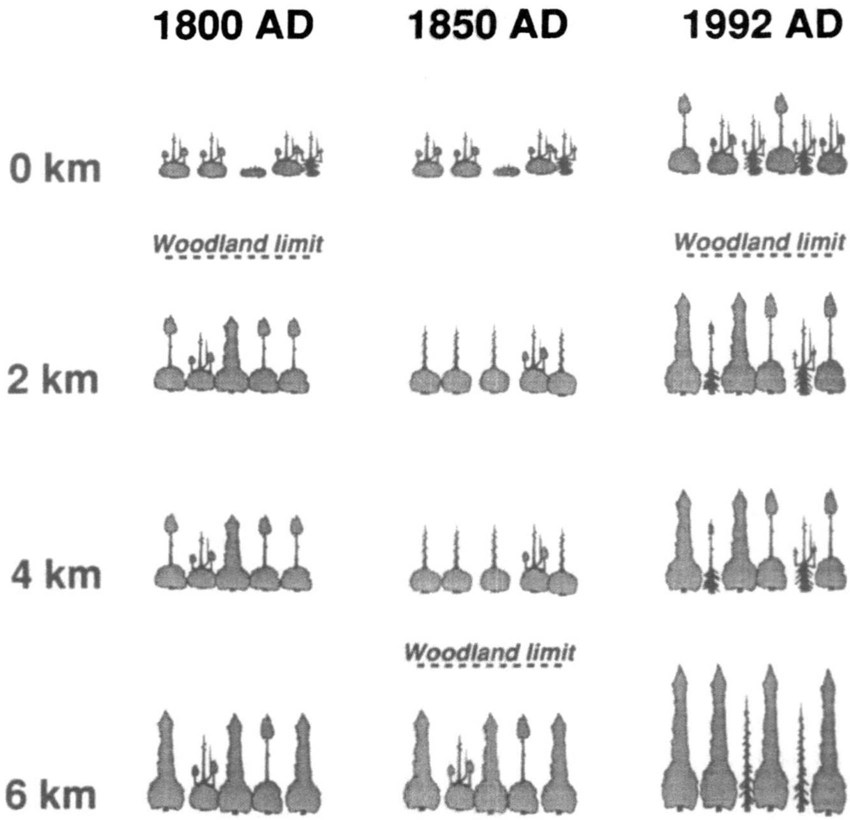

Figure 13.5. The development of black spruce woodlands located at the extreme limit of forest–tundra and 2, 4, and 6 km south during 1800 to 1992. Over this period there was no major change in the spruce krummholz at the tree limit. Marginal woodland trees were severely abraded during the first part of the 1800s, resulting in the loss of stems, and the trees became islands or krummholz. Over this period the woodland effectively retreated some 4 km. Since the late 1800s, the woodland moved 4 km northward as stems developed on the trees in response to the climate warming of the 20th century. During the last 100 to 150 years, at the sites 2 and 4 km within the woodland, shrubby spruces were replaced by stemmed trees. The site 6 km within the woodland was not affected by the harsh winter conditions during the 1800s (after Lavoie and Payette, 1994).

winter (Scott et al., 1993a) so that any condition that affects the quantity of snowfall (Lavoie and Payette, 1992) or the redistribution of snow (Filion and Payette, 1983; Scott et al., 1993c) will have considerable impact initially. During periods of climatic warming, changes in tree form and stem density may occur within decades (e.g., Filion et al., 1985; Lavoie and Payette, 1994;

MacDonald et al., in prep.) but be highly reversible (e.g., Lavoie and Payette, 1992; Payette and Gagnon, 1985).

Stem development in krummholz or tree islands can change local conditions in favor of further tree development and ultimately lead to the formation of woodlands. For example, as ground conditions deteriorate beneath spruce tree islands, the growth becomes poorer and poorer and the density of needles and branches is greatly reduced at snow level. Consequently, the downwind snowdrift will elongate back into the mass of infranival branches and further downwind as well. The moderation in snow distribution can greatly enlarge the region where seedlings establish and where clones begin to develop. In one example documented by Scott et al. (1993c), 12 such islands of white spruce coalesced to become a spruce woodland. Although there is considerable evidence for the reversibility of these processes at each stage including woodlands, there are apparently no studies that document the transition from woodlands to the more stable forest communities.

Documented rapid increases in stem densities on the forest–tundra have been shown to correspond with 20th century warming (Kullman, 1979; Payette and Filion, 1985; Scott et al., 1987b; MacDonald et al., in prep.) and can be reversed in a similar time frame. Once the woodland has been established, a period of climate cooling, for example, can lead to a shift from woodland to krummholz, effectively changing the limit of woodlands. In the Boniface River area of subarctic Québec, a reversion from lichen–spruce woodland to krummholz apparently occurred 150 years ago (Figure 13.5) during one of the coldest parts of the Little Ice Age (Bégin, 1991; Lavoie and Payette, 1992, 1994). Since the late 1800s, the woodlands advanced some 4 km northward, most likely in response to mild winter conditions (Bégin and Payette, 1988; Lavoie and Payette, 1992, 1994; Payette and Delwaide, 1991). The krummholz developed progressively into woodland, because of the change in the growth form—tree height increased some 1 m on average. Although 80% of spruce produced cones (e.g., in 1991), no evidence for a recent spruce establishment in the tundra was recorded, possibly because released seeds have low viability (Luc Sirois, pers. comm.). Indeed, within the woodlands sampled, fewer than 10 seedlings per 900 m^2 quadrat were found. Furthermore, no seedling or sapling was found at 48 well-drained hill sites located up to 16 km north of the extreme limit of forest–tundra. Spruces growing at these sites were all long-lived shrubs that originated from layering (Lavoie and Payette, 1994). Thus, the development and loss of woodlands in the Boniface River area over the last 200 years were the result of in situ shifts in form initially during the cold period that dominated in the 1800s and later during mild winter conditions the have persisted since the late 1800s.

Fossil pollen studies from central Canada also indicate that increases in tree cover within the northern limits of the forest–tundra can occur rapidly and synchronously over wide areas. MacDonald et al. (1993) analyzed sediments from four lakes and showed that the pollen percentages of black spruce more than doubled at all of the lakes at approximately 5000 BP (Figure 13.6). The increase likely reflects a change in crown form and an infilling of forest–tundra, and may have occurred in a period of less than 150 years. The percentages of black spruce pollen then declined again at 4000 BP, resulting in the establishment of the extremely sparse forest–tundra that occurs in the region today. The rate of this decline in tree cover was also extremely rapid at some sites (Figure 13.6).

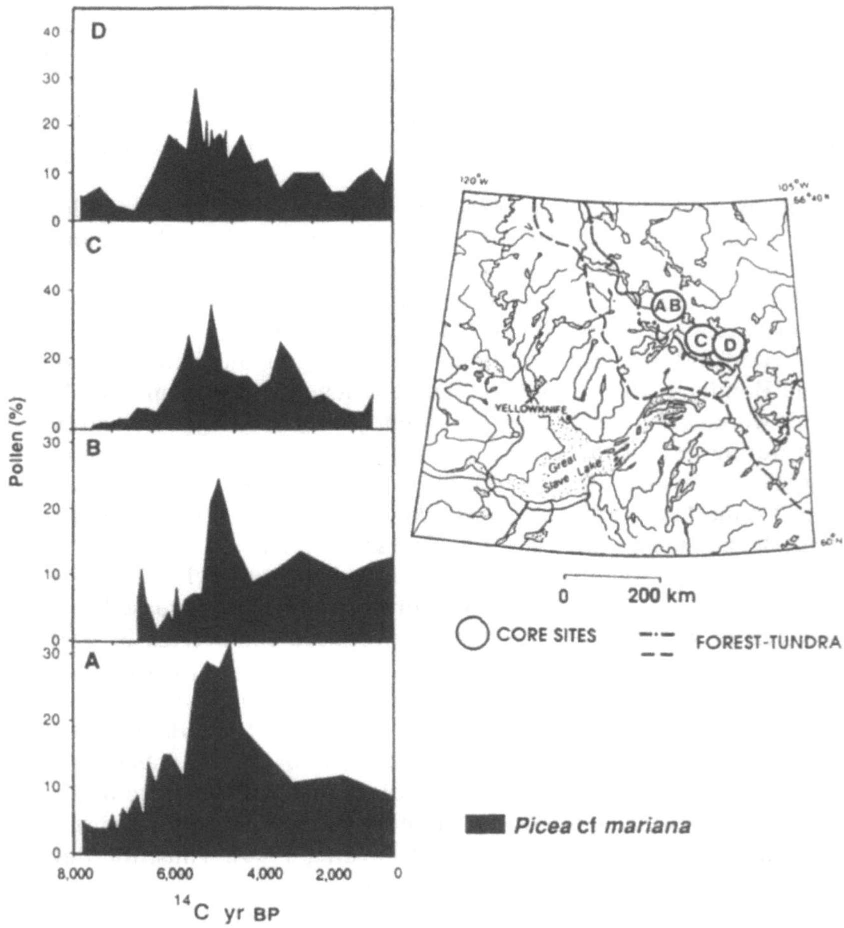

Figure 13.6. Mid-Holocene increases and decreases in *Picea mariana* pollen in lake sediments from tree line in central Canada. The pollen records suggest rapid and synchronous variations in tree cover due to climatic changes. Perhaps pollen productivity and tree form changes enhanced pollen production during this period (after MacDonald et al., 1993).

Evidence for a generative model of forest–tundra development appears to be: (1) the very sparse and gradual establishment of krummholz and particularly of those species that can reproduce asexually; (2) an increase in stem height and density, which can change the local snow distribution to favor successful establishment of trees; (3) the formation of woodlands that successfully reproduce sexually. Tundra communities burn much less frequently than forests (Wein, 1976); these fires are patchy and tree islands in microsites are affected to a limited extent on the landscape scale. Conversely, evidence for a degenerative model of forest–tundra development indicates: (1) an increase in fire frequency either at a stable forest edge or in the tree zone on the forest–tundra; (2) a decline in the regenerative capacity of the trees; (3) eventual loss of trees and the development of shrub zone or tundra.

Future Climate Change and Arctic Tree Line

Climate change will strongly influence the establishment of seedling cohorts, which ultimately determines whether there will be forest or individual trees (i.e., forest–tundra). Once a forest is established it appears adapted to infilling the gaps and clearings caused by disturbances and consequently is largely unresponsive to short-term climate change. The forest–tundra tree forms (i.e., ecotone) are much more responsive to short-term fluctuations where stemless krummholz, for example, can very quickly develop stems, produce flowers, and release large quantities of pollen and seed, and then become asexual stemless forms again. However, lack of suitable soil substrate for germination and seedling survival coupled with high rates of wind abrasion may act against forest development at the northward limit of tree species.

Palaeoecological records from fossil pollen and plant macrofossils provide some insight into the potential for the tree line to change in position or for shifts in tree density in response to climatic change. Stumps of *Picea* have been found in growth position on the tundra of the Tuktoyaktuk Peninsula of northwestern Canada (Ritchie, 1984). Radiocarbon dates from these stumps range from 6620 to 4940 BP (BP = radiocarbon years before AD 1950). Tree ring widths from the stumps indicate rates of growth similar to trees now found in the northern boreal forest over 100 km to the south. This evidence of a northward displacement of the forest border is consistent with evidence from fossil pollen studies of lake sediment in the region that suggest northern boreal forest was present on the peninsula between approximately 9000 and 5000 BP. The northward advance of the forest was in response to warmer summer temperatures that prevailed at that time (Ritchie, 1984; Ritchie et al., 1983). The almost complete absence of even krummholz spruce on the peninsula today indicates that late Holocene climate cooling has been sufficient to overcome the positive feedbacks associated with forest presence and produce deforestation. Similarly, samples of *Larix* wood have been recovered on the eastern Taimyr Peninsula of Siberia—on completely treeless tundra approximately 200 km north of the northern forest limit. This wood has produced radiocarbon ages ranging from 8760 to 5770 BP (Khotinsky, 1983) and suggests a northward displacement of forest in the Taimyr of up to 200 km in the late Holocene.

Evidence for significant displacements of forest is not ubiquitous in the circumpolar tree line zone. For example, although fossil pollen studies indicate that forest–tundra trees in northern Québec were burned off and failed to regenerate, Gajewski et al. (1993) do not provide evidence of changes in the latitudinal limits of forest over the course of the Holocene. Likewise, there is scant evident of change in northern forest limits from fossil pollen studies in Alaska (Brubaker et al., 1983). In the case of Alaska, the Brooks Range may have provided a physiographic barrier to the northward expansion of the forest in the past. Northern Québec was not completely deglaciated until the mid-Holocene, and forest was precluded from taking advantage of maximum northern hemisphere summer insolation that occurred during the early to mid-Holocene (Ritchie et al., 1983). The constancy of the northern forest in Québec may reflect strong positive feedback mitigating against the late Holocene climatic cooling.

Both ecological and palaeoecological studies suggest that care should be taken in predicting the response of arctic tree line to future global warming. It is clear that a number of factors, including air and soil temperatures, moisture, wind and associated snow abrasion, could influence rates of change in tree form and distribution, as well as advance in the northern limit of the tree line species. Past responses at tree line to climate change have been geographically variable, making it difficult to use palaeoecological records as a general approximation of how tree line will respond to future climatic change. However, these records and ecological observations do suggest that changes in the tree density in forest–tundra may respond relatively rapidly to climate warming.

Summary

Northern forest trees function as an interacting system that can moderate the environment by altering or eliminating many direct weather influences. Evidence shows that the moderated conditions in establishing forests corresponds to enhanced tree growth which promotes reproduction and further infilling of tree stands in marginal environments.

In the forest–tundra trees do not establish in sufficient density to moderate constraining environmental factors, and the trees fail to function as a unit. In addition to marginal heat availability that limits tree growth near the forest edge, individual trees experience the full severity of the weather.

This results in continuous damage that further impedes growth. At some point poleward, either damage will be severe enough, or there will be insufficient heat energy to enable growth and repair of the ongoing damage to individual trees. Consequently, there is a gradient on the forest–tundra from sparse groupings of erect wind-abraded trees, to dense multistemmed individuals that form when slow terminal growth is continually destroyed by abrasion at the snow surface, to prostrate stemless mats.

While the forest may develop as tree complexes that create positive feedback for further development, beyond the tree line the trees of the forest–tundra ecotone develop as individuals largely under negative feedback. Palaeoecological records indicate that long-term changes in climate during the Holocene have been sufficient to cause northward extensions and subsequent retreats of northern forest for tens of kilometers or more in northwestern Canada and central Eurasia. Palaeoecological evidence from the forest–tundra in central Canada and Québec indicates that during the Holocene long- and short-term variations of temperature have caused both rapid increases in tree density and similarly rapid decreases. The rapid increase in tree density during short periods of warming suggest that the intensity of the negative feedbacks working against tree establishment in the forest–tundra is continually changing and may be rapidly overcome during a significant warming. In recent examples seedling establishment on forest–tundra has increased in coincidence with the climate warming that has occurred during the middle of this century.

References

Abadie, W.D. 1991. Growth and carbon flux of white spruce at different elevations in the Chugach Mountains, Alaska. M.Sc. thesis, University of Alaska at Anchorage.

Arseneault, D., and Payette, S. 1992. A postfire shift from lichen–spruce to lichen–tundra vegetation at tree line. *Ecology* 73:1067–1081.

Barry, R.G. 1967. Seasonal location of the Arctic Front over North America. *Geogr. Bull.* 9:79–95.

Bégin, C. 1991. Analyse architecturale et dendroécologique d'une pessière à lichens à la limite des forêts. Ph.D. dissertation, Université Laval.

Bégin, Y., and Payette, S. 1988. Dendroecological evidence of lake-level changes during the last three centuries in subarctic Québec. *Quat. Res.* 30:210–220.

Black, R.A., and Bliss, L.C. 1980. Reproductive ecology of *Picea mariana* (Mill.) BSP., at treeline near Inuvik, Northwest Territories, Canada. *Ecol. Monogr.* 50:331–354.

Brubaker, L.E., Garfinkel, H.L., and Edwards, M.E. 1983. A late Wisconsin and Holocene vegetation history from the central Brooks Range: Implications for Alaskan palaeoecology. *Quat. Res.* 20:194–214.

Bryson, R.A. 1966. Air masses, streamlines, and the boreal forest. *Geogr. Bull.* 8:228–269.

Chapin, F.S., III. 1986. Controls over growth and nutrient use by taiga forest trees. In: Van Cleve, K., Chapin, F.S., III, Flanagan, P.W., Viereck, L.A., and Dyrness, C.T. (eds.), *Forest Ecosystems in the Alaska Taiga* (pp. 96–111). New York: Springer-Verlag.

Davis, J., Schober, A., Bahn, M., and Sveinbjörnsson, B. 1991. Soil carbon and nitrogen turnover at and below the elevational treeline in northern Fennoscandia. *Arctic Alpine Res.* 23:279–286.

Day, T.A., Heckathorn, S.A., and DeLucia, E.H. 1991. Limitations of photosynthesis in *Pinus taeda* L. (Loblolly Pine) at low soil temperatures. *Plant Physiol.* 96:1246–1254.

DeLucia, E.H. 1986. Effect of low root temperature on the net photosysnthesis, stomatal conductance and carbohydrate concentration in Engelmann spruce (*Picea engelmannii* Parry *ex* Engelm.) seedlings. *Tree Physiol.* 2:143–154.

DeLucia, E.H., and Smith, W.K. 1987. Air and soil temperature limitations on photosynthesis in Englemann spruce during summer. *Can. J. Forest Res.* 17:527–533.

Elliot, D.L. 1979. The current regenerative capacity of northern Canadian trees, Keewatin, N.W.T., Canada: Some preliminary observations. *Arctic Alpine Res.* 11:243–251.

Emmanuel, W.R., Shugart, H.H., and Stevenson, M.P. 1985. Climatic change and broad-scale distribution of terrestrial ecosystem complexes. *Climatic Change* 7:29–44.

Filion, L., and Payette, S. 1983. Régime nival et végétation chionophile à Poste-de-la-Baleine, Nouveau-Québec. *Nat. Can.* 109:557–571.

Filion, L., Payette, S., and Gauthier, L. 1985. Analyse dendroclimatique d'un krummholz à la limite des arbres, lac Bush, Québec nordique. *Geogr. Phys. Quat.* 39:221–226.

Flanagan, P.W., and Van Cleve, K. 1983. Nutrient cycling in relation to decomposition and organic matter quality in taiga ecosystems. *Can. J. Forest Res.* 13:795–817.

Gajewski, K., Payette, S., and Ritchie, J.C. 1993. Holocene vegetation history at the boreal-forest—shrub-tundra transition in north-western Québec. *J. Ecol.* 81:433–443.

Gill, D. 1975. Influence of white spruce trees on permafrost-table microtopography, MacKenzie River Delta. *Can. J. Earth Sci.* 12:263–272.

Goldstein, G.H. 1981. Ecophysiological and demographic studies of white spruce (*Picea glauca* [Moench] Voss) at treeline in the Central Brooks Range of Alaska. Ph.D. dissertation, University of Washington.

Hadley, J.L., and Smith, W.K. 1983. Influence of wind exposure on needle desiccation and mortality for timberline conifers in Wyoming, U.S.A. *Arctic Alpine Res.* 15:127–135.

Hadley, J.L., and Smith, W.K. 1989. Wind erosion of leaf surface wax in alpine timberline conifers. *Arctic Alpine Res.* 21:392–398.

Hansen-Bristow, K.J. 1986. Influence of increasing elevation on growth characteristics at timberline. *Can. J. Bot.* 64:2517–2523.

Hare, F.K., and Ritchie, J.C. 1972. The boreal bioclimates. *Geogr. Rev.* 62:333–365.

Holtmeier, F.-K., and Broll, G. 1992. The influence of tree islands and microtopography on pedoecological conditions in the forest–alpine tundra ecotone on Niwot Ridge, Colorado Front Range, U.S.A. *Arctic Alpine Res.* 24:216–228.

Hustich, I. 1966. On the forest-tundra and northern treelines. *Ann. Univ. Turku* A. 11:7–47.

Hustich, I. 1979. Ecological concepts and biogeographical zonation in the north: The need for a generally accepted terminology. *Holarctic Ecol.* 2:208–217.

Johnson, E.A. 1975. Buried seed populations in the subarctic forest east of Great Slave Lake, Northwest Territories. *Can. J. Bot.* 53:2933–2941.

Johnson, E.A. 1992. *Fire and Vegetation Dynamics: Studies from the North American Boreal Forest.* New York: Cambridge University Press.

Jones, P.D., Raper, S.C.B., Bradley, R.S., Diaz, H.F., Kelly, P.M., and Wigley, T.M.L. 1986. Northern Hemisphere surface air temperature variations: 1851–1984. *J. Climate Appl. Meteorol.* 25:161–179.

Khotinsky, N.A. 1983. Holocene vegetation history. In: Velichko, A.A. (ed.), *Late Quaternary Environments of the Soviet Union* (pp. 179–200). Minneapolis: University of Minnesota Press.

Kullman, L. 1979. Change and stability in the altitude in the birch tree-limit in the southern Swedish Scandes 1915–1975. *Acta Phytogeogr. Suecicam* 65:1–121.

Kullman, L. 1990. Dynamics of altitudinal tree-limits in Sweden: a review. *Norsk Geografisk Tidsskrift* 44:103–116.

Kullman, L., and Engelmark, O. 1991. Historical biogeography of *Picea abies* (L.) Karst. at its subarctic limit in northern Sweden. *J. Biogeogr.* 18:63–70.

Kullman, L., and Högberg, N. 1989. Rapid natural decline of upper montane forests in the Swedish Scandes. *Arctic* 42:217–226.

Landhäusser, S.M., and Wein, R.W. 1993. Post-fire vegetation recovery and tree establishment at the arctic treeline: Climate-change–vegetation-response hypothesis. *J. Ecol.* 81:665–672.

Larsen, J.A. 1989. *The Northern Forest Border in Canada and Alaska.* New York: Springer-Verlag.

Lavoie, C., and Payette, S. 1992. Black spruce growth forms as a record of a changing winter environment at treeline, Quebec, Canada. *Arctic Alpine Res.* 24:40–49.

Lavoie, C., and Payette, S. 1994. Recent fluctuations of the lichen-spruce forest limit in subarctic Québec. *J. Ecol.* 82:725–734.

MacDonald, G.M., Edwards, T.W.D., Moser, K.A., Pienitz, R., and Smol, J.P. 1993. Rapid response of treeline vegetation and lakes to past climate warming. *Nature* 361:243–246.

Marr, J.W. 1948. Ecology of the forest–tundra ecotone on the east coast of Hudson Bay. *Ecol. Monogr.* 18:117–144.

Masuzawa, T. 1985. Ecological studies on the timberline of Mt. Fuji. I. Structure of plant community and soil development on the timberline. *Bot. Mag. Tokyo* 98:15–28.

Morneau, C., and Payette, S. 1989. Postfire lichen–spruce woodland recovery at the limit of the boreal forest in northern Québec. *Can. J. Bot.* 67:2770–2782.

Odin, H., and Degermark, C. 1990. The spring in the forest terrain at Svartberget, northern Sweden. *Geogr. Annaler* 72A:167–178.

Oechel, W.C., and Van Cleve, K. 1986. The role of bryophytes in nutrient cycling in the taiga In: Van Cleve, K., Chapin, F.S., III, Flanagan, P.W., Viereck, L.A., and Dyrness, C.T. (eds.), *Forest Ecosystems in the Alaska Taiga* (pp. 121–137). New York: Springer-Verlag.

Olsen, S.K. 1993. Simulated fire severity and seedling growth at treeline: A climate change analogue. M.Sc. thesis, University of Alberta, Edmonton.

Payette, S. 1983. The forest–tundra and present treelines of the northern Quebec-Labrador peninsula. *Nordicana* 47:3–23.

Payette, S., and Delwaide, A. 1991. Variations séculaires du niveau d'eau dans le bassin de la rivière Boniface (Québec nordique): Une analyse dendroécologique. *Geogr. Phys. Quat.* 45:59–67.

Payette, S., and Filion, L. 1985. White spruce expansion at the treeline and recent climatic change. *Can. J. Forest Res.* 15:241–251.

Payette, S., and Gagnon, R. 1985. Late Holocene deforestation and tree regeneration in the forest–tundra of Québec. *Nature* 313:570–572.

Payette, S., Deshaye, J., and Gilbert, H. 1982. Tree seed populations at the treeline in Rivière aux Feuilles area, Northern Québec, Canada. *Arctic Alpine Res.* 14:215–221.

Payette, S., Morneau, C., Sirois, L., and Desponts, M. 1989. Recent fire history of the northern Québec biomes. *Ecology* 70:656–673.

Pearce, C.M., McLennan, D., and Cordes, L.D. 1988. The evolution and maintenance of white spruce woodlands on the Mackenzie Delta, N.W.T., Canada. *Holarctic Ecol.* 11:248–258.

Ritchie, J.C. 1984. *Past and Present Vegetation of the Far Northwest of Canada*, Toronto: University of Toronto Press.

Ritchie, J.C., Cwynar, L.C., and Spear, R.A. 1983. Evidence from north-west Canada for an early Holocene Milankovitch thermal-maximum. *Nature* 305:126–128.

Rizzo, B., and Wiken, E. 1992. Assessing the sensitivity of Canada's ecosystems to climatic change. *Climatic Change* 21:37–55.

Rouse, W.R. 1982. Microclimate of low arctic tundra and forest at Churchill, Manitoba. In: *Proceedings of the Fourth Canadian Permafrost Conference* (pp. 68–80). Ottawa: National Research Council.

Rouse, W.R. 1984. Microclimate of the arctic treeline 1. Radiation balance of tundra and forest. *Water Resources Res.* 20:56–66.

Scott, P.A. 1994. Influence of summer heat availability on forest establishment at the treeline near Churchill, Manitoba. Ph.D. dissertation, University of Toronto.

Scott, P.A., Hansell, R.I.C., and Erickson, W.R. 1993a. Influences of wind and snow on treeline environments at Churchill, Manitoba, Canada. *Arctic* 46:316–323.

Scott, P.A., Hansell, R.I.C., and Fayle, D.C.F. 1987b. Establishment of white spruce populations and responses to climatic change at the treeline, Churchill, Manitoba, Canada. *Arctic Alpine Res.* 19:45–51.

Scott, P.A., Staniforth, R.J., and Hansell, R.I.C. 1993c. Responses of white spruce to tundra environments: Woodland development (unpublished manuscript).

Scott, P.A., Bentley, C.V., Fayle, D.C.F., and Hansell, R.I.C. 1987a. Crown forms and shoot elongation of white spruce at the treeline, Churchill, Manitoba, Canada. *Arctic Alpine Res.* 19:175–186.

Scott, P.A., Fayle, D.C.F., Bentley, C.V., and Hansell, R.I.C. 1988. Large scale changes in atmospheric circulation interpreted from patterns of tree growth at Churchill, Manitoba, Canada. *Arctic Alpine Res.* 20:199–211.

Scott, P.A., Hansell, R.I.C., Fayle, D.C.F., and Staniforth, R.J. 1993b. Responses of white spruce to tundra environments: Development of tree islands (unpublished manuscript).

Sirois, L. 1992. The transition between boreal forest and tundra. In: Shugart, H.H., Leemans, R., and Bonan, G.B. (eds.), *Systems Analysis of the Global Boreal Forest* (pp. 196–215). Cambridge: Cambridge Press.

Sirois, L., and Payette, S. 1991. Reduced postfire tree regeneration along a boreal forest—forest-tundra transect in northern Québec. *Ecology* 72:619–627.

Skre, O. 1991. Physiology of plant survival under cold conditions, with particular reference to dark respiration as a factor limiting growth at timberline. A literature review. *Medd. Norsk Inst. Skogforskning* 44:1–34.

Staniforth, R.J., Scott, P.A., and Hansell, R.I.C. 1993. Responses of white spruce to tundra environments: Seed production (unpublished manuscript).

Stevens, G.C., and Fox, J.F. 1991. The causes of treeline. *Annu. Rev. Ecol. System.* 22:177–191.

Sveinbjörnsson, B. 1992. Arctic tree line in a changing climate. In: Chapin, F.S., III, Jefferies, R.L., Reynolds, J.F., Shaver, G.R., Svoboda, J. (eds.), *Arctic Ecosystems in a Changing Climate* (pp. 239–258). San Diego: Academic Press.

Sveinbjörnsson, B., Nordell, O., and Kauhanen, H. 1992. Nutrient relations of mountain birch growth at and below the elevational tree-line in Swedish Lapland. *Funct. Ecol.* 6:213–220.

Sveinbjörnsson, B., Sonesson, M., Nordell, O.K., and Karlsson, S.P. 1993. Performance of mountain birch growth in different environments in Sweden and Iceland: Implications for afforestation. In: Alden, J.N., Mastrantonio, L., and Odum, S. (eds.), *Forest Development in Cold Climates* (pp. 79–88). New York: Plenum Press.

Timoney, K.P., and Wein, R.W. 1991. The aerial pattern of burned tree vegetation in the subarctic region of northwestern Canada. *Arctic* 44:223–230.

Tranquillini, W. 1979. *Physiological Ecology of Alpine Timberline.* Berlin: Springer-Verlag.

Van Cleve, K., and Yarie, J. 1986. Interaction of temperature, moisture, and soil chemistry in controlling nutrient cycling and ecosystem development in the taiga of Alaska. In: Van Cleve, K., Chapin, F.S., III, Flanagan, P.W., Viereck, L.A., and Dyrness, C.T. (eds.), *Forest Ecosystems in the Alaska Taiga* (pp. 160–189). New York: Springer-Verlag.

Van Cleve, K., Barney, R., and Schlentner, R. 1981. Evidence of temperature control of production and nutrient cycling in two interior Alaska black spruce ecosystems. *Can. J. Forest Res.* 11:258–273.

Van Cleve, K., Oechel, W.C., and Hom, J.L. 1990. Response of black spruce (*Picea mariana*) ecosystems to soil temperature modification in interior Alaska. *Can. J. Forest Res.* 20:1530–1535.

Viereck, L.A. 1979. Characteristics of treeline plant communities in Alaska. *Holarctic Ecol.* 2:228–238.

Vowinckel, T. 1975. The effect of climate on the photosynthesis of *Picea mariana* in the subarctic tree line. Ph.D. dissertation, McGill University.

Wein, R.W. 1976. Frequency and characteristics of arctic tundra fires. *Arctic* 29:213–222.

Werren, G.L. 1979. Winter stress in subarctic spruce associations: A Schefferville case study. M.Sc. thesis, McGill University.

Zasada, J.C. 1986. Natural regeneration of trees and tall shrubs on forest sites in interior Alaska. In: Van Cleve, K., Chapin, F.S., III, Flanagan, P.W., Viereck, L.A., and Dyrness, C.T. (eds.), *Forest Ecosystems in the Alaska Taiga* (pp. 44–73). New York: Springer-Verlag.

Zasada, J.C., Sharik, T.L., and Nygren, M. 1991. The reproductive process in boreal forest trees. In: Shugart, H.H., Leemans, R., and Bonan, G.B. (eds.), *Systems Analysis of the Global Boreal Forest* (pp. 85–185). Cambridge: Cambridge University Press.

Zasada, J.C., Foote, M.J., Deneke, F.J., and Parkerson, R.H. 1978. Case history of an excellent white spruce cone crop and seed crop in interior Alaska: Cone and seed production, germination, and seedling survival. USDA Forest Service General Technical Report PNW-65.

14. The Role of Northern Ecosystems in the Global Methane Budget

George L. Vourlitis and Walter C. Oechel

The concentration of atmospheric methane has doubled over the last century from a preindustrial concentration of 0.7 ppmv (parts per million by volume) to the current level of 1.7 ppmv (Bolin et al., 1994; Khalil and Rasmussen, 1987). Blake and Rowland (1988) estimated the globally averaged increase of atmospheric CH_4 between 1978 and 1987 to nearly 16 ppbv (parts per billion by volume) yr^{-1}; however, this increase declined substantially during the late 1980s to around 10 ppbv yr^{-1} (Bolin et al., 1994). Although the trend in CH_4 increase appears to have recently declined (Khalil and Rasmussen, 1992), data from 1993 suggest that atmospheric CH_4 growth rates are on the rise again (Bolin et al., 1994).

Atmospheric CH_4, both directly and indirectly, is an effective greenhouse gas (Cicerone and Oremland, 1988; Lashof and Ahuja, 1990; Ramanathan et al., 1985). Atmospheric CH_4 can directly act as a greenhouse gas by trapping outgoing tropospheric infrared radiation, which acts to warm the earth's surface and near-surface atmosphere and cool the stratosphere (Cicerone and Oremland, 1988). Assuming a 1.5 ppm concentration of atmospheric CH_4, Donner and Ramanathan (1980) calculated that the earth would be approximately 1.3 K higher compared to an atmosphere with zero CH_4. Atmospheric CH_4 oxidation may reduce the direct effects of CH_4-induced radiative forcing, but indirectly affect global climate by producing as much as 340 Tg C yr^{-1} (1 Tg = 10^{12} g) as CO_2, which is equivalent to roughly 6% of the total anthropogenic CO_2 emission (Cicerone and Oremland,

1988). Other reactions, such as the production of tropospheric O_3 in the presence of No_x, and H_2O vapor during CH_4 oxidation, may also indirectly affect radiative forcing as tropospheric O_3 and H_2O are also important greenhouse gases (Cicerone and Oremland, 1988).

Because of the potential effects of increased atmospheric CH_4 on the global energy budget, considerable effort has recently been made to quantify both terrestrial and atmospheric CH_4 sources and sinks. Total CH_4 emissions are estimated to be nearly 535 Tg $CH_4 yr^{-1}$, 70% of which are derived from anthropogenic sources (Bolin et al., 1994; Table 14.1). Major CH_4 sources (60 to 100 Tg yr^{-1}) include wetlands, rice production, and ruminants, whereas minor sources (10 to 60 Tg yr^{-1}) include landfills, coal mining, termites, biomass burning, urban areas, sewage disposal, natural gas leaks, lakes, oceans, and tundra (Khalil et al., 1993). Major CH_4 sinks include tropospheric oxidation of atmospheric CH_4 by OH radicals (e.g., 85%), stratospheric oxidation, and oxidation by methyltrophic bacteria in terrestrial soils (Bolin et al., 1994; Khalil et al., 1993) (Table 14.1). Sources emit nearly 37 Tg $CH_4 yr^{-1}$ more than the combined sinks consume (Table 14.1), resulting in the rise of atmospheric CH_4 (Bolin et al., 1994; Khalil et al., 1993). However, the estimated flux is approximately 20 Tg $CH_4 yr^{-1}$, resulting in an imbalance between the estimated and observed emissions of 17 Tg $CH_4 yr^{-1}$ (Table 14.1). Because anthropogenic CH_4 sources are thought to be well known (Bolin et al., 1994), it is possible that this imbalance is due to an underestimation in the amount of CH_4 lost from natural systems annually, or to an overestimation in the amount of CH_4 oxidized in the atmosphere and soils.

Table 14.1. Estimated Sources, Sinks, and Atmospheric Fluxes (Observed and Estimated) of CH_4 Between 1980 and 1990

	Average Flux (Tg CH_4 yr^{-1})	Range (Tg CH_4 yr^{-1})
Sources		
Natural	160	110–210
Anthropogenic	375	300–450
Total sources	535	410–660
Sinks		
Tropospheric OH	445	360–530
Stratosphere	40	32–48
Soil oxidation	30	15–45
Total sinks	515	430–600
Atmospheric flux		
Observed	37	35–40
Estimated[a]	20	–20–60

Data from Bolin et al. (1994).
[a] Total sources – total sinks.

Natural wetlands are important sources of CH_4 to the global atmosphere, and may be responsible for nearly 30% of the CH_4 produced annually (Table 14.1). Analysis of global CH_4 emissions indicates that more than half of the annual wetland emission, or approximately 50 Tg of C, is from forested and nonforested wetlands from 50 to 70°N (Mathews and Fung, 1987) (Figure 14.1). This high rate of emission is undoubtedly due to the presence of large soil C stocks (Billings, 1987; Gorham, 1991; Miller et al., 1983) and frequently inundated soils. Northern ecosystems contain an estimated 380 Pg of C (10^{15} g), or roughly 25% of the global soil C pool (Schlesinger, 1991). Of this, approximately 92% is composed of dead organic matter present in the seasonally thawed soil active layer and permafrost (Billings, 1987; Gorham, 1991; Miller et al., 1983), and thus represents a potentially huge C source for fermentation and methanogenesis.

Due to the uncertainty in the natural wetland CH_4 sources and sink strengths, and the possibility that these source and sink strengths will be altered by climate change, the aim of this chapter is to review recent estimates of high-latitude CH_4 emissions, discuss how methanogenesis and methyltrophy are controlled by their environment, and discuss how rates of CH_4 production and consumption in high-latitude ecosystems might be affected by climate change.

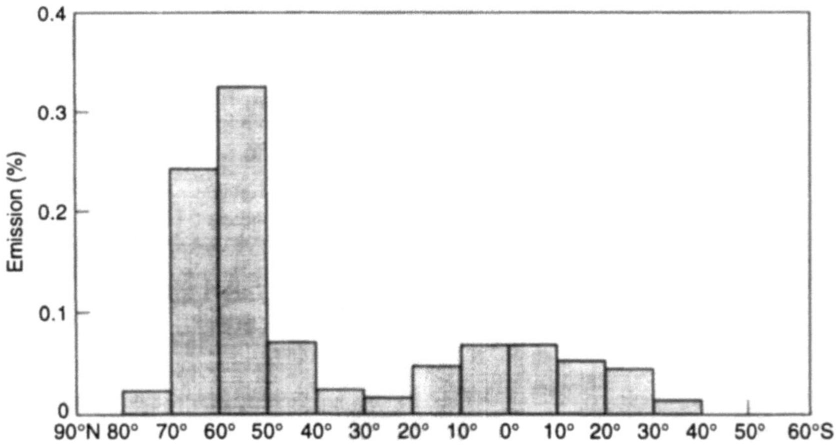

Figure 14.1. Percentage of CH_4 flux from natural wetlands as a function of latitude. Data from Mathews and Fung (1987).

Estimates of CH_4 Flux in Northern Ecosystems

Estimates of the source and sink strength of boreal, subarctic, and arctic wetlands have received considerable attention over the past decade. Recent initiatives such as the 1988 Arctic Boundary Layer Experiment (ABLE-IIIA) and the 1990 Canadian Northern Wetlands Study (NOWES) have focused primarily on the fluxes of CH_4 from subarctic and boreal forest wetlands, and the results of these experiments have been published in dedicated volumes of the *Journal of Geophysical Research.* In addition, a significant amount of CH_4 flux research has been conducted in arctic and subarctic Alaska over the past 5 to 10 years, allowing for more precise estimation of high-latitude CH_4 emissions.

In our analysis, arctic/subarctic wetlands (60 to 80°N; Aselmann and Crutzen, 1989; Bubier et al., 1993; Harriss et al., 1991; Mathews and Fung, 1987) were partitioned into aquatic, moist, and wet sedge tundra groups (Table 14.2). Aquatic ecosystems are defined as ponds and lakes that lack emergent vegetation and have a water table above the soil surface throughout the duration of the thaw season (Whalen and Reeburgh, 1990a). Moist tundra ecosystems are defined as predominantly sedge and low-shrub-dominated peatlands (Whalen and Reeburgh, 1990a), with subsites typically referred to as "black holes," "tussock," and "moss" habitats in the literature (Christensen, 1993; Whalen and Reeburgh, 1992). Soil water content and water table depth of moist tundra ecosystems are generally lower than wet sedge ecosystems, and often these patch types are defined as "upland" habitats (Bartlett et al., 1992; Morrissey and Livingston, 1992). Mathews and Fung (1987) considered tundra to be nonforested bog, whereas here, raised ombrotrophic bogs (e.g., Svensson and Rosswall, 1984) above 60°N latitude are considered to be moist tundra ecosystems. Wet sedge tundra ecosystems were defined as predominantly nontussock-forming graminoid and moss-dominated habitats having saturated soil for a majority of the thaw season (Bartlett et al., 1992; Morrissey and Livingston, 1992; Whalen and Reeburgh, 1990a).

Average arctic and subarctic CH_4 flux estimates range from 12 mg CH_4 $m^{-2} d^{-1}$ for most tundra to 100 mg $CH_4 m^{-2} d^{-1}$ for wet sedge ecosystems (Table 14.2). Aquatic and moist tundra ecosystems are on average comparable sources of between 3 and 31 mg $CH_4 m^{-2} d^{-1}$ (Table 14.2). Wet sedge ecosystems appear to be the dominant arctic and subarctic sources, and are estimated to emit between 30 and 128 mg $CH_4 m^{-2} d^{-1}$ (Table 14.2). The variability associated with these estimates is large, however, indicating the potential for each of these ecosystems to be either significantly greater sources of CH_4 to the atmosphere, or even sinks, depending on the environmental conditions.

Boreal wetlands (45 to 60°N; Harriss et al., 1991) were divided into five wetland types using a classification that has been thoroughly described by

Table 14.2. Estimates (mean ± SD) of CH_4 Flux of Arctic and Subarctic Tundra Ecosystems (60–80°N)

Ecosystem	Reference	CH_4 Flux ($mg\,m^{-2}d^{-1}$)	CV	Location	Sites Used in CH_4 Flux Estimate[a]
Aquatic	Whalen and Reeburgh (1990)	21.0 ± 13.7	0.66	Alaska	AQU sites along a N/S latitudinal transect (*n* = 6).
	Morrissey and Livingston (1992)	3.1 ± 5.2	1.67	Alaska	Open water (*n* = 16 samples).
	Bartlett et al. (1992)	8.4 ± 9.0	1.07	Alaska	Lake Chaos, surrounding area (*n* = 9 samples).
	Hobbie et al. (1980)	14.7 ±		Alaska	
Average		11.8 ± 7.8	0.66		
Moist	Whalen and Reeburgh (1990)	31.0 ± 8.2	0.27	Alaska	MST sites along a N/S latitudinal transect (*n* = 42). July–August 1987.
	Morrissey and Livingston (1992)	7.7 ± 5.3	0.69	Alaska	Habitats with water table below surface, tussock tundra (*n* = 2). August 1987.
	Bartlett et al. (1992)	2.8 ± 4.6	1.64	Alaska	Sites 0, 1, 2, 4 from Bethel, Tower transects (*n* =7). July–August 1988.
	Sebacher et al. (1986)	19.9 ± 27.7	1.39	Alaska	MP 159, 298, 318, 346, 401 sites (*n* = 5). August 1984.
	Whalen and Reeburgh (1992)	27.2 ± 35.5	1.31	Alaska	Black hole, moss, tussock habitats (*n* = 3). Thaw season, 1987–1990.
	Panikov and Zelenev (1991)	5.3 ± 4.7	0.89	Russia	Vorkuta mound and mound base plots (*n* = 2 subsites). July 1990.
	Christensen (1993)	27.4 ± 24.9	0.91	Alaska	Black hole, moss, tussock, elevated habitats (*n* = 4). June–August 1991 and August 1992.
	Torn and Chapin (1993)	2.2 ± 1.1	0.51	Alaska	Moist, wet tussock tundra ecosystems (*n* = 2). July 1991.
	Oechel and Vourlitis (1995)	2.4 ± 2.3	0.99	Alaska	Tussock tundra (*n* = 2 sites). July–August 1990–1991.

	Vourlitis et al. (1994)	8.0 ± 7.2	0.90	Alaska	Moist sedge ($n = 2$ sites). July–August 1990–1991.
	Svensson and Rosswall (1984)	5.1 ± 7.3	0.75	Sweden	Communities I–III ($n = 3$). June, July, September 1974.
Average		12.6 ± 11.3	0.90		
Wet sedge	Whalen and Reeburgh (1990)	90.0 ± 46.2	0.51	Alaska	WET sites along a N/S latitudinal transect ($n = 18$ sites). July–August 1987.
	Morrisey and Livingston (1992)	77.0 ± 25.1	0.33	Alaska	Habitats with water table at or above surface, wet meadow, lakes with emerging vegetation ($n = 4$). August 1987.
	Bartlett et al. (1992)	121.6 ± 68.8	0.57	Alaska	Sites 4, 5 from Bethel, Tower transects, *Carex*, *Arctophila*, *Eriophorum* sites near Lake Chaos ($n = 5$). July–August 1988.
	Sebacher et al. (1986)	142.5 ± 93.2	0.65	Alaska	MP 50, 61, 371, 408, 416 sites ($n = 5$). August 1984.
	Whalen and Reeburgh (1992)	127.9 ± 209.9	1.64	Alaska	*Carex* habitats ($n = 3$ plots). Thaw season, 1987–1990.
	Panikov and Zelenev (1991)	103.7 ± 119.8	1.15	Russia	Vorkuta waterlogged plots ($n = 1$, mean and SD from reported range). July 1990.
	Christensen (1993)	108.6 ± 133.3	1.11	Alaska	*Carex* and depression habitats ($n = 2$). June–August 1991 and August 1992.
	Torn and Chapin (1993)	30.0 ± 31.8	1.06	Alaska	Wet and saturated habitats ($n = 2$). July 1991.
	Vourlitis et al. (1994)	69.7 ± 38.3	0.55	Alaska	Wet and inundated habitats ($n = 2$ sites). July–August 1990–1991.
	Svensson and Rosswall (1984)	111.1 ± 187.7	1.01	Sweden	Communities IV–VI ($n = 3$). June, July, September 1974.
Average		98.2 ± 32.8	0.33		

[a] Site names are those found in references, n = number of measurements, subsites, or sites used in the flux estimate reported here. Dates refer to the time of year that the referenced flux measurements were made.

Mathews and Fung (1987), and later by Aselmann and Crutzen (1989). On average, boreal wetland daily CH_4 flux estimates range between 23 mg $CH_4\,m^{-2}\,d^{-1}$ for forested bogs to 128 mg $CH_4\,m^{-2}\,d^{-1}$ for open bogs (Table 14.3). Again, the variability associated with these estimates is large. For example, open bogs in the Tver Region in Russia are estimated to emit on average (±1 SD) 508.9 ± 885.3 mg $CH_4\,m^{-2}\,d^{-1}$, whereas open bogs in Canadian boreal regions are estimated to emit as little as 21 mg $CH_4\,m^{-2}\,d^{-1}$ (Table 14.3). Forested fen, swamp, and marsh ecosystems are on average estimated to be sources of CH_4 to the atmosphere of about 40 mg $CH_4\,m^{-2}\,d^{-1}$, whereas open fens are estimated to be sources of around 80 mg $CH_4\,m^{-2}\,d^{-1}$ (Table 14.3).

Previously reported estimates of CH_4 flux extrapolated from plot-level arctic, subarctic, and boreal wetlands range (50 to 70°N) between ca. 25 Tg $CH_4\,yr^{-1}$ (Aselmann and Crutzen, 1989), 35 Tg $CH_4\,yr^{-1}$ (Fung et al., 1991; Reeburgh and Whalen, 1992), to as high as 65 Tg $CH_4\,yr^{-1}$ (Mathews and Fung, 1987). Using three different aerial ecosystem coverage estimates found in the literature, our estimates for peatlands at 45 and 80°N range between 25 and 36 Tg $CH_4\,yr^{-1}$ (Table 14.4), which is in good agreement with Aselmann and Crutzen (1989), Reeburgh and Whalen (1992), and Fung et al. (1991). The high estimates by Mathews and Fung (1987) probably reflect the relatively limited data base that was available during the mid-1980s. Of the 25 to 35 Tg CH_4 emitted from northern wetlands annually, approximately 50% is derived from arctic and subarctic wetlands (Table 14.4).

These data, however, should be interpreted with caution. Even cursory examination of the CH_4 emission data displayed in Tables 14.2 and 14.3 reveal that a majority of the flux estimates are from short-term studies that usually were made over a single or partial growing season. A notable exception includes data collected by Whalen and Reeburgh (1992), where CH_4 flux measurements were made at permanent sites over 4 consecutive years (Table 14.2). In addition, 21 out of the 25 CH_4 flux estimates listed for arctic and subarctic ecosystems are from Alaska (Table 14.2), whereas 23 out of the 34 estimates of boreal wetland CH_4 flux are derived from studies of Canadian boreal wetlands (Table 14.3). Hence, there is a potentially large bias in these area-averaged estimates to arctic and boreal ecosystems of North America, and because little work has been done elsewhere, it is unknown how important this bias is. However, comparison between Swedish (Svensson and Rosswall, 1984) and Russian (Panikov and Zelenev, 1991) estimates of net CH_4 flux in moist and wet sedge ecosystems indicate that estimates from Alaskan are comparable (Table 14.2). In contrast, estimates for boreal wetlands in Russia appear to be anomalous to all North American estimates except for forested bogs (Table 14.3), indicating that the North American bias may be relatively more important with the area-averaged boreal wetland estimates. Finally, CH_4 flux estimates for boreal ecosystems usually incorporate only wetlands, which are on average typically found to be sources of CH_4 to the atmosphere (Reeburgh and Whalen, 1992). However, boreal forests have been found to be sinks for atmospheric CH_4 of 0.8 Tg yr^{-1} (Reeburgh and Whalen, 1992; Whalen et al., 1991) to as much as 15 Tg CH_4 yr^{-1}, which would almost completely compensate for the observed wetland CH_4 losses (Born et al., 1990).

Controls on CH_4 Flux in Northern Wetlands

Methane exchange between the ecosystem and atmosphere is the sum of CH_4 production (methanogenesis) and consumption (methanotrophy) (Cicerone and Oremland, 1988; Reeburgh et al., 1994) (Figure 14.2). To understand fluxes of CH_4, and how these fluxes might be altered by climate change, it is important to understand how rates of CH_4 production and oxidation are affected by biotic and abiotic variables. Furthermore, because processes such as production and consumption are antagonistic, affected by similar environmental variables such as water table depth and temperature, and independent of transport processes, simple relationships between CH_4 flux and the soil environment will probably be inadequate for predicting CH_4 fluxes (Valentine et al., 1994; Whiting and Chanton, 1993). Often correlations between net flux and variables that are thought to control CH_4 production are weak and extremely variable (Klinger et al., 1994; Moore et al., 1990; Svennson and Rosswall, 1984; Whalen and Reeburgh, 1990a), underlying the importance of differentiating between production and consumption process, and transport processes.

Soil redox potential is a key control on CH_4 production in terrestrial soils and aquatic and marine sediments, as methanogenesis is a strictly anaerobic process requiring highly reducing conditions (Eh = -300 mV) to occur (Cicerone and Oremland, 1988; Paul and Clark, 1989). Fluctuations in water table depth, which have often been reported to cause variable rates of CH_4 emission (Moore et al., 1990), appear to both directly affect the redox potential of the soil and indirectly affect the substrate quality for methanogenesis, as soil organic carbon commonly varies with depth in the soil profile (Valentine et al., 1994). Although increases in soil water content have been shown to lead to increases in CH_4 production (Freeman et al., 1993; Funk et al., 1994), decreases in soil water content are known to increase the rate of CH_4 oxidation (Whalen et al., 1991). Methanotrophic bacteria, which tend to be most active along oxic–anoxic interfaces (Reeburgh et al., 1993) (Figure 14.2), are primarily limited by slow diffusion of CH_4 when soil water content increases (Whalen et al., 1991).

Provided that anaerobic conditions exist in the soil or sediment environment, substrate quality and abundance appears to be a significant factor in limiting methanogenesis (Delwiche and Cicerone, 1993; Valentine et al., 1994). Methanogenic bacteria are capable of using a limited number of substrates, namely CO_2, H_2, and acetate, which are supplied by the fermen-

Table 14.3. Estimates (mean ± SD, and the CV) of CH_4 Flux of Boreal Forest Wetlands (45–60°N)

Ecosystem	Reference	CH_4 Flux ($mg\,m^{-2}\,d^{-1}$)	CV	Location	Sites Used in CH_4 Flux Estimate[a]
Forested bog	Crill et al. (1988)	86.0 ± 36.5	0.41	Minnesota	Red Lake forested bogs, Marcell sites 1, 2 (*n* = 3). June, August 1986.
	Panikov and Zelenev (1991)	13.9 ± 34.6	2.48	Russia	Tver site 8, Komi site 9 (*n* = 2). July 1990–July 1991 (Tver), July 1990 (Komi).
	Roulet et al. (1992)	5.8 ± 16.7	2.88	Canada	B-T (*n* = 68 measurements). May–October 1990.
	Dise (1993)	24.0 ± 19.8	0.82	Minnesota	S2, collars 3, 4 (*n* = 2). April 1989–April 1990.
	Bubier et al. (1993)	5.1 ± 1.9	0.38	Canada	Sites 9, 14 (*n* = 2). May–October 1991.
	Moore et al. (1994)	47.3 ± 69.7	1.47	Canada	Sites 31, 33, 34 (*n* = 3). June–September 1990.
	Klinger et al. (1994)	4.3 ± 16.7	3.87	Canada	Sites K1–K3 (*n* = 3). July–August 1990.
	Roulet et al. (1994)	1.0 ± 1.1	1.13	Canada	*n* = 2 sites. June–October 1990.
Average		23.4 ± 29.5	1.26		
Open bog	Crill et al. (1988)	189.5 ± 147.8	0.78	Minnesota	Red Lake open bogs, Marcell site 4, Bena bog (*n* = 3). June, August 1986.
	Panikov and Zelenev (1991)	508.9 ± 885.3	1.74	Russia	Tver sites 2–5 (*n* = 4). July 1990–July 1991.
	Roulet et al. (1992)	20.6 ± 27.8	1.34	Canada	B-O (*n* = 65 measurements). May–October 1990.
	Dise (1993)	118.0		Minnesota	Bog S4 (*n* = 1). April 1989–April 1990.
	Bubier et al. (1993)	26.1 ± 35.9	1.38	Canada	Sites 1, 7, 15 (*n* = 3). May–October 1991.
	Moore et al. (1994)	34.7 ± 39.7	1.14	Canada	Sites 24, 25, 36–39 (*n* = 6). June–September 1990.
	Klinger et al. (1994)	54.3 ± 86.2	1.59	Canada	C1, C2, K4 (*n* = 3). July–August 1990.
	Roulet et al. (1994)	70.3 ± 34.5	0.49	Canada	*n* = 5 sites. June–October 1990.
Average		127.8 ± 164.1	1.28		
Forested fen	Crill et al. (1988)	142.0 ± 65.8	0.46	Minnesota	Marcell site 3 (*n* = 12 measurements). June, August 1986.

	Moore et al. (1990)	18.5 ± 22.1	1.19	Canada	Forest margin sites ($n = 6$). June–September 1989.
	Panikov and Zelenev (1991)	0.8 ± 0.8	1.05	Russia	Tver sites 6, 7 ($n = 2$). July 1990–July 1991.
	Bubier et al. (1993)	21.3 ± 35.8	1.68	Canada	Site 4 ($n = 355$ measurements). May–October 1991.
	Moore et al. (1994)	6.3 ± 6.4	1.02	Canada	Sites 6, 16, 10–15 ($n = 7$). June–September 1990.
	Roulet et al. (1994)	2.5 ± 2.1	0.84	Canada	$n = 2$ sites. June–October 1990.
Average		31.9 ± 54.6	1.71		
Open fen	Crill et al. (1988)	348.5 ± 33.2	0.10	Minnesota	Red Lake open fens, Junction fen ($n = 2$). June, August 1986.
	Moore et al. (1990)	37.4 ± 32.6	0.87	Canada	Fen margin, central, flooded, patterned fen, ridges, and pool sites ($n = 17$). June–September 1989.
	Roulet et al. (1992)	3.0 ± 12.3	4.10	Canada	F ($n = 62$ measurements). May–October 1990.
	Dise (1993)	107.5 ± 102.5	0.95	Minnesota	S2 collar 5, Junction fen ($n = 2$). April 1990–April 1991.
	Bubier et al. (1993)	27.6 ± 18.9	0.68	Canada	Sites 3, 5, 6, 13 ($n = 4$). May–October 1991.
	Moore et al. (1994)	18.1 ± 14.9	0.82	Canada	Sites 7–9, 17–22, 26–30 ($n = 14$). June–September 1990.
	Klinger et al. (1994)	14.8 ± 19.7	1.33	Canada	CF.IF, MF ($n = 3$). July–August 1990.
	Roulet et al. (1994)	83.2 ± 77.7	0.93	Canada	$n = 3$ sites. June–October 1990.
Average		80.0 ± 114.3	1.42		
Swamp Marsh	Sebacher et al. (1986)[b]	106.0 ± 12.2	0.12	Alaska	MP 1372 ($n = 7$ measurements), August 1984.
	Roulet et al. (1992)	15.6 ± 28.9	1.85	Canada	CS, MS, TS, M, BP sites ($n = 12$ sites). May–October 1990.
	Bubier et al. (1993)	28.9 ± 37.9	1.31	Canada	Sites 2, 8, 10–12 ($n = 5$). May–October 1991.
	Moore et al. (1994)	15.3 ± 16.8	1.10	Canada	Sites 1–5, 23 ($n = 6$). June–September 1990.
Average		41.5 ± 43.5	1.05		

[a] Site names are those found in references, n = number of measurements, subsites, or sites used in the flux estimate reported here. Dates refer to the time of year that the referenced flux measurements were made.
[b] Site is actually above 60°N latitude, but vegetation is more similar to a boreal marsh.

Table 14.4. Regional Estimates of Net CH_4 Flux from Arctic and Subarctic Tundra (60–80°N) and Boreal Wetland Ecosystems (40–60°N) Using Areal Ecosystem Coverage Estimates

	Miller et al. (1983)[a]	Mathews and Fung (1987)[b]	Aselmann and Crutzen (1989)[c]
Arctic/subarctic	Regional CH_4 Flux (60–80°N) Tg CH_4 yr^{-1}		
Aquatic	0.7 ± 0.5	0.1 ± 0.1	0.4 ± 0.3
Moist tundra	2.9 ± 2.6	1.2 ± 1.0	1.5 ± 1.3
Wet sedge tundra	16.3 ± 23.2	8.3 ± 11.8	16.3 ± 23.2
Σ	19.9	9.6	18.2
Taiga	Regional CH_4 Flux (40–60°N) Tg CH_4 yr^{-1}		
Bogs	7.0 ± 6.6	6.9 ± 0.7	11.1 ± 10.4
Fens	3.3 ± 2.5	7.5 ± 1.5	5.2 ± 3.9
Marsh/swamp	0.7 ± 0.7	1.4 ± 1.5	1.1 ± 1.2
Σ	11.0	15.8	17.4
Σ (40–80°N)	30.9	25.4	35.6

CH_4 flux estimates area averages from data displayed in Tables 14.2 and 14.3. Growing season length was estimated to be on average 130 days for the arctic and subarctic wetlands and 150 days for boreal wetlands.

[a] Areal estimates from Miller et al. (1983). Moist tundra = area of tussock tundra (9×10^{11} m^2) + area of low shrub tundra (1.28×10^{12} m^2); wet sedge tundra = 1×10^{12} m^2. Aquatic estimated to be 0.15% of total tundra area. Bogs, fens, and marsh/swamps were estimated to be 60%, 30%, and 10% (from Aselmann and Crutzen, 1989) of the 1.1×10^{12} m^2 of boreal peatlands (Miller et al., 1983).

[b] Areal estimates from Mathews and Fung (1987). Moist tundra and wet sedge tundra area inferred from vegetation groups 2 and 6G (nonforested bog and tundra, respectively) = 5.06×10^{11} m^2 between 60 and 80°N. Aquatic estimated to be 15% of total tundra area. Bogs = area of forested and nonforested bogs (40–60°N) = 2.2×10^{12} m^2. The area of fens was set equal to the area of bogs, whereas the area of marsh/swamp = area of forested swamp = 2.31×10^{11} m^2.

[c] Areal estimates from Aselmann and Crutzen (1989). Moist tundra was classified as bogs, with the total area between 60 and 80°N = 6.73×10^{11} m^2. Wet sedge tundra was classified as a fen, with the area = 1×10^{12} m^2. Aquatic estimated to be equal to 15% of the total tundra area. Boreal fens and bogs = 5.17×10^{11} and 1.04×10^{12} m^2, respectively, whereas swamp/marsh area was pooled to = 1.81×10^{11} m^2.

tation process (Cicerone and Oremland, 1988; Valentine et al., 1994). Several authors have found that rates of methanogenesis are higher at sites where NPP, plant density, or biomass are high, due primarily to the increase in the supply of labile organic substrates for fermentation (Delwiche and Cicerone, 1993; Sebacher et al., 1985; Valentine et al., 1994; Whiting and Chanton, 1993). These observations are supported by a variety of field and laboratory manipulations, where CH_4 production is typically enhanced by the addition of labile C sources (Valentine et al., 1994), or reduced when plants are removed (Torn and Chapin, 1993).

Soil temperature also represents an important environmental control on CH_4 production (Bartlett et al., 1992; Moore et al., 1994; Morrissey and Livingston, 1992; Svensson, 1984; Svensson and Rosswall, 1984). Methane production typically increases exponentially with added temperature (Svensson, 1984), and reported Q_{10} values range between 1.7 and 4.7 for

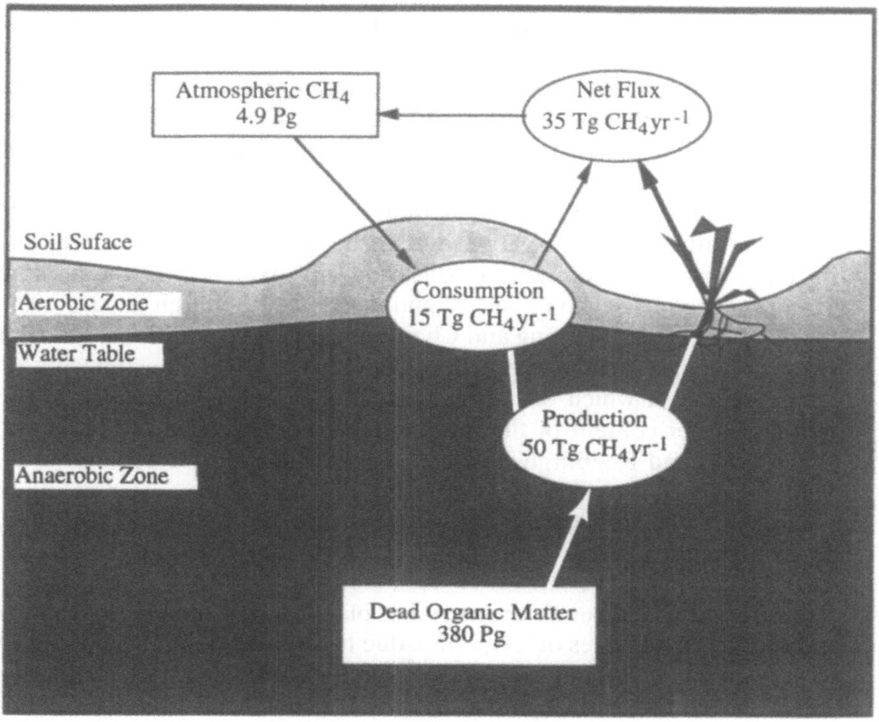

Figure 14.2. Simplified schematic of the CH_4 cycle of northern ecosystems. Quantities in boxes represent C pools, and quantities in ovals represent rates. Data for the dead organic matter pool is from Schlesinger (1991), and for total atmospheric CH_4 burden is from Bolin et al. (1994). Data for total CH_4 production and consumption is from Reeburgh et al. (1993), and data for net CH_4 flux is from this study and Reeburgh et al. (1993). Heavy arows depict relatively larger fluxes.

boreal forest wetlands (Valentine et al., 1994), and 5.5 and 8 for subarctic ecosystems (Svensson and Rosswall, 1984). The variance in these Q_{10} values is thought to be due to differences in substrate quality, as greater temperature sensitivity of methanogens has been observed on higher quality substrates (Valentine et al., 1994; Whiting and Chanton, 1993). These data also suggest that the apparent temperature response of methanogens is due more to an increase in substrate availability through enhanced fermentation rates, rather than a direct effect of temperature on CH_4 production (Valentine et al., 1994; Westerman et al., 1989). Methanotrophy is also found to increase as a function of temperature, but as with the soil moisture response, this is probably due more to the indirect effects of temperature on CH_4 diffusion rates (Whalen et al., 1991).

Transport of CH_4 from the soil to the atmosphere is a function of diffusion through the soil matrix, ebullition of gas bubbles, and diffusion through the arenchyma of vascular plants (Cicerone and Oremland, 1988; Fechner and Hemond, 1992; Sebacher et al., 1985; Whiting and Chanton, 1993). In saturated soils, CH_4 diffuses up through the soil profile and is released to the atmosphere either by diffusion or by ebullition of gas bubbles, and the rate of transfer is simply a function of the CH_4 diffusivity, which is driven by soil structural characteristics, temperature, and water content (Cicerone and Oremland, 1988; Fechner and Hemond, 1992). In soil where the surface layer is unsaturated and relatively more aerated, CH_4 transport to the atmosphere is a function of both diffusion and oxidation in the aerated surface soil zone (Fechner and Hemond, 1992; Reeburgh et al., 1993; Sebacher et al., 1985; Whiting and Chanton, 1993). The additional affect of CH_4 oxidation can substantially reduce the efflux of CH_4 to the atmosphere, and the degree in which oxidation controls CH_4 release depends on the depth and water content of the unsaturated zone (Fechner and Hemond, 1992; Whalen and Reeburgh, 1990b). Estimates made by Reeburgh et al. (1993) suggest that oxidation of diffused CH_4 can reduce rates of efflux by as much as 30% (Figure 14.2), whereas in a West Virginia peat bog, oxidation has been found to diminish CH_4 flux by nearly 55% (Yavitt et al., 1988).

Transport of CH_4 through arenchymous plant stems is thought to result in significantly greater rates of CH_4 efflux due to a reduction in the amount of CH_4 consumed by methanotrophs (Reeburgh and Whalen, 1992; Sebacher et al., 1985; Whiting and Chanton, 1993). Methane release from the soil has been found to increase linearly with plant density (Morrissey and Livingston, 1992) and primary productivity (Whiting and Chanton, 1993). Methane concentrations within arenchymous stems may be significantly higher compared to ambient air concentrations, suggesting that plant tissues are important conduits for CH_4 release (Sebacher et al., 1985). Comparison of the CH_4 exchange rate between the air–water interface and leaves of *Typha latifolia* indicates that CH_4 flux through plant structures accounts for more than 50% of the CH_4 released (Sebacher et al., 1985). In studies where vegetation is removed and replaced with stainless steel straws to simulate plant stems, CH_4 efflux was nearly four times greater in the "straw" plots compared to unvegetated plots, whereas efflux was nearly an order of magnitude greater in unmanipulated control plots compared to unvegetated plots (Torn and Chapin, 1993).

These studies point to the dual role of plants in controlling CH_4 production and release. Plant density and productivity affect CH_4 production and consumption by providing a steady supply of labile C for methanogenesis (Valentine et al., 1994; Whiting and Chanton, 1993) and by influencing the rhizosphere O_2 concentrations through root and rhizome respiration (Fechner and Hemond, 1992). Plants also affect the amount of CH_4 released by providing a conduit for produced CH_4 that may effectively bypass the oxidation process (Reeburgh et al., 1993; Sebacher et al., 1985; Torn and Chapin, 1993), and by the transport of O_2 to the rhizosphere produced through photosynthesis (Fechner and Hemond, 1992; Whalen et al., 1991).

Potential Effects of Climate Change on Northern CH_4 Fluxes

High-latitude climate change may significantly affect future CH_4 emissions by altering both the rate of CH_4 gas emission (Reeburgh and Whalen, 1994) and the extent of wetland surface area (Neilson, 1993; Smith and Shugart, 1993). The increase in the concentration of radiatively active gasses such as CO_2 and CH_4 has the potential to increase surface temperature and affect climate on a global scale (Gates et al., 1992; Watson et al., 1992). The most significant increases are expected to occur in high latitudes (Gates et al., 1992; Manabe and Stouffer, 1993; Meehl et al., 1993; Mitchell et al., 1989), especially in northern latitudes outside the North Atlantic region where the well-mixed ocean tends to buffer atmospheric forcing (Meehl et al., 1993; Schlesinger and Xingjian, 1991). Winter and spring temperatures are expected to increase on the order of 6 to 15°C, due primarily to decreases in snow- and icecover, which have a pronounced effect on surface albedo (Gates et al., 1992; Manabe and Stouffer, 1993; Meehl et al., 1993), whereas summer temperature increases in high-latitude regions are expected to be more modest (e.g., 2 to 4°C).

Warming is expected to affect regional climate patterns, energy balance, and hydrology of northern ecosystems (Gates et al., 1992; Groisman et al., 1994; Hinzman and Kane, 1992; Kane et al., 1990; Kane et al., 1992; Waelbroeck, 1993). Precipitation is predicted to increase significantly (Gates et al., 1992); however, higher temperatures may lead to increased evapotranspiration and reductions in soil moisture during the summer growing season (Hinzman and Kane, 1992). In addition, warmer winter and spring temperatures may result in an earlier spring thaw, which may translate into reductions of soil water content during the summer growing season due to earlier drainage and evapotranspiration of runoff (Hinzman and Kane, 1992; Waelbroeck, 1993). Active layer depth (depth of the unfrozen soil layer that overlies permafrost during the warm summer months) is expected to increase due to high-latitude warming (Hinzman and Kane, 1992; Kane et al., 1990; Kane et al., 1992; Waelbroeck, 1993). Increases in active layer depth will subsequently feed back on the hydrology of these ecosystems in the form of lower water table depth, changes in the soil water holding capacity, and the amount of subsurface flow, and alterations in the soil temperature regime (Hinzman and Kane, 1992; Kane et al., 1992; Waelbroeck, 1993).

High-latitude warming has been recently reported in arctic Alaska, Canada, and the former Soviet Union. Thermal profiles of permafrost and surface temperature records indicate a temperature rise of 2 to 4°C across the north slope of Alaska and throughout northern Canada within the last few decades (Beltrami and Mareshal, 1990; Chapman and Walsh, 1993; Karl et al., 1991; Lachenbruch and Marshall, 1986; Oechel et al., 1993). The annual increase is due primarily to warmer winter and spring temperatures, which have increased on the order of 4.5°C over the last three decades (Chapman and Walsh, 1993). Although increased winter temperatures are probably not significant in terms of biogenic trace gas production, the springtime temperature increase is significant, due to its potential to alter both the timing and amount of runoff from snowmelt (Groisman et al., 1994; Kane et al., 1992). Summer and fall mean temperatures have increased by approximately 2°C over the last three decades (Chapman and Walsh, 1993), which has the potential to alter evapotranspiration rates significantly (Hinzman and Kane, 1992; Waelbroeck, 1993).

The effects of climate change on the CH_4 budget of high-latitude ecosystems are complex. From the previous discussion regarding the controls on CH_4 production and oxidation, the scenario of a warmer and wetter future climate is likely to result in significant increases in CH_4 production and efflux (Morrissey and Livingston, 1992; Roulet et al., 1992). This assumes that increases in precipitation are large enough to counteract increases in evapotranspiration and a drop in water table depth due to permafrost thaw. However, if soil drying occurs in response to increased evapotranspiration and reductions in summer soil water content (Hinzman and Kane, 1992), then CH_4 oxidation, which already plays an important role in the CH_4 budget in dryer boreal forest and tundra ecosystems, will undoubtedly increase (Reeburgh et al., 1994). With other variables such as primary productivity held constant, dryer soil conditions, and warmer temperatures may actually convert boreal and arctic ecosystems from a source of atmospheric CH_4 to a sink (Reeburgh et al., 1994).

Increases in plant productivity are expected in response to high-latitude warming (Shaver et al., 1992), due primarily to a poleward shift in plant species distributions (Billings, 1987; Neilson, 1993; Smith and Shugart, 1993). Because of the strong relationship observed between CH_4 flux and primary productivity (Whiting and Chanton, 1993), the poleward shift in primary productivity may be important in the future CH_4 balance. Under a warmer and wetter climatic regime, greater rates of primary productivity will undoubtedly enhance CH_4 production, and probably efflux as well (Whiting and Chanton, 1993). However, under a warmer and dryer scenario, the possible response in CH_4 flux is not as straightforward. Much depends on the type of vegetation that replaces the extant vegetation, and the magnitude of the soil drying and possible reductions in water table depth. For example, increases in primary productivity will probably enhance CH_4 production (Whiting and Chanton, 1993), but if water table depth drops beyond the plant's root system (or soil drying in the rhizosphere favors oxidation over production), then CH_4 produced in deeper anaerobic soil layers may be completely oxidized before it diffuses up to the root system. Under this scenario, increases in primary productivity may affect actual CH_4 exchange minimally. However, if plants are deeper rooted (or if the reduction in water table depth or soil moisture content in the rhizosphere is not significant), then enhanced primary productivity may counteract increases in oxidation even though surface soil layers are relatively drier, and net CH_4 flux may either increase or remain unchanged. The interactions between altered climatic and biotic regimes are clearly not straightforward, and more process level research must be carried out before changes in CH_4 flux under various climate change scenarios can be accurately predicted.

Techniques Available for Large-Scale CH_4 Flux Measurements

Recent interest in ecological research is expanding from the plot level (square meters) to the landscape (hectares) and regional levels (square kilometers). With the advent of fast response gas analyzers and anemometers, micrometeorological techniques have shown promise in the characterization of trace gas and energy fluxes (Baldocchi et al., 1988; Verma, 1990). Aerodynamic methods, which primarily focus on vertical gradients in wind speed and trace gas concentration, are not limited by sensor time response and have been used for decades for landscape-scale measurements of trace gas and energy fluxes. Eddy correlation methodology has proven reliable using both tower and aircraft platforms, and allows the accurate determination of both landscape (e.g., tower) and regional (e.g., aircraft) level fluxes. Using these methods, questions concerning the scaling of fine-scale phenomena (e.g., trace gas flux from individual patches) to landscape and regional scales can be addressed by calibrating scaled estimates to spatially integrated, landscape- and regional-scale fluxes that are measured directly (Baldocchi et al., 1988; Hicks, 1989; Rastetter et al., 1992; Verma, 1990).

Although large-scale fluxes of CO_2, H_2O vapor, and energy can be routinely measured, large-scale estimates of CH_4 fluxes are rare due to technological difficulties in designing and implementing a sensitive, fast response CH_4 sensor (Hovde et al., in press). Nevertheless, there are recent studies that have successfully measured large-scale CH_4 fluxes using recently developed technology (Fan et al., 1992; Hastie et al., 1983; Hovde et al., in press; Ritter et al., 1992; Verma et al., 1992). Wet herbaceous tundra emit an estimated (mean ± 1 SD) 65.6 ± 36.0 mg $CH_4\,m^{-2}\,d^{-1}$, whereas moist tussock tundra ecosystems release 22.4 ± 15.6 mg $CH_4\,m^{-2}\,d^{-1}$ (Table 14.5). Boreal wetland estimates obtained using eddy correlation techniques range between 14 and 177 mg $CH_4\,m^{-2}\,d^{-1}$ for forested wetlands and open bogs, respectively (Table 14.5).

These large-scale estimates are comparable to chamber flux extimates found in the literature (Tables 14.2 and 14.3), the only exception being subarctic aquatic ecosystems near Bethel, Alaska, which were estimated

Table 14.5. Landscape and Regional Estimates of CH_4 Emission Rates in Arctic and Subarctic Tundra Ecosystems, and in Boreal Forest Ecosystems

Reference	Method	Ecosystem Type	CH_4 Flux ($mg\,m^{-2}\,d^{-1}$)
Arctic and Subarctic Tundra			
Harazono et al., unpublished data[a]	A	wet herbaceous	66.2 ± 34.2
Fan et al. (1992)[b]	EC		29.3
Vourlitis and Hovde, unpublished data[c]	EC		101.3 ± 64.8
$\bar{X} \pm SD$			65.6 ± 36.0
Fan et al. (1992)[b]	EC	moist tussock	11.3
Vourlitis and Hovde, unpublished data[c]	EC		33.4 ± 32.2
$\bar{X} \pm SD$			22.4 ± 15.6
Fan et al. (1992)[b]	EC	aquatic	56.5
Ritter et al. (1992)[d]	EC	subarctic tundra	57.5 ± 20.4
Boreal Wetlands			
Verma et al. (1992)[e]	EC	bog	177.0 ± 50.0
Moore et al. (1994)[f]	EC	fen	49.4
Edwards et al. (1994)[g]	EC	forested	14.5 ± 7.9

Data were collected using aerodynamic (A) and eddy correlation (EC) methods. Data are means ± 1 SD for the measurements made over the indicated duration of the experiment.
[a] Data from Barrow, Alaska, 15–22 August 1993.
[b] Data from Bethel, Alaska, 14 July–12 August 1988.
[c] Data from the North Slope of Alaska, 20–24 July 1994.
[d] Aircraft data from Bethel, Alaska, average of three flights between 28 and 31 July 1988.
[e] Data from Bog Lake Peatland, Chippewa National Forest, north-central Minnesota, 7–14 August 1990.
[f] Data from Schefferville, Quebec, Canada, 30 May–19 August 1990.
[g] Data from Kinosheo Lake, Ontario, Canada, 1–25 July 1990.

using eddy correlation techniques to emit 56 mg $CH_4\,m^{-2}\,d^{-1}$ (Table 14.5). Plot level estimates from strictly aquatic ecosystems such as ponds and lakes, which lack emergent vegetation (Whalen and Reeburgh, 1990), indicate that efflux is on average only 10.8 ± 9.2 mg $CH_4\,m^{-2}\,d^{-1}$ from these ecosystems (Table 14.2). This discrepancy is undoubtedly due to differences in spatial averaging, as landscape-scale estimates integrate not only efflux from the aquatic surface, but from all of the surface types in the upwind tower "footprint" in the pond complex (Gash, 1986; Schuepp et al., 1990).

Regional estimates of CH_4 flux from aircraft eddy correlation flux platforms in the Bethel, Alaska, region (Ritter et al., 1992) compare well with landscape level estimates (Fan et al., 1992). This may be due to the comparable scale of measurement, and to the similarity in the controlling factors that mediate trace gas exchange between the surface and the atmosphere. The integrated regional estimates may "view" the landscape surface heterogeneity on a scale that is similar to the integrated landscape estimates from the eddy flux tower. Furthermore, atmospheric effects, such as the effect of wind speed on trace gas exchange between the surface and the overlying atmosphere, are accounted for in both landscape- and regional-scale measurement techniques. Unfortunately, current landscape- and regional-scale CH_4 flux estimates have been obtained from short-term studies (e.g., weeks), so the seasonal (and annual) dynamics of CH_4 flux in northern latitude landscapes is largely unknown. However, these techniques appear to be amenable to frequent sampling over the long term, and will likely be important tools for monitoring CH_4 fluxes and understanding spatially integrated high-latitude CH_4 sources and sinks.

Summary

Considerable effort has been made over the last two decades to quantify sources and sinks for atmospheric CH_4. Although much progress has been made, there still is significant uncertainty regarding sources and sinks in terrestrial ecosystems, which is highlighted in the imbalance of nearly 17 Tg $CH_4\,yr^{-1}$ between the observed and estimated atmospheric CH_4 flux. One potentially strong source is northern latitude wetlands (50 to 70°N), and assuming that natural sources emit 160 Tg CH_4 into the global atmosphere annually, northern wetlands may be responsible for an estimated 16% (data displayed here) to 45% (Mathews and Fung, 1987) of the total natural source.

Using estimates of CH_4 flux from the current literature, our analysis indicates that average arctic and subarctic CH_4 flux extimates range from 12 mg $CH_4\,m^{-2}\,d^{-1}$ for moist tundra to 100 mg $CH_4\,m^{-2}\,d^{-1}$ for wet sedge ecosystems. Similarly, boreal wetland daily CH_4 flux estimates range between 23 mg $CH_4\,m^{-2}\,d^{-1}$ for forested bogs to 128 mg $CH_4\,m^{-2}\,d^{-1}$ for open bogs. High variability was associated with all flux estimates, indicating considerable spatial heterogeneity in CH_4 emissions both within and between specific patch types, and that similar patch types can alternate between being a source or sink depending on the environmental conditions. Using three estimates of arctic, subarctic, and boreal wetland surface area, northern wetlands are estimated to be sources of between 25 and 36 Tg $CH_4\,yr^{-1}$, which agree well with other recent estimates of the high-latitude wetland source.

Production, consumption, and transport processes must be considered before a complete understanding of the controls on CH_4 exchange can be realized. Methane production is affected by water table depth, temperature, and substrate quality, and increases under anaerobic conditions and increased temperature provided that C quality and supply are relatively high. Methane consumption is typically enhanced under well-aerated conditions and higher temperatures, due to greater rates of CH_4 diffusion in soils. Transport of CH_4 to the atmosphere is due to diffusion, ebullition of gas bubbles, and diffusion through arenchymous plant structures, and much of the CH_4 produced may be oxidized as it diffuses through the anaerobic–aerobic interface. Diffusion of CH_4 through arenchymous plant structures may be extremely important, as this pathway may effectively reduce the amount of CH_4 consumed by methanotrophic microorganisms. Because processes such as production and consumption are antagonistic, and independent of transport processes, simple relationships between CH_4 flux and the soil environment will probably be inadequate for predicting CH_4 fluxes, which may explain why relationships between CH_4 flux and, for example, temperature are strong in some cases and weak in others.

The effects of climate change on CH_4 flux in high-latitude ecosystems are largely unknown, due to the lack of process-oriented models. If climate change leads to warmer and wetter conditions, which is possible given that precipitation is expected to increase significantly, then high-latitude ecosystems are likely to become even greater sources of CH_4 to the global atmosphere. If climate change leads to warmer and drier conditions, which may occur if evapotranspiration rates increase relatively more than precipitation, then oxidation will undoubtedly become even more important than it currently is, and high-latitude ecosystems may become sinks for atmospheric CH_4. One of the many unknown factors in climate change is the plant response to climatic warming, and because CH_4 production appears to be intimately linked to primary productivity, it is possible that the anticipated changes in plant species distributions will have an important, but currently unknown, effect on CH_4 exchange. The uncertainty in the high-latitude response to climate change highlights the need for more process-level experimentation.

Finally, recent technological advances allow for area-averaged estimates of CH_4 flux to be measured directly using eddy correlation techniques. Although fast-response CH_4 sensors are not currently available, several groups using sensor prototypes have measured landscape- and regional-scale CH_4 fluxes that are comparable to the average flux estimates in the literature. Other techniques, such as aerodynamic or gradient techniques, do not require fast-response sensors and have been used for decades in measurements of energy and CO_2 fluxes. More recently, these methods have been used to quantify large-scale CH_4 flux in arctic wet sedge tundra with good results. Because these techniques allow frequent sampling over the long term, they will undoubtedly be important tools for monitoring CH_4 fluxes and understanding spatially integrated high-latitude CH_4 sources and sinks.

References

Aselmann, I., and Crutzen, P.J. 1989. Global distribution of natural freshwater wetlands and rice paddies, their net primary productivity, seasonality and possible methane emissions. *J. Atmos. Chem.* 8:307–358.

Baldocchi, D.D., Hicks, B.B., and Meyers, T.P. 1988. Measuring biosphere-atmosphere exchanges of biologically related gases with micrometeorological methods. *Ecology* 69(5):1331–1340.
Bartlett, K.B., Crill, P.M., Sass, R.L., Harriss, R.C., and Dise, N.B. 1992. Methane emissions from tundra environments in the Yukon-Kuskokwim Delta, Alaska. *J. Geophys. Res.* 97:16,645–16,660.
Beltrami, H., and Mareshal, J.C. 1991. Recent warming in eastern Canada inferred from geothermal measurements. *Geophys. Res. Lett.* 18:605–608.
Billings, W.D. 1987. Carbon balance of Alaskan tundra and taiga ecosystems: Past, present, and future. *Quat. Sci. Rev.* 6:165–177.
Blake, D.R., and Rowland, F.S. 1987. Continuing worldwide increase in tropospheric methane, 1979 to 1987. *Science* 239:1129–1131.
Bolin, B., Houghton, J., and Meira-Filho, L.G. 1994. *Radiative Forcing of Climate Change: The 1994 Report of the Scientific Assessment Working Group of IPCC.* World Meteorological Office.
Born, M., Dörr, H., and Levin, I. 1990. Methane consumption in aerated soils of the temperate zone. *Tellus* 42B:2–8.
Bubier, J.L., Moore, T.R., and Roulet, N.T. 1993. Methane emissions from wetlands in the midboreal region of northern Ontario, Canada. *Ecology* 74:2240–2254.
Chapman, W.L., and Walsh, J.E. 1993. Recent Variations of Sea Ice and Air Temperature in High Latitudes. *Bull. Am. Meteorol. Soc.* 74:33–47.
Christensen, T.R. 1993. Methane emission from arctic tundra. *Biogeochemistry* 21:117–139.
Cicerone, R.J., and Oremland, R.S. 1988. Biogeochemical aspects of atmospheric methane. *Global Biogeochem. Cycles* 2:299–327.
Crill, P.M., Bartlett, K.B., Harriss, R.C., Gorham, E., Verry, E.S., Sebacher, D.I., Madzar, L., and Sanner, W. 1988. Methane flux from Minnesota peatlands. *Global Biogeochem. Cycles* 2:371–384.
Delwiche, C.C., and Cicerone, R.J. 1993. Factors affecting methane production under rice. *Global Biogeochem. Cycles* 7:143–156.
Dise, N.B. 1993. Methane emission from Minnesota peatlands: Spatial and seasonal variability. *Global Biogeochem. Cycles* 7:123–142.
Donner, L., and Ramanathan, V. 1980. Methane and nitrous oxide: Their effects on the terrestrial climate. *J. Atmos. Sci.* 37:119–124.
Fan, S.M., Wofsy, S.C., Bakwin, P.S., Jacob, D.J., Anderson, S.M., Kebabian, P.L., McManus, J.B., Kolb, C.E., and Fitzjarrald, D.R. 1992. Micrometeorological measurements of CH_4 and CO_2 exchange between the atmosphere and subarctic tundra. *J. Geophys. Res.* 97:16,627–16,643.
Fechner, E.J., and Hemond, H.F. 1992. Methane transport and oxidation in the unsaturated zone of a Sphagnum peatland. *Global Biogeochem. Cycles* 6:33–44.
Freeman, C., Lock, M.A., and Reynolds, B. 1993. Fluxes of CO_2, CH_4 and N_2O from a Welsh peatland following simulation of water table draw-down: Potential feedback to climate change. *Biogeochemistry* 19:51–60.
Fung, I., John, J., Lerner, J., Mathews, E., Prather, M., Steele, L.P., and Frases, P.J. 1991. Three-dimensional model synthesis of the global methane cycle. *J. Geophys. Res.* 96:13,033–13,065.
Funk, D.W., Pullman, E.R., Peterson, K.M., Crill, P.M., and Billings, W.D. 1994. Influence of water table on carbon dioxide, carbon monoxide, and methane fluxes from taiga bog microcosms. *Global Biogeochem. Cycles* 8:271–278.
Gash, J.H.C. 1986. A note on estimating the effect of a limited fetch on micrometeorological evaporation measurements. *Boundary-Layer Meteorol.* 35:409–413.
Gates, W.L., Mitchell, J.F.B., Boer, G.J., Cubasch, U., and Meleshko, V.P. 1992. Climate modeling, climate prediction and model validation. In: Houghton, J.T.,

Callander, B.A., and Varney, S.K. (eds.), *Climate Change 1992: The Supplemental Report to the IPCC Scientific Assessment* (pp. 97–135). Cambridge: Cambridge University Press.

Gorham, E. 1991. Northern peatlands: Role in the carbon cycle and probable responses to climate warming. *Ecol. Appl.* 1:182–195.

Groisman, P.Y., Karl, T.R., and Knight, R.W. 1994. Observed impact of snow cover on the heat balance and rise of continental spring temperatures. *Science* 263:198–200.

Harriss, R., Bartlett, K., Frolking, S., and Crill, P. 1991. Methane emissions from northern high-latitude wetlands. In: *Biogeochemistry of Global Change: Radiatively Active Trace Gases* (pp. 449–486). Tenth International Symposium on Environmental Biogeochemistry. New York: Chapman and Hall.

Hastie, D.R., Mackay, G.I., Iguchi, T., Ridley, B.A., and Schiff, H.I. 1983. Tunable diode laser systems for measuring trace gases in tropospheric air. *Environ. Sci. Technol.* 17:352–364.

Hicks, B.B. 1989. Regional Extrapolation: Vegetation-Atmosphere Approach. In: Andreas, M.O., and Schimol, D.S. (eds.), *Exchange of Trace Gases between Terrestrial Ecosystems and the Atmosphere*, (pp. 109–118). New York: John Wiley & Sons.

Hinzman, L.D., and Kane, D.L. 1992. Potential response of an arctic watershed during a period of global warming. *J. Geophys. Res.* 97:2811–2820.

Hobbie, J.E., Traaen, T., Rublee, P., Reed, J.P., Miller, M.C., and Fenchel, T. 1980. Decomposers, bacteria, and microbenthos. In: Hobbie, J.E. (ed.), *Limnology of Tundra Ponds* (pp. 340–387). Stroudsburg, PA: Dowden, Hutchinson, and Ross.

Hovde, D.C., Meyers, T.P., Stanton, A.C., and Matt, D.R. Methane emissions from a landfill measured by eddy correlation using a fast response diode laser sensor. *J. Atmos. Chem.* In press.

Kane, D.L., Gieck, R.E., and Hinzman, D.L. 1990. Evapotranspiration from a small arctic watershed. *Nordic Hydrol.* 21:253–272.

Kane, D.L., Hinzman, L.D., Woo, M.K., and Everett, K.R. 1992. Arctic Hydrology and Climate Change. In: Chapin, F.S., III, Jefferies, R., Shaver, G., Reynolds, J., and Svoboda, J. (eds.), *Physiological Ecology of Arctic Plants: Implications for Climate Change* (pp. 35–57). New York: Academic Press.

Karl, T.R., Kukla, G., Razuvayev, V.N., Changery, M.J., Quayle, R.G., Heim, R.R., Easterling, D.R., and Fu, C.B. 1991. Global warming: Evidence for assymetric diurnal temperature change. *Geophys. Res. Lett.* 18:2253–2256.

Khalil, M.A.K., and Rasmussen, R.A. 1987. Atmospheric methane: trends over the last 10,000 years. *Atmos. Environ.* 21(11):2445–2452.

Khalil, M.A.K., and Rasmussen, R.A. 1992. Decline in the trend of atmospheric methane. In: Ferguson, E.E., and Rosson, R.M. (eds.), U.S. Department of Commerce, National Oceanic and Atmospheric Administration, Climate Monitoring and Diagnostics Laboratory Summary Report No. 20 (pp. 105–107). Boulder, CO.

Khalil, M.A.K., Rasmussen, R.A., Shearer, M.J., Ge, S., and Rau, J.A. 1993. Methane from coal burning. *Chemosphere* 26:473–477.

Klinger, L.F., Zimmerman, P.R., Greenberg, J.P., Heidt, L.E., and Guenther, A.B. 1994. Carbon trace gas fluxes along a successional gradient in the Hudson Bay lowland. *J. Geophys. Res.* 99:1469–1494.

Lachenbruch, A.H., and Marshall, B.V. 1986. Changing Climate: Geothermal Evidence from Permafrost in the Alaskan Arctic. *Science* 234:689–696.

Lashof, D.A., and Ahuja, D.R. 1990. Relative contributions of greenhouse gas emissions to global warming. *Nature* 344:529–531.

Manabe, S., and Stouffer, R.J. 1993. Century-scale effects of increased atmnospheric CO_2 on the ocean-atmosphere system. *Nature* 364:215–218.

Mathews, E., and Fung, I. 1987. Methane emission from natural wetlands: Global distribution, area, and environmental characteristics of sources. *Global Biogeochem. Cycles* 1:61–86.

Meehl, G.A., Washington, W.M., and Karl, T.R. 1993. Low-frequency variability and CO_2 transient climate change: Part 1. Time averaged differences. *Clim. Dynam.* 8:117–133.

Miller, P.C., Kendall, R., and Oechel, W.C. 1983. Simulating carbon accumulation in northern ecosystems. *Simulation* 40:119–131.

Mitchell, J.F.B., Senior, C.A., and Ingram, W.J. 1989. CO_2 and climate: A missing feedback? *Nature* 341:132–134.

Moore, T.R., Heyes, A., and Roulet, N.T. 1994. Methane emissions from wetlands, southern Hudson Bay lowland. *J. Geophys. Res.* 99:1455–1467.

Moore, T.R., Roulet, N., and Knowles, R. 1990. Spatial and temporal variations of methane flux from subarctic/northern boreal fens. *Global Biogeochem. Cycles.* 4:29–46.

Morrissey, L.A., and Livingston, G.P. 1992. Methane emissions from Alaska arctic tundra: An assessment of local spatial variability. *J. Geophys. Res.* 97:16,661–16,670.

Neilson, R.P. 1993. Vegetation redistribution: A possible biosphere source of CO_2 during climatic change. *Water, Soil, Air Pollut.* 70:659–673.

Oechel, W.C., Hastings, S.J., Vourlitis, G.L., Jenkins, M., Riechers, G., and Grulke, N. 1993. Recent change of arctic tundra ecosystems from a net carbon dioxide sink to a source. *Nature* 361:520–523.

Oechel, W.C., and Vourlitis, G.L. 1995. Effects of global change on carbon storage in cold soils. In: Lal, R., Kimbel, J., Levine, E., and Stewart, B.A. (eds.), *Advances in Soil Science: Soils and Global Change* (pp. 117–129). Boca Raton, FL: Lewis Publishers.

Panikov, N., and Zelenev, V. 1991. Methane and carbon dioxide production and uptake in some boreal ecosystems of Russia. In: Vinson, T.S., and Kolchugina, T.P. (eds.), *Carbon Cycling in Boreal Forests and Sub-arctic Ecosystems* (pp. 125–138). Washington, DC: United States Environmental Protection Agency Report EPA/600R-93/084, Office of Research and Development.

Paul, E.A., and Clark, F.E. 1989. Soil Microbiology and Biochemistry. San Diego: Academic Press.

Ramanathan, V., Cicerone, H.B., Singh, H.B., and Kiehl, J.T. 1985. Trace gas trends and their potential role in climate change. *J. Geophys. Res.* 90:5547–5566.

Rastetter, E.B., King, A.W., Cosby, B.J., Hornberger, G.M., O'Neill, R.V., and Hobbie, J.E. 1992. Aggregating fine-scale ecological knowledge to model coarse scale attributes of ecosystems. *Ecol. Appl.* 2:55–70.

Reeburgh, W.S., and Whalen, S.C. 1992. High-latitude ecosystems as CH_4 sources. *Ecol. Bull.* 42:62–70.

Reeburgh, W.S., Roulet, N.T., and Svensson, B.H. 1994. Terrestrial biosphere-atmosphere exchange in high latitudes. In: Prinn, R.G. (ed.), *Global Atmospheric-Biospheric Chemistry* (pp. 165–178). New York: Plenum Press.

Reeburgh, W.S., Whalen, S.C., and Alperin, M.J. 1993. The role of methyltrophy in the global methane budget. In: Murrell, J.C., and Kelley, D. (eds)., *Microbial Growth on C_i Compounds* (pp. 1–14). Andover, UK: Intercept.

Ritter, J.A., Barrick, J.D.W., Sachse, G.W., Gregory, G.L., Woerner, M.A., Watson, C.E., Hill, G.F., and Collins, Jr, J.E. 1992. Airborne flux measurements of trace species in an arctic boundary layer. *J. Geophys. Res.* 97:16,601–16,625.

Roulet, N.T., Ash, R., and Moore, T.R. 1992. Low boreal wetlands as a source of atmospheric methane. *J. Geophys. Res.* 97:3739–3749.

Roulet, N.T., Jano, A., Kelley, C.A., Klinger, L.F., Moore, T.R., Protz, R., Ritter, J.A., and Rouse, W.R. 1994. Role of the Hudson Bay lowland as a source of atmospheric methane. *J. Geophys. Res.* 99:1439–1454.

Schlesinger, M.E., and Xingjian, J. 1991. Revised projection of future greenhouse warming. *Nature* 350:219–221.

Schlesinger, W.H. 1991. *Biogeochemistry: An Analysis of Global Change*. San Diego, CA: Academic Press.

Schuepp, P.H., Leclerc, M.Y., MacPherson, J.I., and Desjardins, R.L. 1990. Footprint prediction of scalar fluxes from analytical solutions of the diffusion equation. *Boundary-Layer Meteorol.* 50:355–374.

Sebacher, D.I., Harriss, R.C., and Bartlett, K.B. 1985. Methane emissions to the atmosphere through aquatic plants. *J. Environ. Qual.* 14:40–46.

Sebacher, D.I., Harriss, R.C., Bartlett, K.B., Sebacher, S.M., and Grice, S.S. 1986. Atmospheric methane sources: Alaskan tundra bogs, an alpine fen, and a subarctic boreal marsh. *Tellus* 38B:1–10.

Smith, T.M., and Shugart, H.H. 1993. The transient response of terrestrial carbon storage to a perturbed climate. *Nature* 361:523–526.

Svensson, B.H., and Rosswall, T. 1984. In situ methane production from acid peat in plant communities with different moisture regimes in a subarctic mire. *Oikos* 43:341–350.

Svensson, B.N. 1984. Different temperature optima for methane formation, when enrichments from acid peat are supplemented with acetate or hydrogen. *Appl. Environ. Microbiol.* 48:389–394.

Torn, M.S., and Chapin, F.S. III. 1993. Environmental and biotic controls over methane flux from arctic tundra. *Chemosphere* 26:357–368.

Valentine, D.W., Holland, E.A., and Schimel, D.S. 1994. Ecosystem and physiological controls over methane production in northern wetlands. *J. Geophys. Res.* 99:1563–1571.

Verma, S.B. 1990. Micrometeorological methods for measuring surface fluxes of mass and energy. *Rem. Sens. Rev.* 5(1):99–115.

Verma, S.B., Ullman, F.G., Billesbach, D., Clement, R.J., Kimm, J., and Verry, E.S. 1992. Eddy correlation measurements of CH_4 flux in a northern peatland ecosystem. *Boundary-Layer Meteorol.* 58:289–304.

Vourlitis, G.L., Oechel, W.C., Hastings, S.J., and Jenkins, M.A. 1994. The effect of soil moisture and thaw depth on CH_4 flux from wet coastal tundra ecosystems on the North Slope of Alaska. *Chemosphere* 28:R1–R3.

Waelbroeck, C. 1993. Climate-soil processes in the presence of permafrost: a systems modeling approach. *Ecol. Model.* 69:185–225.

Watson, R.T., Meira Filho, L.G., Sanhueza, E., and Janetos, A. 1992. Sources and sinks. In: Houghton, J.T., Callander, B.A., and Varney, S.K. (eds.), *Climate Change 1992: The Supplemental Report to the IPCC Scientific Assessment* (pp. 24–46). Cambridge: Cambridge University Press.

Westermann, P., Ahring, B.K., and Mah, R.A. 1989. Temperature conpensation in Methanosarcina barkeri by modulation of hydrogen and acetate affinity. *Appl. Environ. Microbiol.* 55:1262–1266.

Whalen, S.C., and Reeburgh, W.S. 1990. A methane flux transect along the trans-Alaska pipeline haul road. *Tellus* 42B:237–249.

Whalen, S.C., and Reeburgh, W.S. 1990b. Consumption of artmospheric methane by tundra soils. *Nature* 346:160–162.

Whalen, S.C., and Reeburgh, W.S. 1992. Interannual variations in tundra methane emission: A 4-year time series at fixed sites. *Global Biogeochem. Cycles* 6:139–160.

Whalen, S.C., Reeburgh, W.S., and Kizer, K.S. 1991. Methane consumption and emission by Taiga. *Global Biogeochem. Cycles* 5:261–273.
Whiting, G.J., and Chanton, J.P. 1993. Primary production control of methane emission from wetlands. *Nature* 364:794–795.
Yavitt, J.B., Lang, G.E., and Downey, D.M. 1988. Potential methane production and methane oxidation rates in peatland ecosystems of the Appalachian Mountains. *Global Biogeochem. Cycles* 2:253–268.

15. The Responses of Vegetation to Past and Future Climate Changes

Brian Huntley

One area of consensus among forecasts of potential future climate changes resulting from anthropogenically increased levels of the so-called "greenhouse gases" is the amplification of these changes at higher latitudes, especially in the northern hemisphere and in winter (Houghton et al., 1990, 1992). Arctic and Boreal ecosystems thus may experience the greatest impacts of "global warming." Palaeoenvironmental evidence, however, suggests that past global changes also may have been rapid in these regions (Dansgaard et al., 1989; GRIP Members, 1993). The palaeoecological record from high northern latitudes can provide invaluable evidence of the response of Arctic and Boreal ecosystems to these past changes; this evidence can help predict the future response of these ecosystems.

Numerous palynological studies have been made of Pleistocene and especially of Holocene sediments. In addition, the plant macrofossil content of many such deposits has been investigated. Information from such studies is especially abundant for Europe and North America, with lesser quantities of data available for other regions. In Arctic and Boreal regions, data for the Holocene are numerous; the history of glaciation and ice-sheet development, however, has removed most of the sediments of earlier Pleistocene warm stages in these regions so that the evidence from these stages is sparse. However, it is not the purpose of this chapter to provide any comprehensive review of palaeovegetation data from Arctic and Boreal areas, nor would this author feel qualified to attempt such a task; rather, the aim of this chapter is to draw attention to some of the general conclusions arising from such work that are relevant to considerations of the response of Arctic and Boreal ecosystems to predicted future climate changes. Illustrative examples will be drawn principally from Boreal and Arctic regions of Europe and western Russia, the area for which the author does have some personal experience. Furthermore, the Holocene vegetation history of the Arctic as a whole recently was reviewed by Lamb and Edwards (1988) whereas Ritchie (1987) provided a synthesis of the evidence from Canada. Anderson et al. (1991) provided a synthesis of modern pollen–vegetation–climate relationships in northern North America and western Greenland; Huntley and Birks (1983) and Huntley and Prentice (1993) together provided parallel information for Europe and western Russia. Huntley and Birks (1983), Huntley (1988), and Huntley and Prentice (1993) also provided syntheses of the late-Quaternary vegetation and environment history of Europe and western Russia. Similar syntheses or regional studies are available for other Arctic and Boreal areas, including, for example, Alaska (Anderson and Brubaker, 1993, 1994), southern Québec (Richard, 1994), northeastern Russia (Lozhkin et al., 1993), and eastern North America (Prentice et al., 1991; Webb et al., 1993). Late-Weichselian vegetation and environment history of several Arctic regions have been the subject of recent research (Birks et al., 1994—Norway and Svalbard; Ingólfson and Norddahl, 1994—Iceland). The history of environmental change in Iceland also was discussed in a series of papers in Maizels and Caseldine (1991). An overview of the last interglacial in the Arctic was recently presented (LIGA Members, 1991a) along with a discussion of the linkages between the various components of the global system at that time (LIGA Members, 1991b); more recently a series of papers has provided updated information for the last interglacial from several regions, notably northeast Siberia (Lozhkin and Anderson, 1995) and northwest Alaska (Brigham–Grette and Hopkins, 1995).

Holocene palynological data for the European Arctic were included by Huntley and Birks (1983) in their compilation. Data from the northern part of the former USSR were compiled by Peterson (1983). Huntley (1988) combined these data sets and mapped the data for some Arctic and Boreal taxa in northern Europe and the northwestern part of Russia so as to illustrate changes in the position of the Arctic tree line during the Holocene as well as changes in the composition of the forests forming this tree line at different times during the Holocene. Subsequently, Huntley (1990a, 1990b) analyzed this combined data set of Holocene pollen data from Europe and the former USSR and presented maps of the vegetation of this region at various times since the last glacial.

In the sections that follow a brief review will be presented of the principal results obtained from Quaternary palynological and plant macrofossil investigations, in the Arctic and Boreal regions of Europe and western Russia especially, that are relevant to considering how the vegetation of these regions responds to climatic change. The emphasis will be upon the Holocene because of the greater quantity of data available from this period. The results will be considered in relation to three general topics:

1. past distribution patterns of individual plant taxa and how these have changed through migration;
2. past vegetation patterns, changes in the composition of vegetation units, and changes in the positions and orientation of major ecotones; and
3. past climate changes as deduced both from the palaeobotanical evidence and from independent indicators of past environmental conditions.

The principal conclusions that emerge from these data with respect to the response of Arctic and Boreal vegetation to climate change then will be discussed.

Upon the basis of this evidence of responses to past climate changes it will be argued that the extent of the lowland tundra biome is likely to be greatly reduced in the event of any significant climatic warming, especially upon "mainland" areas of northern Europe, northern Russia, and North America. Similarly, the montane tree line is likely to advance upwards in those mountain regions where wind is not the principal determinant of the maximum elevation to which trees can grow; this will reduce the extent of montane "tundra" in many regions. The latter impact is likely to be felt more rapidly than any extensive impact upon lowland tundra that may be limited by the rate at which trees can migrate in response to altered climate conditions. Whereas past climate conditions in the Arctic apparently were warmer than today both during the early Holocene and especially during the last interglacial (Lozhkin and Anderson, 1995), the magnitude of this warming was less than that forecast for the near future. Furthermore, at least in northeast Siberia (Lozhkin and Anderson, 1995), the greatest magnitude of warming during the last interglacial was in summer, with winter conditions apparently cooler than today; this is consistent with the enhanced solar radiation in summer at that time (Bartlein, 1988) and contrasts with the essentially aseasonal forcing provided by elevated levels of greenhouse gases. These past warmer conditions nonetheless led both to elevated montane tree lines and to a reduced extent of lowland tundra; these in turn likely led to population crises for many Arctic organisms in some parts of their former ranges, in some cases leading to their local or even global extinction. Future impacts are likely to be of greater magnitude than the impacts of these past changes.

Past Distribution Patterns and Migration

Compilations of palynological data from numerous individual localities enable the past patterns of distribution and abundance of taxa to be mapped by means of isopoll maps (Szafer, 1935). Such maps may be drawn for individual pollen taxa and for particular points in time. The migratory response of a taxon to past environmental changes then is revealed by a time-series of maps for that taxon (Bernabo and Webb, 1977; Huntley, 1988; Huntley and Birks, 1983; Huntley and Prentice, 1993; Huntley and Webb, 1989; Webb, 1988; Webb et al., 1993), whereas the composition and patterns of past vegetation at a given time can be deduced from a comparison of the maps for a series of taxa at that time (Huntley, 1988, 1990a, 1990b; Huntley and Birks, 1983; Jacobson et al., 1987).

A time-series of isopoll maps for *Picea*, a major component of the Boreal forests of Europe, Asia, and North America, has been drawn for both western Eurasia and eastern North America (Huntley and Webb, 1989) (Figure 15.1). These maps illustrate both the changing distribution and the changes in extent and abundance of *Picea* since the last glacial stage. In western Eurasia the most striking result is the demonstration that the Boreal population of *P. abies* was initially distributed principally in northwestern Russia—to the southeast of the Fennoscandian ice sheet—and remained absent from most of Scandinavia until as recently as 4000 years ago. Furthermore, this Boreal population apparently was, until recent millennia, geographically isolated from the population of *P. abies* that formed a dominant component of many of the montane forests of eastern and central Europe. Both populations have shown an overall expansion of range and a westward shift of their center of abundance during the Holocene. In the case of the Boreal population, the initial center of abundance 10,000 years ago has become an area of low abundance in recent millennia as the center of abundance has shifted into Fennoscandia.

In eastern North America, *Picea* has a quite different postglacial history. In contrast to Europe, *Picea* was abundant across an extensive region to the south of the Laurentide ice sheet during the last glacial maximum; this is reflected by the pattern mapped for 12,000 years ago. Subsequently, it declined in both area of occurrence and abundance as a result of the component taxa of the broad-leaved forests extending their ranges northward more rapidly during the early Holocene than the remnant Laurentide ice sheet decayed (Jacobson et al., 1987; Webb, 1988). A minimum was reached about 8000 years ago, following which *Picea* spread rapidly northward into the area formerly occupied by the ice sheet. In both continents the southern boundary of the range of *Picea* has apparently shifted southward in recent millennia. This coincides with marked climatic cooling, especially during the Little Ice Age of the last millennium, and is likely a response to this climatic trend (Grimm and Jacobson, Jr., 1992; Huntley, 1992).

Other pollen taxa that represent components of the Boreal forests, of the sub-Arctic woodlands, and of the tundra itself also can be mapped to reveal changes in their distribution and abundance. Huntley (1988) drew a series of such maps to illustrate the patterns of occurrence and abundance 6000 years ago in contrast to those at the present day for four Boreal/Arctic

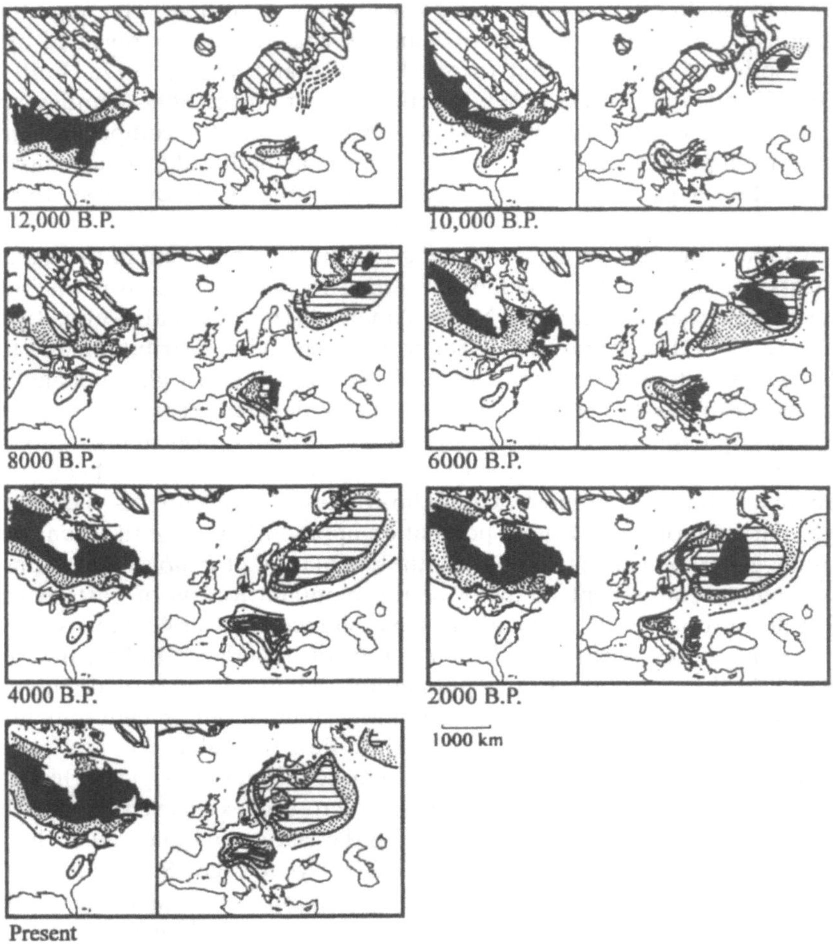

Figure 15.1. *Picea* isopoll map sequences for western Eurasia and eastern North America; maps are drawn at 2000-year intervals from 12,000 years before present (BP) to the present day. Areas of diagonal shading indicate the extent of the residual ice sheets of the last glacial stage. Isopolls are drawn at 2%, 5%, 10%, and 20% in western Eurasia and at 1%, 5%, and 20% in eastern North America. Redrawn from Huntley and Webb (1989).

pollen taxa (Figure 15.2). The first map shows the two pollen taxa distinguishable within the genus *Pinus. Pinus* (Diploxylon) in this case represents *P. sylvestris*, the only north European species in this subgenus. This pollen taxon extended almost to the north coast of Norway 6000 years ago, whereas it showed a steep decline in abundance as the Baltic coast and southern Scandinavia were approached. At the present day it is reduced in abundance throughout the far north and does not reach the north coast of Norway; in contrast, however, it is much more extensively abundant in southern Scandinavia. *Pinus* (Haploxylon) also represents a single species within the region under consideration, in this case *P. sibirica*. This is one of the most cold-tolerant trees of the region and today its pollen is found throughout northwestern and central Siberia (Peterson, 1983, 1993), coin-

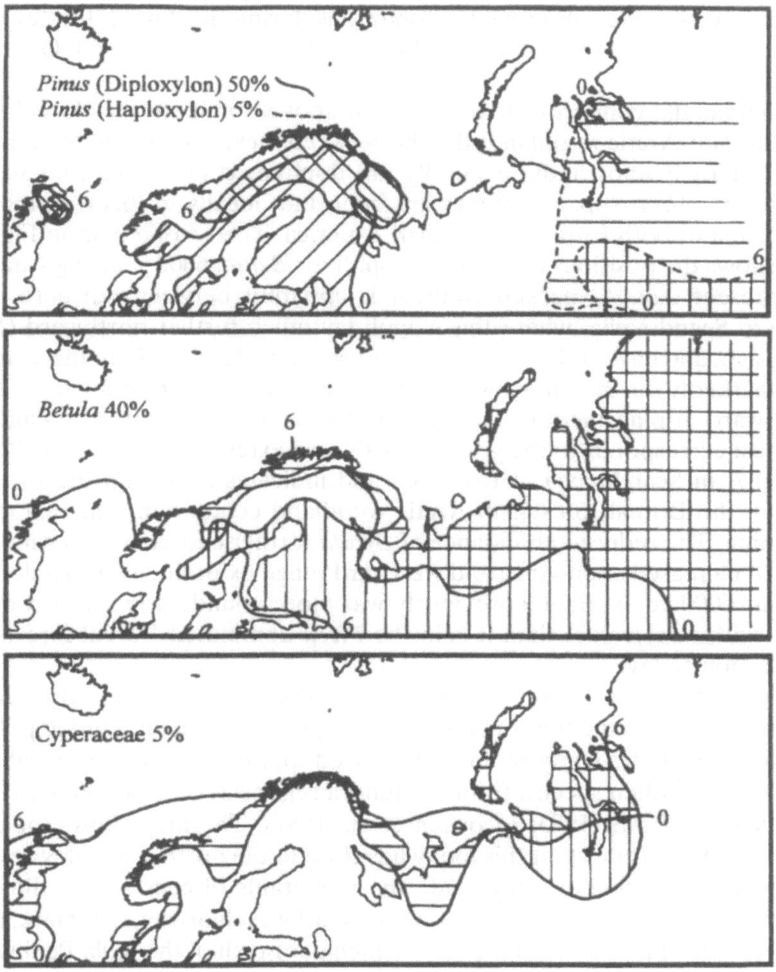

Figure 15.2. The Arctic tree line in Europe and western Russia. Isopolls for four taxa at 6000 BP and the present day superimposed to illustrate changes between the mid-Holocene and the present in the position of the Arctic tree line and in the composition of the boreal forests and woodlands forming that tree line. Redrawn from Huntley (1988).

ciding with the area of distribution of the tree in the Boreal forests of that region (Meusel et al., 1965). Six thousand years ago, however, its pollen was sparse or absent in this region and reached abundance levels of 5% or more only further south towards the continental interior of Siberia (Peterson, 1983, 1993). At that time the north Siberian Boreal forests had much greater abundance of *Picea* than today (Peterson, 1983, 1993) (Figure 15.1); *P. sibirica* apparently has to some extent replaced *P. abies* in this region during the second half of the Holocene, perhaps as a consequence of progressively colder, drier winter conditions during this time (Huntley and Prentice, 1993; Kutzbach and Gallimore, 1988; Kutzbach and Guetter, 1986).

Whereas the coniferous taxa discussed above dominate the Boreal forests, the sub-Arctic woodlands that lie between these forests and the tundra in northern Fennoscandia generally are dominated by *Betula pubescens*; elsewhere, *Alnus* spp. sometimes are important locally in this transitional ecosystem in both Eurasia and North America. The second map in Figure 15.2 shows the positions of the 40% isopoll for *Betula* 6000 years ago and at the present day. A marked contrast is apparent between northern and western Scandinavia, where this isopoll lay much further northward 6000 years ago, and eastern Fennoscandia and western Russia, where this isopoll lies markedly further northward and eastward at the present day. The northward displacement of the area of abundant *Betula* 6000 years ago in the west corresponds to the greater northward extent of the range of *Pinus sylvestris* in Scandinavia at that time and indicates an overall northward shift in the Boreal forest–sub-Arctic woodland ecotone in that region at that time. The reduced abundance of *Betula* further east coincides with the area of increased abundance and westward range extension of *Picea* during the last 6000 years; *Betula* previously was more abundant in mixed woodlands with *P. sylvestris* than it is in the *Picea*-dominated Boreal forests of that region today.

The third map in Figure 15.2 illustrates a pollen taxon that is characteristic principally of the tundra biome; although Cyperaceae is a broad pollen taxon that includes many ecologically varied species, it is extensively abundant only in pollen spectra from the tundra regions (Anderson et al., 1991; Huntley and Birks, 1983; Jacobson et al., 1987). The map shows the 5% isopoll and illustrates that this today lies close to the north coast of Norway, although also extending to include the mountains of central Scandinavia and the extensively deforested regions of southwest Norway, Denmark, and northern Britain. Eastward the isopoll includes much of the Kola Peninsula and a belt of northern Russia varying in a southward extent. Six thousand years ago this isopoll included only the extreme north of Britain and of Norway, but also a large area of northwest Siberia that today lies outside the isopoll. Once again, this evidence is consistent with the ecotones being shifted northward in Scandinavia at that time relative to the present; the extent of tundra in northern Scandinavia was apparently very small. The displacement northward in northwest Siberia during the last 6000 years is also consistent with the reduced abundance of *Picea* in this region and the northward extension of the range of *P. sibirica*; it seems that the tree line in this region has shifted northward during the last 6000 years as a result of a climatic shift towards conditions that favor *P. sibirica* at the expense not only of *P. abies*, but also of the tundra biome.

These examples serve to illustrate the information that syntheses of palynological data from broad geographical areas can provide. They show also how the distribution and abundance patterns of taxa have changed in response to past environmental changes, the principal response of taxa to a changing environment being one of migration (Good, 1931; Huntley, 1991; Huntley and Webb, 1989). The rates at which these migrations can be achieved by major forest tree taxa have been estimated generally to be between 200 and 2000 $m\,yr^{-1}$ (Davis, 1976; Huntley, 1988, 1989; Huntley and Birks, 1983). Some examples of these rates for Boreal woody taxa are given in Table 15.1. The consistency observed between different geographical areas in such estimates of migration rates strongly suggests that they may be close to the maximum achievable by the taxa in question (Huntley, 1989, 1991).

Such extensive migrations, however, are not restricted to the woody taxa about which the pollen record is generally most informative at the broad geographical scale. Studies of plant macrofossils provide evidence that low-growing herbaceous and dwarf-shrub taxa, bryophytes, and aquatic plants all have shown similar large-scale range shifts in response to past environmental changes. The coincidence of the central and thickest accumulations of the Fennoscandian and Alpine ice sheets with the present areas of distribution of many European Arctic–Alpine taxa in itself necessitates that these taxa have migrated to occupy their present ranges since the end of the last glacial period. The extensive records of macrofossil remains of such taxa from deposits representing the last glacial period in unglaciated areas of northern Europe and southern Britain indicate that these taxa migrated away from the centers of ice accumulation, often occupying localities many hundreds or even more than 1000 km beyond their present range limits. The

Table 15.1. Migration Rates of Some Boreal and Arctic Woody Taxa Estimated from Pollen Data

Taxon	Range of Observed Migration Rates ($m\,yr^{-1}$)
Alnus	500–2000
Picea	80–240–500 (–2000)[a]
Pinus (Diploxylon)	1500

From Huntley and Birks (1983).
[a] Higher rate for *Picea* reported by Ritchie and MacDonald (1986).

subsequent migration of taxa to occupy their present ranges, in Scandinavia especially, was of a similar magnitude because land areas marginal to the Fennoscandian ice sheet lay at least several hundred kilometers from areas such as the Dovrefjell in Norway, where many Arctic–Alpine taxa are today concentrated. The discovery of macrofossil remains of seeds of one such species, *Papaver radicatum*, in Scotland in sediments of late-glacial age (Huntley, 1994) necessitates a migration of at least 750 km from the closest present locality; the actual distance moved is likely to be considerably greater because of the barrier to dispersal of the intervening area of ocean even at the glacial maximum.

Past Vegetation Patterns and Composition

Whereas much can be learned from isopoll maps about past distribution patterns of individual taxa and about their migration rates, as discussed above, palaeovegetation patterns and composition are more effectively revealed by multivariate analysis of data for a large number of the major pollen taxa. Huntley (1990a) analyzed data for 44 pollen taxa in >6400 "samples" derived from >500 pollen stratigraphically studied localities and >1000 surface pollen samples in western Eurasia. The samples analyzed represented successive points in time at millennial intervals from 13,000 years BP to the present. The results were presented as a series of 14 palaeovegetation maps that illustrate how both the vegetation patterns and the composition of major vegetation units ("formations") have changed since the end of the last glacial.

Focusing upon the Arctic and Boreal regions of Europe, the most extensive vegetation unit at the present day is characterized by pollen of *Picea*, *Pinus*, *Betula*, and *Alnus*; this unit corresponds closely in extent to the "Boreal Forest," although it also extends southward to include areas of mixed conifer and hardwood forest. To the north of this unit, the most extensive unit is characterized by pollen of *Picea*, *Pinus*, and *Betula*; *Alnus* pollen values are on average only one fifth of those in the Boreal Forest unit. This unit represents the northernmost Taiga and the eastern sub-Arctic woodlands. The western sub-Arctic woodlands, in contrast, have pollen spectra heavily dominated by *Betula* and with little or no *Picea*; *Alnus* is sometimes more abundant in these units than further east but not as abundant as in the Boreal Forest unit. Nowhere on the mainland of western Eurasia were pollen spectra characteristic of tundra vegetation extensive enough at the present day to register in this broad-scale analysis; Iceland, however, had pollen spectra characterized by *Betula* accompanied by shrub and dwarf-shrub taxa (*Salix*, *Juniperus*, and Ericales).

Thirteen thousand years ago most of the Arctic and Boreal region of western Eurasia was covered by the Fennoscandian ice sheet. Small areas of far northern Norway and of the far west were, however, already ice free; the pollen spectra and macrofossil assemblages from these regions were of a "cold steppe" or even "polar desert" character, indicating extremely arid conditions in these periglacial areas at that time (Birks et al., 1994). Southeast of the ice sheet the pollen spectra were similar to those found in the sub-Arctic woodlands today. Subsequently, however, during the rapid environmental changes associated with deglaciation, the composition of the major vegetation units changed, as did the orientation of their boundaries; vegetation patterns did not settle into something resembling the present pattern, at least in terms of the orientation of major ecotones, until about 8000 years BP.

During the early millennia of the Holocene, the pollen spectra indicate that the vegetation of the region under consideration was dominated by *Betula*, often accompanied by *Pinus*. Vegetation units most similar to the modern sub-Arctic woodlands were extensive but so too was a unit that is not present today; this unit has pollen spectra dominated jointly by *Betula* and *Pinus* and apparently represents a form of northern Taiga lacking *Picea*. This unit has varied in extent throughout the Holocene, although it has tended to be eastern in distribution and to lie northward of the gradually more extensive Boreal Forest unit with abundant *Picea*.

Pollen spectra representing tundra vegetation units were extensive in mainland areas of Europe at the end of the last glacial but were displaced rapidly by woodland and forest during the early Holocene; thereafter they principally characterize Iceland. The southern part of Scandinavia has pollen spectra characteristic of broad-leaved forests over a much larger area during the mid-Holocene than at the present day. Only during the second half of the Holocene has the present Boreal Forest characterized by *Picea* extended westward as a wedge between these broad-leaved forests to the south and sub-Arctic woodlands to the north.

The present broad-scale vegetation pattern is thus, even in Quaternary geological terms, very recent in origin. The individualistic response of taxa to their changing environment (Davis, 1976, 1981, 1983; Huntley, 1988, 1991; Huntley and Birks, 1983; Huntley and Webb, 1989; Webb, 1981; Webb et al., 1983a, 1983b) has resulted in changes in the composition and extent of major vegetation units ("formations") as well as in changes in the character and orientation of major ecotones. Although the evidence for similar changes in composition of smaller scale vegetation units ("associations") is less clear, principally because absence of taxa in the fossil record cannot be taken as positive evidence of their absence from the vegetation at that time, there is circumstantial evidence both from studies of present vegetation and of the fossil record that indicates changes in composition in response to environmental changes. Geographical gradients in species composition within associations are widely discussed in the phytosociological literature (see, e.g., McVean and Ratcliffe, 1962); such gradients generally are related to climatic gradients along which component taxa show individualism both with respect to their abundance optima and to their range limits. The late-Quaternary fossil record often includes assemblages of species that have nonoverlapping geographical ranges at the present day, although their ecological characteristics and/or habitat requirements are similar (Graham and Grimm, 1990); here too it seems most probable that such taxa were components of past associations that differ in composition from their modern counterparts.

Past Climate Changes

Past climate conditions at particular localities may be reconstructed either from the fossil evidence of past biota (see, e.g., Atkinson et al., 1987; Huntley, 1994; Kelly and Huntley, 1991) or from independent physical and/or chemical records that can be obtained from various stratified deposits (see, e.g., Dansgaard et al., 1989; GRIP Members, 1993). Syntheses of such reconstructions from many localities can provide synoptic views of past climate patterns (see, e.g., Bartlein and Webb, 1985; Huntley and Prentice, 1988; Prentice et al., 1991). In addition, atmospheric general circulation models (GCMs) may be used to simulate global palaeoclimate patterns for particular times in the past for which the necessary boundary conditions and forcing factors are known (see, e.g., Kutzbach and Guetter, 1986).

GCM simulations and other evidence lead to an estimate of ca. 5°C shift in mean *global* temperature during the last glacial termination (ca. 15,000 to 10,000 years BP) (Schneider, 1989). Reconstructions from individual localities, however, often indicate much larger and/or more rapid shifts either in mean annual temperature (Dansgaard et al., 1989) or in seasonal temperature conditions (Atkinson et al., 1987; Field et al., 1994; Huntley, 1994). Such large (≥20°C change in mean temperature of the coldest month) (Atkinson et al., 1987; Field et al., 1994; Huntley, 1994) and rapid (ca. 10°C increase in mean annual temperature within ca. 100 years) (Dansgaard et al., 1989) localized changes do not conflict with the estimated global changes; rather they serve to highlight what also is apparent from the GCM simulations, namely, that some regions experience changes that are much larger than the global mean whereas in other regions conditions may be almost unchanged (COHMAP Members, 1988; Kutzbach and Guetter, 1986). Furthermore, evidence of such large, rapid changes to date has come principally from the North Atlantic region and from northern and western Europe, areas that are indicated by the GCM simulations to be especially sensitive to changes in global conditions (COHMAP Members, 1988; Kutzbach and Gallimore, 1988; Kutzbach and Guetter, 1986).

Whereas glacial climate conditions were globally cooler, and locally much colder, than present conditions, the interglacial optima of the present (Holocene) and last [Eemian (Europe), Ipswichian (British Isles), Mikulino (Russia), Sangamon (North America)] interglacials globally were only marginally warmer than the present. Globally, the Holocene optimum was no more than 1°C warmer than today and may have been no warmer (Webb and Wigley, 1985), although the last interglacial generally is regarded as having been ca. 2°C warmer than recent times (Folland et al., 1990). Some reconstructions indicate that these temperature optima may have been amplified at high latitudes. Huntley and Prentice (1988) reconstructed July temperatures for Europe from pollen data for 6000 years BP; their results showed >2°C higher summer temperatures in far northern Scandinavia. The magnitude of the winter warming has not yet been quantitatively reconstructed; however, the changes in vegetation recorded by the pollen data and discussed above indicate that winter temperatures were likely warmer than today by a somewhat larger amount than were summer temperatures. This inference is consistent with the modeling results of Kutzbach and Gallimore (1988). Lake-level evidence from Finnish Lapland (Hyvärinen and Alhonen, 1994) indicates drier conditions also during the early Holocene. Mean annual temperatures for the last interglacial have been estimated to have been as much as 5°C warmer than present at high latitudes (Folland et al., 1990; LIGA Members, 1991a). Lozhkin and Anderson (1995) recently have inferred summer temperatures 4 to 8°C higher than today in northeast Siberia during the last interglacial, although with a contrasting marked cooling of up to 12°C in winter close to the Arctic coast.

Discussion

The brief review presented above has outlined some of the evidence relating to the past distribution patterns of organisms and their migratory response to environmental changes, to past vegetation patterns and the composition of past formations and associations, and to past changes in climate. A number of conclusions emerge from these various lines of evidence.

- Both herbaceous and woody taxa have achieved extensive migrations in response to the environmental changes between the alternating glacial and interglacial stages of the Quaternary. Range boundaries frequently have shifted by >1000km even among Arctic and Boreal taxa; some trees of the temperate deciduous forests have exhibited shifts of >2000km in their northern boundaries. In many cases the glacial and interglacial geographical ranges of Arctic taxa may not have overlapped; even where such overlaps have occurred the area of optimum abundance generally does not coincide between glacial and interglacial stages.
- Complementing the evidence of extensive migrations is that of a lack of any significant or persistent macroevolutionary responses elicited by the considerable environmental changes that have taken place throughout the Quaternary. Bennett (1990) points out that the selective pressures exerted by the environment have changed continually throughout the

Quaternary so that traits selected for under glacial conditions may become unfavorable during the next warm period and vice versa. Huntley et al. (1989) showed that the environmental tolerances of two geographically isolated species within the genus *Fagus* were more similar to each other than to other taxa found in each of the two geographical regions; they concluded that the physiological traits that determine these tolerances had been conserved during >5 million years of isolation. This is consistent with the "Theory of Tolerance" advanced by Good (1931) and whereas similar systematic comparisons have not yet been made for other taxa, subjective examination of the climate response surfaces (Anderson et al., 1991; Bartlein et al., 1986; Huntley, 1993; Prentice et al., 1991) for equivalent pollen taxa in Europe and North America shows that they generally are similar.

- In order to survive the changing combinations of conditions to which they have been exposed at different phases of the glacial–interglacial cycle, however, species are likely to have shown microevolutionary changes in allele frequencies and combinations at polymorphic loci. Furthermore, they must have evolved the genetic "competence," expressed in terms of such polymorphisms and/or in terms of phenotypic plasticity, to cope with these changing combinations of conditions; without this competence they would not have survived Quaternary environmental changes. Not only have climate conditions unlike any today been experienced at some times and places during the Quaternary, but the climate has combined with other factors such as atmospheric CO_2 concentration, day length, and seasonal radiation patterns in ways that are not matched today.
- Average migration rates of 200 to 500 m yr^{-1} frequently have been attained in response to the rapid environmental changes of deglaciation, even by long-lived trees. Such rates were sustained over several millennia leading to range boundary shifts of 1000 to 2000 km. In extreme cases migration rates of 1 to 2 km yr^{-1} have been attained; such rates are relatively more frequent amongst Boreal tree taxa than those of the temperate deciduous forests.
- The consistency among the estimates of migration rates made for different geographic rergions (Huntley, 1989), the possibility that vegetation responses may at some times have lagged behind climatic changes (Davis, 1989), as well as the correspondence between expected and observed rates (Huntley, 1991), leads to the conclusion that these migration rates likely approach the maximum rates of which trees, and perhaps most higher plants, are capable (Huntley, 1991).
- These migratory responses have been individualistic, with no two taxa examined showing the same patterns of change. As a consequence, taxa that at one time may have broadly coincident ranges with extensive overlap have at other times ranges that overlap only partially or not at all. Furthermore, the glacial and interglacial ranges of any one taxon often do not overlap and for many Arctic taxa their interglacial range coincides

with areas of major ice accumulations during glacial stages. These observations have important implications with respect both to considerations of the present biogeography of Arctic and Boreal taxa and to investigations into the evolution of vicariant taxa, including the intraspecific evolution of races or subspecies. For example, the disjunct ranges of several plant species in the mountains of Scandinavia (e.g., *Papaver radicatum*, *Rhododendron lapponicum*, *Saxifraga hieracifolia*, and *Silene wahlenbergella*) can be explained only in terms of postglacial history because at its maximum extent the last glacial Fennoscandian ice sheet (see, e.g., Denton and Hughes, 1981) covered the entire region of the Scandinavian mountians and reached a maximum depth that apparently precludes any hypothesis of survival of these species on "nunataks" *within* the mountainous regions where they presently occur. This does not, of course, preclude their survival in the lowland areas of northern and northwestern Fennoscandia that apparently were unglaciated at the time of the last glacial maximum (Birks et al., 1994) or even on "nunataks" emerging through the relatively thinner marginal areas of the ice sheet in western Norway. Similarly, the majority of the 10 subspecies of *Papaver radicatum* that are recognized from Norway/Sweden, as well as the additional three from Iceland and that from the Faeroe Islands (Mossberg et al., 1992), today occur solely in areas that were occupied by the last glacial ice sheets. Some of these vicariant subspecies occupy areas that are separated from one another by only a few tens of kilometres. Whereas it seems impossible now to conceive of these taxa as "glacial relicts" in the conventional sense, because they could not have occupied their present areas of distribution until the early postglacial, the evidence of the occurrence of *Papaver radicatum* in regions where it is now extinct (e.g., Huntley, 1994) demonstrates the capacity of the species to dynamically adjust its range to changing climatic conditions. The present vicariant subspecies may have evolved only during the postglacial period as a consequence of isolation in "interglacial refuges"; their evolutionary history may only be deduced by critical studies of the genetics of each taxon rather than by inference from their present distribution patterns.

- As a further consequence of this individualism, past vegetation units, whether at the level of formations or of associations, frequently have differed in composition from any modern community. This in turn has led to differences in the nature as well as in the position and orientation of major ecotones. At the broad geographical scale, the present patterns of Boreal and Arctic formations have existed for only a few millennia. At finer scales, the macrofossil contents of many peat accumulations indicate that the present species composition of the associations that form their surface vegetation differs, often quite markedly, from that of the associations occupying the mire surface only one or a few millennia ago (e.g., the former extensive occurrence of *Sphagnum imbricatum* as a component of the surface vegetation of ombrogenous mires in the British Isles in contrast to its present rarity of occurrence and absence from most such habitats).

- Whereas the change in global mean temperature between glacial and interglacial conditions was ca. 5°C, the optima of the present and last interglacials were no more than 1°C and ca. 2°C warmer than the present, respectively.
- Localized changes of temperature were often rapid and large in magnitude, especially around the North Atlantic and in northern and western Europe. Although the maximum reconstructed rate of temperature increase is ca. $0.1°C\,yr^{-1}$ for mean annual temperature at a locality in Greenland (Dansgaard et al., 1989), this is exceptional and the most rapid rates reconstructed from terrestrial records in Europe are ca. $0.03°C\,yr^{-1}$ for mean temperature of the coldest month (Atkinson et al., 1987; Field et al., 1994; Huntley, 1994).

When viewed in the context of predicted future climate changes as a consequence of anthropogenic increases in greenhouse gases in the atmosphere, a number of these conclusions become of particular importance. Firstly, the forecast rise in global mean temperature for a doubling of the radiative effect of greenhouse gases (the so-called $2 \times CO_2$ scenario) is in the range 1.5 to 4.5°C (Mitchell et al., 1990) with a "best guess" of ca. 2.5°C. This latter value is approximately half the magnitude of the change in global mean temperature estimated between glacial maximum conditions and the Holocene (Schneider, 1989), whereas the upper limit on the range of forecasts is of the same magnitude as this past warming. The forecast warming is also of a greater magnitude, perhaps as much as twice as great, as the warming estimated for the last interglacial and may result in global conditions warmer than any during the Quaternary (the last ca. 2.4 Myr).

Although the largest forecast increases in regional temperatures are not as great as those reconstructed at some localities for the last glacial to Holocene transition, the increase of 6°C, and locally perhaps as much as 10°C, in winter temperature in high northern latitudes in a period of only 50 to 70 years (Rowntree, Chapter 4) represent rates of increase that are comparable to or exceed by a factor of two the most rapid reconstructed past changes. These latter changes are reconstructed from stable isotope records in ice cores (Dansgaard et al., 1989; GRIP Members, 1993) and themselves exceed by three to seven times the most rapid temperature changes (in the mean temperature of the coldest month) reconstructed from terrestrial evidence (Atkinson et al., 1987; Field et al., 1994; Huntley, 1994). Furthermore, caution is needed when considering these forecasts, firstly because they are for a $2 \times CO_2$ scenario and secondly, because current GCMs have a relatively coarse resolution and omit several important feedback mechanisms, the majority of which are expected to be positive feedbacks (Houghton et al., 1990; Schimel, 1990). These limitations of the GCMs may hide changes of even greater magnitude, such as those seen in the palaeoclimate records from ice cores.

Greenhouse gas levels are forecast to continue rising beyond the equivalent of 2 × CO_2 and may reach the equivalent of 4 × CO_2 by the 22nd century; such continuing and accelerating increases inevitably will lead to further rapid increases in global mean temperature and associated increases in regional temperatures. Manabe and Stouffer (1993) have simulated a transition to a 4 × CO_2 scenario in 140 years, with that level then maintained for a further 360 years; the resulting global mean temperature increase is 7°C, with increases of as much as 16°C in the Arctic region. The rapid changes in temperature at individual localities in Europe and Greenland during the last glacial termination generally are accepted to be related to rapid changes in the deep ocean circulation that is driven especially by the thermohaline circulation in the North Atlantic (Broecker and Denton, 1989) and the resulting formation of North Atlantic Deep Water. Broecker and Denton (1989) have argued that the extreme climate sensitivity around the North Atlantic may relate to the existence of "on" and "off" modes of the thermohaline circulation, a hypothesis supported by modeling results that indicate two alternate stable states for global ocean circulation (Manabe and Stouffer, 1988). In their simulation for 4 × CO_2, Manabe and Stouffer (1993) report the almost complete extinction of the thermohaline circulation. They note also that they have not included the effect of ice sheet melting in their simulation and suggest that the impact of the fresh meltwater input so generated might lead to the loss of the thermohaline circulation even at 2 × CO_2. They also point out the potentially very important implications for various biogeochemical cycles, not least the carbon cycle, were such an extinction of the ocean thermohaline circulation to occur. The possibility that such a change could occur and that, through feedback mechanisms that are not included in present GCMs, it might lead to climate changes as large as or even larger than those in the palaeoclimate record cannot be ignored.

Given the likelihood of climate changes in the near future, the conclusions from the palaeovegetation record lead to the prediction that species will respond by individualistic migrations. Such migrations may not, however, keep pace with the changing climate except where altitudinal adjustments rather than spatial adjustments of range limits are involved or where scattered krummholz individuals already are present as much as 100 km beyond the tree line, providing the potential for an initial, although not a sustained, very rapid response (MacDonald et al., 1993). The impacts upon advancing and retreating range boundaries are likely to be qualitatively different. Some species may be unable to tolerate changed conditions at this "warm" and/or "dry" range limit and mature individuals may die as a result. In other cases regeneration may fail; this will lead to progressive wasting of the population with the possibility of natural or human disturbance triggering catastrophic decline. Along their "cool" range limit, species may expand at rates of 20 to 200km per century, by analogy with the palaeoecological record; this, however, may be as much as 10 to 40 times slower than the rate of advance of their potential range margin. As a consequence of the potential for rapid decline at "warm" range boundaries and the lag in adjustment of "cool" range limits, large areas may be colonized by early successional or "opportunist" species. In the Arctic this may result in a broadening of the zones with tall shrubs and, in the Eurasian sector at least, of *Betula* woodlands between the conifer-dominated Boreal forests and the tundra. Thus, it is likely that mainland tundras, at least, rapidly will diminish in extent. The Boreal forests themselves also are likely to be reduced in extent; in the Eurasian sector *Picea* is likely to fail to regenerate near its southern and western boundaries. The likelihood is that a formation dominated by *Pinus sylvestris* and *Betula*, similar to that which was extensive during the early Holocene in Fennoscandia, might opportunistically occupy areas from which *Picea* disappears but into which the nemoral broad-leaved species will migrate only relatively more slowly. Thus, the extent, and even the number, of formations in northern Eurasia may change, at least temporarily, during the centuries of adjustment necessary before any equilibrium with the new conditions can be reached. It is likely that similar changes will occur in other regions of the Arctic.

Where tundra occurs at higher elevations upon sub-Arctic mountains, the pace of change is likely to be even greater. Tree lines may advance upslope at a pace that allows equilibrium to be maintained with the changing climate, although in those mountain areas where wind is a principal determinant of the tree line elevation such upslope movements are likely to be less pronounced. Many areas of alpine vegetation will be reduced in extent and it is likely that species will be lost from some mountain areas (Grabherr et al., 1994). Such elevation of tree lines occurred at the time of the Holocene optimum, as did a maximal northward extent of forest and woodland vegetation in the Arctic lowlands. It is likely that these past advances led to local extinctions of some tundra and/or alpine zone species. Where, as may often have been the case, these species have subsequently failed to recolonize the areas from which they were eradicated, then initially there may be little apparent impact of the newly advancing tree line upon the biodiversity of the herbaceous commumities. A noticeable impact is to be expected only when the tree line extends beyond its Holocene maximum position and this may not occur for at least some decades. Thereafter, many herbaceous and other tundra species may face crises of population reduction as a consequence of reduced habitat availability and some at least likely will become locally extinct in some areas.

Only the Arctic islands are likely to suffer limited initial impact. Their relative isolation from sources of immigrant species (apart from those that humans may accidentally and/or deliberately introduce) and their freedom from the impact of advancing woody taxa likely will allow the shade-intolerant tundra species to persist. However, even in these situations some species likely will be sensitive to the changed and perhaps less stable climate conditions (Molau, Chapter 7) and may be unable to persist and/or to regenerate. In such cases the composition of the tundra communities will change and some species may even be lost from some islands. Those species most likely to persist in such circumstances are those with a wide range of genetic polymorphisms that enable them to occupy a range of habitats and thus to adapt to changing conditions (Crawford and Chapman, 1995).

Acknowledgments. I am grateful to the organizers of the *Global Change and Arctic Terrestrial Ecosystems* conference for the invitation to prepare this paper; they and the University of Durham provided financial support that enabled my participation in this meeting. Discussions and interactions with numerous colleagues have contributed to the development of the ideas presented here, as have previous opportunities to address related issues. Particular thanks go to Wolfgang Cramer and Jarle Holten for providing the opportunity to become more familiar with Arctic ecosystems, and to Jacqui James and Alexina Huntley who have shared several trips to explore the European Arctic. A number of other contributors to the conference presented ideas in their papers that stimulated new ways of thinking about the problem of climate change impacts upon Arctic vegetation; they too must be thanked for their unconscious contributions to the final form of this text. Wolfgang Cramer provided helpful critical comments upon an initial draft of the paper; Jacqui Huntley carefully and critically read the subsequent revised version and two anonymous reviewers also made numerous constructive criticisms that have aided in the preparation of the final version.

References

Anderson, P.M., Bartlein, P.J., Brubaker, L.B., Gajewski, K., and Ritchie, J.C. 1991. Vegetation–pollen–climate relationships for the arcto-boreal region of North America and Greenland. *J. Biogeogr.* 18:565–582.

Anderson, P.M., and Brubaker, L.B. 1993. Holocene vegetation and climate histories of Alaska. In: Wright, H.E. Jr., Kutzbach, J.E., Webb, T. III, Ruddiman, W.F., Street-Perrott, F.A., and Bartlein, P.J. (eds.), *Global Climates Since the Last Glacial Maximum* (pp. 386–400). Minneapolis: University of Minnesota Press.

Anderson, P.M., and Brubaker, L.B. 1994. Vegetational history of northcentral Alaska: A mapped summary of Late Quaternary pollen data. *Quat. Sci. Rev.* 13:71–92.

Atkinson, T.C., Briffa, K.R., and Coope, G.R. 1987. Seasonal temperatures in Britain during the past 22,000 years reconstructed using beetle remains. *Nature* 325:587–593.

Bartlein, P.J. 1988. Late-Tertiary and Quaternary palaeoenvironments. In: Huntley, B., and Webb, T. III (eds.), *Vegetation History* (pp. 113–152). Dordrecht: Kluwer Academic.

Bartlein, P.J., Prentice, I.C., and Webb, T. III 1986. Climatic response surfaces from pollen data for some Eastern North American taxa. *J. Biogeogr.* 13:35–57.

Bartlein, P.J., and Webb, T. III 1985. Mean July temperature estimates for 6,000 yr BP in eastern North America; Regression equations for estimates from fossil-pollen data. In: Harrington, C.R. (ed.), *Climatic Change in Canada* (pp. 301–342). *Syllogeus* 55.

Bennett, K.D. 1990. Milankovitch cycles and their effects on species in ecological and evolutionary time. *Paleobiology* 16:11–21.

Bernabo, J.C., and Webb, T. III 1977. Changing patterns in the Holocene pollen record of northeastern North America: a mapped summary. *Quat. Res.* 8:64–96.

Birks, H.H., Paus, A., Svendsen, J.I., Alm, T., Mangerud, J., and Landvik, J.Y. 1994. Late Weichselian environmental change in Norway, including Svalbard. *J. Quat. Sci.* 9:133–145.

Brigham-Grette, J., and Hopkins, D.M. 1995. Emergent marine record and palaeoclimate of the last interglaciation along the northwest Alaskan coast. *Quat. Res.* 43:159–173.

Broecker, W.S., and Denton, G.H. 1989. The role of ocean-atmosphere reorganizations in glacial cycles. *Geochim. Cosmochim. Acta* 53:2465–2501.

COHMAP Members 1988. Climatic changes of the last 18,000 years: Observations and model simulations. *Science* 241:1043–1052.

Crawford, R.M.M., and Chapman, H.M. 1995. Climatic change and species polymorphism in the Arctic. In: *Global Change and Arctic Terrestrial Ecosystems* (pp. 115–121). Ecosystems research report 10, European Commission, Luxembourg.

Dansgaard, W., White, J.W.C., and Johnsen, S.J. 1989. The abrupt termination of the Younger Dryas climate event. *Nature* 339:532–534.

Davis, M.B. 1976. Pleistocene biogeography of temperate deciduous forests. *Geosci. Man* 13:13–26.

Davis, M.B. 1981. Quaternary history and the stability of forest communities. In: West, D.C., Shugart, H.H., and Botkin, D.B. (eds.), *Forest Succession: Concepts and Application* (pp. 132–153). New York: Springer-Verlag.

Davis, M.B. 1983. Holocene vegetational history of the eastern United States. In: Wright, H.E. Jr. (ed.), *Late Quaternary Environments of the United States, Volume 2: The Holocene* (pp. 166–181). Minneapolis: University of Minnesota Press.

Davis, M.B. 1989. Lags in vegetation response to greenhouse warming. *Climate Change* 15:75–82.

Denton, G.H., and Hughes, T.J. (eds.). 1981. *The Last Great Ice Sheets*. New York: John Wiley and Sons.

Field, M.H., Huntley, B., and Müller, H. 1994. Eemian climate fluctuation observed in a European pollen record. *Nature* 371:779–783.

Folland, C.K., Karl, T.R., and Vinnikov, K.Ya. 1990. Observed climate variations and change. In: Houghton, J.T., Jenkins, G.J., and Ephraums, J.J. (eds.), *Climate Change: the IPCC Scientific Assessment* (pp. 195–238). Cambridge: Cambridge University Press.

Good, R. 1931. A Theory of Plant Geography. *New Phytol.* 30:11–171.

Grabherr, G., Gottfried, M., and Pauli, H. 1994. Climate effects on mountain plants. *Nature* 369:448.

Graham, R.W., and Grimm, E.C. 1990. Effects of global climate change on the patterns of terrestrial biological communities. *Trends Ecol. Evol.* 5:289–292.

Grimm, E.C., and Jacobson, G.L., Jr. 1992. Fossil-pollen evidence for abrupt climate changes during the past 18,000 years in eastern North America. *Clim. Dynam.* 6:179–184.

GRIP Members 1993. Climate instability during the last interglacial period recorded in the GRIP ice core. *Nature* 364:203–207.

Houghton, J.T., Callander, B.A., and Varney, S.K. 1992. *Climate Change 1992: The Supplementary Report to the IPCC Scientific Assessment.* Cambridge: Cambridge University Press.
Houghton, J.T., Jenkins, G.J., and Ephraums, J.J. 1990. *Climate Change: The IPCC Scientific Assessment.* Cambridge: Cambridge University Press.
Huntley, B. 1988. Glacial and Holocene Vegetation History: Europe. In: Huntley, B., and Webb, T. III (eds.), *Vegetation History* (pp. 341–383). Dordrecht: Kluwer Academic.
Huntley, B. 1989. European post-glacial vegetation history: a new perspective. In: Ouellet, H. (ed.), *Proceedings of the XIX International Ornithological Congress, Volume 1* (pp. 1060–1077). Ottawa: University of Ottawa Press.
Huntley, B. 1990a. European vegetation history: palaeovegetation maps from pollen data—13,000 yr BP to present. *J. Quat. Sci.* 5:103–122.
Huntley, B. 1990b. European post-glacial forests: compositional changes in response to climatic change. *J. Vegeta. Sci.* 1:507–518.
Huntley, B. 1991. How plants respond to climate change: Migration rates, individualism and the consequences for plant communities. *Ann. Bot.* 67 (Suppl. 1):15–22.
Huntley, B. 1992. Rates of change in the European palynological record of the last 13,000 years and their climatic interpretation. *Clim. Dynam.* 6:185–191.
Huntley, B. 1993. The use of climate response surfaces to reconstruct palaeoclimate from Quaternary pollen and macrofossil data. *Philos. Trans. R. Soc. Lond. Series B* 341:215–224.
Huntley, B. 1994. Late-Devensian and Holocene palaeoecology and palaeoenvironments of the Morrone Birkwoods, Aberdeenshire, Scotland. *J. Quat. Sci.* 9:311–336.
Huntley, B., Bartlein, P.J., and Prentice, I.C. 1989. Climatic control of the distribution and abundance of beech (*Fagus* L.) in Europe and North America. *J. Biogeogr.* 16:551–560.
Huntley, B., and Birks, H.J.B. 1983. *An Atlas of Past and Present Pollen Maps for Europe: 0–13,000 BP.* Cambridge: Cambridge University Press.
Huntley, B., and Prentice, I.C. 1988. July temperatures in Europe from pollen data 6000 years before present. *Science* 241:687–690.
Huntley, B., and Prentice, I.C. 1993. Holocene vegetation and climates of Europe. In: Wright, H.E. Jr., Kutzbach, J.E., Webb T. III, Ruddiman, W.F., Street-Perrot, F.A., and Bartlein, P.J. (eds.), *Global Climates since the last glacial maximum* (pp. 136–168). Minneapolis: University of Minneapolis Press.
Huntley, B., and Webb, T. III 1989. Migration: species response to climatic variations caused by changes in the earth's orbit. *J. Biogeogr.* 16:5–19.
Hyvärinen, H., and Alhonen, P. 1994. Holocene lake-level changes in the Fennoscandian tree-line region, western Finnish Lapland: diatom and cladoceran evidence. *The Holocene* 4:251–258.
Ingólfson, O., and Norddahl, H. 1994. A review of the environmental history of Iceland, 13,000–9000 yr BP. *J. Quat. Sci.* 9:147–150.
Jacobson, G.L. Jr., Webb, T. III, and Grimm, E.C. 1987. Patterns and rates of vegetation change during the deglaciation of eastern North America. In: Ruddiman, W.F., and Wright, H.E. Jr. (eds.), *North America and Adjacent Oceans during the Last Deglaciation* (pp. 277–288). Boulder, CO: Geological Society of America.
Kelly, M.G., and Huntley, B. 1991. An 11,000-year record of vegetation and environment from Lago di Martignano, Latium, Italy. *J. Quat. Sci.* 6:209–224.
Kutzbach, J.E., and Gallimore, R.G. 1988. Sensitivity of a coupled atmosphere/mixed-layer ocean model to changes in orbital forcing at 9000 years BP. *J. Atmos. Sci.* 93:803–821.

Kutzbach, J.E., and Guetter, P.J. 1986. The influence of changing orbital parameters and surface boundary conditions on climate simulations for the past 18,000 years. *J. Atmos. Sci.* 43:1726–1759.
Lamb, H.F., and Edwards, M.E. 1988. Glacial and Holocene Vegetation History: The Arctic. In: Huntley, B., and Webb, T. III (eds.), *Vegetation History* (pp. 519–555). Dordrecht: Kluwer Academic.
LIGA Members 1991a. Report of the first discussion group: The last interglacial in high latitudes of the northern hemisphere: terrestrial and marine evidence. *Quat. Int.* 10–12:9–28.
LIGA Members 1991b. Report of the second discussion group: Inter-relationships and linkages between the land, atmosphere and oceans during the last interglacial. *Quat. Int.* 10–12:29–48.
Lozhkin, A.V., and Anderson, P.M. 1995. The last interglaciation in northeast Siberia. *Quat. Res.* 43:147–158.
Lozhkin, A.V., Anderson, P.M., Eisner, W.R., Ravako, L.G., Hopkins, D.M., Brubaker, L.B., Colinvaux, P.A., and Miller, M.C. 1993. Late quaternary lacustrine pollen records from southwestern Beringia. *Quat. Res.* 39:314–324.
MacDonald, G.M., Edwards, T.W.D., Moser, K.A., Plenitz, R., and Smol, J.P. 1993. Rapid response of treeline vegetation and lakes to past climate warming. *Nature* 361:243–246.
Maizels, J.K., and Caseldine, C. 1991. *Environmental Change in Iceland: Past and Present.* Dordrecht: Kluwer Academic.
Manabe, S., and Stouffer, R.J. 1988. Two stable equilibria of a coupled ocean-atmospheric model. *J. Climatol.* 1:841–866.
Manabe, S., and Stouffer, R.J. 1993. Century-scale effects of increased atmospheric CO_2 on the ocean-atmosphere system. *Nature* 364:215–218.
McVean, D.N., and Ratcliffe, D.A. 1962. *Plant Communities of the Scottish Highlands.* London: HMSO.
Meusel, H., Jäger, E., and Weinert, E. 1965. *Vergleichende Chorologie der Zentraleuropäischen Flora: Vol. 1 (Karten).* Jena: Gustav Fischer Verlag.
Mitchell, J.F.B., Manabe, S., Tokioka, T., and Meleshko, V. 1990. Equilibrium Climate Change. In: Houghton, J.T., Jenkins, G.J., and Ephraums, J.J. (eds.), *Climate Change: The IPCC Scientific Assessment* (pp. 131–172). Cambridge: Cambridge University Press.
Mossberg, B., Stenberg, L., and Ericsson, S. 1992. *Den Nordiska Floran.* Stockholm: Wahlström and Widstrand.
Peterson, G.M. 1983. *Holocene Vegetation and Climate in the Western U.S.S.R.* Ph.D. thesis. Madison: University of Wisconsin.
Peterson, G.M. 1993. Vegetational and climatic history of the western former Soviet Union. In: Wright, H.E. Jr., Kutzbach, J.E., Webb, T. III, Ruddiman, W.F., Street-Perrott, F.A., and Bartlein, P.J. (eds.), *Global Climates Since the Last Glacial Maximum* (pp. 169–193). Minneapolis: University of Minnesota Press.
Prentice, I.C., Bartlein, P.J., and Webb, T. III 1991. Vegetation and climate change in eastern North America since the last glacial maximum. *Ecology* 72:2038–2056.
Richard, P.J.H. 1994. Postglacial palaeophytogeography of the eastern St. Lawrence river watershed and the climatic signal of the pollen record. *Palaeogeog. Palaeoclimat. Palaeoecol.* 109:137–161.
Ritchie, J.C. 1987. *Postglacial Vegetation of Canada.* Cambridge: Cambridge University Press.
Ritchie, J.C., and MacDonald, G.M. 1986. The patterns of post-glacial spread of white spruce. *J. Biogeogr.* 13:527–540.

Schimel, D. 1990. Biogeochemical Feedbacks in the Earth System. In: Leggett, J. (ed.), *Global Warming: The Greenpeace Report* (pp. 68–82). Oxford: Oxford University Press.

Schneider, S. 1989. The changing global climate. *Sci. Am.* 261(3):38–47.

Szafer, W. 1935. The significance of isopollen lines for the investigation of trees in the postglacial period. *Bulletin de l'Academie Science Polonaise* B1935:235–239.

Webb, T. III 1981. 11,000 years of vegetational change in eastern North America. *BioScience* 31:501–506.

Webb, T. III 1988. Glacial and holocene vegetation history: Eastern North America. In: Huntley, B., and Webb, T. III (eds.), *Vegetation History* (pp. 385–414). Dordrecht: Kluwer Academic Publishers.

Webb, T. III, Bartlein, P.J., Harrison, S.P., and Anderson, K.H. 1993. Vegetation, lake levels, and climate in eastern North America for the past 18,000 years. In: Wright, H.E. Jr., Kutzbach, J.E., Webb, T. III, Ruddiman, W.F., Street-Perrott, F.A., and Bartlein, P.J. (eds.), *Global Climates Since the Last Glacial Maximum* (pp. 415–467). Minneapolis: University of Minnesota Press.

Webb, T. III, Cushing, E.J., and Wright, H.E. Jr. 1983a. Holocene changes in the vegetation of the Midwest. In: Wright, H.E. Jr. (ed.), *Late-Quaternary Environments of the United States, Volume 2* (pp. 142–165). Minneapolis: University of Minnesota Press.

Webb, T. III, Richard, P.J.H., and Mott, R.J. 1983b. A mapped history of Holocene vegetation in Southern Québec. *Syllogeus* 49:273–336.

Webb, T. III, and Wigley, T.M.L. 1985. What past climates can indicate about a warmer world. In: MacCracken, M.C., and Luther, F.M. (eds.), *Projecting the Climatic Effects of Increasing Carbon Dioxide* (pp. 239–257). Washington, DC: U.S. Department of Energy.

16. Modeling the Possible Impact of Climate Change on Broad-Scale Vegetation Structure: Examples from Northern Europe

Wolfgang Cramer

Climate has long been recognized as a major determinant of vegetation structure and composition (Woodward, 1987). Changes in climate, such as ambient temperature or the availability of light and moisture, are therefore likely to change vegetation on a relatively short time scale, such as years to decades. In fact, the connection between climate and vegetation has been considered so close that knowledge of past vegetation composition, derived from pollen, macrofossils, or other sources, could be used to reconstruct historical climatic patterns by analysis of the correlation between the presence of certain species and the climatic conditions. The key concepts that were developed for such paleoecological reconstruction have provided valuable contributions to the more recent development of *predictive* modeling of climate–vegetation interactions (see Huntley, Chapter 15).

The problems connected to the prediction of possible impacts of future climate change require the development of climate–vegetation models that are capable of approximating, firstly, the potential direction and magnitude of the ecosystem response as well as its associated feedbacks to the atmosphere, and secondly, the time scales that are required until the new conditions have been established. A wide range of simulation models exists for the first purpose, with different capabilities and with different areas of application. Here, some of the current techniques for broad-scale ecosystem modeling will be reviewed, with particular reference to Arctic and Subarctic ecosystems. The second purpose, the estimation of the transient nature of ecosystem change, will be discussed, and it will be shown that a realistic prediction of broad-scale transient change of ecosystems currently is beyond the reach of the scientific community. Some approaches exist, however, to move towards this goal.

The term "ecosystem model" is used in its broadest meaning here, implying almost any scientific method that appears to be capable of predicting relevant structural characteristics of ecosystems at a broad regional scale. Using this definition, it is often considered useful to distinguish between statistical (empirical) and mechanistic models. In this distinction, statistical models are based on the correlation between observations of ecosystems and climate, and mechanistic models rely on understanding of fundamental climate-driven ecosystem processes. It is important to note, however, that this distinction rapidly breaks down when the existing models are analyzed more closely. The more successful biogeographic models, which are mainly based on spatial correlation between climate and vegetation, can be shown to perform particularly well if there is a certain degree of ecophysiological reasoning behind the definition of the variables and their response functions. An example for this is the use of moisture balance estimates, which have improved the predictive power of global biome models considerably by making a more explicit description of processes of evapotranspiration. Similarly, models that are classified as being "mechanistic" or "process models" inevitably rely on correlations between subprocesses and climate, because it seems to be impossible to cover very many levels of process resolution simultaneously in one model.

For Arctic ecosystems, climate is a particularly significant factor, and there is rather low direct impact of human land use in most areas. Although nutrient availability is a crucial factor controlling the local distribution of ecosystem types in Arctic regions, it is usually a function of the ecosystem processes themselves (e.g., the activity of soil micro-organisms or sea birds). At the regional scale, climate dominates—this makes Arctic and Subarctic ecosystems a useful test case for the development of broad-scale ecosystem models. Correlations between climate and vegetation have already been shown to give particularly powerful predictive models of the distribution of Arctic plant communities (Elvebakk, 1985; Kverndal et al., 1990; Sjörs, 1963). Arctic ecosystems are also expected to experience one of the strongest regional signals from global climate change, and there is therefore a particular focus on models that allow the estimation of impacts in this area (Woodwell and Houghton, 1993). A further reason is the potentially strong feedback that is likely to occur due to methane released from thawing permafrost soils—an indication of the magnitude of this feedback could be gained if the areas experiencing such change were better known (Nelson and Anisimov, 1993).

So far, both validation and application of terrestrial ecosystem models have been limited by their coarse spatial resolution. Developments in data gathering and interpolation techniques, as well as in computer storage and processing power, now allow the lower limits of spatial resolution to be stretched more than earlier, and the results have shown increasingly more realistic ecosystem distributions. This application shown here demonstrates some of the techniques that can be used to produce higher resolution (≈5 km) mapped model output that might provide greater potential for validation and hence be more adequate for climate change impact analysis in selected regions.

Background

Early Global Vegetation Models

The history of climate–vegetation models dates back to the beginnings of explorations of the natural history of the earth. The works of the great explorers of the 18th and 19th centuries used methods that are still in use in large-scale biogeographic modeling, such as bioclimatic indicators and map comparison techniques (Cramer and Leemans, 1993; Von Humboldt, 1807). Among the earliest *mathematical* formulation of a broad-scale climate–vegetation model are the studies by W. Köppen, L. Holdridge, and C.W. Thornthwaite, who all postulated systematical relationships between regional ecosystem types and the regional climate (Holdridge, 1947; Holdridge, 1967; Köppen, 1884; Köppen, 1936; Thornthwaite, 1948). Interestingly, the primary goal of Holdridge's much-quoted work was to "predict" climate from vegetation, without measurements from weather stations, based on the observation of vegetation types alone. He proposed that there was a well-defined systematic relationship that subdivides climate space, consisting of temperature and rainfall, into a regular pattern of zones with similar climate. These zones were described using biogeographic, rather than climatological terminology ("life zones" sensu Merriam, 1898), so that they would be recognizable from simple observation of the vegetation. If a sample of vegetation was related successfully to a life zone, then the model predicted a fairly narrow range of the climatic conditions, without requiring any measurements to be made. Although the model was first formulated for the vegetation of Central America, it was rapidly extended to include all life zones occurring on earth.

The implicit assumption, which was the basis for the first application of the Holdridge model to a climate change scenario, almost 40 years after the original publication, is that spatial shifts in climate imply corresponding shifts of the life zones (Emanuel et al., 1985). This assumption, however, must be rejected on the basis of paleoecological studies, many of which indicate that previous times have seen combinations of species that do not exist today (Huntley and Webb III, 1988; Huntley, Chapter 15). The implication is that correlations between climatic zones and vegetation types cannot be extrapolated into the past or into the future, because vegetation types can be different in different environments. Therefore, the species should be treated in an individualistic manner (Gleason, 1926). They may respond in different ways to changing climate and may therefore form different associations under different environmental conditions. However, from the point of view of climate change impact research, the only alternative to the "Clementsian" approach of species responding in groups to climatic zones would be to develop separate models for all or for the majority of the individual species. The global application of this idea is simply limited by the vast number of species that would need to be considered.

The Functional Types Concept

A compromise between the individualistic and the community approach is to lump species into groups with characteristics that are so similar that they can be considered functionally identical. A pioneering approach for this was the life form concept developed by Raunkiær (1907). It is based on the notion that most plants essentially are limited by their ability to survive the most stressful period of the seasonal cycle, no matter whether this period is the low-temperature situation of winter or the low-moisture situation of the dry season. All of Raunkiær's life forms are described by the way the meristems are protected from the influence of the stressful climate, such as in the seeds of annuals or in the well-protected buds of rosette plants. Combining the climatic classification of Holdridge and the life form concept of Raunkiær gave the core of a more complex equilibrium model of climate–vegetation relationships developed by Box (1981). Box defined approximately 90 plant types, based primarily on physiognomic characteristics, but with implicit functional definitions not unlike those of Raunkiær (see Table 16.1 for a list of the plant types defined for Arctic regions). For each of these plant types, which can be seen as analogues to the life forms, simple mathematical step functions were parameterized in such a way that a bioclimatic "envelope" describes the range of climatic conditions under which the plant can survive. In contrast to Holdridge, the types here actually can form different assemblages, and therefore be interpreted as resembling vegetation types. These assemblages are not fixed as a part of the model definition, but rather result from the range of climatic conditions that occur at real locations.

An interesting feature of the plant assembly rules in this model is that competitive relationships between different plant types are modeled in a relatively simple manner by estimating the numerical proximity of each existing plant type to that one of its environmental limits to which it is closest. A simple comparison of these numbers for all local types yields a dominance hierarchy that resembles reality quite well. The other important difference to previous work is that the bioclimatic variables that were used as input into the model were much more directly related to the functional

Table 16.1. Subset of Arctic Plant Types From the Global Vegetation Model of Box Together With the Threshold Values for the Bioclimatic Variables

Plant Type	T_{max}	T_{min}	D_T	P	MI	P_{max}	P_{min}	P_{Tmax}
Subpolar broad-leaved evergr. small trees	7–15	2–10	2–15	>500	>2	>150	>80	>80
Dwarf-needle small trees	18–30	−5–18	10–40	>200	0.45–3	25–300	<100	5–200
Needle-leafed treeline krummholz	10–13	−35–3	5–70	>75	>0.6	>25	2–100	>15
Cold-winter xeromorphic shrubs	15–32	−20–6	10–60	50–800	0.1–0.9	15–200	<30	<60
Summergreen tundra dwarf-shrubs	6–18	−60–3	5–80	>100	>0.7	>20	>10	>10
Xeric cushion-herbs	5–30	−60–25	<100	40–1000	0.2–1	5–300	<50	<300
Mat-forming thallophytes	0–30	−100–30	<130	>20	>0.8	>10	>1	>1
Xeric thallophytes	−2–40	−100–30	<150	>1	0.01–2	>1	<100	>0

From Box (1981). T_{max} = temperature of warmest month (°C), T_{min} = temperature of coldest month (°C), D_T = range between T_{max} and T_{min} (°C), P = total annual precipitation, MI = moisture index (P/PET), P_{max} = precipitation of wettest month (mm), P_{min} = precipitation of driest month (mm), P_{Tmax} = precipitation of warmest month (mm).

relationships between plants and their environment than the ones used by Holdridge (see below). The Box model has been used for several applications to climate change scenarios (Cramer and Leemans, 1993; Sargent, 1988).

A limitation of Box's model is that in spite of the impressive literature review produced by Box (1981), the relatively large numbers of plant types (90) and of bioclimatic variables (8) make it practically impossible to parameterize each response function with global validity (Cramer and Leemans, 1993). The actual distributions of even the few Arctic plant types listed in Table 16.1, for example, are too sparsely known to be used in a correlation analysis with large amounts of climate observations. In a more recently developed model (BIOME, see below) this list has therefore been simplified considerably (Prentice et al., 1992).

Ecophysiological Basis for Species-to-Climate Response Functions

Another implicit assumption of Holdridge's work is that the main climatic variables that plants are affected by are average mean annual precipitation and "biotemperature" (an estimate of the amount of warmth received by the plants through an average annual cycle). The work of the Russian climatologist Köppen and that of many later biogeographers showed, however, that other variables in the climatological record are better suited for the characterization of the suitability of climate zones for plant growth, such as the temperature of the coldest month, or an index of available soil moisture, rather than average rainfall (Köppen, 1936). These considerations are not based on significance estimates of correlation analysis, but on the ecophysiological factors behind the limits to plant distributions. The temperature of the coldest month, for example, has been shown to be highly correlated with the absolute minimum temperature at most locations (Prentice et al., 1992), and this variable (which is less frequently available in compilations of climatic normals) is in turn closely linked to basic processes in the leaves of many plants (Woodward, 1987). Similarly, the estimation of the climatic soil water balance from radiation, temperature, precipitation, and basic soil characteristics, pioneered by Thornthwaite (1948), changed the view of water availability limitations for plant growth into a more mechanistic concept, particularly when more physically sound estimations of evapotranspiration were included (Priestley and Taylor, 1972).

For modern biogeographic modeling, the conclusion of this is that current static models are based on more than empirical correlations between arbitrarily chosen climatic variables, but rather on "mechanistic" considerations of the nature of the limitations for plant growth that are experienced in harsh climates (Walter, 1960). The formulations for these limitations may be based on ecophysiology or biophysics in one or both of two ways: in the derivation of more appropriate climatic indicators from the observed climatological record (such as the formulation of growing degree days above a certain threshold value), and in the determination of the response functions for plant performance in relation to these variables.

Most of these considerations are dealt with by the BIOME model. It is based on 14 functional plant types only, and these are related to only three major (and two supplementary) bioclimatic variables. Another important difference to earlier models is that the response functions were parameterized in a "minimalist" way (i.e., by defining only those climatic limits that were essential to achieve a geographically meaningful result). The original model (Prentice et al., 1992) has recently been updated for greater resolution of plant and vegetation types (version BIOME 1.1) (Prentice et al., 1993b). Both have been applied to a variety of climate change scenarios; version 1.0 has also been used in coupling experiments with a general circulation model (Claußen, 1994).

Data Requirements for Broad-Scale Vegetation Models

If large-scale, spatial shifts of community boundaries or composition are to be predicted from vegetation models, then geographic comprehensiveness and explicitness of the underlying data bases are essential. This is true both for the driving variables (climate, soils and other environment factors) and for the description of existing vegetation used for validation of the model. Details of the development of climatic data bases for global modeling are described elsewhere (Cramer and Fischer, in press; Leemans and Cramer, 1991). Convenient and appropriate spatial resolutions of global data bases used in vegetation change scenarios are currently in the range of 5 to 50 km (5 to 30° longitude and latitude). For areas with sparse networks of meteorological stations, suitable techniques are available that allow the interpolation of data to be based on topographical features. These techniques are capable of producing sufficiently representative results for equilibrium models (Hutchinson and Bischof, 1983; Hutchinson and Gessler, 1994). Significant data shortages for global coverages with similar resolutions exist for soils and land use data, although activities are under way to improve the existing data bases. Satellite data can in many cases help, because sufficiently long observation series and appropriate data analysis methods are now available (Fischer et al., in press).

Limitations of Current Models

The major limitations to the validity of the current models predicting either climate or vegetation from spatially or temporally organized information about the other are the following.

- Climate and vegetation, although influencing each other, need not be in equilibrium at every point in time. Particularly in the case of rapid climate change or of heavy disturbance of the existing vegetation, the early successional communities that may now occupy the site are usually not in equilibrium with climate, but will eventually be replaced by other communities. Migrational processes may be a part of this process and may therefore delay the development of any equilibrium for long periods of time. Dynamic models of climate–vegetation interactions describe the processes of plants replacing each other individually, such as establishment, growth, death, and dispersal, in a much more detailed and mechanistic way. A suite of models has been developed for this purpose, but virtually all of them are parameterized for defined species, rather than types (a recent noticeable exception is by Bugmann, in press). Their applicability is therefore limited to certain biomes, or even only certain vegetation types. Good examples for such models exist for the boreal forest, where the competition of the major tree species for light and other environmental variables can be described successfully (Leemans and Prentice, 1989; Prentice, et al., 1993a; Shugart, 1984; Shugart et al., 1992; Sirois et al., 1994).
- Atmospheric chemistry, most notably the content of CO_2 in the lower atmosphere, may have significant influence on the performance of plants. One of the best known examples for this is the change in water use efficiency caused by increasing CO_2 concentrations, which has been observed for a number of species (Eamus and Jarvis, 1989; Owensby, 1993).

The impacts of toxic or otherwise damaging substances, or of nutrients deposited from atmospheric pollution, belong to the same group of nonclimatic effects. By definition, they cannot be captured by a model that is based on climate variables alone. A separate group of models describe climate- and CO_2-dependent aspects of productivity (nutrient cycling) or atmospheric trace gas fluxes through the ecosystem, for example, the Terrestrial Ecosystem Model (TEM) (Melillo et al., 1993) or the related General Ecosystem Model (GEM) (Rastetter et al., Chapter 22). The particular value of these models is that the interface with the atmosphere usually can be described fairly explicitly in terms of exchange of gases (e.g., carbon release or sequestration). Most of them require initialization with a given description of the ecosystem, and they can therefore only respond to changing environmental conditions by changes in productivity of the existing vegetation. Boundary shifts of biomes and changes in the composition of ecosystems are not reflected. It is possible, however, to initialize a biogeochemical model with an ecosystem structure model (Esser and Lautenschlager, 1994; Plöchl and Cramer, 1995).

- Human land use is a major determinant of ecosystem structure and composition and, in fact, is the second most important component of global change. Most present climate–vegetation models are only valid for the prediction of the potential natural vegetation. Some attempts have been made using a simple concept of changing availability of agricultural land (Cramer and Solomon, 1993), and others have studied the realized land use of the world in a more integrated approach (Alcamo, 1994). None of the current models, however, can claim to be capable of estimating the future distribution of global land cover in a convincing way.

In the light of these limitations, why has the development of climate-driven vegetation models received such great attention recently? In the first place, vegetation is one of the most significant components of the immediate human environment, as well as the basis of food production and other essential life functions. Changes in potential vegetation may relatively easily be converted into potential changes of the existing land cover (e.g., by constructing risk maps). Secondly, vegetation composition and structure are readily observed and serve well as an integrative indicator for the environmental conditions of a given locality, without requiring measurements or other long-term observations. Thirdly, vegetation is involved in major feedback relationships with the atmosphere, due to its ability to change the composition (and therefore radiative properties) of the atmosphere. These changes are most directly notable for the cycles of water and CO_2, but can also be observed for other chemical components. Globally, a large fraction of the terrestrial surface of the earth is relatively close to its natural state, and the broad-scale feedbacks can therefore be studied with this class of models.

Large-Scale Changes in Structure and Composition of Arctic Ecosystems

Historical reconstructions show that major changes of Arctic and Subarctic ecosystems have occurred in the recent past, and that they are not adequately described by simple north and southward shifts (Gajewski et al., 1993; Huntley, Chapter 15). Even when considering the general limitations described above, it is nevertheless worthwhile applying the existing vegetation models in a climate change context, for several reasons.

- The analysis of climatic scenarios and their effect on potential vegetation distribution provides a sensitivity analysis for the vegetation models.
- Although equilibrium models cannot predict the transient nature of climate-driven vegetation change, the beginning (current potential vegetation) and the endpoint (the likely composition of future potential vegetation) can be identified. The length of time needed to go through the associated successional sequence can then roughly be estimated from studies of observed chronosequences.
- Although future land use cannot be modeled in a physically deterministic way, a risk analysis for the current land use systems can be performed if the potential new vegetation is known.
- A summary of the overall changes resulting from these scenarios can help to improve the biosphere–atmosphere interface description needed for general circulation models. There is growing evidence that general circulation models might provide more realistic simulations of atmospheric dynamics when terrestrial vegetation can respond adequately to changes in the atmosphere, on both short and longer time scales (Henderson-Sellers and McGuffie, 1995).

Application of the BIOME Model to the European Arctic

The global application of the BIOME 1.0 model proved to be successful in recovering the major geographic boundaries of the tundra and taiga biomes, even in relatively mountainous terrain. This will be further explored here using an improved, higher resolution data base of the regional climate and the version 1.1 of BIOME. In northern regions, the model distinguishes six major types of plants (Table 16.2), the occurrence of which is distinguished from unvegetated areas by requiring at least 100 growing degree days over a base of 0°C. Application of the aggregation rules yields six vegetation types from these plants (Table 16.3).

The application of a geographically explicit, spatially comprehensive model like BIOME requires driving variables to be provided for each grid cell. In this case, the critical variables are the bioclimatic variables, which, because of the shortage of climate stations in the area, to a large extent must be generated using interpolation techniques. For this application, a modi-

Table 16.2. Subset of Northern Plant Types from the BIOME 1.1 Global Vegetation Model Together with Their Bioclimatic Envelopes

Plant Functional Type	No.	Bioclimatic Variable				
		GDD_0	GDD_5	T_{min}	T_{max}	AET/PET
Southern cold evergreen trees	1		>500	−35–−2		>0.75
Northern cold evergreen trees	2		350–500	−35–−2		>0.75
Southern cold deciduous trees	3		>500	<5		>0.65
Northern cold deciduous trees	4		350–500	<5		>0.65
Cold shrubs (S tundra)	5	>100	>250			>0.33
Cold grasses (N tundra)	6	>100	<250			>0.33

From Prentice et al. (1993b). GDD_0 = growing degree days above 0°C, GDD_5 = growing degree days above 5°C, T_{min} = temperature of the coldest month (°C), T_{max} = temperature of the warmest month (°C), AET/PET = ratio of actual versus potential evapotranspiration. Plant functional type numbers are given here only for comparison with Table 16.3.

fied version of the database by (Leemans and Cramer, 1991) was used.[1] The main modifications were the following:

- A large number of new stations were merged into the data base, many of them originating from the states formed from the previous USSR.
- A new, improved interpolation technique was used, based on a 3-D thin-plate spline algorithm that generated realistic elevational trends for all climate variables (Hutchinson and Bischof, 1983; Hutchinson and Gessler, 1994).
- The interpolated data were generated for a finer spatial grid, using an elevation data base with a 5° lon/lat resolution (Row et al., 1995).

Table 16.3. Resulting Vegetation Types for Northern Boreal and Arctic Regions in the BIOME 1.1 Model

Biome Type	Plant Functional Type No.					
	1	2	3	4	5	6
Northern tundra						×
Southern tundra					×	
Northern taiga		×		×		
Southern taiga	×		×			
Northern cold deciduous forest				×		
Southern cold deciduous forest			×			

From Prentice et al. (1993b).

[1] This data base had been used for the original version of the BIOME model (Prentice et al., 1992).

For each grid cell, temperature, precipitation, and sunshine data were generated. From these, the bioclimatic variables for the BIOME model were calculated using the same methods and software as described in the original BIOME publication. The modified algorithm (for version 1.1) was then used to screen the data for each grid cell for the potential availability of the plant functional types, as well as their aggregation into biome types. The resulting map of Northern Europe is shown in Figure 16.1(A). No statistical comparison with an existing vegetation map was made in this case, but the resulting biome map can be said to be in fairly good agreement (Leemans et al., in press; Monserud and Leemans, 1992) with classic studies of the biogeography of Northern Europe (e.g., Sjörs, 1963), or the globe (e.g., Olson et al., 1985).

Application of the Model for a Climate Change Scenario

There is a general agreement about the likelihood of significant temperature increases at high latitudes due to increased concentrations of atmospheric CO_2, although there is currently no observational evidence that would prove the detection of this simulation result (Houghton et al., 1990; Houghton et al., 1992). In the following, a possible ecosystem change scenario following a regional warming trend is illustrated using data from a general circulation model. The model simulation was performed using the general circulation model of the Max Planck Institute for Meteorology, Hamburg, Germany, at T42 resolution (i.e., a grid size of several hundreds of kilometers) (Perlwitz, 1992). The scenario is not used as a *prediction* of future conditions, but as one of several possible futures and as a basis for the sensitivity analysis of the BIOME 1.1. model. To generate a scenario at the appropriate spatial resolution for the ecosystem model, the GCM output was first converted to anomaly values (relative differences between the "normal CO_2" simulation and a "$3 \times CO_2$" simulation) at the center points of the GCM grid cells. Then these anomalies were smoothed (temperature anomalies above 5°C were reduced to 5°C) and interpolated to the same 5′ grid as was used for the observed climate. Finally, the anomalies were applied to high-resolution climate data, assuming that, in the absence of a regionalized climate simulation, an unchanged regional pattern of climate is more likely to be realistic than the absence of any pattern.

The result of the application of this scenario to the data base underlying BIOME 1.1 is shown in Figure 16.1(B). The northward shift of practically all vegetation zones is clearly visible and not fundamentally different from other estimates at similar or coarser resolution, for example, by Sargent (1988) or Emanuel et al. (1985). Table 16.4 lists the areas covered by northern biomes in Northern Europe (between 30°W and 80°E, and north of 55°N). The largest net losses under this climate scenario would occur in the Northern boreal forests ("N taiga"), which have a potential area less than half of their current area (while a large portion of the new areas would be gained outside the present range, thereby requiring extensive migration to take place). The estimated loss of "ice/polar desert" clearly cannot be taken as a realistic scenario, because it would require melting of large areas that are currently glaciated, but it nevertheless illustrates the magnitude of the warming that results from this scenario. Not shown in the table are the more southern biome types that are likely to replace the boreal forests.

Potential and Limitation of the High-Resolution Scenario Analysis

Different from previous analyses is the relatively high spatial resolution of the maps in Figure 16.1. A prime advantage of the high-resolution simulation presented here is that comparison with the distribution of ecosystem types at regional scale is easier now, and this type of validation is currently under way. However, these maps should be interpreted with caution, like any other climate changes scenario. The purpose of the higher resolution here is not to suggest a higher "accuracy" than any other method, because these simulations do not currently have the potential of an "accurate" prediction. The goal was rather to develop and demonstrate the appropriate algorithms for the assimilation of all information about regional patterns in a coherent spatial framework. From an ecosystem point of view, the aggregation of all data to the resolution of the coarsest data set (here the GCM scenarios) is not appropriate, because response patterns are known at much finer scales. What is required is transparency of the *regional implications* of coarse-scale trends of changing climate, and the present technique allows this.

It is important to reemphasize the main limitations of this approach. As mentioned earlier, the ecosystem model used here is an equilibrium model, which predicts potential natural vegetation after succession and migration have continued long enough to reach a steady-state equilibrium with the local climate conditions. No emission scenario is currently predicting such an equilibrium even for the dominant chemical components of the atmosphere. Secondly, several major constraints on vegetation processes are not accounted for. At these latitudes, one of the more significant constraints is the photoperiod adaptation of the plants, which is likely to require a different genetic adaptation after the climatically suitable area for the plants has been shifted to more northern regions.

Summary

Global vegetation models are able to predict broad-scale equilibrium distributions of major ecosystems at a fairly high spatial resolution, given

Table 16.4. Total and Relative Areas Covered by Northern Biomes in Northern Europe (Between 30°W and 80°E, and North of 55°N) Under Current and a Possible Future Climate, Predicted by the BIOME 1.1 Model

	Current Climate		3 × CO_2 MPI T42	
	($10^3 km^2$)	(%)	($10^3 km^2$)	(%)
Ice/polar desert	475.2	5.3	4.6	0.1
N tundra	985.7	10.9	434.9	4.8
N taiga	596.7	6.6	246.8	2.7
N cold deciduous forest	34.6	0.4	17.9	0.2
S tundra	439.5	4.9	81.8	0.9
S taiga	3502.3	38.7	2320.0	25.6
S cold deciduous forest	53.1	0.6	29.3	0.3

Relative areas are expressed in proportion to the total area, which also contains more southern biomes.

that appropriate data bases for the main environmental factors exist. To use them for scenario analysis of possible climate change impacts, however, several improvements are required, most of all a capability of predicting possible rates of changes. These might be used to characterize both potential delays in vegetation response or sudden events, such as large-scale dieback with the associated impacts on the global carbon balance. So far, the magnitude of the anticipated changes in climate are well illustrated by the magnitude of potential vegetation response, and this information could be useful to the analysis of possible feedback mechanisms, as well as for the regional identification of potential risk areas.

Acknowledgments. I thank the organizers of the conference on Global Change and Arctic Ecosystems, Jarle I. Holten and Walter Oechel, for the kind invitation to Oppdal, Norway. I also thank Harald Bugmann, Martin Claußen, Alberte Fischer, Jarle I. Holten, Brian Huntley, Michael Hutchinson, Rik Leemans, Matthias Plöchl, I. Colin Prentice, Al Solomon, two anonymous reviewers, and others for various contributions to this work. The linguistic revision was made by Mrs. Alison Schlums. Much of the background material has been produced with the support of Trondheim University, Trondheim, Norway, the Norwegian Research Council for Arts and the Humanities, Oslo, Norway, and the Potsdam Institute for Climate Impact Research, Potsdam, Germany. This work contributes to and has benefited from the IGBP core project "Global Change and Terrestrial Ecosystems" (GCTE).

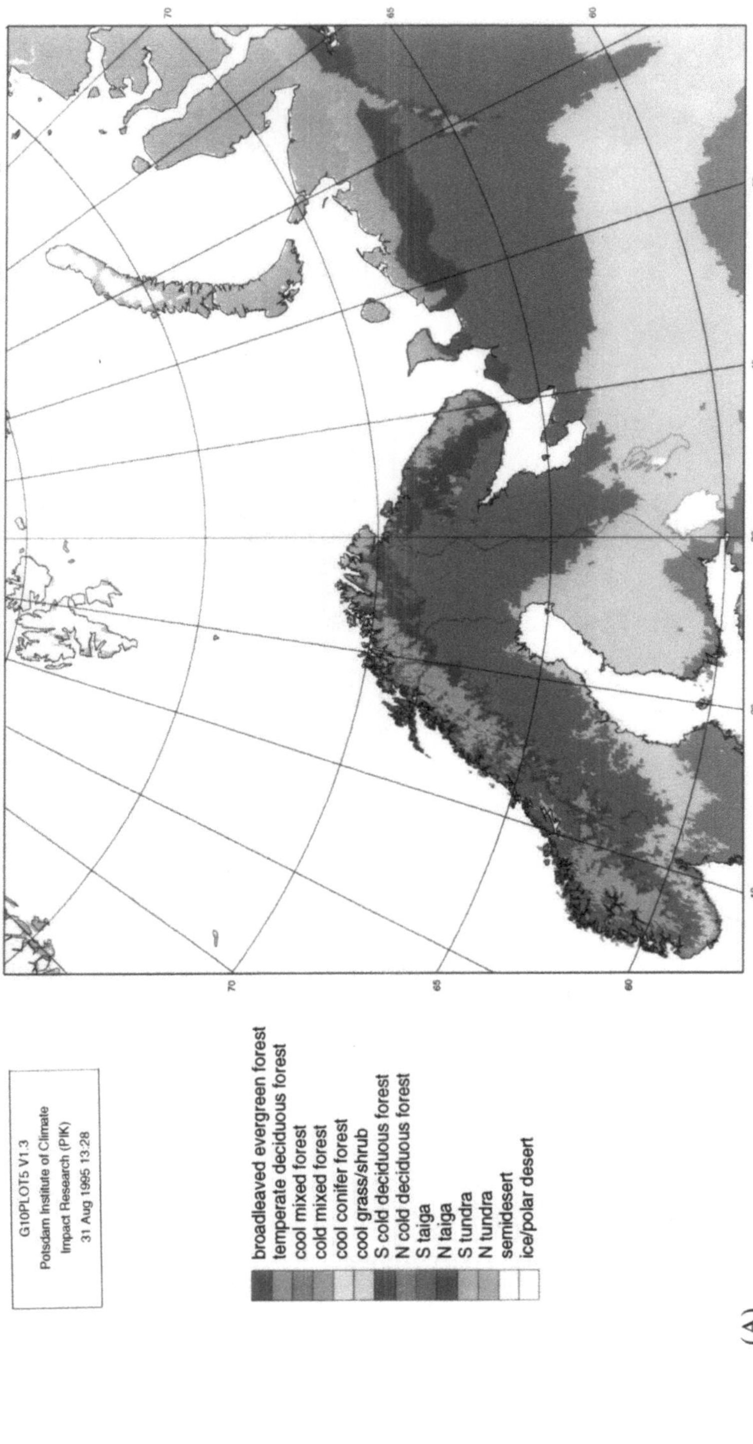

(A)

Figure 16.1. Distribution of potential natural vegetation in Northern Europe as simulated by the BIOME 1.1 model (Prentice et al., 1993b), using a modified version of the climate data base of Leemans and Cramer (1991). (A) Current climate.

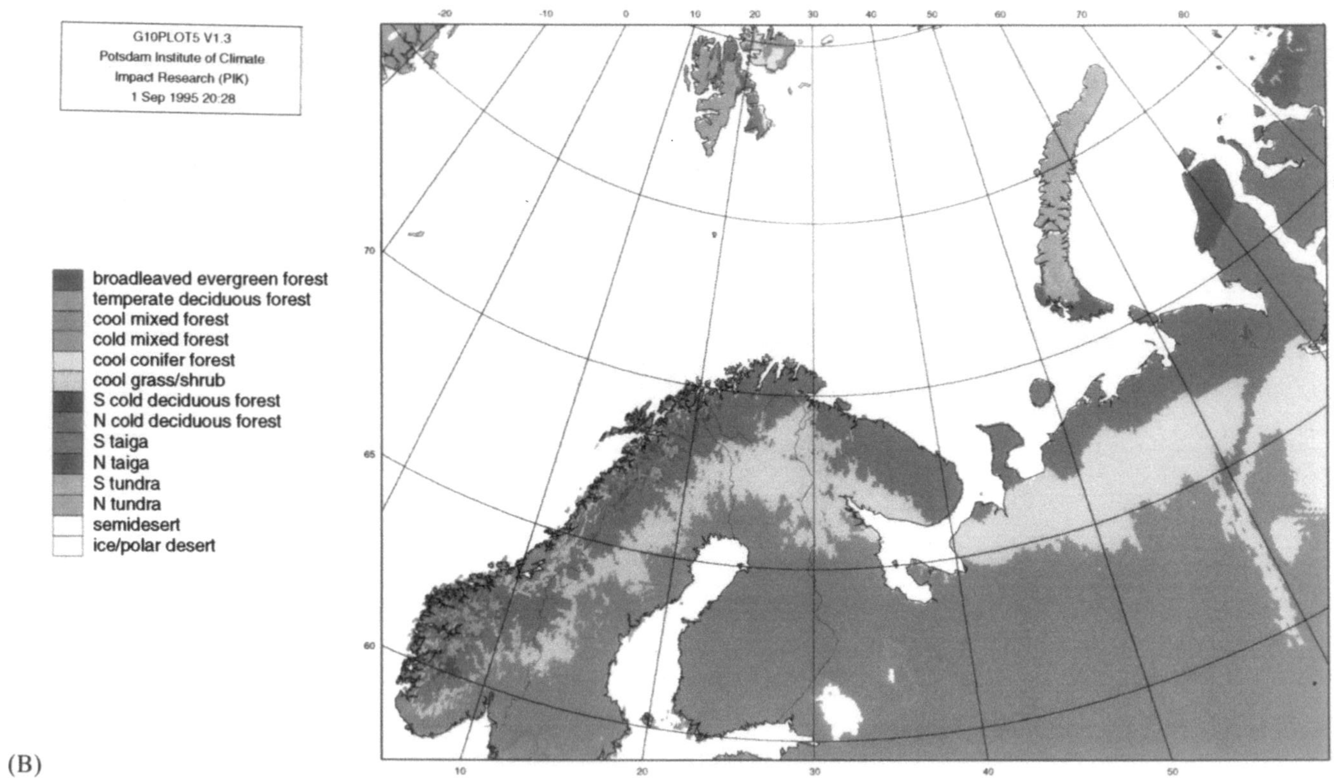

Figure 16.1. *Continued* (B) current climate modified using a scenario derived from comparing a control and a $3 \times CO_2$ climate simulation made with the T42 general circulation model of the Max Planck Institute for Meteorology, Hamburg, Germany (Perlwitz, 1992).

References

Alcamo, J. (ed.) 1994. *IMAGE 2.0: Integrated Modeling of Global Climate Change.* Dordrecht: Kluwer Academic.

Box, E.O. 1981. *Macroclimate and Plant Forms: An Introduction to Predictive Modeling in Phytogeography*. The Hague: Dr. W. Junk.

Bugmann, H. In press. Functional types of trees in temperate and boreal forests: Classification and testing. *J. Vegetat. Sci.*

Claußen, M. 1994. On coupling global biome models with climate models. *Clim. Res.* 4(3):203–221.

Cramer, W., and Fischer, A. In press. Data requirements for global terrestrial ecosystem modelling. In: Walker, B.E.A. (ed.), *Invited Papers to the 1st Science Conference of IGBP-GCTE.*

Cramer, W., and Leemans, R. 1993. Assessing impacts of climate change on vegetation using climate classification systems. In: Solomon, A.M., and Shugart, H.H. (eds.), *Vegetation Dynamics and Global Change* (pp. 190–217). New York: Chapman and Hall.

Cramer, W., and Solomon, A.M. 1993. Climatic classification and future redistribution of global agricultural land. *Clim. Res.* 3:97–110.

Eamus, D., and Jarvis, P.G. 1989. The direct effects of increase in the global atmospheric CO_2 concentration on natural and commercial temperate trees and forests. *Adv. Ecol. Res.* 19:1–55.

Elvebakk, A. 1985. Higher phytosociological syntaxa on Svalbard and their use in subdivision of the Arctic. *Nord. J. Bot.* 5(3):273–284.

Emanuel, W.R., Shugart, H.H., and Stevenson, M.P. 1985. Climatic change and the broad-scale distribution of terrestrial ecosystems complexes. *Climatic Change* 7(1):29–43.

Esser, G., and Lautenschlager, M. 1994. Estimating the change of carbon in the terrestrial biosphere from 18,000 BP to present using a carbon cycle model. *Environ. Pollut.* 83(1–2):45–53.

Fischer, A., Louahala, S., Maisongrande, P., Kergoat, L., and Dedieu, G. Satellite data for monitoring, understanding and modelling of ecosystem functioning. In: Walker, B.E.A. (ed.), *Invited Papers to the 1st Science Conference of IGBP-GCTE.* In press.

Gajewski, K., Payette, S., and Ritchie, J.C. 1993. Holocene vegetation history at the boreal-forest—shrub-tundra transition in North-Western Quebec. *J. Ecol.* 81(3):433–443.

Gleason, H.A. 1926. The individualistic concept of the plant association. *Bull. Torrey Bot. Club* 57:7–26.

Henderson-Sellers, A., and McGuffie, K. 1995. Global climate models and dynamic vegetation changes. *Global Change Biol.* 1(1):63–75.

Holdridge, L.R. 1947. Determination of world plant formations from simple climatic data. *Science* 105:367–368.

Holdridge, L.R. 1967. *Life Zone Ecology*, rev. ed. San José, Costa Rica: Tropical Science Center.

Houghton, J.T., Callander, B.A., and Varney, S.K. (eds.) 1992. *Climate Change 1992: The Supplementary Report to the IPCC Scientific Assessment.* Cambridge: Cambridge University Press.

Houghton, J.T., Jenkins, G.J., and Ephraums, J.J. (eds.) 1990. *Climate Change: The IPCC Scientific Assessment.* Cambridge: Cambridge University Press.

Huntley, B., and Webb, T. III. (eds.) 1988. *Vegetation History*. Dordrecht: Kluwer Academic.

Hutchinson, M.F., and Bischof, R.J. 1983. A new method for estimating the spatial distribution of mean seasonal and annual rainfall applied to the Hunter Valley, New South Wales. *Aust. Meteorol. Mag.* 31:179–184.

Hutchinson, M.F., and Gessler, P.E. 1994. Splines—more than just a smooth interpolator. *Geoderma* 62:45–67.

Köppen, W. 1884. Die Wärmezonen der Erde, nach der Dauer der heissen, gemässigten und kalten Zeit und nach der Wirkung der Wärme auf die organische Welt betrachtet. *Meteorol. Z.* 1:215–226 (+map).

Köppen, W. 1936. Das geographische System der Klimate. In: Köppen, W., and Geiger, R. (eds.), *Handbuch der Klimatologie* (pp. 1–46). Berlin: Gebrüder Bornträger.

Kverndal, A.-I., Elvebakk, A., Jaworowski, Z., and Hansson, R. 1990. *Virkninger av klimaendringer i polarområdene—bidrag til den interdepartementale klimautredningen*. Norsk Polarinstitutt. 82-90307-66-7.

Leemans, R., and Cramer, W. 1991. *The IIASA Database for Mean Monthly Values of Temperature, Precipitation and Cloudiness of a Global Terrestrial Grid*. International Institute for Applied Systems Analysis (IIASA), Laxenburg, Austria. Research Report.

Leemans, R., Cramer, W., and Van Minnen, J.G. In press. Prediction of global biome distribution using bioclimatic equilibrium models. In: Melillo, J.M., and Breymeyer, A. (eds.), *C Cycling in Grassland and Forested Ecosystems*. New York: SCOPE, John Wiley and Sons.

Leemans, R., and Prentice, I.C. 1989. *FORSKA, A General Forest Succession Model*. Uppsala, Sweden: Department of Plant Ecology, Uppsala University.

Melillo, J.M., McGuire, A.D., Kicklighter, D.W., Moore, B. III, Vörösmarty, C.J., and Schloss, A.L. 1993. Global climate change and terrestrial net primary production. *Nature* 363:234–240.

Merriam, C.H. 1898. Life zones and crop zones of the United States. *Bulletin US Department of Agriculture, Division Biological Survey* 10.

Monserud, R.A., and Leemans, R. 1992. Comparing global vegetation maps with the Kappa statistic. *Ecol. Model.* 62:275–293.

Nelson, F.E., and Anisimov, O.A. 1993. Permafrost zonation in Russia under anthropogenic climatic change. *Permafrost Periglacial Proc.* 4:137–148.

Olson, J., Watts, J.A., and Allison, L.J. 1985. *Major World Ecosystem Complexes Ranked by Carbon in Live Vegetation: A Database*. Carbon Dioxide Information Center.

Owensby, C.E. 1993. Potential impacts of elevated CO_2 and aboveground and belowground litter quality of a tallgrass prairie. *Water, Soil, Air Pollut.* 70(1–4):413–424.

Perlwitz, J. 1992. Preliminary results of a global SST anomaly experiment with a T42 GCM. In: *VII General Assembly of the European Geophysical Society*. Edinburgh, UK, April 6–10, 1992.

Plöchl, M., and Cramer, W. 1995. Coupling global models of vegetation structure and ecosystem processes—An example from Arctic and Boreal ecosystems. *Tellus Series B—Chem. Phys. Meteorol.* 47(1/2):240–250.

Prentice, I.C., Cramer, W., Harrison, S.P., Leemans, R., Monserud, R.A., and Solomon, A.M. 1992. A global biome model based on plant physiology and dominance, soil properties and climate. *J. Biogeogr.* 19(2):117–134.

Prentice, I.C., Sykes, M.T., and Cramer, W. 1993a. A simulation model for the transient effects of climate change on forest landscapes. *Ecol. Model.* 65(1–2):51–70.

Prentice, I.C., Sykes, M.T., Lautenschlager, M., Harrison, S.P., Denissenko, O., and Bartlein, P.J. 1993b. Modelling global vegetation patterns and terrestrial carbon storage at the last glacial maximum. *Global Ecol. Biogeogr. Lett.* 3(3):67–76.

Priestley, C.H.B., and Taylor, R.J. 1972. On the assessment of surface heat flux and evaporation using-large scale parameters. *Monthly Weather Rev.* 100(2):81–92.

Raunkiær, C. 1907. *Planterigets Livsformer.* Copenhagen/Kristiania: Gyldendalske Boghandel and Nordisk Forlag.

Row, L.W. III, Hastings, D.A., and Dunbar, P.K. 1995. *TerrainBase—Wordwide Digital Terrain Data, Documentation Manual CD-ROM Release 1.0.* National Geophysical Data Center. NGDC Key to Geophysical Records Documentation: 30.

Sargent, N.E. 1988. Redistribution of the Canadian boreal forest under a warmed climate. *Climatol. Bull.* 22(3):23–34.

Shugart, H.H. 1984. *A Theory of Forest Dynamics: The Ecological Implications of Forest Succession Models.* New York: Springer-Verlag.

Shugart, H.H., Leemans, R., and Bonan, G.B. (eds.) 1992. *A Systems Analysis of the Global Boreal Forest.* Cambridge: Cambridge University Press.

Sirois, L., Bonan, G.B., and Shugart, H.H. 1994. Development of a simulation model of the forest tundra transition zone of northeastern Canada. *Can. J. Forest Res. (Journal Canadien de la Recherche Forestière)* 24(4):697–706.

Sjörs, H. 1963. Amphi-atlantic zonation, nemoral to arctic. In: Löve, A., and Löve, D. (eds.), *North Atlantic Biota and their History* (pp. 109–125). Oxford: Pergamon Press.

Thornthwaite, C.W. 1948. An approach toward a rational classification of climate. *Geogra. Rev.* 38(1):55–94.

Von Humboldt, A. 1807. *Ideen zu einer Geographie der Pflanzen neben einem Naturgemälde der Tropenländer.* Tübingen, Germany: Lotta.

Walter, H. 1960. *Einführung in die Phytologie. Grundlagen der Pflanzenverbreitung.* Stuttgart: Gustav Fischer.

Woodward, F.I. 1987. *Climate and Plant Distribution.* Cambridge: Cambridge University Press.

Woodwell, G.M., and Houghton, R.A. 1993. Warming the north: What happens? In: Vinson, T., and Kolchugina, T.P. (eds.), *Carbon Cycling in Boreal Forest and Subarctic Ecosystems: Bio-Spheric Responses and Feedbacks To Global Climate Change* (pp. 1–8). Corvallis, Oregon: Environmental Research Laboratory, U.S. Environmental Protection Agency.

17. Past, Present, and Future Potential Sources Contributing to Artificial Radionuclides in the Arctic Region

Brit Salbu and Per Strand

The presence of artificially produced radionuclides in the arctic region is due to both accidental and authorized releases from the nuclear weapon cycle, including atmospheric nuclear weapon tests, and the nuclear fuel cycle, especially reprocessing facilities. Among radionuclides released to the environment, the long-lived ^{137}Cs, ^{90}Sr, and transuranics (Pu isotopes, ^{241}Am) are of major radiological concern. In addition, the contribution of mobile species of ^{99}Tc and ^{129}I is of relevance for marine ecosystems. For dose assessments, however, the contribution from naturally occurring radionuclides (e.g., U and Th isotopes and decay products, ^{40}K, ^{14}C, ^{3}H) should also be taken into account.

Global fallout, especially during the 1950s to 1960s, is by far the most important source of radioactive contamination for Arctic terrestrial ecosystems. The terrestrial food chain, including fresh water fish, contains ^{137}Cs levels at orders of magnitude higher than the marine food chain, and hence constitutes the major doses to man in the Nordic region (Dahlgaard et al., 1995). For the Arctic region the Chernobyl fallout represents an additional source, showing large spatial and temporal variations. For the Arctic marine environment, in particular the Barents and Kara Seas, the following sources have contributed significantly to enhanced levels of artificially produced radionuclides in the past:

From 1948 — Transport, especially by the Ob and Yenisey Rivers, of radionuclides released from nuclear installations in the catchment areas.

1950s to 1960s	Global fallout from atmospheric nuclear weapon testing.
1957 to 1961	Underwater nuclear detonations southeast of Novaya Zemlya.
1970s to 1980s	Marine transport of radionuclides in discharges from European reprocessing plants (Sellafield, UK and La Hague, France).
After 1986	Atmospheric deposition and marine transport of Chernobyl fallout.

In addition, dumping at sea of liquid and solid radioactive waste, including reactors with spent fuel, during the 1960s to 1990s at Novaya Zemlya has a local impact. Accidental releases associated with waste disposal (Kyshtym accident), satellite (Canada), aircraft (Thule), and submarine accidents (Komsomolets, Bear Island) have so far been of little general significance for the Arctic Seas.

Due to both the reduced input of radionuclides (cessation of nuclear weapon testing and dumping of radioactive waste at sea, reduced releases from reprocessing plants) and the physical decay of radionuclides already released, including the global fallout, the present level of radioactive contamination in the Arctic region is lower today than during the last 30 to 40 years. Results of the joint Russian–Norwegian expeditions to the Kara Sea performed annually since 1992 demonstrate that the concentrations of radionuclides in waters and sediments are in general very low (JRNC, 1993). However, enhanced levels of radionuclides due to leakage from the dumped radioactive waste were localized to near the dumped objects (JRNC, 1994; Salbu et al., 1995).

As no further nuclear weapon testing or dumping operations are expected, potential future sources include increased leakage from the solid radioactive waste dumped at sea or stored in ships or at the shore, and accidental scenarios associated with civilian and military installations. Furthermore, a change in the distribution of effluent radionuclides from the Sellafield reprocessing plant is expected following the installation of the new actinide separation systems. In most cases the estimation of the potential source terms (i.e., list of inventory, activity level, activity ratio, and particularly the physicochemical forms of released radionuclides) is extremely difficult. Thus, assessment of future consequences of potential releases to the environment suffers from considerable uncertainties. Furthermore, unknown sources may still exist.

Releases in the Past

An overview of the past sources influencing the Barents and Kara Seas is given in Table 17.1.

Table 17.1 Estimated Integrated Input of ^{137}Cs and ^{90}Sr to the Barents and Kara Seas (JRNC, 1994)

Sources	Barents Sea		Kara Sea	
	^{137}Cs (PBq)	^{90}Sr (PBq)	^{137}Cs (PBq)	^{90}Sr (PBq)
Global fallout	2.1	1.5	1.4	1.0
River transport				
1949–52			0.06	0.6, 11.1
Ob[a]				
1961–89				
Penchora, Onega	0.02	2.0		
Severnaga Dvina				
1961–89				
Ob, Yenisey			0.11	1.1
Marine transport				
1956–92				
Sellafield[a]	8.2	1.8	0.16	0.03
1966–85				
La Hague[a]	0.18	0.22	0.004	0.004
Chernobyl fallout[a]	1–5		0.02–0.10	
Waste disposal[a]				
Reactors			totally 85 + 3.7	
SNW	totally 0.01		totally 0.6	
LNW	totally 0.8			

[a] At the time of deposition or disposal.

Transport by the Ob and Yenisey Rivers

Radionuclides released from nuclear installations situated in the catchments of Ob and Yenisey Rivers have been transported to the Kara Sea. At the nuclear installation Mayak PA, Chelyabinsk, reactor and reprocessing effluents associated with nuclear weapon production were discharged directly into the Techa River, a tributary to the Ob River (Techa → Iset → Tobol → Irtysh → Ob) during 1948 to 1951. Later, effluents were discharged into Lake Karachay. In 1957, a waste tank explosion (Kyshtym accident) at Mayak PA contaminated a terrestrial area of about 10,000 km^2 within the Ob catchment (Karabolka → Sinara → Iset → Tobol → Irtyrsh → Ob). Furthermore, highly contaminated sediments were transported by wind erosion (tornado) from Lake Karachay to the surrounding area in 1967. Additional contributions are due to discharges and accidental releases from the Siberian Chemical Plant, Tomsk-7 (Romashka → Tom → Ob). The main source of radionuclide transport by the Yenisey River previously is direct releases from nuclear installations at Krasnoyarsk (Vakulovsky, 1995).

Totally about 120 PBq are assumed to have been discharged from Mayak into Lake Karachay from 1947/1948 to date (Romanov, 1995). Discharges into Lake Karachay have also led to severe contamination of underlying groundwaters seeping into the hydrological system of the Techa. Most of the activity in the cascade lakes upstream from the Techa River is associated with sediments. Futhermore, the floodplain of the Techa River is heavily contaminated from the previous releases from Mayak. The total input of radionuclides to the rivers, especially for the discharges to the Techa during 1948 to 1951, is difficult to estimate as data in the open literature are inconsistent. The distribution of Cs- and Pu-isotopes in sediment profiles from the Ob estuary reflects global fallout only (Panteleyev et al., 1995). Based on measurements during 1961 to 1989, the transport by Ob and Yenisey Rivers to the Kara Sea has been estimated to about 1.1 PBq ^{90}Sr and probably 0.1 PBq ^{137}Cs (Vakulovsky et al., 1993). Furthermore, about 0.5 PBq ^{90}Sr and probably 0.05 PBq ^{137}Cs have been transported by the Pechora, Severnaya, Dvina, Lena, and Indigirka Rivers to the Arctic Seas (Chumichev, 1995). However, as more than 90% of suspended materials may be transported by rivers during high flow episodic events, fluxes based on monthly sampling and annual flow data may be severely underestimated. The ongoing Joint Russian–Norwegian Collaboration on the contamination of radionuclides from the Mayak nuclear installation is expected to provide valuable information on previous and present releases as well as on potential sources in the area of importance for the assessment of future consequences for the Kara Sea.

Nuclear Weapon Tests

Since 1945, more than 2000 nuclear weapon tests have been carried out worldwide, of which 513 were atmospheric tests (Norris and Arkin, 1994). The total release of ^{137}Cs and ^{90}Sr during the atmospheric testings is estimated to be about 910 and 600 PBq, respectively (UNSCEAR, 1993). From 1955 to 1964, 87 atmospheric tests were performed at Novaya Zemlya whereas in 1955, 1957, and 1961, underwater nuclear tests were carried out southeast of the island. During 1961 to 1990, 45 underground nuclear weapon tests were also carried out at Novaya Zemlya.

The major deposition of global fallout from the atmospheric nuclear weapon tests occurred during 1955 to 1966. Reminiscences of the global fallout are still recognized. The integrated deposition was at its highest in the Northern Hemisphere and at its lowest in the polar regions (UNSCEAR, 1982). The underwater tests are assumed to have a short-term impact on waters and a long-term impact on sediments. No significant impact on the level of contamination is expected from the underground nuclear tests, even though volatile fission products occasionally have been identified in Scandinavia (Bjurman et al., 1990). In addition, civilian nuclear detonations (e.g., for mining purposes) have been carried out.

Discharges from European Nuclear Installations

Releases from the reprocessing plants at Sellafield, UK (formerly Windscale) and La Hague, France, situated respectively at the Irish Sea and at the English Channel, represent an important source of contamination to the Arctic marine environment. The transit time from Sellafield to the Barents and Kara Seas is estimated to be 4 to 6 years whereas it is significantly shorter from La Hague (2 to 3 years) (Dahlgaard, 1995; Guegueniat et al., 1995). Based on recent estimates, the Sellafield reprocessing plant has discharged, by means of authorized releases, to the Irish Sea about 41 PBq ^{137}Cs and 6.2 PBq ^{90}Sr during 1952 to 1992, with a maximum annual release during 1974 to 1978 (Kershaw and Baxter, 1993). During 1966 to 1985 about 0.9 PBq ^{137}Cs and 0.8 PBq ^{90}Sr have been discharged into the English Channel from La Hague. Mobile fission products (^{137}Cs, ^{90}Sr, ^{99}Tc) are transported into the North Sea, along the Norwegian coast to the Arctic Seas. Radionuclides associated with particles are mainly deposited in sediments close to the site. The discharges from Sellafield and La Hague have been significantly reduced in recent years. However, mobilization from sediments contaminated from previous effluents may represent a diffuse source in the future (Tronstad et al., 1995).

Discharges and Dumping of Radioactive Waste

According to the Governmental Commission of the Russian Federation on sea dumping of radioactive waste (Whitebook No. 3, 1993), six submarine reactors and one reactor assembly from the icebreaker Lenin containing spent fuel, and 11 reactors without fuel have been dumped in Stepovogo Bay, Tsivolky Bay, and Abrosimov Bay at the east coast of Novaya Zemlya and in the designated area of the open Kara Sea within the Novaya Zemlya Trough (Figure 17.1). Based on estimates given by the Russian Commission, the total activity at the time of dumping was about 85 PBq for the fueled reactors and 3.7 PBq for the unloaded reactors (White Book No. 3, 1993). For most of the fueled reactors' furfural-based material has been used as protection barriers.

According to the Whitebook No. 3 (1993) about 0.45 and 0.32 PBq of liquid radioactive waste and about 574 and 1.5 TBq of low- and medium-level solid radioactive waste have been discharged or dumped into designated areas of the Barents and Kara Seas, respectively (Figure 17.1). Totally about 6500 containers, 155 large objects, and 17 vessels containing low- or intermediate-level radioactive waste have been dumped. The inventory of the dumped reactors is presently being revised by the International Arctic Sea Assessment Project (IASAP). Preliminary results indicate lower inventory level than given in the Whitebook.

Fallout from the Chernobyl Accident

During the Chernobyl accident, April 1986, a large number of radionuclides were released, including an estimated amount of 100 PBq ^{137}Cs (Borovoy et al., 1992; IAEA, 1986). Even though the major part of the fallout was deposited in Belarus, Russia, and Ukraine, most countries in Europe were affected by the fallout. Due to the height of the plume and the initial wind directions, Central Scandinavia, including the Baltic Sea, was significantly contaminated. Compared to global fallout, the contamination of Cs isotopes in heavily affected areas was a factor of 10 to 100 higher. In Northern Scandinavia, the Chernobyl and global fallout levels were of the same order of magnitude (e.g., 1 to 2 kBq/m^3 in Norway). The distribution of global fallout was relatively homogeneous within large areas, whereas the Chernobyl fallout was characterized by rather uneven distributions and local hot spots. Close to the source the Chernobyl fallout was characterized by radionuclides associated to uranium fuel particles (Loshchilov et al., 1991; Salbu et al., 1994a). Fuel particles were also observed at a far distance from the source (e.g., Scandinavia) (Devell et al., 1986; Salbu, 1988). The association with particles reduced the mobility and biological uptake of radionuclides (Hansen and Hove, 1991; Oughton et al., 1992; Salbu et al., 1994b).

Based on whole-body measurements of individuals from the Sami population in Norway involved in the reindeer breeding for the years 1965 to 1992, the annual effective dose has been calculated (Figure 17.2). The increase shown in 1987 measurements is caused by the fallout from the Chernobyl accident in 1986. The annual dose of the reindeer breeding Sami population south of the polar circle is also shown during the years 1987 to 1992. In this part of the country the fallout levels were considerably higher than during the nuclear weapons testing period. Thus, a range of different countermeasures was implemented to reduce the intake of radiocaesium to man. Without the use of countermeasures (Figure 17.2), the levels would have been 7 to 10 times higher (Strand et al., 1992).

The importance of the lichen–reindeer–man pathway in the Arctic was identified during the global fallout period to be a very vulnerable pathway because of the high transfer of radiocaesium. After the Chernobyl accident other pathways were of concern, especially those related to natural food products or the use of seminatural ecosystems for domestic animals grazing.

Nuclear Accidents

Accidents of potential relevance for the Arctic region are: the Kyshtym (1957) and Lake Karashay (1967) accidents in Chelyabinsk, Ural; the B-52 aircraft crash at Thule, Greenland, in 1968 (four nuclear weapons); the crash of Soviet satellite COSMOS 954 in Canada in 1978 (nuclear reactor); and accidents at sea, for example, the sunken Russian submarine Komsomolets at 1658 m depth close to Bear Island in 1989 (nuclear reactor and two warheads). In contrast to the Chernobyl accident, relatively smaller and more remote territories were affected by the Cosmos and Kyshtym accidents. Local impacts only were recognized after the Lake Karachay and Thule accidents; so far no significant leakage has been observed from the Komsomolets.

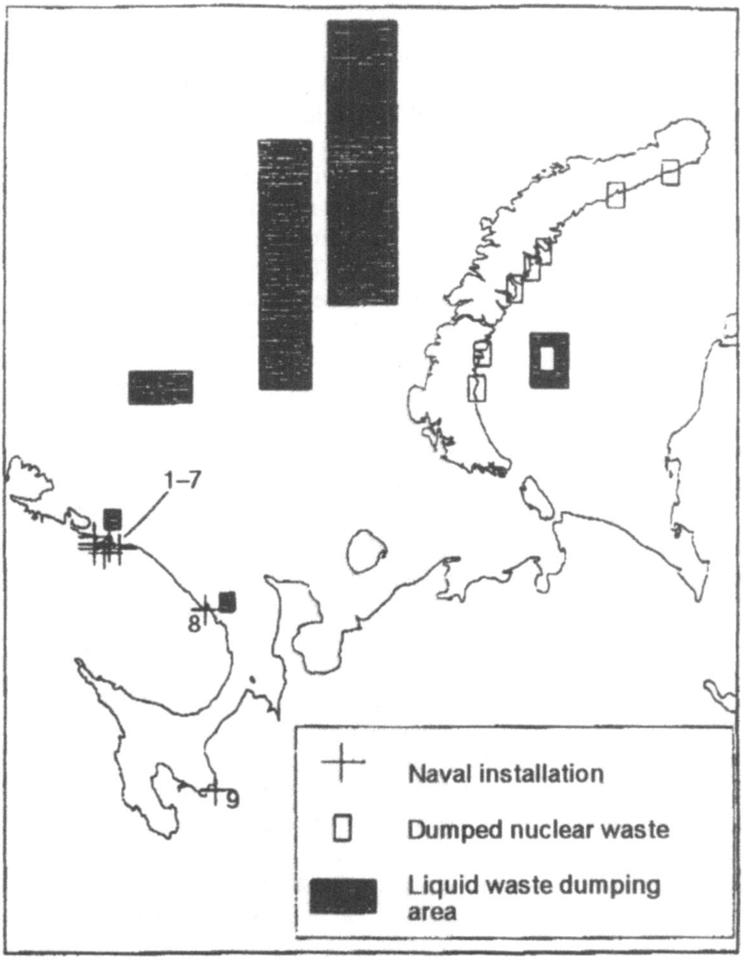

(a)

Figure 17.1. (a) Estimated total activity (TBq) of dumped solid radioactive waste in the Kara Sea at the time of dumping (Whitebook No. 3, 1993). The reactors were dumped in bays at the East coast of Novaya Zemlya. The sites investigated include Abrosimov Bay (southern site, 44 PBq), Stepovogo Bay (North of Abrosimov, 7.4 PBq), and Tsikolky Bay (North of Stepovogo, 5.5 PBq). In addition, the investigations included the dumping site at the open Kara Sea (Novaya Zemlya Trough, 29.6 PBq). The numbers refer to naval facilities hosting nuclear-powered vessels or submarines.

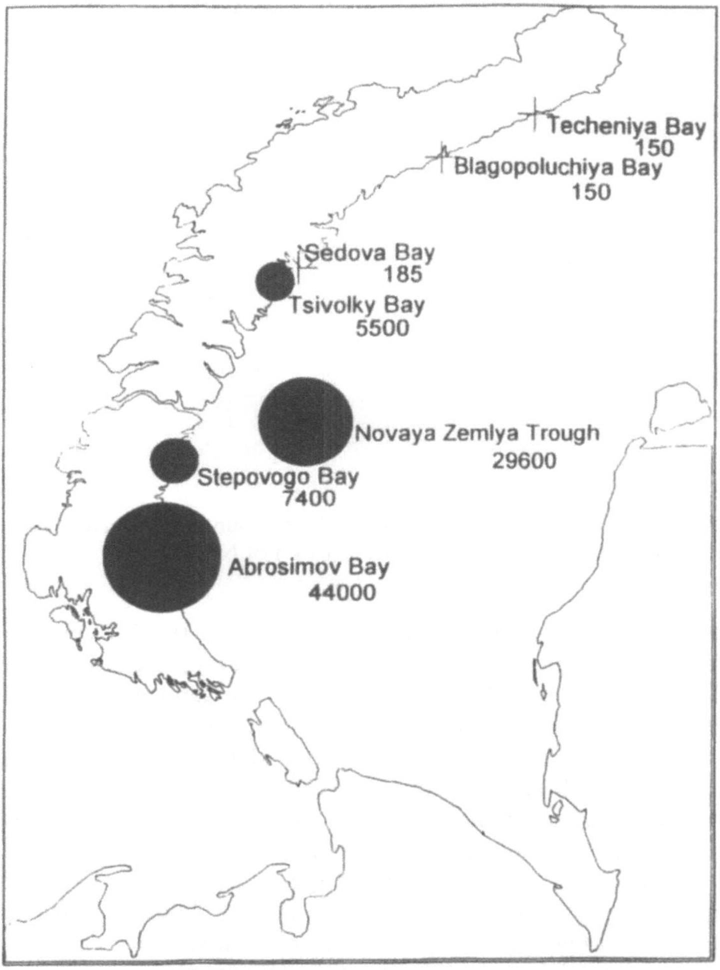

(b)

Figure 17.1. *Continued* (b) Estimated total activity of the solid radioactive waste at the time of dumping in the Kara Sea (Whitebook No. 3, 1993).

Present Level of Radioactivity

Artificially produced radionuclides can be identified in all compartments of the terrestrial and marine ecosystems in the Arctic region. The concentrations, especially in the marine system, are low and are assumed to have a low impact on the environment and health in the Arctic.

Global fallout is the major source contributing to the overall background contamination level in both terrestrial and marine Arctic ecosystems. At present, the total integrated deposition on ground surfaces at the latitude of northern Scandinavia/Novaya Zemlya is about 1.5 to 2.6 GBq/km^3 for ^{137}Cs

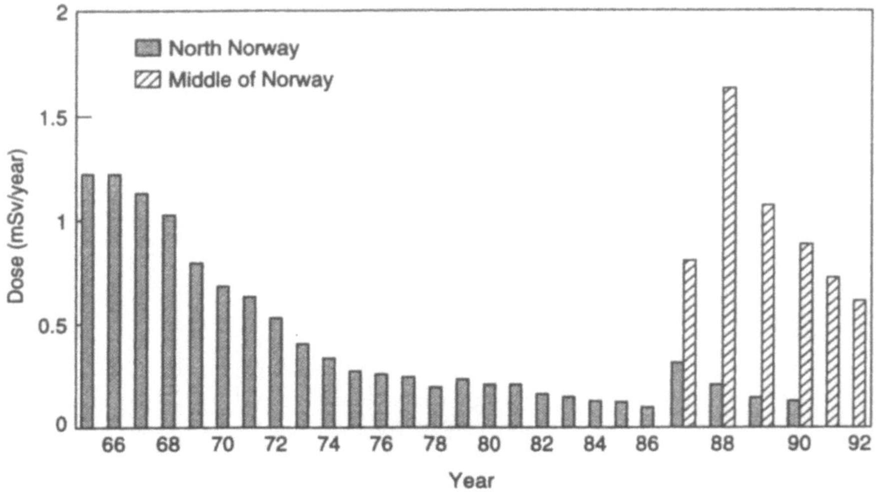

Figure 17.2. Annual effective dose (whole-body counting) received by the Sami population living in the North and central region of Norway.

and about 1 to 2 GBq/km^3 for ^{90}Sr (Boltneva et al., 1977). Global fallout is also transferred to marine systems by rivers draining large catchment areas (e.g., Ob, Yenisey, Lena, McKensey), runoff from land, and by sea currents (e.g., from the Atlantic Ocean to the Arctic Seas).

In the Arctic terrestrial ecosystem, a considerable fraction of deposited radionuclides is still retained in soils, lichens, and mosses. The ecological half-lives seem to be significantly longer in seminatural alpine ecosystems than in cultivated areas situated in temperate zones (Hove and Strand, 1990). Reindeer feeding on lichens is the key pathway of radionuclides (^{137}Cs) to the Sami population. Based on whole-body measurements of selected groups of people in Northern Norway, maximum concentrations of ^{137}Cs were observed during the atmospheric nuclear weapons testing (mid-1960s) even though an additional contribution was due to the Chernobyl accident (Figure 17.2).

Due to resuspension and runoff, radionuclides are transferred to fresh water systems where a major fraction is retained in sediments, but is also transferred in measurable amounts to fish. Especially during high flow (e.g., snowmelt) the flux of radionuclides to river systems may be substantial. A major fraction of Chernobyl ^{137}Cs in snowmelt input waters to an alpine lake was associated with colloids and about 50% was retained in the lake, whereas ^{90}Sr was transported through the lake in a mobile form (Salbu et al., 1992). High uptake of ^{137}Cs in fish from oligotrophic lakes (low conductivity waters) and reduced excretion at low temperature observed during the global fallout period have been confirmed after the Chernobyl accident. The half-lives are known to decrease with decreasing temperature, and increase with high water residence time and weight-specific metabolism for the actual fish species (Brittain, 1991). Furthermore, Cs-isotopes are concentrated from lower to higher trophic levels.

Radionuclides in the marine environment are transferred into the food chain, and seaweed (Fucus) has proved to be a suitable bioindicator. The concentration of ^{137}Cs in fish in the North Sea, Norwegian Sea, and Barent Sea has declined from the 1980s and is presently low (average less than 1 Bq/kg). The intervention level for Cs isotopes in fish applied in Europe after the Chernobyl accident is significantly higher (i.e., 600 Bq/kg). Thus, the terrestrial food chain, including fish, contains levels of ^{137}Cs orders of magnitude higher than the marine food chain in the Nordic region.

Joint Russian–Norwegian Expeditions to the Kara Sea

Information on radionuclides in the Kara Sea has previously been based on reports from Russian institutes. The Joint Russian–Norwegian 1992 expedition to the open Kara Sea was the first investigation, involving participation of Western scientists in this area. The expedition in 1993 included investigations at sites close to dumped radioactive waste in the Tsivolky Bay and the Stepovogo Bay at the east coast of Novaya Zemlya. In 1994, investigations were carried out in Abrosimov Bay and additional investigations were performed in Stepovogo Bay.

Based on results from the 1992 expedition, the concentration of ^{137}Cs varied within the range 3 to 20 Bq/m^3, ^{90}Sr within 3 to 11 Bq/m^3, and $^{239,240}Pu$ within 2 to 16 mBq/m^3 in sea waters from the Kara Sea (JRNC, 1993; Strand et al., 1994). The concentration in the upper 10-cm sediments ranged within 120 to 500 Bq/m^2 for ^{137}Cs and within 2 to 24 Bq/m^2 of $^{239,240}Pu$. Significantly higher levels were observed outside the Kara Gate (Figure 17.3), probably due to previous underwater detonations. Based on the spatial distribution of the concentrations of ^{137}Cs, ^{90}Sr, and Pu isotopes in waters and sediments and activity ratios, global fallout was found to be the major source of contamination (JRNC, 1993; Strand et al., 1994). In addition, transport especially of ^{90}Sr by the Ob and Yenisey rivers could be identified by the increase in water concentrations in areas influenced by low salinity river waters (Figure 17.4). These rivers are draining large catchment areas affected by global fallout as well as areas contaminated with radionuclides from releases from nuclear facilities. Furthermore, marine transport of radionuclides was observed, especially Cs isotopes from the Sellafield nuclear reprocessing plant and from fallout from the Chernobyl accident. Based on the isotopic ratio ($^{134}Cs/^{137}Cs$) about 30% of the Cs contribution in the waters could be attributed to the Chernobyl fallout. Compared to the radioactivity level in waters, sediments, and fish from other European marine systems (e.g., the Irish Sea, Baltic Sea, North Sea, and Norwegian Sea), the levels observed in the Barents and Kara Seas are extremely low. No impact of dumped radioactive waste (one reactor with

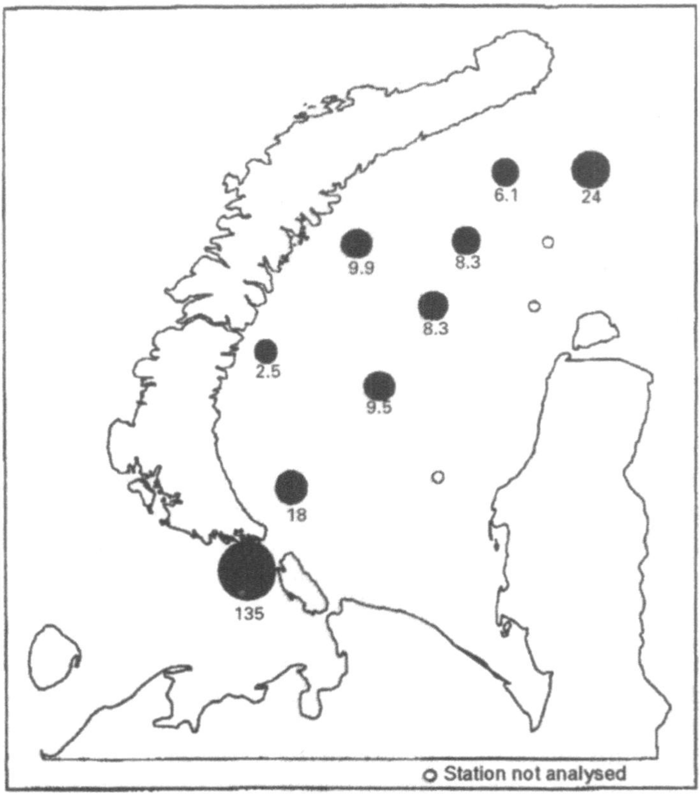

Figure 17.3. Distribution of $^{239,240}Pu$ in upper 10-cm sediments (Bq/m^2) in the Kara Sea, 1992.

fuel was dumped at 300m depth in the Novaya Zemlya Trough in 1972) could be observed.

Expeditions to bays of Novaya Zemlya took place in 1993 and 1994, using sonars, an underwater camera, underwater detectors, as well as carrying out an extensive sampling program for waters, sediments, and biota (Nikitin et al., 1995). Based on the results from the 1993 joint expedition, the levels of ^{137}Cs and ^{90}Sr in water from the Tsivolky Bay were similar to that of the open Kara Sea. However, the presence of ^{60}Co (up to 20Bq/kg) in upper sediment layers at the dumping sites reflected leakages from dumped radioactive waste (JRNC, 1994). In this area, reactor shielding assembly with fuel (60% of the fuel) and three reactors without fuel from the icebreaker Lenin were dumped in 1967 (Whitebook No. 3, 1993).

In the Stepovogo Bay one submarine having two reactors with fuel was dumped in 1981 (Whitebook No. 3, 1993). The levels of ^{137}Cs and ^{90}Sr in surface waters from the Stepovogo Bay were similar to that of the open

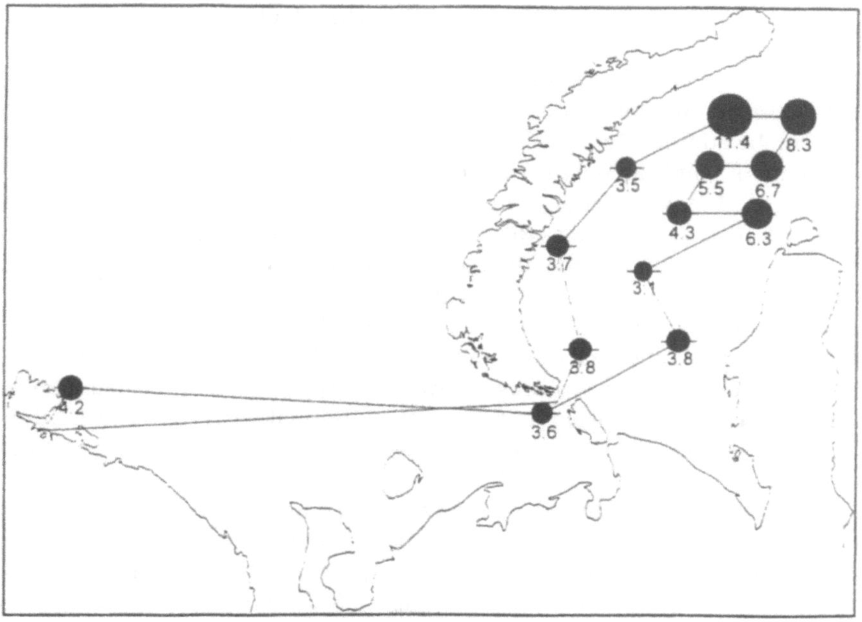

Figure 17.4. Distribution of ^{90}Sr in surface waters (Bq/m^3) in the Kara Sea, 1992.

Kara Sea, whereas enhanced levels (factor of 5) were observed in bottom waters close to the dumped submarine (Table 17.2). Furthermore, the presence of ^{60}Co as well as traces of ^{152}Eu and ^{154}Eu in sediments close to the hull of the sunken submarine (JRNC, 1994) and close to dumped containers (Salbu et al., 1995) reflected leakages.

In the Abrosimov Fjord three reactors with fuel and three reactors without fuel were dumped during 1965 to 1966 (Whitebook No. 3, 1993). A large number of containers and several large objects, including vessels, were also observed during the 1994 expedition. Enhanced levels of ^{137}Cs, ^{90}Sr, ^{60}Co, and $^{239,240}Pu$ were observed close to all localized dumped objects, especially in the areas where containers were dumped (Salbu et al., 1995).

Table 17.2. Concentrations of ^{137}Cs and ^{90}Sr in Sea Waters (Bq/m^3) from the Stepovogo Bay (JRNC, 1994)

Station		^{137}Cs	^{90}Sr
5	Surface	5–6	4–6
	Bottom	6–8	4–5
6	Surface	5–7	5–6
	Bottom	22–32	24–26
7	Surface	4–6	5–7
	Bottom	6–8	4–5

Potential Sources of Future Releases

No additional contribution from atmospheric nuclear weapon testing, underwater detonations, and probably underground detonations are expected. In addition, no further discharges of liquid radioactive waste or dumping of solid radioactive waste are expected in the future.

Enhanced leakage from the dumped radioactive waste, including reactors with fuel, is expected as the corrosion proceeds. In order to assess long-term consequences of the dumped waste, information on source term (inventories) and behavior of the different radionuclides under Arctic climate conditions is needed. The source term will be reassessed in 1996 by the International Arctic Sea Assessment Project (IASAP). Leakage of radionuclides from land-based storage of waste or waste stored intermediately in ships at the Kola Peninsula represents an additional potential source. Information is still needed to assess the source terms for the land-based intermediate waste.

Additional contribution from the Ob and Yenisey rivers may occur from flooding events (e.g., transport of contaminated sediments) or accidents (e.g., dam rupture at Mayak PA). Due to previous direct releases into the Techa River from Mayak, sediments, especially in the several cascade lakes and floodplains of the Techa, are still highly contaminated. Furthermore, large terrestrial areas are contaminated from accidental releases. Episodic events (runoff, flooding) representing a risk of soil and sediment transport may therefore lead to a substantial flux of associated radionuclides.

Releases from new actinide separation systems at Sellafield may contribute to an increase of certain mobile nuclides (e.g., ^{99}Tc), whereas the total activity releases are expected to continue to decline.

Additional contribution in the future may arise due to accidental releases associated with:

- nuclear installations, especially the Kola reactors,
- military activities (e.g., nuclear installations, handling, storage, and transportation of nuclear weapons),
- handling and storage, especially of spent fuel and high activity waste (on board ships, coastal areas, installations situated in catchment areas of Ob and Yenisey rivers),
- transportation by devices driven by nuclear reactors (submarines, icebreakers, satellites, etc.),
- lighthouses containing 10 to 15 PBq ^{90}Sr each.

Summary

Global fallout is the key source contaminating Arctic terrestrial ecosystems, although contamination of the marine environment can be attributed to several sources. The concentrations of artificially produced radionuclides in the Arctic region are low when compared to most areas in Europe. Especially due to the high retention of radionuclides in lichens (reindeer) and high uptake of ^{137}Cs in freshwater fish, the terrestrial food chain represents orders of magnitude higher levels of ^{137}Cs than the marine food chain, hence contributing to the major doses to man in the Nordic region. No impact on the general levels of radionuclides in the Barents and Kara Seas can be seen from dumped radioactive waste. The leakage from the waste is still low and localized to areas close to the dumped objects. Potential sources include enhanced leakage from dumped radioactive wastes with proceeding corrosion and different accidential scenarios associated with civilian and military facilities.

References

Bjurman, B., De Geer, L.-E., Vintersved, I., Rudjord, A.L., Ugletveit, F., Aaltonen, H., Sinkko, K., Rantavaara, A., Nielsen, S.P., and Aarkrog, A. 1990. The detection of radioactive material from a venting underground nuclear explosion. *J. Environ. Radioact.* 11:1–14.

Boltneva, L.I., Israel, U.A., Ionov, V.A., and Nazarov, L.N. 1977. Global contamination by the ^{137}Cs and ^{90}Sr and sources to external contamination on the territory of the USSR. *Atomic Energy* 42:355–361.

Borovoy, A.A., Demin, V.F., and Blinova, L.D. 1992. Radioactive releases originating from the Chernobyl accident. In: Kryshev, I.I. (ed.), *Radioecological Consequences of the Chernobyl Accident* (pp. 9–20). Moscow: Nuclear Society International.

Brittain, J.E. 1991. Radiocesium in brown trout (*Salmo trutta*) from a subalpine lake ecosystem after the Chernobyl reactor accident. *J. Environ. Radioact.* 14:81–191.

Chumichev, V. 1995. Sr-90 discharge with the main rivers of Russia into the Arctic Ocean during 1961–1993. Oslo: Proc. Int. Conf. on Environ. Radioact. in the Arctic. 21–25 August 1995. In press.

Dahlgaard, H. 1995. Transfer of European coastal pollution to the Arctic: Radioactive tracers. *Marine Pollut. Bull.* In press.

Dahlgaard, H., Notter, M., Brittain, J.E., Strand, P., Rantavaara, A., and Holm, E. 1994. General Summary and Conclusions. In: Dahlgaard, H. (ed.), *Studies in Environmental Science 62. Nordic Radioecology* (pp. 7–21). Amsterdam: Elsevier.

Devell, L., Tovedal, M., Bergstrom, U., Appelgren, A., Chussler, J., and Andersson, L. 1986. Initial observations of fallout from the reactor accident at Chernobyl. *Nature* 321:817–819.

Gueguenіat, P., Bailly de Bois, P., and Gandon, R. 1995. Estimation of La Hague contribution to the artificial radioactivity of Norwegian waters (1992–1995) and Kara Sea (1995–1998). Oslo: Proc. Int. Conf. on Environ. Radioact. in the Arctic. 21–25 August 1995. In press.

Hansen, H.S., and Hove, K. 1991. Radiocaesium bioavailability: transfer of Chernobyl and tracer radiocaesium to goat milk. *Health Phy.* 60:665–673.

Hove, K., and Strand, P. 1990. Predictions for the duration of the Chernobyl radiocaesium problem in non-cultivated areas based on a reassessment of the behaviour of fallout from nuclear weapon tests. In: Flitton, S., and Katz, E.W. (eds.), *Environmental Contamination Following a Major Nuclear Accident* (pp. 215–223). Vienna: IAEA-SM-306 1.

IAEA, 1986. Report of the USSR State Committee on the Utilization of Atomic Energy to the IAEA Meeting on the Chernobyl Accident, 25–29 August, Vienna.

JRNC, 1993. A survey of artificial radionuclides in the Kara Sea. Results from the Russian–Norwegian 1992 expedition to the Barents and Kara Seas. Report ISBN 82-993079-0-2. Norwegian Radiation Protection Authority, Norway.

JRNC, 1994. Radioactive contamination at dumping sites for nuclear waste in the Kara Sea. Results from the Russian–Norwegian 1993 expedition to the Kara Sea. Report ISBN 82-993079-2-9. Norwegian Radiation Protection Authority, Norway.

Kershaw, P.J., and Baxter, A.J. 1993. Sellafield as a source of radioactivity to the Barents Sea. Proc. Int. Conf. Environmental Radioactivity in the Arctic and the Antarctic, Kirkenes, August 23–27, 1993.

Loshchilov, N.A., Kashparov, V.A., Yudin, Ye. B., Protsak, V.P., Zhurba, M.A., and Parshakov, A.E. 1991. Physical–chemical forms of the radioactive fallout from the Chernobyl reactor accident. 1992. In: *The Radiobiological Impact of Hot Beta-Particles from the Chernobyl Fallout: Risk Assessment.* Kiev, Ukraine, August 1991. IAEA, Vienna, Part I, 34–39.

Nikitin, A.I., Salbu, B., Strand, P., Christensen, G.C., Chumichev, V.B., Lind, B., Fjelldal, H., Bjerk, T.O., Selnæs, T.D., Rudjord, A.L., Sickel, M., Valetova, N.K., and Føyn, L. 1995. Joint Russian-Norwegian collaboration on radioactive contamination from dumped nuclear waste in the Kara Sea—3 years with expeditions to the dumping sites: Goals and investigations. Proc. Int. Conf. on Environ. Radioact. in the Arctic. 21–25 August 1995, Oslo. In press.

Norris, R.S., and Arkin, W.M. 1994. Nuclear Notebook Bulletin of the Atomic Scientists, May/June, p. 9.

Oughton, D.H., Salbu, B., Riise, G., Lien, H.N., Østby, G., and Nøren, A. 1992. Radionuclide mobility and bioavailability in Norwegian and Soviet Soils. *The Analyst* 117:481–486.

Panteleyev, G.P., Livingston, H.D., Sayles, F.L., and Medkova, O.N. 1995. Deposition of plutonium isotopes and cesium-1376 in sediments of the Ob delta from the beginning of the nuclear age. Proc. Int. Conf. on Environ. Radioact. in the Arctic. 21–25 August 1995, Oslo. In press.

Romanov, G. 1995. Environmental impact of "Mayak" PA activity, Proc. Int. Conf. on Environ. Radioact. in the Arctic. 21–25 August 1995, Oslo. In press.

Salbu, B. 1988. Radionuclides associated with colloids and particles in rainwaters, Oslo, Norway. In: von Philipsborn, H., Steinhausler, F. (eds.), *Hot Particles from the Chernobyl Fallout.* Proceedings of an International Workshop, Theuern: Bergbau- und Industriemuseums, Theuern, Vol. 16, 83–84.

Salbu, B., Bjørnstad, H.E., and Brittain, J.E. 1992. Fractionation of Cs-isotopes and ^{90}Sr in snow-melt runoff and lake waters from contaminated areas in Norway. *J. Radioanal. Nucl. Chem.* 156(1):7–20.

Salbu, B., Krekling, T., Oughton, D.H., Østby, G., Kashparov, V.A., and Day, J.P. 1994a. Hot particles in accidental releases from nuclear installations. *The Analyst* 119:125–130.

Salbu, B., Nikitin, A.I., Strand, P., Christensen, G.C., Chumichev, V.B., Lind, B., Fjelldal, H., Selmæs, T.D., Sickel, M., Valetova, N.K., and Føyn, L. 1995. Joint Russian–Norwegian collaboration on radioactive contamination from dumped nuclear waste in the Kara Sea. 1. Three years with expeditions to the dumping sites: goals and investigations 2. Results from the 1994 expedition to Abosimov and Stepovogo fjords. Proc. Int. Conf. on Environ. Radioact. in the Arctic. 21–25 August 1995, Oslo. In press.

Salbu, B., Oughton, D.H., Ratnikov, A.V., Zhigareva, T.L., Kruglov, S.V., Petrov, K.V., Averin, B.V., Firzakova, S.K., Astasheva, N.P., Loshchilov, N.A., Hove,

K., and Strand, P. 1994b. The mobility of ^{137}Cs and ^{90}Sr in agricultural systems Ukraine, Belarus and Russia, 1991. *Health Phys.* 67:518–528.

Strand, P., Nikitin, A.I., Rudjord, A.L., Salbu, B., Christensen, G.C., Føyn, L., Kryshev, I.I., Chumishev, V.B., Dahlgaard, H., and Holm, E. 1994. Survey of artificial radionuclides in the Barents Sea and Kara Sea. *J. Environ. Radioact.* 25:99–112.

Strand, P., Selnes, T.D., Boe, E., and Andersson-Sorlie, A. 1992. Internal doses to different parts of the Norwegian population after the Chernobyl accident and the effect of dietary advice. *Health Physics* 64:385–392.

Tronstad, E., Nøren, A., Oughton, D.H., Salbu, B., Skipperud, L., Clacher, A., Strand, P., and Christensen, G.C. 1995. Mobilization of ^{90}Sr from Mayak sediments and soils: present and potential sources of contamination in the Techa River. Int. Conf. on Environ. Radioact. in the Arctic. 21–25 August 1995, Oslo. In press.

Vakulovsky, S.M. 1995. An estimation of the impact of the radioactive waste discharges into the Yenisey River on the Kara Sea radioactive contamination. Proc. Int. Conf. on Environ. Radioact. in the Arctic. 21–25 August 1995, Oslo. In press.

Vakulovsky, S.M., Nikitin, A.I., Chumichev, V.B., and Malyshev, S.V. 1993. Radioactive contamination of the Barents and Kara Seas Proc. International meeting on assessment of Actual and Potential Consequences of Dumping of Radioactive waste into Arctic Seas, Oslo, February 1–5.

Whitebook No. 3 1993. Yablokov, A.V., Karasev, V.K., Rumyantsev, V.M., Kokeev, M.E., and Petrov, O.J. 1993. Facts and problems related to radioactive waste disposal in seas adjacent to the territory of the Russian Federation. (Materials from Report by the Govermental Commission on Matters related to Radioactive Waste Disposal at Sea, created by Decree No. 613 of the Russian Federation President, October 24, 1992.) Moscow: Small World Publisher.

Woodhead, D. 1995. Preliminary estimates of fluxes of Cs-137 in northern waters from recent measurements. Proc. Int. Conf. on Environ. Radioact. in the Arctic. 21–25 August 1995, Oslo. In press.

UNSCEAR, 1982. *Ionizing Radiation: Sources and Biological Effects. United Nations Scientific Committee on the Effects of Atomic Radiation. UNSCEAR 1982 Report.* New York: United Nations.

UNSCEAR, 1993. *Sources and Effects of Ionizing Radiation. United Nations Scientific Committee on the Effects of Atomic Radiation. UNSCEAR 1993 Report.* New York: United Nations.

4. Effects of Anticipated Global Change on Ecosystem Function and Feedbacks to the Atmosphere

18. Potential Impacts of Climate Change on Nutrient Cycling, Decomposition, and Productivity in Arctic Ecosystems

Knute J. Nadelhoffer, Gaius R. Shaver, Anne Giblin, and Edward B. Rastetter

Short thaw seasons, low soil temperatures, high moisture content, and low rates of evapotranspiration through much or all of growing seasons in the Arctic combine to slow both litter decomposition and soil organic matter turnover. As a result, most arctic soils are overlain by mats consisting of plant litter and partially decomposed organic matter. These organic mats serve to retain moisture, impede the progression of seasonal soil thawing, and maintain low soil temperatures. The consequently cold, wet soil conditions serve to severely constrain microbially mediated processes such as decomposition and nutrient mineralization and create an ecosystem "bottleneck" (Chapin et al., 1980) by lowering rates of nutrient supply to plant roots. As a result, rates of plant growth and nutrient cycling between plants and soils are exceedingly low in arctic ecosystems. This view is consistent with fertilization studies in various tundra types showing consistent increases in plant growth and net primary production (NPP—the amount of plant biomass produced annually in an ecosystem) in response to nitrogen (N), phosphorus (P), or N plus P additions (e.g., Chapin and Shaver, 1985a; Haag, 1974; McCown, 1978; McKendrick, 1980; Ulrich and Gersper, 1978). Such studies suggest that the effects of low nutrient availability limit plant growth in the Arctic more than do the direct effects of cold conditions on plant processes.

Because soil processes are sensitive to variations in temperature and moisture, and because arctic soils are typically both cold and wet, climatic warming in the Arctic (Maxwell, Chapter 1) will likely have important effects on soil carbon and nutrient stocks, nutrient cycling patterns, and plant–soil interactions in arctic ecosystem. However, predicting the long-term effects of regional temperature increases on nutrient cycling and plant productivity will require an improved understanding of how temperature, moisture, and the quality of plant litter inputs to soils interact to control decomposition and nutrient mineralization in soils. In this chapter we compare soil characteristics and nutrient cycling patterns along natural and experimental environmental gradients to provide insights into how plant–soil interactions will respond to climate warming in the Arctic.

Tundra Ecosystem Carbon and Nutrient Contents

Soils

The combined effects of moisture and temperature on both decomposition and primary production influence the accumulation of organic matter in arctic soils (French, 1977; Heal et al., 1981). Because moisture balances and temperature regimes vary at both regional and local scales in the Arctic, the characteristics of surface organic mats also vary at these scales. For example, soils in relatively dry regions such as extreme northern Siberia and the Canadian High Arctic (Andreev and Aleksandrova, 1981; Bliss, 1981) have thin and sometimes discontinuous mats, whereas soils in wet, low-lying regions such as northern Fennoscandia and northern coastal Alaska (Gersper et al., 1980; Hinneri et al., 1975) have thick, saturated mats that typically extend into permanently frozen soil. Similarly, local variations in geomorphology and drainage patterns within small areas can lead to predictable variations in tundra soils with organic mat thickness increasing along gradients of well to poorly drained sites (Oechel, 1989; Shaver et al., 1990; Sonneson et al., 1975).

We previously characterized such gradients as consisting of "dry heath," "moist tussock," and "wet sedge" tundras (Nadelhoffer et al., 1992). Dry heath tundra has well-drained soils with thin organic mats (<5 cm), deep thaw depths, and vegetation consisting of lichens, heath, dwarf shrubs, or cushion plants. It is characteristic of ridge tops, steeply sloping land, or windswept areas where little snow accumulates in winter. Moist tussock tundra typically occupies gently sloping land, has dense organic mats of intermediate thickness (15 to 40 cm), thaws into the mineral soil during most summers, and is dominated by tussock forming sedges with admixtures of dwarf shrubs, mosses, lichens, and forbs. Wet sedge tundra occupies poorly drained lowlands, is characterized by peat accumulations (>40 cm, often below permafrost), and is dominated by sedges and mosses. We use results of recent studies of ecosystem function in the foothills region of Alaska's North Slope to compare soil properties and element stocks of these three tundra types. We also evaluate responses of tundra ecosystems in this region to experimental warming and fertilization in an attempt to predict the impacts of climatic warming on nutrient cycles and plant production in tundra ecosystems.

Differences in C, N, and P contents in seasonally thawed soils of dry heath, moist tussock, and wet sedge tundras near the Sagavanirktok River in northern Alaska (Figure 18.1) reflect differences in soil microclimate and nutrient cycling among tundra types. Soil C stocks ranged from about $6\,kg\,m^{-2}$ in dry heath to about $25\,kg\,m^{-2}$ in moist tundra. It should be noted that although the organic mat was thicker and depth of thaw was greater in the wet sedge than in the moist tussock tundra, C stocks were lower in wet sedge (about $20\,kg\,m^{-2}$). This is due, in part, to higher water content and lower bulk density in wet sedge than moist tussock tundra soils (below). However, because thaw depths do not reach mineral soil horizons at the wet sedge site and because we did not sample into permafrost, our estimates of C and nutrient stocks do not include the entire organic horizon. Unknown and presumably large amounts of organic C are permanently frozen in the wet sedge. Soil N contents ranged from about $0.3\,kg\,m^{-2}$ in dry heath to about $1.3\,kg\,m^{-2}$ in moist tussock and to about $1.4\,kg\,m^{-2}$ in wet sedge tundra.

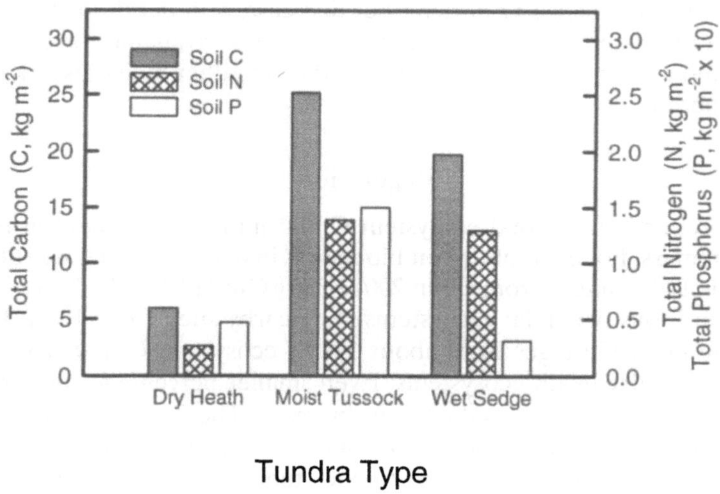

Figure 18.1. Soil C, N, and P contents in three tundra types located along a mesotopographic gradient near the Sagavanirktok River on Alaska's North Slope. Soils were sampled to 10 cm depth in the dry heath tundra where rocky soils prevented sampling to permafrost. Soils were sampled to maximum depths of thaw in the wet sedge (40 cm) and moist tussock (35 cm) tundras. Organic mat thicknesses averaged 4 cm in dry heath, 30 cm in moist tussock, and >40 cm in wet sedge tundras. Soil element contents are calculated from data in Giblin et al. (1991) and Nadelhoffer et al. (1991).

Soil P content ranged from about 0.03 $kg\,m^{-2}$ in wet sedge to about 0.15 $kg\,m^{-2}$ in moist tussock tundra.

More detailed comparisons of soil element stocks within tundra ecosystems (Table 18.1) provide further insights into patterns of soil organic matter distribution and nutrient cycling in tundra ecosystems. First, although large proportions of organic carbon stocks occur in organic horizons, considerable amounts of organic C can reside in mineral horizons. For example, at the Sagavanirktok River site almost 15% of organic C mass in the seasonally thawed layers of tussock tundra and almost 50% of organic C in dry heath soil occurs in 5 to 6 cm of mineral soil immediately beneath the organic mats. Although low C concentrations in mineral soil horizons are low, high bulk densities lead to relatively high C contents. Increasing thaw depths in a warmer climate would be expected to increase both the relative proportions and the total amounts of C in seasonally thawed mineral soils. Distributions of soil N within tundra types are roughly equivalent to soil C distributions due to relatively small changes in soil C/N ratios, especially within organic horizons. Soil P contents are dependent upon the presence of mineral horizons in thawed soils. Organic soil in the wet sedge tundra, where mineral horizons are below thaw depth, had the lowest P contents and had much higher C/P and N/P ratios than organic horizons in either of the other tundra ecosystems. Moreover, these ratios increase with depth in the wet sedge O horizon but decrease with depth in the tussock tundra O horizon, suggesting that P is much less available for uptake by plants and microbes in wet sedge tundra than in either moist tussock or dry heath tundra.

Vegetation

Small proportions of total ecosystem C and nutrients are held in tundra plant biomass. For example, plant biomass C in dry, moist, and wet tundras at Toolik Lake ranged from about 200 to 570 $g\,C\,m^{-2}$ (Table 18.2) versus 6 to 25 $kg\,C\,m^{-2}$ soils of similar ecosystems at a nearby site (Table 18.1). Therefore, the plant C ranges from about 1% of ecosystem C in wet sedge to about 3% in dry heath ecosystems. Even smaller percentages of nutrients in these ecosystems are contained in biomass. The 3 $g\,N\,m^{-2}$ in dry heath plant biomass is only about 1% of total ecosystem N, whereas the 5 to ~11 $g\,N\,m^{-2}$ in wet sedge and moist tussock tundra are less than 1% of total ecosystem N. Plants contain less than 1% of total ecosystem P in dry heath and moist tussock tundra and about 3% of ecosystem P in wet sedge where soil P stocks are extremely low.

Carbon/nutrient ratios of plant biomass differ dramatically from soil ratios. For example, although soil C/N ratios ranged from about 15 to 23 in tundra soils (Table 18.1), vegetation ratios were about three times greater within each ecosystem type, ranging from 44 to 66 (Table 18.2). Vegetation C/N ratios increased as the proportion of woody shrubs in plant biomass

Table 18.1. Selected Soil Properties and Element Stocks in Dry, Moist, and Wet Tundra Ecosystems Along a Mesotopographic Gradient Near the Sagavanirktok River in Northern Alaska

Tundra Type	Horizon	Thickness (cm)	Db ($g\,cm^{-3}$)	C (%)	N (%)	P (%)	C ($g\,m^{-2}$)	N ($g\,m^{-2}$)	P ($g\,m^{-2}$)	C/N (g/g)	C/P (g/g)	N/P (g/g)
Dry heath	Oe + Oa	3.7	0.22	37.9	1.38	0.11	3,085	112	9	27	345	13
	C	6.3	0.75	6.1	0.32	0.08	2,882	151	38	19	76	4
						Profile:	5,967	264	47	23	128	6
Moist tussock	Oe	15.0	0.13	36.9	1.84	0.10	7,196	359	20	20	369	18
	Oa	15.0	0.55	17.5	1.07	0.11	14,438	883	91	16	159	10
	B2g	5.0	1.00	7.1	0.29	0.08	3,550	145	40	24	89	4
						Profile:	25,183	1,387	150	18	168	9
Wet sedge	O1	15.0	0.11	40.1	2.54	0.07	6,617	419	12	16	573	36
	O2	25.0	0.13	40.4	2.69	0.06	13,130	874	20	15	673	45
						Profile:	19,747	1,293	31	15	636	42

Horizon thicknesses, bulk densities (Db), and element concentrations are from Giblin et al. (1991), except for moist tussock B2g where Db is estimated and element concentrations are from Nadelhoffer et al. (1991).

Table 18.2. Biomass C, N, and P Stocks in Three Tundra Types Near Toolik Lake, Alaska

Tundra Type	Biomass C ($g\,m^{-2}$)	Biomass N ($g\,m^{-2}$)	Biomass P ($g\,m^{-2}$)	C/N (g/g)	C/P (g/g)	N/P (g/g)
Dry heath	203	3.05	0.25	66	812	12
Moist tussock	570	10.53	1.12	54	509	9
Wet sedge	253	5.78	0.98	44	258	6

Data are from Shaver and Chapin (1991) and include element stocks in aboveground biomass (vascular plus nonvascular species) and rhizomes.

increased, with the lowest ration in wet sedge where shrubs are rare and the highest ratio in dry heath where shrubs are the dominant plant growth form. Plant C/P ratios also increase as the proportions of shrub growth forms increased. Plant C/P ratios were also much higher than soil ratios in dry and moist tundra, but plant C/P ratios were lower than soil ratios in wet sedge. This reflects the extremely low P availability in these wet sedge soils and indicates that productivity is likely to be more P limited in wet sedge than in drier tundra ecosystem types.

Decomposition and Microbial Activity

In the Arctic, as elsewhere, temperature, moisture, and substrate quality interact to control microbially mediated processes such as litter and organic matter decomposition. When moisture content is less than about 20% of dry mass, respiration rates are extremely low and relatively independent of temperature. Above this level, temperature sensitivity increases until moisture content reaches 200% or more, with less moisture required for optimum decomposition as temperature increases (Bunnell and Tait, 1974; Heal et al., 1981). Decomposition is slowed dramatically under moisture-saturated conditions due to the effects of low oxygen on the rates and pathways of microbial metabolism. Overall, the independent and interactive effects of moisture and temperature probably account for at least 70% of the variation in decomposer activity among arctic ecosystems (Svensson, 1980). Cold, wet conditions lead to relatively incomplete as well as slow decomposition. For example, we reported previously (Nadelhoffer et al., 1992) that lignocellulose indices [LCI = lignin/(cellulose + lignin, sensu Melillo et al., 1989] of arctic organic soil horizons are low (0.35 to 0.55) compared to LCIs of temperate forest and grassland soils (0.7 to 0.8). Low LCI values in arctic soils indicate that organic C in arctic soils is functionally more similar to fresh litter than to well-decomposed humus. A more detailed view of C chemistry in tundra soils (Figure 18.2) reveals that arctic organic mats have high concentrations of extractive and acid-soluble C and relatively low concentrations of acid-insoluble (humus plus lignin) C. Organic C in both wet sedge and moist tussock soils at the Sagavanirktok site was about 35% acid insoluble, 50% acid soluble, and about 15% extractive (soluble in hot water and nonpolar solvents). In contrast, organic soils in well-drained temperate soils typically contain at least 50% acid-insoluble C and lower percentages of acid-soluble and extractive C than either moist or wet tundra.

The quality, or decomposability, of plant litter also plays an important role in controlling soil C and nutrient cycling. Litter quality has been linked to functional characteristics such as nutrient contents, N/lignin ratios, leaf longevity, and photosynthetic capacity (Aber and Melillo, 1982; Reich et al., 1992; Swift et al., 1979). Therefore, differences in plant functional types among ecosystems are likely correlated with differences in nutrient cycling. Litter quality is probably particularly important in the Arctic, where soils are characterized by steep temperature and moisture gradients and where small differences in decomposability at the surface can produce large differences in the proportion of litter that is transferred to depths where decomposition is slowed by cold, wet conditions (Heal et al., 1981).

Different combinations of soil temperature and moisture regimes, together with differences in the quality of litter inputs to soils, can lead to large differences in the quality as well as the amounts of organic matter in arctic ecosystems. For example, O horizons from our dry heath, moist tussock, and wet sedge tundra ecosystems along a mesotopographic gradi-

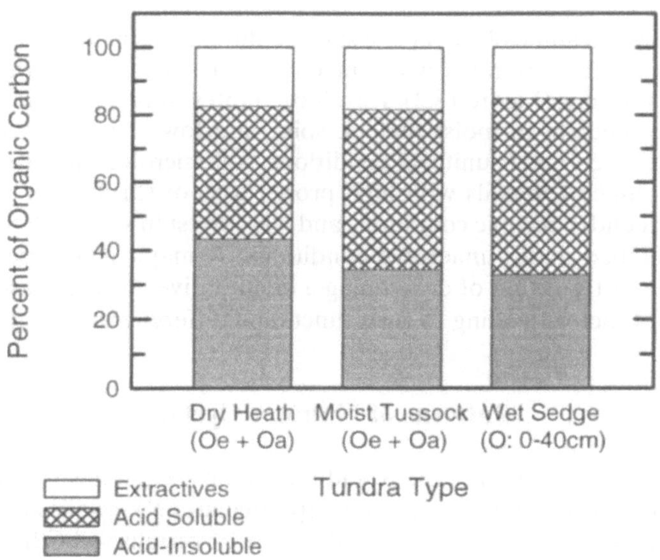

Figure 18.2. Soil C fractions in organic horizons from dry, moist, and wet tundra ecosystems near the Sagavanirktok River in northern Alaska.

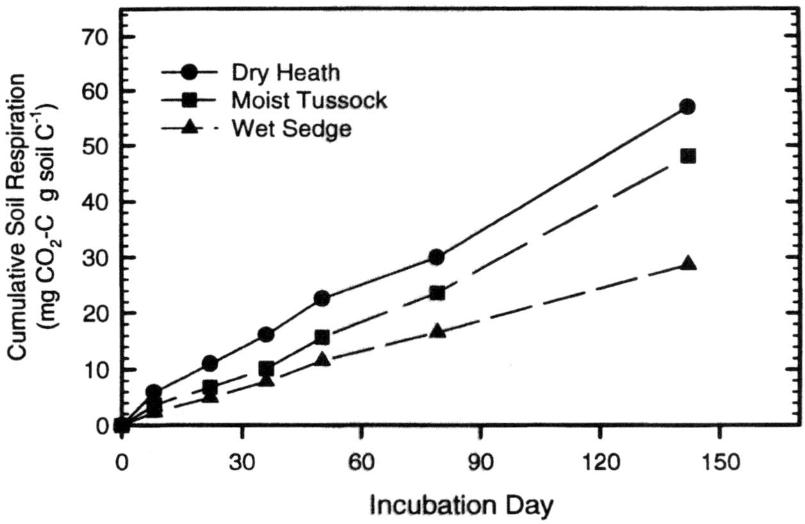

Figure 18.3. Cumulative releases of CO_2 from organic horizons collected from dry, moist, and wet tundra ecosystems and incubated under uniform conditions (9°C and –0.06 MPa moisture). Incubated materials were Oe + Oa horizons from the dry heath and moist tussock tundra and O1 (0–15 cm) from wet sedge. Soils for incubations were collected from the same sites as soils described in Figure 18.1 and Table 18.1.

ent at the Sagavanirktok River site showed differences in microbial respiration and C turnover rates when incubated at similar temperatures and moisture tensions (Figure 18.3). Respiration rates were highest in dry heath soils, intermediate in moist tussock soils, and lowest in wet sedge soils. Therefore, even under uniform conditions, soil microbial activity (respiration) was highest in soils with high proportions of shrub litter input that developed under aerobic conditions and was lowest in soils with few shrubs that developed under anaerobic conditions. A major challenge to arctic biogeochemistry is that of developing a quantitative, process-based understanding of factors leading to such functional differences in arctic soils.

Patterns of Nutrient Cycling

Because more than 99% of the N and P in arctic ecosystems resides in soils and because growth-limiting nutrients (particularly N) in soils are bound as organic compounds, the amounts and quality of various soil C fractions are key determinants of nutrient availability to plants (above). Soil C quality interacts with soil microclimate and microbial activity to determine the amounts and seasonality of nutrient availability to arctic plants. High

C/nutrient ratios (Table 18.1) and large amounts of soluble C in arctic soils (Figure 18.2) lead to high microbial demands for nutrients. These high microbial demands result in high ratios of nutrient immobilization to gross nutrient mineralization and to extremely low net nutrient mineralization ratios. For example, comparisons of in situ net N mineralization in arctic soils versus soils of other ecosystems (Nadelhoffer et al., 1992) show that net N mineralization in the Arctic ranges from about 0.1 to $0.5\,g\,N\,m^{-2}\,yr^{-1}$ whereas rates in boreal, temperate, and tropical ecosystems range from 2 to $20\,g\,N\,m^{-2}\,yr^{-1}$. Even though net N mineralization rates are considerably lower in the Arctic than elsewhere, mineralization rates do vary among arctic ecosystem types. For example, up to fivefold variations in annual net N mineralization rates have been reported among dry, moist, and wet tundra in alpine (Hart and Gunther, 1989) and arctic (Giblin et al., 1991) sites in Alaska.

The seasonality of nutrient uptake by plants can vary considerably in arctic ecosystems and nutrient availability can be relatively low during the peak growing season. Exchangeable NH_4^+ concentrations in organic soils are typically greater just after soil thaw or before fall freezing than in midseason (Gersper et al., 1980). Midsummer net N mineralization (July) in a moist tussock tundra on Alaska's North Slope was only slightly greater than zero in a watertrack site and was negative in an adjacent non-watertrack site (Chapin et al., 1988). Results of in situ soil incubations in dry, moist, and wet tundra ecosystems reported by Giblin et al. (1991) showed that only half of the annual net N mineralization in moist tussock and dry heath ecosystems occurred in July to August 1985, and that mineralization was negative in a wet sedge ecosystem during this period. Such patterns of relatively high N availability early and late in the growing season and low availabilities in midsummer probably result from lower microbial demands when soil temperatures are cooler. Even if gross N mineralization rates are highest when soils are warmest, net N mineralization can be lowered by high rates of microbial N immobilization. Nitrification (below), in contrast to net N mineralization, is more likely to occur during the warmest months of the year in ecosystem types where it is important (Giblin et al., 1991).

The dominant forms of N available for uptake by plants may vary in arctic landscapes. Although cold, wet conditions and low ammonium concentrations in arctic soils might not be considered favorable for nitrification, nitrate may be an important form of plant-available N in some arctic ecosystems. Indirect evidence for nitrification includes reports of inducible nitrate reductase activities and $^{15}NO_3$ uptake by some arctic plant species (Atkin and Cummins, 1994; Atkin et al., 1993). Direct evidence for nitrification in arctic ecosystems includes reports of nitrate arctic plant tissues and solutions (Haag, 1974; Ulrich and Gersper, 1978) and of nitrification in arctic soils (Chapin et al., 1988; Giblin et al., 1991).

Phosphorus dynamics, in contrast to N dynamics, are subject to physical–chemical as well as biological controls. In soils of lower latitudes, the chemical release of phosphate from weatherable minerals is an important source of plant-available P (Williams and Walker, 1969). Weathering is less important in arctic ecosystems, however, because plant roots are often isolated from mineral soil horizons for much or all of the growing season and because low temperatures slow weathering reactions (Hill and Tedrow, 1961). In moist and wet tundra ecosystems, P inputs to organic horizons are small, coming mainly from precipitation. Therefore, P recycling from organic soils must supply much or all of the P taken up by plants in these ecosystems (Shaver et al., 1990). In dry tundra ecosystems, where organic horizons are thin, P availability depends more on properties of mineral soil horizons. Phosphorus availability can be high in young soils with large pools of weatherable P in primary minerals, but low in more intensively weathered soils, especially if phosphates are tightly held to secondary minerals such as aluminum or iron oxides (Sayers and Walker, 1969).

Phosphorus is tightly conserved in arctic ecosystems, even when it is not the most limiting element (Chapin et al., 1980). The C/P ratios often increase with soil depth in arctic peats where mineral horizons lie below the maximum depth of thaw, whereas C/N ratios generally remain constant or decrease with depth (Malmer and Nihlgård, 1980; Table 18.1), suggesting that strong biological demands for P in active soil layers minimize P burial in accumulating peats.

Phosphorus availability in arctic soils can vary markedly across growing seasons. As with mineral N, soluble and exchangeable phosphate pools in organic horizons are greatest at snowmelt and decline as soil temperatures increase (Dowding et al., 1981). As a result, tundra plants probably compete best for soil P shortly after surface thawing in early summer, just before freezing in fall, or during other periods when microbial populations decline (Chapin and Bloom, 1976; Chapin et al., 1978). Laboratory incubations of tundra soils also indicate that P availability is highest at temperatures just above freezing, when microbial demands for P are relatively low (Nadelhoffer et al., 1991).

Plant–Soil Interactions in a Warmer Arctic

Even relatively modest increases in growing season temperatures could have important effects on C turnover and nutrient cycling in tundra soils and on net primary production and biomass accumulation in tundra ecosystems. This is because microbial respiration and net N mineralization in tundra soils are temperature sensitive, particularly at the upper end of the temperature ranges experienced by tundra soils under current climates. For example, we reported previously that microbial respiration in soils from six tundra ecosystems located along an arctic toposequence increased about twofold between 10 and 15°C under laboratory conditions (Nadelhoffer et al., 1991). Increases in net N mineralization in these soils were even larger, increasing by four- to fivefold across the same temperature range. These results suggest that increases in growing season temperatures in the Arctic would increase both C and N turnover in tundra soils, but that increases in N turnover would be greater. Therefore, ratios of N mineralized to C respired from arctic soils will likely increase if soil temperatures increase. Also, deeper thawing and longer growing seasons that could accompany a warmer climate would increase both the volume of active soil and the length of time that microbes are active, thereby increasing annual C and nutrient mineralization.

Climatic warming would likely also alter seasonal patterns and forms of plant nutrient availability in arctic soils. If warmer soils lead to decreases in relatively labile C fractions, rates of microbial nutrient immobilization would likely decrease, thereby increasing plant nutrient availability during the peak growing season. The likelihood of nitrification in many soils will increase because of the direct effects of increases in temperature and oxygen and because higher nitrogen mineralization will increase the pools of substrate (NH_4^+) available to nitrifiers.

Soil and plant processes will probably be most affected in moist tundra ecosystems that have the greatest potential for changes in nutrient availability in a warmer climate (Table 18.3). Higher air temperatures and increases in evapotranspiration will likely lead to increases in soil temperatures and rates of organic matter decomposition. Increases in decomposition rates should, in turn, lead to thinner organic mats that will provide less thermal insulation in summer and will feed back to increase soil temperatures and depths of thaw. Soil C/nutrient ratios will likely decrease and rates of net N mineralization should increase in warmer, drier soils. This in turn would increase the amount of ammonium available as substrate for nitrifiers and increase the likelihood of nitrification. Increased nitrification would have important implications for primary productivity and for community composition in the Arctic, both because plants metabolize nitrate differently from

Table 18.3. Possible Effects of a Warmer Climate on Ecosystem Properties and Processes in Dry Heath, Moist Tussock, and Wet Sedge Tundras

Ecosystem Property	Tundra Type		
	Dry heath	Moist tussock	Wet sedge
Decomposition and soil C turnover	= to +	+++	= to +++ f(drainage)
N mineralization	= to +	+++	= to ++
Nitrification	more probable	more probable	more probable with better drainage
P availability	f(mineralogy)	+++	− to +++
Organic mat thickness	= to −	− to −−−	− or + f(drainage)
Thaw depth	+	+++	++
Drainage	=	++ to +++	= to +++ f(topography)
Soil moisture	=	− to −−−	= to −−−
Net primary production	= to +	++ to +++	= to ++

ammonium and because species preferences for these forms differ (Smirnoff and Stewart, 1985). Phosphorus availability will probably also increase as organic matter stocks decline and turn over more rapidly. Increases in P availability will also occur if mineral horizons thaw more deeply and plant roots become less isolated from mineral soils, especially where mineral horizons are not excessively weathered. Long-term (9 year) manipulations of air temperature (using plastic greenhouses) in moist tussock tundra near Toolik Lake, Alaska (Chapin et al., 1995), are consistent with our predictions. Chapin and colleagues reported that warming (about 5°C during growing seasons) increased growing season soil temperatures by about 2°C and availabilities of ammonium, nitrate, and phosphate in soils. Although increases in net primary production and biomass were not detectable after 9 years of warming, growth and biomass of vascular plants increased at the expense of nonvascular plants. However, fertilization (N + P) and fertilization plus warming increased primary production and led to dominance of moist tundra by deciduous shrubs at the expense of nonvascular plants, graminoids, and evergreen shrubs. Therefore, we expect that sustained increases in nutrient availability resulting from warming could lead to increases in primary production and deciduous shrub biomass over the longer term in moist tundra.

In wet tundra ecosystems, variations in local topography and drainage will determine responses of soil microbial processes and nutrient cycling to climate warming (Table 18.3). In locations where water inputs from well-drained uplands are not attenuated, soil organic matter stocks will probably remain high or continue to accumulate. This is because microbial respiration and nutrient mineralization rates of saturated, anaerobic peats will not increase greatly with soil temperature and thaw depths. In wet tundra soils that become better drained because of local perturbations (e.g., decreased inputs from uplands or thawing of ice dams), soil responses could be similar to those in moist tundra ecosystems. Phosphorus availability should remain low where summer thawing does not extend to mineral soils. Increases in P availability are possible, however, in those wet tundras where peats are thin enough to allow thawing and contact of roots with mineral soils.

Dry tundra ecosystems will probably be the least responsive to a warmer arctic climate. Soil temperatures, thaw depths, and nutrient availability should all increase somewhat with air temperature in these ecosystems (Table 18.3). Under current conditions, however, summer soil temperatures are high and thawing is deep relative to moister ecosystems due to the presence of relatively thin organic mats. Therefore, increases in soil respiration and decreases in organic matter stocks would not feed back to further increase soil temperatures as strongly as in wetter tundra ecosystems. Relative increases in N mineralization rates would probably be less than in wetter ecosystems. Dramatic increases in N availability in dry tundra would probably require greater input via N fixation or precipitation. Changes in P availability are highly dependent on soil mineralogy. Phosphorus weathering and P availability could increase in dry tundra ecosystems on relatively young soils with large amounts of primary minerals. Large increases in P availability are unlikely, however, on highly weathered mineral soil, where P is largely bound with iron and aluminum in secondary minerals (Giblin et al., 1991).

Finally, although NPP in ecosystems worldwide is typically nutrient limited (Vitousek and Howarth, 1991), comparisons of plant nutrient requirements versus soil nutrient availability along global gradients show that nutrient mineralization/NPP ratios decrease markedly from low (tropical) to high (arctic) latitudes (Nadelhoffer et al., 1992). As a result, arctic plant species rely heavily on either nutrient retranslocation (Chapin and Shaver, 1985b), N fixation (Chapin and Bledsoe, 1992), or direct uptake of organic N from soils (Keilland, 1994) to supply their demands for growth-limiting nutrients. We expect that increases in soil nutrient availability resulting from climatic warming will favor species that rely more on direct uptake of mineral N from soils and less on these alternative mechanisms for meeting their growth requirements for nutrients.

Summary

Short thaw seasons, low soil temperatures, and extremes in moisture availability all serve to slow litter decomposition and soil organic matter turnover in terrestrial ecosystems of arctic regions. As a result, rates at which growth-limiting nutrients are mineralized in arctic soils are exceedingly low and primary productivity is strongly nutrient limited.

Although the supply rates of nutrients in arctic soils are consistently low in relation to plant nutrient demands, the relative degree of nutrient limitation and the specific nutrients most limiting to growth are quite variable arctic ecosystem types. Fertilization studies and measures of soil nutrient availability indicate that N and P are the most important growth-limiting elements, with P limitation tending to be more important in wet ecosystems and other systems where mineral soils remain continuously frozen. Nitrogen is generally more limiting where mineral horizons are close to the soil surface, such as on hilltops or in riparian zones. Productivity in extensive areas of moist tundra dominated by the tussock-forming sedge, *Eriophorum vaginatum*, can be limited by either N, P, or both elements.

Warming and climatic factors that affect soil moisture in the Arctic will likely have major implications for nutrient cycling, productivity, and plant growth forms. Warming will generally increase decomposition, soil nutrient availability, and primary production. Effects will likely vary, however, for different ecosystem types. Moist tussock ecosystems will likely be most responsive. As soils warm, decomposition and N mineralization rates will increase in surface organic horizons and P availability will increase as mineral horizons thaw more deeply. Greater nutrient availability in moist ecosystems will likely increase primary productivity and favor growth of woody shrubs over graminoids and mosses. Wet sedge ecosystems will be less altered unless local disturbances or a drier climate serve to improve drainage or lower water tables. Responses of dry, well-drained ecosystems with low organic carbon contents will be influenced more by factors that determine overall N storage than by changes in N mineralization.

References

Aber, J.D., and Melillo, J.M. 1982. Nitrogen immobilization in decaying hardwood leaf litter as a function of initial nitrogen and lignin content. *Can. J. Bot.* 60:2263–2269.

Andreev, V.N., and Aleksandrova, V.D. 1981. Geobotanical division of the Soviet Arctic. In: Bliss, L.C., Heal, O.W., Moore, J.J. (eds.), *Tundra Ecosystems: A Comparative Analysis* (pp. 25–34). Cambridge: Cambridge University Press.

Atkin, O.K., and Cummins, W.R. 1994. The effect of nitrogen source on growth, nitrogen economy and respiration of two high arctic plant species differing in relative growth rate. *Functional Ecol.* 8:389–399.

Atkin, O.K., Villar, R., and Cummins, W.R. 1993. The ability of several high arctic plant species to utilize nitrate nitrogen under field conditions. *Oecologia* 96:239–245.

Bliss, L.C. 1981. North American and Scandinavian tundras and polar deserts. In: Bliss, L.C., Heal, O.W., and Moore, J.J. (eds.), *Tundra Ecosystems: A Comparative Analysis* (pp. 38–46). Cambridge: Cambridge University Press.

Bunnell, F.L., and Tait, D.E.N. 1974. Mathematical simulation models of decomposition processes. In: Holding, A.J., Heal, O.W., MacLean, S.F. Jr., and Flanagan, P.W. (eds.), *Soil Organisms and Decomposition in Tundra* (pp. 207–226). Stockholm: Tundra Biome Steering Committee.

Chapin, D.M., and Bledsoe, C.S. 1992. Nitrogen fixation in arctic plant communities. In: Chapin, F.S. et al. (eds.), *Physiological Ecology of Arctic Plants: Implications for Climate Change* (pp. 301–320). New York: Academic Press.

Chapin, F.S. III, and Bloom, A.J. 1976. Phosphate absorption: Adaptation of tundra graminoids to a low-temperature, low-phosphorus environment. *Oikos* 26:111–121.

Chapin, F.S. III, and Shaver, G.R. 1985a. Individualistic growth response of tundra plant species to manipulation of light, temperature, and nutrients in a field experiment. *Ecology* 66:564–576.

Chapin, F.S. III, and Shaver, G.R. 1985b. The physiological ecology of arctic plants. In: Chabot, B.F., and Mooney, H.A. (eds.), *Physiological Ecology of North American Plant Communities* (pp. 16–40). London: Chapman and Hall.

Chapin, F.S. III, Barsdate, R.J., and Barèl, D. 1978. Phosphorus cycling in Alaskan coastal tundra: A hypothesis for the regulation of nutrient cycling. *Oikos* 31:189–199.

Chapin, F.S. III, Miller, P.C., Billings, W.D., and Coyne, P.I. 1980. Carbon and nutrient budgets and their control in coastal tundra. In: Brown, J., Miller, P.C., Tieszen, L.L., and Bunnel, F.L. (eds.), *An Arctic Ecosystem: The Coastal Tundra at Barrow, Alaska. US/IBP Synthesis Series Vol. 12* (pp. 458–482). Stroudsburg, PA: Dowden, Hutchinson and Ross.

Chapin, F.S. III, Fetcher, N., Keilland, K., Everett, K.R., and Linkins, A.E. 1988. Productivity and nutrient cycling of Alaskan tundra enhanced by flowing soil water. *Ecology* 69:693–702.

Chapin, F.S. III, Shaver, G.R., Giblin, A.E., Nadelhoffer, K.J., and Laundre, J.A. 1995. Responses of Arctic tundra to experimental and observed changes in climate. *Ecology* 76:694–711.

Dowding, P., Chapin, F.S., Wielgolaski, F.E., and Kilfeather, P. 1981. Nutrients in tundra ecosystems. In: Bliss, L.C., Heal, O.W., and Moore, J.J. (eds.), *Tundra Ecosystems: A Comparative Analysis* (pp. 647–683). Cambridge: Cambridge University Press.

French, D.D. 1977. Multivariate characteristics of IBP Tundra Biome site characteristics. In: Bliss, L.C., Heal, O.W., and Moore, J.J. (eds.), *Tundra Ecosystems: A Comparative Analysis* (pp. 47–75). Cambridge: Cambridge University Press.

Gersper, P.L., Alexander, V., Barkley, S.A., Barsdate, R.J., and Flint, P.S. 1980. The soils and their nutrients. In: Brown, J., Miller, P.C., Tieszen, L.L., and Bunnel, F.L. (eds.), *An Arctic Ecosystem: The Coastal Tundra at Barrow, Alaska. US/IBP Synthesis Series Vol. 12* (pp. 219–254). Stroudsburg, PA: Dowden Hutchinson and Ross.

Giblin, A.E., Nadelhoffer, K.J., Shaver, G.R., Laundre, J.A., and McKerrow, A.J. 1991. Biogeochemical diversity along a riverside toposequence in arctic Alaska. *Ecol. Monogr.* In press.

Haag, R.W. 1974. Nutrient limitations to plant production in two tundra communities. *Can. J. Bot.* 52:103–116.

Hart, S.C., and Gunther, A.J. 1989. In situ estimates of annual net nitrogen mineralization and nitrification in a subarctic watershed. *Oecologia* 80:284–288.

Heal, O.W., Flanagan, P.W., French, D.D., and MacLean, S.F., Jr. 1981. Decomposition and accumulation of organic matter in tundra. In: Bliss, L.C., Heal, O.W., and Moore, J.J. (eds.), *Tundra Ecosystems: A Comparative Analysis* (pp. 587–633). Cambridge: Cambridge University Press.

Hill, D.E., and Tedrow, J.C.F. 1961. Weathering and soil formation in the arctic environment. *Am. J. Sci.* 259:84–101.

Hinneri, S., Sonesson, M., and Veum, A.K. 1975. Soils of Fennoscandian IBP tundra ecosystems. In: Wielgolaski, F.E. (ed.), *Fennoscandian Tundra Ecosystems. Part 1, Plants and Microorganisms* (pp. 31–40). Berlin: Springer-Verlag.

Keilland, K. 1994. Amino acid absorption by arctic plants: Implications for plant nutrition and nitrogen cycling. *Ecology* 75:2362–2372.

Malmer, N., and Nihlgård, B. 1980. Supply and transport of mineral nutrients in a subarctic mire. In: Sonesson, M. (ed.), *Ecology of a Subarctic Mire* (pp. 63–95). Stockholm: Ecological Bulletins.

McCown, B.H. 1978. The interactions of organic nutrients, soil nitrogen and plant growth and survival in the arctic environment. In: Tieszen, L.L. (ed.), *Vegetation and Production Ecology of An Alaskan Arctic Tundra* (pp. 435–456). New York: Springer-Verlag.

McKendrick, J.D., Batzli, G.O., Everett, K.R., and Swanson, J.C. 1980. Some effects of mammalian herbivores and fertilization on tundra soils and vegetation. *Arctic Alpine Res.* 12:565–578.

Melillo, J.M., Aber, J.D., and Muratore, J.M. 1982. Nitrogen and lignin control of hardwood leaf litter decomposition dynamics. *Ecology* 63:621–626.

Melillo, J.M., Aber, J.D., Linkins, A.E., Ricca, A., Fry, B., and Nadelhoffer, K.J. 1989. Carbon and nitrogen dynamics along the decay continuum: Plant litter to soil organic matter. In: Clarholm, M., and Bergstrom, L. (eds.), *The Ecology of Arable Land* (pp. 53–62). Dordrecht: Kluwer Academic.

Nadelhoffer, K.J., Giblin, A.E., Shaver, G.R., and Laundre, J.L. 1991. Effects of temperature and substrate quality on element mineralization in six arctic soils. *Ecology*. In press.

Nadelhoffer, K.J., Giblin, A.E., Shaver, G.R., and Linkins, A.E. 1992. Microbial processes and plant nutrient availability in arctic soils. In: Chapin, F.S., et al. (eds.), *Physiological Ecology of Arctic Plants: Implications for Climate Change*. (pp. 281–300). New York: Academic Press.

Oechel, W.C. 1989. Nutrient and water flux in a small arctic watershed: An overview. *Holarctic Ecol.* 12:229–237.

Reich, P.B., Walters, M.B., and Ellsworth, D.S. 1992. Leaf life-span in relation to leaf, plant, and stand characteristics among diverse ecosystems. *Ecol. Monogr.* 62:365–392.

Sayers, J.K., and Walker, T.W. 1969. Phosphorus transformations in a chronosequence of soils developed on a windblown sand in New Zealand. I. Total and organic phosphorus. *J. Soil Sci.* 23:50–64.

Shaver, G.R., and Chapin, F.S., III. 1991. Production: Biomass relationships and element cycling in contrasting arctic vegetation types. *Ecol. Monogr.* 61:1–31.

Shaver, G.R., Nadelhoffer, K.J., and Giblin, A.E. 1990. Biogeochemical diversity and element transport in a heterogeneous landscape, the North Slope of Alaska. In: Turner, M., and Gardner, R. (eds.), *Quantitative Methods in Landscape Ecology*. New York: Springer-Verlag.

Smirnoff, N., and Stewart, G.R. 1985. Nitrate assimilation and translocation by higher plants: Comparative physiology and ecological consequences. *Physiol. Plantar.* 64:133–140.

Sonesson, M., Wielgolaski, F.E., and Kallio, P. 1975. Description of Fennoscandian tundra ecosystems. In: Wielgolaski, F.E. (ed.), *Fennoscandian Tundra Ecosystems. Part 1, Plants and Microorganisms* (pp. 3–28). Berlin: Springer-Verlag.

Svensson, B.H. 1980. Carbon dioxide and methane fluxes from the ombrotrophic parts of a subarctic mire. In: Sonesson, M. (ed.), *Ecology of a Subarctic Mire* (pp. 235–250). Stockholm: Ecological Bulletins.

Swift, M.J., Heal, O.W., and Anderson, J.M. 1979. *Decomposition in Terrestrial Ecosystems* (pp. 276–291). Berkeley, CA: University of California Press.

Ulrich, A., and Gersper, P.L. 1978. Plant nutrient limitations of tundra plant growth. In: Tieszen, L.L. (ed.), *Vegetation and Production Ecology of An Alaskan Arctic Tundra* (pp. 457–482). New York: Springer-Verlag.

Vitousek, P.M., and Howarth, R.W. 1991. Nitrogen limitation on land and in the sea: How can it occur? *Biogeochemistry* 13:87–115.

Williams, J.D.H., and Walker, T.W. 1969. Fractionation of phosphate in a maturity sequence of New Zealand basaltic soil profiles. *Soil Sci.* 107:213–219.

19. Buffering of Arctic Plant Responses in a Changing Climate

Sven Jonasson

Responses of plants to predicted global climate change are complex, and will depend on the magnitude and interaction of changes among a suite of environmental variables (e.g., atmospheric CO_2 concentration, air and soil temperature, amount and seasonal distribution of precipitation, radiation regime, etc).

In addition to being influenced by changes in the environment, all stages in the plants' life cycles are also sensitive to biotic interactions that, in turn, often are indirectly controlled by the environment. For instance, the strength of competition may vary with the resource levels (Grime, 1973) or the balance between various environmental resources (Tilman, 1982). Mutualism, such as the interaction between mycorrhiza and the host plant and the activity of nutrient-mineralizing microorganisms, is influenced by the water and nutrient content of the soil (Heal and Block, 1987; Koide, 1991). Reproductive success is influenced by climatic factors, which regulate the activity and behavior of the pollinators, etc.

Even though our knowledge of plant responses to single or a few environmental factors and to biotic interactions, seen in isolation, is reasonably good, complications arise when we try to predict or model the response to the integrated effect from a large suite of interacting factors at the community or ecosystem level (Körner, 1993). Due to the scarcity of data from these levels, predictions are often based on extrapolations from known responses to environmental influences on lower levels such as populations, single plants, or even from physiological responses at tissue levels. Such extrapolations are risky because they do not take the possible interactions into account, and the processes act on time scales that often are much shorter than those acting on higher levels of organization (Oechel and Billings, 1992). We must therefore accept a high level of uncertainty when predicting community and ecosystem responses to a changing climate until they are properly tested by in situ experiments at the appropriate levels of biological organization.

At present, it appears that the predicted changes in the environment will lead to increased biological activity at the lower organization levels. For instance, increased CO_2, temperature, and nutrient levels generally lead to increased growth (Chapin et al., 1992). By simple summing of the responses, it appears that the presumed changes in the environment will cause dramatic effects in arctic plant communities. Several recent observations of experimental perturbations at community level, however, suggest that the systems may have a considerable resistance to change, and that known reactions from lower levels of organization cannot be extrapolated uncritically to communities (Körner, 1993). For instance, the well-known effect of increased photosynthesis and growth of most C3 plants exposed to CO_2 levels above ambient was not observed under field conditions in an in situ experiment on an Alaskan tussock tundra (Tissue and Oechel, 1987). The low response was probably because the dominant species in the community, *Eriophorum vaginatum*, was more strongly limited by nutrients than by CO_2 in its natural environment. Hence, increased growth by enhancement of CO_2 should not be expected unless the nutrient level also increases.

This chapter aims at focusing on a selection of possible reasons why arctic plant communities may show higher resistance or "buffering capacity" against major climatic changes during a transitional period than generally anticipated. As with buffers in chemical systems, it seems inevitable, however, that the resistance eventually will break down and the systems will undergo large, and possibly also fast, changes if the climate continues to change in the predicted way. I have chosen examples of probable buffering mechanisms from various levels of organization, including physiological adaptations, life history traits, community organization, and stabilizing influences from other trophic levels. Where possible, I have attempted to find support for my examples from observations at community or ecosystem levels.

Life History and Clonality as Buffering Agents

Arctic plants are typically perennial, an obvious advantage in the Arctic because one growing season is too short for "normal sized" plants to complete the entire life cycle of vegetative growth and reproduction (Callaghan and Emanuelsson, 1985). In addition, between-year storage of resources in perennial plants buffers against temporarily adverse conditions for resource assimilation in certain years and ensures survival even if resource acquisition is strongly impeded or even entirely prevented (Jonasson and Chapin, 1985).

Clonality, vegetative reproduction and vegetative spread are other typical attributes, particularly among plants in the low Arctic (Callaghan et al., 1992; Savile, 1972; Sonesson and Callaghan, 1991). Clonality adds advantages to the perennial life cycle because connections between ramets permit partitioning of resources between modules of different ages and types (Callaghan, 1984; Jónsdóttir and Callaghan, 1990) so that resources can be translocated to different parts of the clone to buffer spatial heterogeneity of the habitat (Callaghan et al., 1992; Pitelka and Ashmun, 1985). Clonality also reduces risks associated with the younger life stages in nonclonal, seed-recruited plants. In spite of often low recruitment from seeds in clonal plants (Callaghan and Emanuelsson, 1985), those with long rhizome connections can spread and dominate the local space in a manner comparable to seed-dispersed plants, and in some species a single genet can even dominate larger areas of several hectares (Oinonen, 1967).

Taken together, all these factors, which are functions of long life span of the genet and the integration of ramets, contribute to stabilize arctic plant communities and make them less susceptible to moderate changes in the environment over time scales of decades or centuries (McGraw and Fetcher, 1992).

However, the same traits may be a disadvantage under more rapidly changing conditions. The generally low frequency of establishment after sexual reproduction, typical at least in stable vegetation (Jonasson, 1992), implies that genetic variation is low within the communities. This is likely to decrease the potential of the species to respond to drastic environmental change that requires adjustments of the plants beyond their phenotypic plasticity (Callaghan et al., 1992). Species in habitats with higher frequency of establishment from seeds may be less sensitive to such changes. These plants apparently exhibit a high degree of ecotypic specialization in response to local differences in climate or topography, even within distances of meters (Crawford et al., 1993; McGraw, 1987; McGraw and Fetcher, 1992). Hence, if conditions change beyond the limit of tolerance for some ecotypes in some sites, there should still be a possibility that other ecotypes in the neighborhood can spread into that site. For example, if the climate becomes warmer and drier, ecotypes adapted to wet and cold microsites (Crawford et al., 1993) can be replaced by ecotypes presently occurring nearby on dry and warm hilltops. Hence, such spatial genetical specialization could act as a buffer against extinction of species but not against extinction of genotypes.

Buffering in the Soil–Plant Interface

Arctic ecosystems are generally nutrient limited (Chapin, 1987; Haag, 1974; Jonasson, 1992), which implies that the direct effect of the current increase of atmospheric CO_2 may not give a plant growth response of the same magnitude as has been predicted in more fertile ecosystems (Tissue and Oechel, 1987). Instead, increased temperature and changed soil moisture regime, two of the most frequently predicted indirect effects of raised atmospheric CO_2 level (Chapin et al., 1992), could have greater effects than the increasing level of CO_2 per se. It is known that both factors exert strong control over the rate of microbial activity, decomposition of soil organic matter, and thereby nutrient mineralization (Melillo et al., 1990). Except in already dry soils a temperature increase is likely to increase decomposition rates (Rastetter et al., 1993) and enhance nutrient mineralization both as a direct response of increased microbial activity in warmed soils and as a result of decreased waterlogging of the soil.

Recent data suggest, however, that the rate of decomposition and nutrient mineralization may not increased as anticipated, for several reasons. Firstly, several observations from air heating experiments across an array of arctic ecosystems, ranging from the subarctic to the high arctic (Havström et al., 1993; Jonasson et al., 1993; Wookey et al., 1993) have shown proportionally low temperature enhancements in the soil when the air temperature is raised 3 to 5°C, or in one case even a slight cooling of the soil (Wookey et al., 1993).

Although different soil and site conditions probably give different responses in soil temperature to increases in air temperatures, it appears that there is a considerable time lag before an increase in atmospheric temperature has given full effect on soil temperature. Modeling from an Alaskan tundra showed that a 4°C heating of the air during a 50-year period gave a likely summer soil temperature enhancement of 2 to 3°C at 10-cm depth (i.e., in the root zone of the plants) (Kane et al., 1992). Hence, an air warming cannot be translated uncritically to a soil warming of the same magnitude. This, together with observed modest enhancement of microbial nutrient mineralization in tundra soils warmed a few centigrades above ambient (Nadelhoffer et al., 1991; see below), implies that an atmospheric temperature increase could have a much higher immediate effect on aboveground plant processes than on the soil biota and on plant and other biological processes belowground. It cannot be excluded, however, that the allocation pattern between shoots and belowground plant parts changes in response to the changes in atmospheric gas composition and climate aboveground.

Data from both Alaska and Scandinavia suggest that the effects on mineralization may be lower than previously thought, even if the soil is warmed to 4 to 5°C above ambient. Nadelhoffer et al. (1991, 1992) found, in a laboratory experiment that included measurement of respiration and net nutrient mineralization of soil plugs from a number of Alaskan tundra sites at 3, 9, and 15°C, that increased heating generally increased both the respiratory loss of CO_2 and the mineralization rate. The response within the range of 3 to 9°C was modest, however, and in some cases net N and P mineralization even decreased due to a probable increased nutrient immobilization in soil microorganisms. In contrast, further heating up to 15°C, which is far above the present soil temperature in most arctic areas (Jonasson et al., 1993; Wookey, 1993), resulted in a rapid step increase of both respiration and net mineralization.

These observations agree with reports on mineralization in situ. An increase of the air temperature of 4 to 5°C in two arctic soils from a relatively warm heath and a cold, high altitude fellfield (Figure 19.1) resulted in 0.9 to 1.9°C enhancement of the soil temperature, but did not increase net N and P mineralization (Jonasson et al., 1993). In fact, regardless of the temperature enhancement, there was no significant seasonal net N mineralization, but significant P mineralization in the heath, and no measurable net P mineralization but strong N mineralization in the fellfield. This indicates that N in the heath and P in the fellfield were strongly immobilized. Temperature increase of 4°C after transplantation of fellfield soil to the warmer heath site did not cause any increased mineralization of P, the most strongly immobilized element; net mineralization of N, which was released in a considerable rate at ambient temperatures (Figure 19.1), was more pronounced than at its original, cooler site (Figure 19.2).

The experiment suggests two possible buffering properties of the soil and the soil biota. First, the low response of soil temperature enhancement to atmospheric warming, discussed above, resulted in a lower response in nutrient mineralization than expected. Second, the result shows that some nutrients are vulnerable to microorganisms, which have the potential to immobilize them efficiently. If these nutrients limit plant growth, there is probably a strong competition for nutrients between microorganisms and plants (Marion et al., 1982). In fact, it has been suggested recently that only nutrients that are "left over" by the microorganisms when they have maximized the allocation of mineralized nutrients to their own biomass growth are available to plants (Harte and Kinzig, 1993). Under such circumstances, even if gross mineralization increases, immobilization buffers against enhanced nutrient availability to the plants, and they will only benefit from the increased turnover of the organic matter if the microbial biomass declines periodically and their cell content is released to the plant available soil nutrient pool (Chapin et al., 1978; Jonasson and Chapin, 1991). The buffering will continue until the combination of strong nutrient immobilization and enhanced respiratory loss of carbon from the soil organic matter lower the carbon to nutrient ratio of the soil to the extent of releasing the microorganisms from nutrient limitation.

A high buffering through microbial immobilization of nutrients appears contradictory to the strong responses to fertilization that have been shown

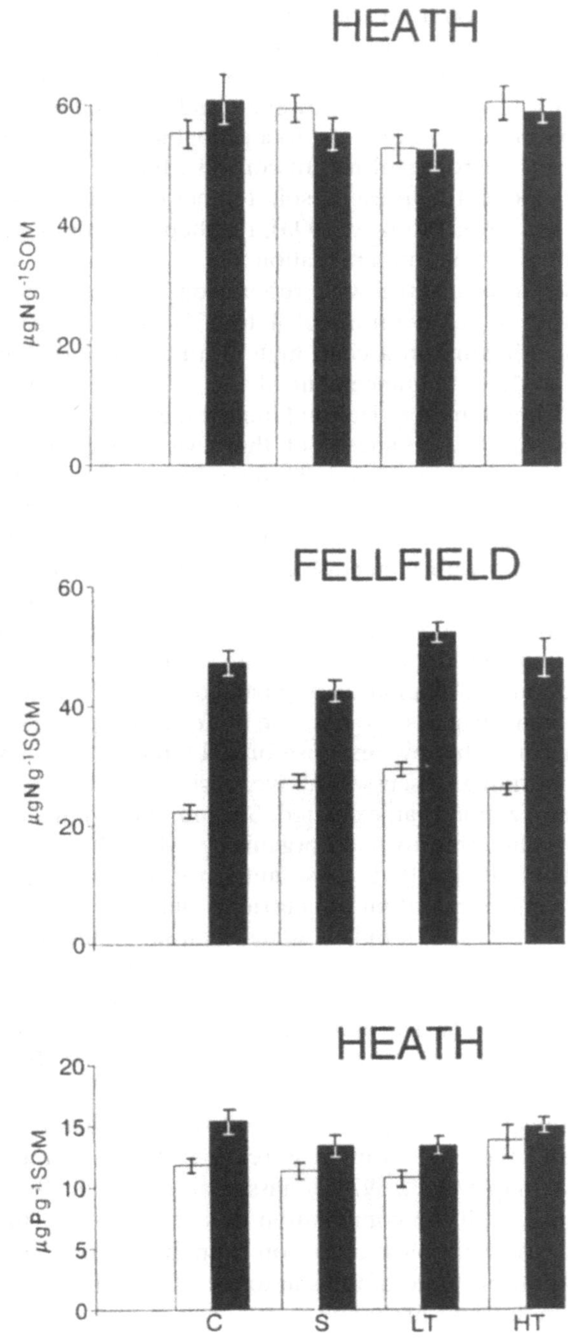
HEATH
60
40
20
0
μgNg⁻¹SOM
FELLFIELD
60
40
20
0
μgNg⁻¹SOM
HEATH
20
15
10
5
0
μgPg⁻¹SOM
C
S
LT
HT

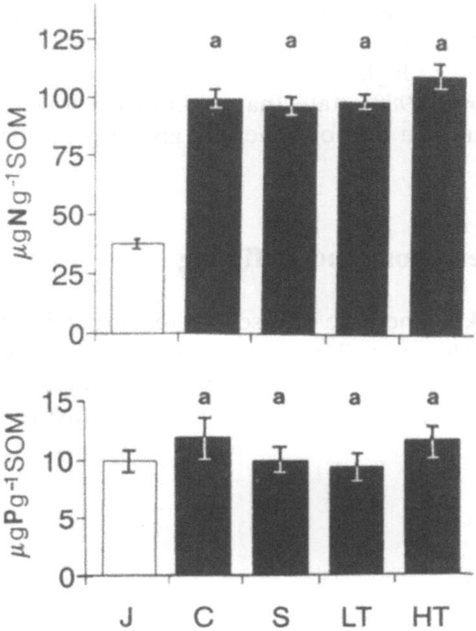

Figure 19.2. Concentrations of nitrogen and phosphorus per gram soil organic matter of fellfield soil in June at the time of transplantation and incubation (white bars; J) and in September after transplantation and incubation at a 4°C warmer heath (black bars). For further information see Figure 19.1. Note, the soil type was different from that in Figure 19.1 (adapted from Jonasson et al., 1993).

in almost all nutrient addition experiments throughout the Arctic. It should be borne in mind, however, that fertilization gives a high instantaneous pulse of nutrients and a release from the possible nutrient competition between microorganisms and plants. These conditions are quite different from the slow and continuous release of nutrients from soil organic matter that will occur if microbial activity increases in a warmer environment. In fact, results from nutrient amendments have indicated that microbial immobilization is high. Marion et al. (1982) reported that less than 20% of ^{15}N added to the soil was taken up by the vegetation, and the remaining amount was immobilized in the soil. Jonasson and Chapin (1991) found similar low uptake and high immobilization, on average 87%, of ^{32}P added in low amounts to *Eriophorum vaginatum* tussocks in a subarctic muskeg. Virtu-

Figure 19.1. Concentrations of nitrogen and phosphorus per gram soil organic matter (SOM) before (white bars) and after (black bars) soil incubations in situ buried bags at a low-altitude heath (450 m) and high-altitude fellfield (1150 m), Abisko, Swedish Lapland. The differences between white and black bars represent seasonal net mineralization from late June to early September in control (C), shaded (S), low air temperature enhancement (2–3°C; LT) and high air temperature enhancement (4–5°C; HT) plots. Phosphorus was not possible to measure with confidence at the fellfield because of extremely low concentrations (adapted from Jonasson et al., 1993).

ally all uptake took place within a few days after the application and no additional uptake could be detected during the following 4 to 5 weeks. In fact, microbial immobilization may be high even if nutrients are added in large amounts. Shaver and Chapin (1980) state that additions below 10 times the annual plant nutrient uptake do not generally give any strong response in plant growth.

Competitive Interactions and Buffering

It is generally assumed that the likely increase of essential resources, both CO_2 and nutrients, in a changing climate will cause increased competitive exclusion. A few species with great ability to respond to the changes by enahnced resource assimilation and growth will be competitively favored

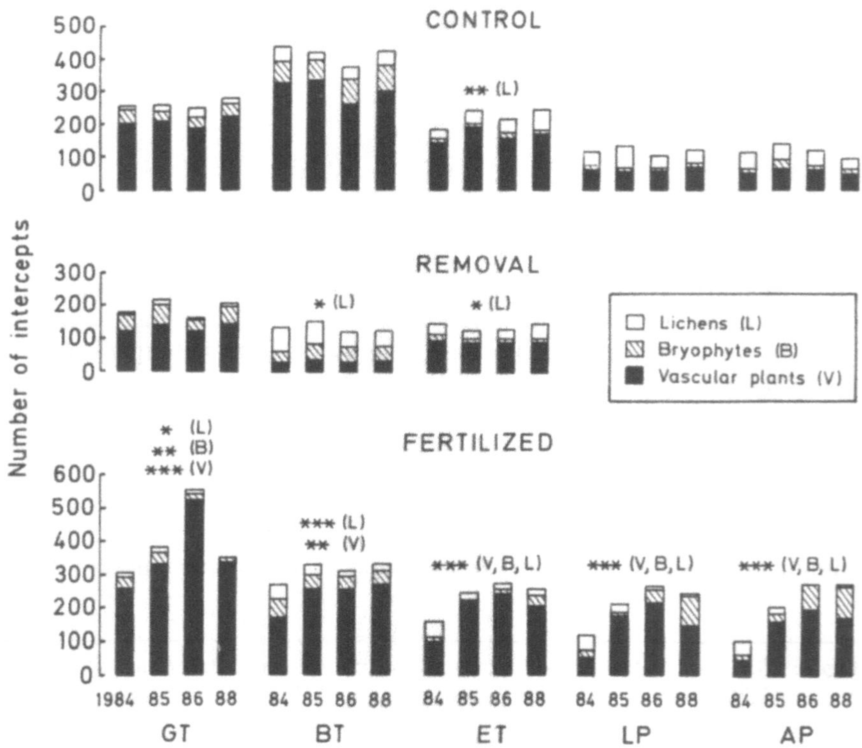

Figure 19.3. Vegetation response (measured as number of point intercepts) to removal of dominant plants and fertilization on undisturbed graminoid (GT), *Betula* (BT), and *Empetrum* tundra (ET) and on polygon patterned ground of low (LP) and high (AP) frost-heaving activity during 4 years following the treatment (from Jonasson, 1992).

Table 19.1. Shannon Diversity Values

Site	Vascular Plants			Vascular Plants plus Cryptogams		
	Control	Removal	Fertilized	Control	Removal	Fertilized
GT	0.78	0.94	0.87	1.00	1.21	1.11
	0.82	0.94	0.93	1.04	1.19	1.00
BT	0.48	0.54	0.45	0.82	1.09	0.89
	0.47	0.49	0.75	0.84	1.07	0.98
ET	0.56	0.56	0.72	0.91	1.02	1.15
	0.63	0.56	0.71	1.07	1.10	0.94
LP	0.87	—	0.80	1.21	—	1.07
	0.91	—	0.73	1.22		0.79
AP	1.03	—	1.10	1.25		1.26
	1.08	—	0.64	1.21	—	0.64

Shannon diversity immediately after (upper rows), and 4 years after (lower rows) removal of dominant shrubs on undisturbed grass tundra (GT), *Betula nana* tundra (BT), and *Empetrum* tundra (ET), and diversity changes before and 4 years after fertilization of the undisturbed tundra and frost-heaved fields of polygon patterned ground of low (LP) and high (AP) frost heaving within the *Betula* tundra (Jonasson, 1992).

relative to species that rely on restriction of losses (Berendse et al., 1987). There are few in situ experiments on tundra that attempt to show this effect. McGraw and Chapin (1989), however, reported that growth of the cotton sedge, *Eriophorum vaginatum*, which is the main component of the Alaskan tussock tundra, declined when grown together with the more nutrient demanding *E. scheuchzeri* at high nutrient levels, indicating that *E. scheuchzeri* was a stronger competitor for nutrients than *E. vaginatum* under fertile soil conditions. Likewise, Chapin and Shaver (1985) found that coexisting species on an Alaskan tundra responded in a species-specific manner to different levels of various environmental resources, and that growth of individual species varied from year to year. They interpreted the annual fluctuations as responses to climatic variations between years and suggested that the observed niche differences in growth and resource use could have been a result of specialization in response to competition.

In contrast, other studies suggest that competition is low among tundra plants. Jonasson (1992) could not demonstrate competitive exclusion of vascular plant species on a number of nondisturbed north Swedish shrub tundra sites within 4 years after addition of NPK fertilizer. Although growth of a few species increased dramatically (Figure 19.3), no species with weak, or no growth response declined in abundance. Furthermore, Shannon diversity generally increased instead of decreased (Table 19.1, sites GT, BT and ET), mainly as a function of increased evenness. The abundance and diversity of cryptogams, on the contrary, declined in general when the canopy of vascular plants grew denser (Table 1). Almost an identical increase in diversity and evenness among vascular plants after NPK addition was reported by Fox (1992) from fellfields in an Alaskan alpine area.

Reduced diversity due to increased dominance of a few species is the expected response to nutrient increase, judging from responses to similar fertilization experiments in most nonarctic communities (DiTommaso and Aarssen, 1989). The lack of any marked competitive exclusion in tundra was probable, firstly, because the plants are clonal and seedlings and young plants are virtually absent (Jonasson, 1992). Hence, the increase of a few species did not influence the recruitment of other species; seedlings in these communities are much more susceptible to competitive effects than established plants (Del Moral, 1983). Secondly, the species are of about the same height so that expansion of a few species into the open spaces did not cause any dramatic increase of shading among nonexpanding species.

In the same communities, removal of structurally dominant shrubs, resulting in a 30 to 50% reduction of plant cover, did not increase the abundance of remaining plant species (Figure 19.3). An increase was predicted if the shrubs exerted a strong competitive dominance above- or belowground on the other species.

In contrast to the response on the stable tundra, fertilization of polygon patterned ground, particularly on sections that were disturbed by strong frost heaving (Figure 19.3, site AP), caused a decline of both species richness and Shannon diversity (Table 19.1). These disturbed sites have a species composition that is fundamentally different from that of stable ground (Jonasson, 1986; Jonasson and Callaghan, 1992; Jonasson and Sköld, 1983). Seed-recruited, low-growing forbs are common and the abundance of seedlings and juvenile plants is high. The nutrient addition resulted in a spread of a few species, which rapidly covered and outcompeted seedlings, juveniles, low growing vascular plants, and lichens, with a decline of species richness and diversity as a logical consequence.

Low effects of competition in environmentally stressed vegetation was also demonstrated by Del Moral (1983). Observations of seedling establishment and survival of transplanted adult plants after dominant removals and fertilization within a range of plant communities in the Rocky Mountains demonstrated higher competitive interference in highly productive meadows than in environmentally "stressed" communities.

All these observations taken together suggest that even drastic changes in nutrient levels do not necessarily induce any rapid effect on the species composition of vascular plants on resource-poor tundra dominated by clonal plants, even though biomass is likely to increase and the relative abundance of life forms is likely to shift towards higher abundance of life forms that respond rapidly and strongly to enhanced nutrient availability (e.g., graminoids). The resource-poor tundra types are probably buffered against depletion of the vascular species, largely because the low abundance of seedlings and juvenile plants, together with the poorly developed horizontal layering of the vegetation, preclude competitive exclusion by a shading canopy. In contrast, slow-growing, late successional moss species and lichens that form a distinct layer beneath the vascular plants appear sensitive to shading when the density of vascular plants increases (Jonasson, 1992). Changes in cryptogamic cover can occur both as a result of increased competition from phanerogams and as a direct result of climatic change (e.g., increased drought or changed frequency of rainfall). The effects on ecosystem level of such possible change need more attention because cryptogams are the main plant component over vast tundra areas (Bliss and Matveyeva, 1992). They are important regulators of heat and water fluxes between the atmosphere and the soil (Tenhunen et al., 1992), and N-fixing blue–green algae (cyanobacteria) that occur as free-living organisms or as the algal symbiont in many arctic lichens are important contributors to the flux of N from the atmosphere to the tundra soils (Chapin and Bledsoe, 1992).

Except for the effects on cryptogams, the generally small changes in species composition with nutrient enrichment, demonstrated in nutrient-deficient communities, will probably contrast with pronounced competitive exclusion of certain vascular plants on segments of the tundra with high abundance of low-growing, seed-recruited vascular plants, such as on the patterned ground. Also, lush vegetation with pronounced differences in plant height will probably be more sensitive to any change that induces spread of opportunistic, tall-growing species with high competitive ability, as judged from the responses of the alpine meadow vegetation referred to above.

Herbivory as a Buffer Against Changes

The sections above assume that the plants themselves, or soil processes exert the major control over tundra vegetation. A different view is that the vegetation is controlled by herbivory (e.g., Hairston et al., 1960), at least within certain intervals of productivity (Fretwell, 1977). Based on an exploitation model of Fretwell (1977), Oksanen and coworkers (e.g., Oksanen, 1980, 1983, 1990; Oksanen et al., 1981) suggest that the vegetation in most terrestrial ecosystems is controlled by competition among the primary producers in areas of very low and high primary productivity, but that herbivory from mammalian herbivores exerts the major direct control on vegetation within a range of medium productivity. According to the authors, this range, from about 50 to about $500\,g\,m^{-2}\,yr^{-1}$, coincides with the productivity range of most tundra communities (Oksanen, 1983) and equals the approximate interval where primary productivity is high enough to sustain herbivory, but not high enough to permit predators to control the density of herbivores to any major extent. The lack of predator regulation allows the herbivores to exert a grazing pressure on the system, to the extent of controlling the standing stock of biomass and species composition.

The model predicts that as the productivity of the ecosystem increases towards the upper limit of the "medium productivity range," the herbivore population density also increases. However, standing stock of plant biomass will change little because the increased grazing pressure tends to reduce the biomass across the entire productivity interval to a common equilibrium level (see Figure 1 in Oksanen, 1990).

If this model is applied to the global change scenarios, it implies that any increase of primary production due to direct or indirect effects of climate change would not primarily be reflected in any change in biomass or species composition, but rather in the standing stocks of mammalian herbivores such as voles or lemmings. Hence, this scenario suggests that any major changes in tundra productivity will not, primarily, appear as visible increases in plant biomass, or in changes of plant species composition, except at the present upper level of the medium productivity range. The changes will instead appear at higher trophic levels, which act as buffers against major structural changes in the plant communities.

There is no general agreement on these theories, however, although it has been known for a long time (e.g., Thompson, 1955) that herbivory and exclusion of herbivory from arctic tundra (Andersson and Jonasson, 1986) can cause dramatic changes in the plant biomass, cover, and the relative abundance of plant species.

Summary

Several properties in physiological traits, life history, and population development of arctic plants, together with interactions within and among trophic levels, suggest that tundra vegetation may be buffered against gradual environment change, but that dramatic changes may take place over longer time periods. For instance, the long life span of most arctic plants and their generally low dependence on sexual propagation, at least in low-latitude tundra, are efficient buffers acting over decades. The time span of responses by decrease of abundance and extinction of species must therefore be viewed in relation to be life span of the species, unless the changes are great enough to rapidly override the limits of phenotypical adjustment in the present flora. On a longer term, continuous climatic change could result in extinction of ecotypes or vegetatively reproducing species with low genetic variation.

There is increasing evidence for strong buffering against anticipated heavy nutritional impacts on arctic plant communities, associated with temperature-induced increases in decomposition. The buffering is related to both physical factors (relatively low soil heating when the atmospheric temperature increases) and biotic factors (immobilization of extra released nutrients in soil microorganisms). The buffering will probably be effective until the carbon to nutrient ratio of the soil organic matter is considerably decreased. Increased knowledge of soil processes, nutrient turnover, and nutrient utilization in plants is urgently needed, however, because nutrient limitation appears to be one of the strongest regulating factors on plant growth in tundra, acting both directly on plant growth and indirectly by regulating plant photosynthesis.

Knowledge of interactions between plants in arctic communities is insufficient. It appears, however, that competition between vascular plants in communities dominated by vegetatively reproducing species generally is relatively low because there is space available for spatial expansion of the plants, shading is relatively low, and there is a minimum of antagonism between adult plants and seedlings. In contrast, competition in fertile and disturbed sections of the tundra with high abundance of plants recruited from seeds, and competition between vascular plants and cryptogams is probably higher. A long-term biomass increase in response to enhanced CO_2 or increased soil nutrient levels is therefore likely to change the present balance between sexually and vegetatively recruited vascular plants and between cryptogams and vascular plants. How competitive interactions may change if new species invade is still poorly known, however, and can at present only be open to speculation.

The role of herbivory in assessing the impact of changed climatic conditions has attracted less interest than the direct impact of changing abiotic conditions and mutualistic (or antagonistic) interactions between plants and soil microorganisms. Herbivores exert strong grazing pressure in several tundra ecosystems and may act as buffers against increasing standing stock of biomass if climate changes.

In conclusion, arctic ecosystems, like most other ecosystems, probably have a considerable resistance against predicted changes in the environment on a short-term basis. Hence, the anticipated continuous change in the environment is not expected to lead to dramatic initial changes in composition, or in other properties of the plant communities. Observed dramatic responses to experimental perturbations, including those simulating environmental change, have most often been results of large step changes of conditions to levels expected only after decades or centuries of continuous environmental change.

Several studies of competitive interactions, community response to increased atmospheric CO_2 levels, microbial mineralization and immobilization, and theories of herbivory as controlling mechanism in tundra vegetation have indicated that the earliest changes may show up in relatively productive sections of the tundra.

On a longer time scale it seems inevitable, however, that continuous changes in the environment would lead to drastic changes in plant composition and community organization once the buffering systems break down. Such changes can probably take place within a relatively short period of time.

References

Andersson, M., and Jonasson, S. 1986. Rodent cycles in relation to food resources on an alpine heath. *Oikos* 46:93–106.

Berendse, F., Oudhof, H., and Bol, J. 1987. A comparative study on nutrient cycling in wet heathland ecosystems. I. Litter production and nutrient losses from the plants. *Oecologia* 74:174–184.

Bliss, L.C., and Matveyeva, N.V. 1992. Circumpolar arctic vegetation. In: Chapin, F.S. III, Jefferies, R.L., Reynolds, J.F., Shaver, G.R., and Svoboda, J. (eds.), *Arctic Ecosystems in a Changing Climate. An Ecophysiological Perspective* (pp. 59–89). San Diego: Academic Press.

Callaghan, T.V. 1984. Growth and translocation in a clonal southern hemisphere sedge *Uncinia meridensis*. *J. Ecol.* 72:529–546.

Callaghan, T.V., Carlsson, B.Å., Jónsdóttir, I.S., Svensson, B.M., and Jonasson, S. 1992. Clonal plants and environmental change: introduction to the proceedings and summary. *Oikos* 63:341–347.

Callaghan, T.V., and Emanuelsson, U. 1985. Population structure and processes of tundra plants and vegetation. In: White, J. (ed.), *The Population Structure of Vegetation* (pp. 399–439). Dordrecht: Junk Publishers.

Chapin, D.M., and Bledsoe, C.S. 1992. Nitrogen fixation in arctic plant communities. In: Chapin, F.S. III, Jefferies, R.L., Reynolds, J.F., Shaver, G.R., and Svoboda, J. (eds.), *Arctic Ecosystems in a Changing Climate. An Ecophysiological Perspective* (pp. 301–319). San Diego: Academic Press.

Chapin, F.S. III. 1987. Environmental control over growth of tundra plants. *Ecol. Bull.* 38:69–76.

Chapin, F.S. III, Barsdate, R.J., and Barel, D. 1978. Phosphorus cycling in Alaskan coastal tundra: A hypothesis for the regulation of nutrient cycling. *Oikos* 31:189–199.

Chapin, F.S. III, Jefferies, R.L., Reynolds, J.F., Shaver, G.R., and Svoboda, J. 1992. Arctic plant physiological ecology in an ecosystem context. In: Chapin, F.S. III, Jefferies, R.L., Reynolds, J.F., Shaver, G.R., and Svoboda, J. (eds.), *Arctic Ecosystems in a Changing Climate. An Ecophysiological Perspective* (pp. 441–451). San Diego: Academic Press.

Chapin, F.S. III, and Shaver, G.R. 1985. Individualistic response of tundra plant species to environmental manipulations in the field. *Ecology* 66:564–576.

Crawford, R.M.M., Chapman, H.M., Abbot, R.J., and Balfour, J. 1993. Potential impact of climate warming on arctic vegetation. *Flora* 188:367–381.

Del Moral, R. 1983. Competition as a control mechanism in subalpine meadows. *Am. J. Bot.* 70:232–245.

DiTommaso, A., and Aarssen, L.W. 1989. Resource manipulation in natural vegetation: a review. *Vegetatio* 84:9–29.

Fox, J.F. 1992. Repsonses of diversity and growth-form dominance to fertility in Alaskan tundra fellfield communities. *Arctic Alpine Res.* 24:233–237.

Fretwell, S.D. 1977. The regulation of plant communities by food chains exploiting them. *Perspect. Biol. Med.* 20:169–185.

Grime, J.P. 1973. Competitive exclusion in herbaceous vegetation. *Nature* 242:344–347.

Haag, R.W. 1974. Nutrient limitations to plant production in two tundra communities. *Can. J. Bot.* 52:103–116.

Hairston, N.G., Smith, F.E., and Slobodkin, L.B. 1960. Community structure, population control, and competition. *Am. Nat.* 94:421–425.

Harte, J., and Kinzig, A.P. 1993. Mutualism and competition between plants and decomposers: Implications for nutrient allocation in ecosystems. *Am. Nat.* 141:829–846.

Havström, M., Callaghan, T.V., and Jonasson, S. 1993. Differential growth responses of *Cassiope tetragona*, an arctic dwarf shrub, to environmental perturbations among three contrasting high- and subarctic sites. *Oikos* 66:389–402.

Heal, O.W., and Block, W. 1987. Soil biological processes in the North and South. *Ecol. Bull.* 38:47–57.

Jonasson, S. 1986. Influence of frost heaving on soil chemistry and on the distribution of plant growth forms. *Geograf. Annaler* 68A:185–195.

Jonasson, S. 1992. Plant responses to fertilization and species removal in tundra related to community structure and clonality. *Oikos* 63:420–429.

Jonasson, S., and Callaghan, T.V. 1992. Root mechanical properties related to disturbance and stress in tundra. *New Phytol.* 122:179–186.

Jonasson, S., and Chapin, F.S. III. 1991. Seasonal uptake and allocation of phosphorus in *Eriophorum vaginatum* L. measured by labelling with ^{32}P. *New Phytol.* 118:349–357.

Jonasson, S., Havström, M., Jensen, M., and Callaghan, T.V. 1993. In situ mineralization of nitrogen and phosphorus of arctic soils after perturbations simulating climate change. *Oecologia* 95:179–186.

Jonasson, S., and Sköld, S-E. 1983. Influence of frost-heaving on vegetation and nutrient regime of polygon-patterned ground. *Vegetatio* 53:97–112.

Jónsdóttir, I.S., and Callaghan, T.V. 1990. Intraclonal translocation of ammonium and nitrate in *Carex bigelowii* using ^{15}N and nitrate reductase assays. *New Phytol.* 114:419–428.

Kane, D.L., Hinzman, L.D., Woo, M.-K., and Everett, K.R. 1992. Arctic hydrology and climate change. In: Chapin, F.S. III, Jefferies, R.L., Reynolds, J.F., Shaver, G.R., and Svoboda, J. (eds.), *Arctic Ecosystems in a Changing Climate. An Ecophysiological Perspective* (pp. 35–57). San Diego: Academic Press.

Koide, R.T. 1991. Nutrient supply, nutrient demand and plant responses to mycorrhizal infection. *New Phytol.* 117:365–386.

Körner, C. 1993. CO_2 fertilization: The great uncertainty in future vegetation development. In: Solomon, A.M., and Shugart, H.H. (eds.), *Vegetation Dynamics and Global Change* (pp. 53–70). New York, London: Chapman and Hall.

Marion, G.M., Miller, P.C., Kummerow, J., and Oechel, W.C. 1982. Competition for nitrogen in a tussock tundra ecosystem. *Plant Soil* 66:317–327.

McGraw, J.B. 1987. Experimental ecology of *Dryas octopetala* ecotypes. IV. Fitness response to reciprocal transplanting in ecotypes with differing plasticity. *Oecologia* 73:465–468.

McGraw, J.B., and Chapin, F.S. III. 1989. Competitive ability and adaptation to fertile and infertile soils in two *Eriophorum* species. *Ecology* 70:736–749.

McGraw, J.B., and Fetcher, N. 1992. Response of tundra plant populations to climatic change. In: Chapin, F.S. III, Jefferies, R.L., Reynolds, J.F., Shaver, G.R., and Svoboda, J. (eds.), *Arctic Ecosystems in a Changing Climate. An Ecophysiological Perspective* (pp. 359–376). San Diego: Academic Press.

Melillo, J.M., Callaghan, T.V., Woodward, F.I., Salati, E., and Sinha, S.K. 1990. Effects on ecosystems. In: Houghton, J.T., Jenkins, G.T., and Ephraums, J.J. (eds.), *Climate Change: The IPCC Scientific Assessment* (pp. 282–310). Cambridge: Cambridge University Press.

Nadelhoffer, K.J., Giblin, A.E., Shaver, G.R., and Laundre, J.L. 1991. Effects of temperature and substrate quality on element mineralization in six arctic soils. *Ecology* 72:242–253.

Nadelhoffer, K.J., Giblin, A.E., Shaver, G.R., and Linkins, A.E. 1992. Microbial processes and plant nutrient availability in arctic soils. In: Chapin, F.S. III, Jefferies, R.L., Reynolds, J.F., Shaver, G.R., and Svoboda, J. (eds.), *Arctic Ecosystems in a Changing Climate. An Ecophysiological Perspective* (pp. 281–300). San Diego: Academic Press.

Oechel, W.C., and Billings, W.D. 1992. Effects of global change on the carbon balance of arctic plants and ecosystems. In: Chapin, F.S. III, Jefferies, R.L., Reynolds, J.F., Shaver, G.R., and Svoboda, J. (eds.), *Arctic Ecosystems in a Changing Climate. An Ecophysiological Perspective* (pp. 139–168). San Diego: Academic Press.

Oinonen, E. 1967. Sporal regeneration of bracken in Finland in the light of the dimensions and age of its clones. *Acta. Forest. Fenn.* 83:3–96.

Oksanen, L. 1980. Abundance relationships between competitive and grazing-tolerant plants in productivity gradients in Fennoscandian mountains. *Ann. Bot. Fenn.* 17:410–429.

Oksanen, L. 1983. Trophic exploitation and arctic phytomass patterns. *Am. Nat.* 122:45–52.

Oksanen, L. 1990. Predation, herbivory, and plant strategies along gradients of primary productivity. In: Grace, J.B., and Tilman, D. (eds.), *Perspectives on Plant Competition* (pp. 445–474). San Diego: Academic Press.

Oksanen, L., Fretwell, S.D., Arruda, J., and Niemelä, P. 1981. Exploitation ecosystems in gradients of primary productivity. *Am. Nat.* 118:240–261.

Pitelka, L.F., and Ashmun, J.W. 1985. Physiology and integration of ramets in clonal plants. In: Jackson, J.B.C., Buss, L.W., and Cook, R.E. (eds.), *Population Biology and Evolution of Clonal Organisms* (pp. 399–435). New Haven, CT: Yale University Press.

Rastetter, E.B., McKane, R.B., Shaver, G.R., Nadelhoffer, K.J., and Giblin, A.E. 1993. Potential changes in carbon storage by moist arctic tundra resulting from global change: Constraints imposed by carbon-nitrogen interactions. In: Gilmanov, T., Holten, J.I., Maxwell, B., Oechel, W.C., and Sveinbjörnsson, B. (eds.), *Abstracts. Global Change and Arctic Terrestrial Ecosystems.* Norway: Oppdal.

Savile, D.B.O. 1972. *Arctic Adaptations in Plants.* Canada Dept. Agric. Research Branch. Monograph No. 6.

Shaver, G.R., and Chapin, F.S. III. 1980. Response to fertilization by various plant growth forms in an Alaskan tundra: Nutrient accumulation and growth. *Ecology* 61:662–675.

Sonesson, M., and Callaghan, T.V. 1991. Strategies of survival in plants of the Fennoscandian tundra. *Arctic* 44:95–105.

Tenhunen, J.D., Lange, O.L., Hahn, S., Siegwolf, R., and Oberbauer, S.F. 1992. The ecosystem role of poikilohydric tundra plants. In: Chapin, F.S. III, Jefferies, R.L., Reynolds, J.F., Shaver, G.R., and Svoboda, J. (eds.), *Arctic Ecosystems in a Changing Climate. An Ecophysiological Perspective* (pp. 213–237). San Diego: Academic Press.

Thompson, D.Q. 1955. The role of food and cover in population fluctuations of the brown lemming at Point Barrow, Alaska. *Trans. N. Am. Wildl. Conference* pp. 166–176.

Tilman, D. 1982. *Resource Competition and Community Structure.* Princeton: Princeton University Press.

Tissue, D.T., and Oechel, W.C. 1987. Response of *Eriophorum vaginatum* to elevated CO_2 and temperature in the Alaskan tussock tundra. *Ecology* 68:401–410.

Wookey, P.A., Parsons, A.N., Welker, J.M., Potter, J.A., Callaghan, T.V., Lee, J.A., and Press, M.C. 1993. Comparative responses of phenology and reproductive development to simulated climate change in subarctic and high arctic plants. *Oikos* 67:490–502.

20. Climate Change in Northern Latitudes: Alterations in Ecosystem Structure and Function and Effects on Carbon Sequestration

Walter C. Oechel and George L. Vourlitis

The concentration of radiatively active gases such as CO_2 and CH_4 have increased substantially since the beginning of the industrial revolution. Atmospheric CO_2 has increased from the preindustrial level of 270 ppm to the current level of 355 ppm, and is expected to double over the next 50 to 75 years (Tans et al., 1990; Watson et al., 1992). Atmospheric CH_4, which is about 20 times more reactive than CO_2 as a greenhouse gas (Lashof and Ahuja, 1990), is increasing at a rate of 0.8% to 2.0% per year (Khalil and Rasmussen, 1987), due primarily to increased population growth and associated anthropogenic emissions. Recently, however, the trend in CH_4 increase has declined, indicating either a decline in CH_4 emissions or that the concentration of tropospheric OH^-, the primary atmospheric sink for CH_4, is increasing (Khalil and Rasmussen, 1992; Khalil et al., 1993).

The increase in the concentration of these gases has the potential to increase surface temperature and affect climate on a global scale (Gates et al., 1992; Watson et al., 1992). General circulation models (GCMs) predict mean annual global surface temperature increases of approximately 1.3 to 2.3°C with a doubling of atmospheric CO_2, but the most significant increases are expected to occur in high latitudes (Gates et al., 1992; Manabe and Stouffer, 1993; Meehl et al., 1993; Mitchell et al., 1989). Specifically, warming is expected to occur in northern latitudes outside the North Atlantic region where the well-mixed ocean tends to buffer atmospheric forcing (Meehl et al., 1993; Schlesinger and Xingjian, 1991). This warming, however, is expected to occur asymmetrically with the greatest increases in temperature occurring during the winter and spring months (6 to 15°C), due primarily to decreases in snow- and icecover, which have a pronounced effect on surface albedo (Gates et al., 1992; Manabe and Stouffer, 1993; Meehl et al., 1993). Summer temperature increases in high-latitude regions are expected to be more modest (e.g., 2 to 4°C) under doubled atmospheric CO_2 (Gates et al., 1992).

Northern latitudes are considered more sensitive to radiative forcing, due to changes in atmospheric and surface properties that act synergistically with increases in surface temperature. The atmosphere over polar regions contains relatively little water vapor (Hinzman and Kane, 1992) and as a result is poor at trapping outgoing tropospheric radiation. Small changes in cloudiness, which are predicted for high-latitude regions under CO_2-induced climate change (Mitchell et al., 1989), would therefore have a relatively large effect on the surface temperature (Chapman and Walsh, 1993; Gates et al., 1992; Groisman et al., 1994; Mitchell et al., 1989). Reductions in the existence and/or distribution of polar ice- and snowcover also are expected to positively feed back on the increase in surface temperature by reducing the surface albedo and thus increasing the amount of absorbed shortwave radiation (Gates et al., 1992; Groisman et al., 1994; Hinzman and Kane, 1992; Mitchell et al., 1989).

Warming is expected to affect regional climate patterns, energy balance, and hydrology of northern ecosystems (Gates et al., 1992; Groisman et al., 1994; Hinzman and Kane, 1992; Kane et al., 1990; Kane et al., 1992; Waelbroeck, 1993). Precipitation is predicted to increase at high latitudes (Gates et al., 1992). However, higher temperatures may lead to increased evapotranspiration and reductions in soil moisture during the summer growing season (Hinzman and Kane, 1992). In addition, warmer winter and spring temperatures may result in an earlier spring thaw, which may translate into reductions of soil water during the summer growing season due to earlier drainage and evapotranspiration and runoff. Regardless of precipitation changes, active layer depth (depth of the unfrozen soil layer that overlies permafrost during the warm summer months) is expected to increase due to high-latitude warming (Hinzman and Kane, 1992; Kane et al., 1990; Kane et al., 1992; Waelbroeck, 1993). Increases in active layer depth will subsequently feed back on the hydrology of these ecosystems in the form of lower water table depth (assuming summer precipitation remains unchanged), changes in the soil water holding capacity, and the amount of subsurface flow, alterations in the soil temperature regime, and ultimately, changes in the rates of soil and plant metabolic processes (Billings et al., 1982; Gorham, 1991; Hinzman and Kane, 1992; Oechel and Billings, 1992; Silvola, 1986; Waelbroeck, 1993).

High-latitude ecosystems are particularly vulnerable to climate change due to the large C stocks in northern latitude soils and the predominance of permafrost. Northern ecosystems (arctic, boreal forest, and northern bogs) represent approximately 11% of the total global land area; however, they contain approximately 25 to 33% of the total world soil C pool in the permafrost and seasonally thawed soil active layer (Billings, 1987; Gorham, 1991; Miller et al., 1983; Oechel and Billings, 1992). Arctic tundra ecosystems alone contain approximately 12% of the global soil pool even though they only make up about 5% of the total land area (Billings, 1987; Miller et al., 1983; Schlesinger, 1991).

High-latitude warming has been recently reported in arctic Alaska, Canada, and the Former Soviet Union (FSU). Thermal profiles of permafrost indicate a temperature rise of 2 to 4°C across the north slope of Alaska and throughout northern Canada within the last century, or possibly even during the last few decades (Beltrami and Mareshal, 1991; Lachenbruch and Marshall, 1986). Northern latitude weather records indicate a similar increase in annual surface temperature (Chapman and Walsh, 1993; Karl et al., 1991; Oechel et al., 1993). The annual increase is due primarily to warmer winter and spring temperatures, which have increased on the order of 4.5°C over the last three decades (Chapman and Walsh, 1993). Although it is difficult to decouple low-frequency, natural climatic variability from greenhouse gas-induced climate change (Meehl et al., 1993), the observed trends are consistent with simulations based solely on atmospheric forcing by increased trace gas concentrations (Gates et al., 1992; Manabe and Stouffer, 1993; Meehl et al., 1993).

An increasing trend in growing season length over the last 65 years has been observed in several regions throughout Alaska (Sharratt, 1992), which corresponds with (at least over the last two decades) reductions in the extent of snowcover throughout North America, Europe, and Asia (Groisman et al., 1994). Similarly, increases in mean annual precipitation for North America and the former Soviet Union, on the order of those predicted by greenhouse gas driven GCMs, have been observed over the last 120 years (Bradley et al., 1987), with much of the increase observed during the winter and spring months.

The response of high-latitude ecosystems to increased temperature and associated climate change will largely be dependent on how plants and ecosystems respond to elevated CO_2, how soil microbial processes and plant productivity are affected by changes in temperature and hydrology, and how community dynamics are affected by changing thermal, nutrient, and competitive regimes (Chapin et al., in press; Melillo et al., 1990; Oechel and Billings, 1992; Oechel et al., 1994; Oechel et al., 1993; Oechel and Vourlitis, 1994; Shaver et al., 1992). In turn, these plant and ecosystem responses have the potential to feed back (both positively and negatively) on climate change, due for example, to changes in albedo in response to changes in vegetative cover (negative feedback; Bonan et al., 1992) or to increases in the net loss of CO_2 to the atmosphere following a period of warming and soil drying (positive feedback; Oechel et al., 1993) or to increase in CO_2 storage due to increased organic matter accumulation (negative feedback; Shaver et al., 1992).

Our goal in this discussion is to outline the current knowledge regarding the plant, community, and ecosystem response to elevated CO_2 and associated climate change in high-latitude ecosystems. We focus primarily on the arctic because a great deal of both long- and short-term research has been carried out, which will facilitate the analysis of how high-latitude ecosystems respond to changes in trace gas concentration and climate. We also focus on the effects of climate change on trace gas flux (CO_2 and CH_4) because measurements of gas flux (primarily CO_2) integrate both above- and belowground processes, and alterations in the net gaseous flux can feed back on climate change.

Direct Effects of Elevated CO_2

On the positive side, increases in atmospheric CO_2 have the potential to increase photosynthetic rates and biomass production on a global scale (Idso et al., 1991). Over short time scales, leaf biochemistry and physiology control photosynthetic rates, and for terrestrial plants, atmospheric CO_2 concentration appears to universally limit short-term photosynthesis rates (Drake, 1992; Stitt, 1991; Strain, 1992). However, the long-term photosynthetic response to elevated atmospheric CO_2 concentration is much more variable, due to genetic (Graham et al., 1991; Oechel and Billings, 1992), anatomical (Woodward, 1987) growth, demographic, community, and ecosystem level controls and feedbacks (Oechel et al., 1993). These adjustments can be homeostatic; that is, they can tend to maintain ecosystem photosynthetic rates at similar levels, or they can further exaggerate the effect of elevated CO_2 on photosynthesis by increasing the magnitude of the plant or ecosystem response.

Initial exposure to elevated CO_2 results in significant increases in photosynthesis rates of many arctic species; however, this stimulation is relatively short lived. In short-term (weeks to months) growth chamber and in situ studies, homeostatic adjustment of photosynthetic rates to elevated CO_2 occurred in as little as 3 weeks (Tissue and Oechel, 1987) and within 3 months (Oberbauer et al., 1986). Prolonged enhancement of photosynthesis by elevated CO_2 generally requires that nutrients are abundant, which translates directly into increased plant growth (Chapin et al., 1980) and sink activity. Sink limitations, which appear to be the mechanism for the rapid homeostatic adjustment in photosynthetic rates, feed back to limit the amount of ribulose bis-phosphate (RuBP), resulting in a concomitant reduction in photosynthesis (Arp, 1991; Tissue and Oechel, 1987).

Elevated CO_2 also appears to have a minimal effect on biomass accumulation, although increases in the biomass allocation to belowground tissue have been observed (Oberbauer et al., 1986; Tissue and Oechel, 1987).

Root growth, which represents a significant sink for fixed C under ambient CO_2 concentration (Billings et al., 1977), may represent an even larger pool for excess carbohydrates produced under elevated CO_2 (Larigauderie et al., 1988). Changes in relative growth rate of plants exposed to elevated CO_2 are minimal, as are changes in leaf area (Oberbauer et al., 1986; Tissue and Oechel, 1987). Although elevated CO_2 may not increase the leaf area of mature tillers, it apparently acts to increase the total number of tillers produced (Tissue and Oechel, 1987), which could feed back on the competitive dynamics of arctic communities by increasing the mutual shading and possible competitive exclusion of intertussock plants (Körner, 1993).

Early simulation models indicate that C accumulation would increase minimally in a doubled CO_2 environment over the next 50 years, and by 3 to 4% under elevated CO_2 and a 4°C increase in temperature (Miller et al., 1983). In a more recent simulation, however, C accumulation under elevated CO_2 alone is expected to increase by 12% over the next 50 years (Rastetter et al., 1992). The cause for the increase in C accumulation is thought to be due to greater N-use efficiency of plants exposed to elevated CO_2, so more C should be sequestered per unit N mineralized from decomposition (Oechel and Billings, 1992; Rastetter et al., 1992; Shaver et al., 1992). Other simulations, however (e.g., Terrestrial Ecosystem Model), indicate that there will be no stimulation of Net Primary Productivity (NPP) by CO_2 without concomitant increases in nutrient (especially N) availability (Melillo et al., 1993).

In spite of the predicted increases in C accumulation by arctic ecosystems under elevated CO_2, growth chamber and in situ studies clearly show that ecosystem C accumulation is only briefly stimulated by elevated CO_2, and of limited extent (Billings et al., 1984; Grulke et al., 1990; Hilbert et al., 1987; Oechel et al., 1994; Oechel and Vourlitis, in press; Prudhomme et al., 1982). In a 3-year in situ study where most tussock tundra ecosystems were exposed to double-ambient CO_2 concentrations (Grulke et al., 1990; Oechel et al., 1992; Oechel et al., 1994; Oechel and Vourlitis, in press), the initial stimulation in net ecosystem productivity declined during the first year, and by the third year, ambient and high CO_2 grown plots had similar net ecosystem fluxes (Figure 20.1). The rapid, ecosystem-level downregulation of photosynthesis is thought to be due to the overriding effects of nutrient limitation that cause reduced growth and sink formation (Grulke et al., 1990; Oechel et al., 1994). As with the downregulation of leaf-level photosynthesis, whole ecosystem downregulation can occur rapidly as well (Figure 20.1) in situations where productivity is limited more by other resources, especially nutrients. In the arctic, plants and whole ecosystems are likely to be limited by nutrients (Chapin et al., in press; Shaver and Chapin, 1980; Shaver and Chapin, in press), and as a result, the direct effects of CO_2 over the short and medium term will likely be minimal (Grulke et al., 1990; Oechel and Billings, 1992; Oechel et al., 1994).

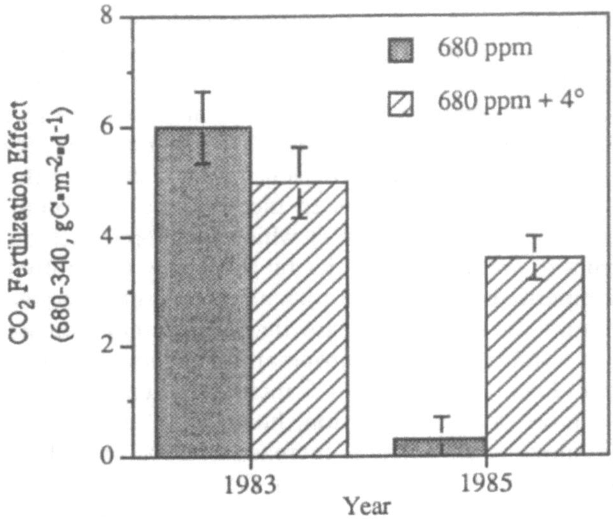

Figure 20.1. The CO_2 fertilization effect coupled with a 4°C increase in ambient temperature (response at 680, response at 340) on the net CO_2 flux of moist tussock tundra ecosystems during the first and last year of a 3-year in situ elevated CO_2 experiment at Toolik Lake, Alaska. Data are means ± 1 SE, $n = 3$ chambers per treatment (from Oechel et al., 1994).

Our knowledge of the long-term (decade to century) response to elevated CO_2 stems largely from observations of low- and high-altitude plants (e.g., Körner, 1993; Körner and Diemer, 1987) and from recent work near cold, CO_2-emitting springs in Iceland (Cook and Oechel, unpublished data). Elevational studies (Körner and Diemer, 1987) are based on the observation that low-elevation plants experience higher partial pressures of CO_2 compared to high-elevation plants. In general, estimates by Körner and Diemer (1987) indicate that an additional $100 \mu l \, l^{-1}$ of atmospheric CO_2 would increase net photosynthesis by approximately 21% and 31% in low- and high-elevation plants, respectively, whereas an additional $200 \mu l \, l^{-1}$ would result in an even smaller relative photosynthetic gain. These results suggest that the reduction in photosynthetic capacity observed under short-term exposure to elevated CO_2 is apparent even after centuries.

Preliminary results from cold CO_2 springs are similar (Cook and Oechel, unpublished data). One recently studied spring is near Olafsvik, Iceland, where plants have presumably been exposed to elevated levels of CO_2 for centuries. Photosynthetic measurements of the dominant C_3 grass, *Nardus stricta*, indicate a reduction in the photosynthetic capacity of high CO_2 grown plants, due primarily to reductions in Rubisco content, final Rubisco activity, and level of chlorophyll concentration (Cook and Oechel, unpublished data). Differences in growth and phenology were also apparent, as plants growing near the CO_2 spring had a lower leaf area index, produced fewer flowers, had slower flower development, and experienced earlier senescence than plants exposed to long-term levels of ambient CO_2 (Cook and Oechel, unpublished data).

At the community level, long-term exposure to elevated CO_2 may be affecting species composition as well. Using the Braun-Blanquet method to compare species abundances in small plots, *Viola palustris*, *Salix herbacea*, *Carex nigra*, and *Carex bigelowii* were all slightly more abundant adjacent to the spring, whereas *Thalictrum alpinum*, *Galium uliginosum*, and *Cerastium fontanum* were more abundant away from the spring (Cook, Sveinbjornsson, and Oechel, unpublished data).

In light of these results, it is unlikely that C accumulation will increase markedly due to the direct effects of CO_2 alone. Rather, an increase in nutrient availability will have to occur to allow the sink required for the utilization of the additional carbohydrates produced under elevated CO_2 (Grulke et al., 1990; Melillo et al., 1993; Miller et al., 1983; Tissue and Oechel, 1987). In the in situ manipulations at Toolik Lake, tussock tundra exposed to elevated CO_2 and a 4°C increase in ambient temperature were net sinks during the entire 3-year study, with little or no homeostatic adjustment of ecosystem C assimilation observed in response to CO_2 enrichment and elevated temperature (Figure 20.1). These results are consistent with model simulations using the TEM model coupled with GCM scenarios of high-latitude climate change; however, the increase in NPP observed in the simulations was more a function of increased temperature and subsequent increases in N mineralization, and not due to the direct effects of elevated CO_2 (Melillo et al., 1993). Increased temperature may directly increase sink strength and activity, as growth and respiration tend to increase at moderately high temperatures in some arctic species (Chapin and Shaver, 1985; Kummerow and Ellis, 1984; Limbach et al., 1982). More importantly, elevated soil temperatures caused by the 4°C increase in ambient temperature may have led to increased mineralization and nutrient availability (Nadelhoffer et al., 1991), resulting in greater sink activity, thereby allowing the utilization of additional photosynthate produced under elevated CO_2 (Grulke et al., 1990; Tissue and Oechel, 1987).

Effects of Climate Change on Net Ecosystem CO_2 Flux

Net primary productivity in northern ecosystems was thought to exceed heterotrophic respiration in the recent and historical geologic past due to the predominance of cold, wet soils (Gorham, 1991; Marion and Oechel, 1993; Oechel and Billings, 1992; Schell, 1983). As a result, arctic ecosystems were estimated to be net sources to atmospheric C of about 0.1 to $0.3\,Pg\,C\,yr^{-1}$ ($1\,Pg = 10^{15}\,g$) (Gorham, 1991; Marion and Oechel, 1993; Miller et al., 1983; Oechel and Billings, 1992; Schell, 1983). Plot level measurements of CO_2 flux made over the past 5 years in arctic Alaska using a passive, closed system chamber sampling technique (Vourlitis et al., 1993) indicate that tussock and wet sedge tundra are now sources of CO_2 to the atmosphere (Grulke et al., 1990; Oechel et al., 1993; Oechel and Vourlitis, 1994; Oechel et al., in press). Tussock tundra ecosystems are now losing on average $112\,g\,C\,m^{-2}\,yr^{-1}$ during the growing season (Figure 20.2). Similarly, wet sedge ecosystems are now either slight sinks for atmospheric CO_2 or roughly in balance (Oechel et al., 1993; Oechel and Vourlitis, 1994; Oechel et al., in press; Figure 20.2; Oechel and Vourlitis, in press). Compared to earlier estimates of C balance using harvest, cuvette, and aerodynamic techniques (Chapin et al., 1980; Coyne and Kelley, 1975; Miller et al., 1983), the current sink strength of tussock and coastal wet sedge ecosystems on the North Slope of Alaska has diminished by approximately 135 and

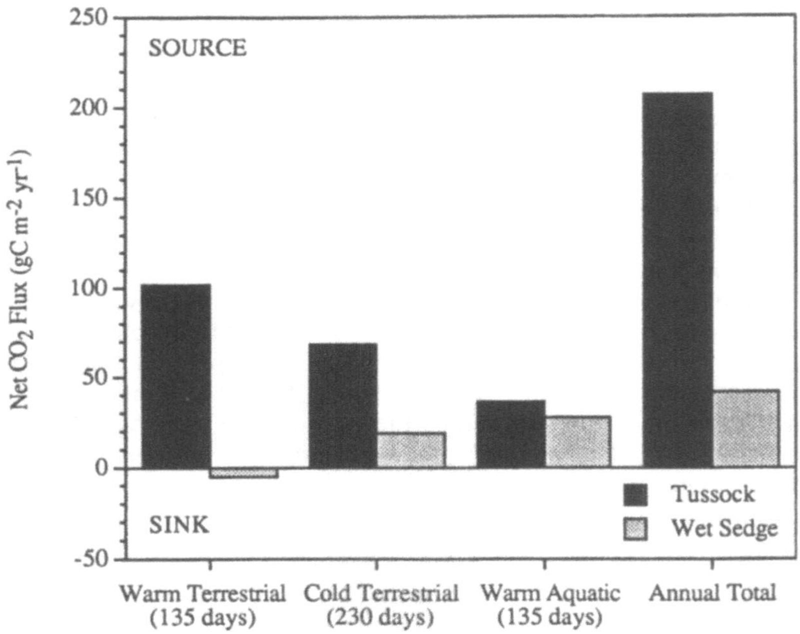

Figure 20.2. Net CO_2 flux during the warm (mid-May to mid-September) and cold seasons (mid-September to mid-May) of terrestrial and aquatic tussock (Toolik Lake) and coastal wet sedge (Prudhoe Bay, West Dock) tundra ecosystems on the North Slope of Alaska. Warm season terrestrial data are from the 1990, 1991, 1993, and 1994 field seasons and are integrated over an average growing season length of 135 days, where the length of the growing season is estimated between the date of snowmelt in the spring and the date of soil freeze-up in the fall (Oechel et al., 1993). Cold season terrestrial data are from the 1993–94 cold season (Oechel et al., unpublished data), and warm season aquatic data are from Kling et al. (1991).

$73\,g\,C\,m^{-2}\,y^{-1}$, respectively, over the last two decades (Oechel and Vourlitis, 1994; Oechel and Vourlitis, in press; Oechel et al., in press). An additional $24\,g\,C\,m^{-2}\,y^{-1}$ may be lost from lakes and streams (Kling et al., 1991), and 19 to $68.5\,g\,C\,m^{-2}$ during the cold season (October to May; Zimov et al., 1993; Oechel et al., unpublished data), thus increasing the actual amount of C lost from arctic ecosystems annually (Figure 20.2). Given the large C stores in arctic soils, this loss of C may represent a significant positive feedback to atmospheric CO_2 concentration and concomitant global change.

The change in net C balance of arctic ecosystems is attributed to a change in water balance associated with the recently reported increase in high-latitude surface temperature, and not from the direct effects of temperature alone (Oechel et al., 1993; Oechel and Vourlitis, 1994; Oechel and Vourlitis, in press; Oechel et al., in press). These observations are supported by a variety of laboratory and in situ experiments (Billings et al., 1982; Freeman et al., 1993; Funk et al., 1994; Silvola, 1986). In laboratory manipulations with wet sedge tundra microcosms subjected to a 10-cm decrease in water table depth, net ecosystem CO_2 incorporation decreased by $212\,g\,C\,m^{-2}$, resulting in a net CO_2 efflux of $84\,g\,C\,m^{-2}$ (Billings et al., 1982). Other experiments where water table depth is lowered, or when soil cores are subjected to drying cycles, indicate that the rate of CO_2 efflux is two to five times greater during drained conditions (Freeman et al., 1993; Funk et al., 1994; Glenn et al., 1993; Hogg et al., 1992).

Similarly, draining of northern peatlands has been shown to convert these ecosystems from a CO_2 sink of approximately $25\,g\,C\,m^{-2}\,yr^{-1}$ to a source of about $150\,g\,C\,m^{-2}\,yr^{-1}$ to the atmosphere (Silvola, 1986). Other in situ experiments in wet sedge tundra ecosystems where water table depth is lowered on average by 10 cm and temperature is elevated indicate that a lowering of the water table results in a seasonal (June to September) loss of between 25 and $45\,g\,C\,m^{-2}$ in drained plots, whereas control plots were found to be net sinks of approximately $25\,g\,C\,m^{-2}\,season^{-1}$ (Figure 20.3). Much of this loss was found to occur in the early portion of the growing season, when primary production is still low, and again during the late portion of the growing season when active layer depth is at a seasonal maximum (Oechel et al., unpublished data). Elevated temperature, in and of itself, was found to be statistically less important in controlling net CO_2 flux than water table, although elevated temperature appeared to accentuate the amount of net CO_2 loss in drained plots (Figure 20.3).

The importance of soil moisture in controlling net CO_2 flux of northern ecosystems and soils is well known, but the effect of temperature on net CO_2 flux is complex because soil temperature is not independent of soil moisture. Soil decomposition is affected by temperature (Moorhead and Reynolds, 1993; Nadelhoffer et al., 1991) whereas photosynthesis is only weakly influenced by temperature (Chapin, 1983; Tieszen, 1973). The temperature sensitivity of soil microorganisms appears to vary according to the soil moisture status. In general, decomposition rates are slow and

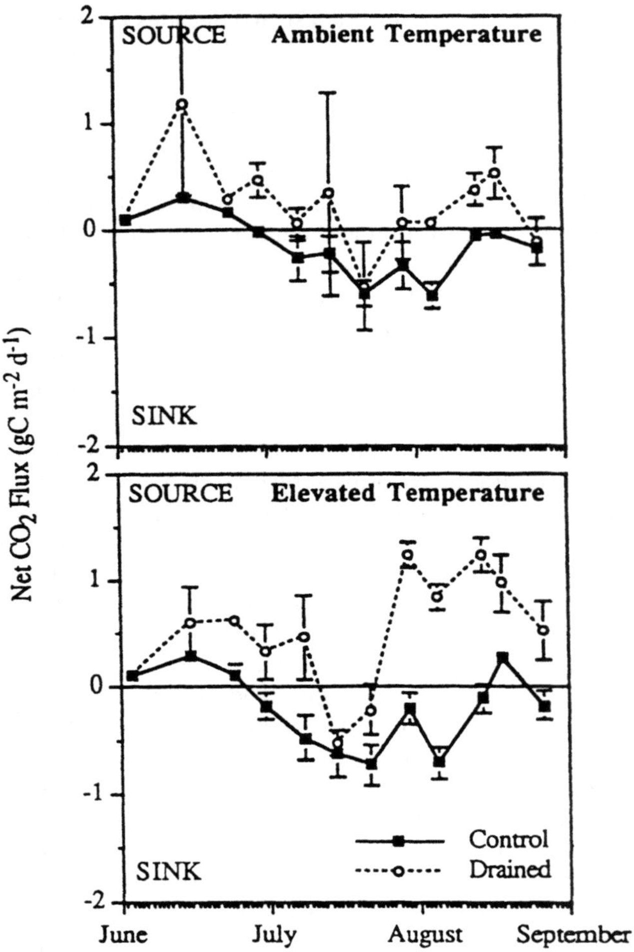

Figure 20.3. The effect of draining and elevated soil temperature on the net CO_2 flux of coastal, wet sedge tundra ecosystems at Prudhoe Bay, West Dock, on the North Slope of Alaska during the 1994 growing season (June to September). Open circles correspond to plots subjected to a 10-cm lowering of the water table, and closed squares correspond to control plots subjected to ambient water table depths. Data are means ± 1 SE, $n = 2$ plots per treatment.

relatively inresponsive to temperature under conditions of low soil moisture content (≤20% of dry mass), whereas temperature sensitivity increases until soil moisture content reaches approximately 200% (Flanagan and Veum, 1992; Nadelhoffer et al., 1991). Under saturated conditions, however, the temperature sensitivity of soil microbial populations is low, and respiration rates appear to be limited more by poor soil aeration (Billings et al., 1982; Funk et al., 1994; Moorhead and Reynolds, 1993; Nadelhoffer et al., 1992).

Increased nutrient mineralization is also expected as arctic soils become warmer and drier (Marion and Black, 1987; Marion and Miller, 1982; Nadelhoffer et al., 1991). In laboratory incubations of tussock and wet sedge tundra soils, N mineralization increased by approximately 45% with a 12°C increase in soil temperature (3 to 15°C; Nadelhoffer et al., 1991). Increased mineralization is expected to increase plant growth, as results from fertilization experiments of arctic plants indicate that increased nutrient availability acts to stimulate tissue production rather than increase leaf photosynthesis. Because essentially all of the N taken up by arctic plants is supplied by mineralization of organic matter, increased N availability through climate change is likely to stimulate whole ecosystem productivity and C accumulation, at least in the long term (Chapin and Shaver, 1985; Shaver et al., 1992).

However, in the short term (<5 years) decomposition and net mineralization are not closely coupled in tundra soils. These two processes of decomposition and mineralization usually increase or decrease together but at different rates so the ratio of respiration to the net N mineralization is quite unpredictable (Moorhead and Reynolds, 1993; Nadelhoffer et al., 1991). Rates of N mineralization appear to be insensitive to variations in soil temperature and soil moisture status over the naturally occurring range of conditions observed; however, microbial respiration is considerably more sensitive to changes in temperature and soil moisture regimes (Flanagan and Bunnell, 1980; Marion and Black, 1987; Marion and Miller, 1982; Moorhead and Reynolds, 1993; Nadelhoffer et al., 1991). The relative independence of changes in soil respiration and net N mineralization is attributed to immobilization of mineralized N by soil microorganisms, as the rate of net N mineralization is more a function of litter quality and the microbial C/N ratio than the overall rate of soil decomposition (Moorhead and Reynolds, 1993). In simulations where soil moisture, temperature, and growing season length were varied over a range of 800%, 4.5°C, and 90 days, respectively, immobilization accounted for nearly all the net N mineralized (Moorhead and Reynolds, 1993). Although soil C/N ratios are reported to be relatively low (15:1; Giblin et al., 1991; Nadelhoffer et al., 1991; Shaver et al., 1992), ratios of C mineralized/N mineralized range between 800:1 and 5000:1, again due to rapid immobilization of mineralized N (Moorhead and Reynolds, 1993).

This phenomenon might easily explain why a concomitant increase in NPP has not been observed with the recently reported change in net CO_2 flux, at least in the short term (Grulke et al., 1990; Oechel et al., 1993; Oechel and Vourlitis, 1994; Oechel and Vourlitis, in press; Oechel et al., in press). Because net CO_2 flux is the difference between gross ecosystem production (GEP) and whole ecosystem respiration (WER), the change in net CO_2 flux is undoubtedly due to a relatively greater increase in the rate of WER compared to GEP. Results from in situ experiments support this hypothesis, as WER is found to increase by 86% in response to a 10-cm reduction in water table depth and elevated temperature, whereas GEP changes by only 30% (Oechel et al., unpublished data). Presumably, any additional N mineralized due to the reduction in the water table depth and elevated temperature is rapidly immobilized by soil microorganisms, resulting in an available N pool that is insufficient to significantly stimulate GEP.

Alternatively, in reciprocal transplant experiments Fetcher and Shaver (1990) found that although populations of *Eriophorum vaginatum* are highly plastic in their growth responses to changes in temperature (degree days), southern populations were significantly more plastic than northern populations. Because of the lower temperature sensitivity of northern populations, productivity of northern populations may respond more slowly to climatic warming than their southern counterparts (Fetcher and Shaver, 1990). This result may further explain why the productivity of arctic moist tussock tundra ecosystems has not been observed to increase following the recently reported trends in high-latitude warming, thus explaining the recent change in net CO_2 flux (Grulke et al., 1990; Oechel et al., 1993; Oechel and Vourlitis, 1994; Oechel and Vourlitis, in press; Oechel et al., in press). Although this loss may be transient in nature (Oechel et al., 1993), the future CO_2 balance undoubtedly will be a function of how nutrient availability, and the subsequent plant response, are affected by climate change and the time scales over which these changes occur (Oechel and Billings, 1992; Oechel and Vourlitis, 1994; Smith and Shugart, 1993).

Effects of Climate Change on CH_4 Flux

Quantifying sources and sinks of CH_4 is of primary concern, as CH_4 is approximately 20 times more reactive than CO_2 as a greenhouse gas (Lashof and Ahuja, 1990; Ramanathan et al., 1985). Major CH_4 sources (60 to 100 Tg yr^{-1}; 1 Tg = 10^{12} g) include wetlands, rice production, and ruminants, whereas minor sources (10 to 60 Tg yr^{-1}) include landfills, coal mining, termites, biomass burning, urban areas, sewage disposal, natural gas leaks, lakes, oceans, and tundra (Khalil et al., 1993).

Natural wetlands are important sources of CH_4 to the global atmosphere (Mathews and Fung, 1987), and may be responsible for nearly 30% of the CH_4 produced annually. Analysis of global CH_4 emissions indicates that more than half of the annual wetland emission, or approximately 50 Tg of C, is from forested and nonforested wetlands from 50 to 70° N (Mathews and Fung, 1987). This high rate of emission is undoubtedly due to the presence of large soil C stocks (Billings, 1987; Gorham, 1991; Miller et al., 1983) and frequently inundated soils.

High-latitude climate change may significantly affect future CH_4 emissions by altering both the rate of CH_4 gas emission and the extent of wetland surface area (Moreas and Khalil, 1993). Changes in soil temperature and moisture can have significant implications for the production of CH_4 and subsequent emission. Methane efflux in northern ecosystems is often positively correlated with soil moisture and water table depth (Bubier et al., 1993; Funk et al., 1994; Moore et al., 1994; Sebacher et al., 1986; Svensson and Rosswall, 1984; Torn and Chapin, 1993; Vourlitis et al., 1994), as soil water content directly affects soil O_2 content and redox potential (Paul and Clark, 1989). This response, however, is often variable, as linear (e.g., Vourlitis et al., 1994) and curvilinear relationships between soil water content and O_2 redox potential (e.g., Sebacher et al., 1986) have been reported. Methane efflux has been shown to decrease markedly in experiments where water table depth is decreased (Funk et al., 1994), and increase as water table depth is returned to within its normal range (Freeman et al., 1993). In other experiments where water table depth was manipulated, CH_4 efflux appeared to be independent of water table depth (Glenn et al., 1993; Hogg et al., 1992).

It appears that there is no single variable that explains the variance observed in CH_4 flux (Reeburgh et al., 1994). Although soil moisture is important, other factors such as soil temperature and substrate quality undoubtedly interact with soil moisture to influence rates of CH_4 emissions (Reeburgh et al., 1994; Valentine et al., 1994). Soil temperature represents an important environmental control on CH_4 flux (Bartlett et al., 1992; Moore et al., 1994; Morrissey and Livingston, 1992; Svensson and Rosswall, 1984; Svensson and Rosswall, 1984); however, correlations between CH_4 efflux and temperature are often weak (Valentine et al., 1994). In addition, plant stems may be important conduits for soil CH_4 release to the atmosphere, as CH_4 efflux through plant stems may bypass oxidation at the soil surface (Morrissey and Livingston, 1992; Sebacher et al., 1985). The amount of CH_4 efflux through hydrophytes is a function of plant surface area, density, and ecosystem productivity (Morrissey and Livingston, 1992; Whiting and Chanton, 1993).

The effects of climate change of the CH_4 budget of high-latitude ecosystems are complex. If soil moisture increases due to higher summer precipitation (Gates et al., 1992), then the source strength of these ecosystems is likely to increase (Morrissey and Livingston, 1992; Reeburgh et al., 1993; Roulet et al., 1992). However, if soil drying occurs in response to increased evapotranspiration and reductions in summer precipitation (Hinzman and Kane, 1992), then CH_4 oxidation, which already plays an important role in the CH_4 budget in dryer boreal forest and tundra ecosystems, will undoubtedly increase (Reeburgh et al., 1994). In fact, Reeburgh et al. (1993) estimate that CH_4 oxidation currently accounts for 30% of the CH_4 produced in high-latitude wetlands, and dryer soil conditions may actually convert boreal and arctic ecosystems from a source of atmospheric CH_4 to a sink (Reeburgh et al., 1994).

Effects of Climate Change on the Distribution of High-Latitude Ecosystems

Climate change scenarios indicate that temperature regimes will shift northward (Gates et al., 1992; Manabe and Stouffer, 1993; Meehl et al., 1993), whereas net radiation received at the earth's surface may either increase or decline due to increased cloudiness (Schlesinger and Xingjiang, 1991) and changes in albedo (Bonan et al., 1992). Current boreal forest distributions are thought to coincide with long-term averages in, particularly, temperature and radiation (Bonan and Shugart, 1989). As temperature bands shift northward, model simulations indicate that vegetation boundaries and distributions may shift northward as well (Bonan et al., 1992; Melillo et al., 1990; Neilson, 1993; Smith and Shugart, 1993). Paleobotanical evidence support this claim, as pollen records indicate that the vegetation in central and northern Alaska during the warmer mid-Holocene was composed primarily of a *Picea*-, *Betula*-, and *Alnus*-dominated forest, whereas during the cooler late-Holocene, the vegetation became more similar to the modern flora (Billings, 1987).

Although there is general agreement between long-term average climatic parameters and the distribution of boreal forest and arctic ecosystems, it is likely that extreme, or anomalous events such as the short-term fluctuations in climate and fire frequency are more important at limiting plant processes, especially at the boreal forest/arctic ecotone (Bonan and Shugart, 1989; Landhäusser and Wein, 1993). Climate change scenarios indicate that the semi-decadal variability is expected to be high as well due to natural climate variability (Meehl et al., 1993). In addition, increases in fire frequency are expected due to soil drying associated with higher temperature (Hengeveld, 1991), and indeed, data collected by the Alaska Fire Service during the last 37 years indicate that an increase in fire frequency in Alaska is occurring (Figure 20.4). Landhäusser and Wein (1993) found that fire can significantly accelerate the conversion of previously unforested subarctic tundra ecosystems to forested boreal ecosystems by removal of the insulating surface organic layer and subsequent exposure of mineral soil, which promotes the thawing on permafrost and provides favorable conditions for the germination and establishment of some boreal tree species.

The extent to which these vegetation changes affect the C sequestration of tundra and boreal ecosystems is largely unknown, due to the inherent long-term nature of vegetation change. Results from simulation models are mixed. Smith and Shugart (1993) suggest that ecosystems could represent a significant source of CO_2 to the atmosphere following a step change in

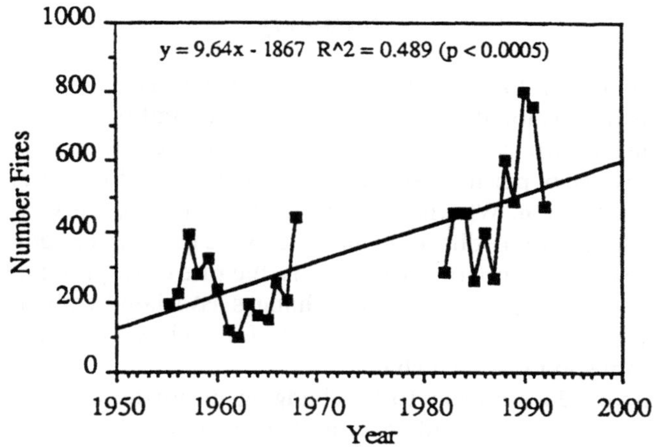

Figure 20.4. Number of reported wildfires per year in Alaska between 1955 and 1992. Missing data between 1969 and 1981 included false alarms and were omitted from this analysis. Data provided by B. Cella of the National Park Service.

climate, due to the more rapid response in the rate of decomposition relative to rates of succession and immigration of vegetation. The redistribution of vegetation toward the poles is predicted to result in a decrease in the extent (area) of tundra and boreal forest (Melillo et al., 1990; Neilson, 1993). Arctic tundra is estimated to decline by an estimated 51 to 75%, whereas boreal forest is estimated to decline on the order of 38 to 64% (Neilson, 1993). The conversion of tundra to boreal forest is estimated to increase C sequestration; however, the conversion of boreal forest to, most likely, boreal and temperate savanna is thought to result in a significant net loss of C to the atmosphere (Neilson, 1993).

Observations of soil cores taken from the North Slope of Alaska indicate that the long-term C accumulation of wet sedge tundra ecosystems should increase in response to high-latitude warming. For example, C accumulation during the mid-Holocene (5500 to 4170 years before present, BP) was significantly greater compared to the cooler late-Holocene (4170–0 years BP; Billings, 1987; Marion and Oechel, 1993). Although it is impossible to determine if the increased C accumulation observed in the middle-Holocene was due to climate and/or vegetation (Marion and Oechel, 1993), or to the normal progression of soil formation and primary succession following deglaciation of the Laurentide Ice Sheet (Harden et al., 1992), it is striking to note that changes in species composition coincide with the change in climate observed between the middle- and late-Holocene (Billings, 1987). These data indicate that following the transient losses of C in the long term, changes in vegetation should result in an increase in C sequestration in northern latitudes.

Summary

Climate change is likely to be amplified in high-latitude regions due to alterations in albedo, atmospheric composition, and permafrost. Northern ecosystems, in turn, are sensitive to climate change due to the large soil C stocks, presence of permafrost, and the importance of hydrology in controlling C sequestration. Whether the arctic and boreal forest will represent a positive or negative feedback on global climate change is largely a function of how drastic the temperature change is, the rate of response of vegetation and soils to the complex, interacting changes associated with an increase in temperature, and the time scales required for each level of biological organization to respond to such changes.

The climate does appear to be changing in northern regions, and this change has apparently resulted in a change in ecosystem function where arctic ecosystems are now net sources of CO_2 to the atmosphere. Although this change is thought to be transient, there is currently no evidence as to how long it will last. Thus far, plant productivity appears to be relatively unresponsive to the observed warming. Changes in the structure of high-latitude ecosystems due to the northward migration of vegetation are also expected in the long term. However, the time scale for this migration is extremely long compared to the relatively instantaneous response of soil microorganisms and the extant vegetation, and results from current simulations based on climate change scenarios are mixed in regard to whether C sequestration will increase or decrease within the next 10 to 20 years.

We are performing an unreplicated, uncontrolled experiment on a global scale. Although much research has been conducted over the past 20 years on the structure and function of arctic ecosystems, we are still are far away from being able to predict accurately the response of northern ecosystems to climate change. Much of this ignorance stems from the mulitvariate nature of natural systems, and the multitude of interactions and feed backs that are difficult to define with field observations, experimentation, and even modeling approaches.

References

Arp, W.J. 1991. Effects of source-sink relations on photosynthetic acclimation to elevated CO_2. *Plant, Cell Environ.* 14:869–875.

Bartlett, K.B., Crill, P.M., Sass, R.L., Harriss, R.C., and Dise, N.B. 1992. Methane emissions from tundra environments in the Yukon-Kuskokwim Delta, Alaska. *J. Geophys. Res.* 97:16645–16660.

Beltrami, H., and Mareshal, J.C. 1991. Recent warming in eastern Canada inferred from geothermal measurements. *Geophys. Res. Lett.* 18:605–608.

Billings, W.D. 1987. Carbon balance of Alaskan tundra and taiga ecosystems: Past, present, and future. *Quat. Sci. Rev.* 6:165–177.

Billings, W.D., Luken, J.O., Mortensen, D.A., and Peterson, K.M. 1982. Arctic tundra: A source or sink for atmospheric carbon dioxide in a changing environment? *Oecologia* 53:7–11.

Billings, W.D., Peterson, K.M., Luken, J.D., and Mortensen, D.A. 1984. Interaction of increasing atmospheric carbon-dioxide and soil nitrogen in the carbon balance of tundra microcosms. *Oecologia* 65:26–29.

Billings, W.D., Peterson, K.M., Shaver, G.R., and Trent, A.W. 1977. Root growth, respiration, and carbon dioxide evolution in an arctic tundra soil. *Arctic Apline Res.* 9:129–137.

Bonan, G.B., Pollard, D., and Thompson, S.L. 1992. Effects of boreal forest vegetation on global climate. *Nature* 359:716–718.

Bonan, G.B., and Shugart, H.H. 1989. Environmental factors and ecological processes in boreal forests. *Annu. Rev. Ecol. System.* 20:1–28.

Bradley, R.S., Diaz, H.F., Eischeid, J.K., Jones, P.D., Kelly, P.M., and Goodess, C.M. 1987. Precipitation fluctuations over northern hemisphere land areas since the mid-19th century. *Science* 237:171–175.

Bubier, J.L., Moore, T.R., and Roulet, N.T. 1993. Methane emissions from wetlands in the midboreal region of northern Ontario, Canada. *Ecology* 74:2240–2254.

Chapin, F.S. III. 1983. Direct and indirect effects of temperature on arctic plants. *Polar Biol.* 2:47–52.

Chapin, F.S. III, Miller, P.C., Billings, W.D., and Coyne, P.I. 1980. Carbon and nutrient budgets and their control in coastal tundra. In: Brown, J., Miller, P.C., Tieszen, L.L., and Bunnell, F.L. (eds.), *An Arctic Ecosystem: The Coastal Tundra at Barrow, Alaska* (pp. 458–482). Stroudsburg, PA: Dowden, Hutchinson, and Ross.

Chapin, F.S. III, and Shaver, G. 1985. Individualistic growth response of tundra plant species to environmental manipulations in the field. *Ecology* 66:564–576.

Chapman, W.L., and Walsh, J.E. 1993. Recent variations of sea, ice, and air temperature in high latitudes. *Bull. Am. Meteorol. Soc.* 74:33–47.

Coyne, P.I., and Kelley, J.J. 1975. CO_2 exchange over the Alaskan arctic tundra: Meteorological assessment by an aerodynamic method. *J. App. Ecol.* 12:587–611.

Drake, B.G. 1992. The impact of rising CO_2 on ecosystem production. *Water, Soil, Air Pollu.* 64:25–44.

Flanagan, P.W., and Bunnell, F.L. 1980. Microflora activities and decomposition. In: Brown, J., Miller, P.C., Tieszen, L.L., and Bunnell, F.L. (eds.), *An Arctic Ecosystem: The Coastal Tundra at Barrow, Alaska*. Stroudsburg, PA: Dowden, Hutchinson, and Ross.

Freeman, C., Lock, M.A., and Reynolds, B. 1993. Fluxes of CO_2, CH_4 and N_2O from a Welsh peatland following simulation of water table draw-down: Potential feedback to climate change. *Biogeochemistry* 19:51–60.

Funk, D.W., Pullman, E.R., Peterson, K.M., Crill, P.M., and Billings, W.D. 1994. Influence of water table on carbon dioxide, carbon monoxide, and methane fluxes from taiga bog microcosms. *Global Biogeochem. Cycles* 8:271–278.

Gates, W.L., Mitchell, J.F.B., Boer, G.J., Cubasch, U., and Meleshko, V.P. 1992. Climate modeling, climate prediction and model validation. In: Houghton, J.T., Callander, B.A., and Varney, S.K. (eds.), *Climate Change 1992: The Supplemental Report to the IPCC Scientific Assessment* (pp. 97–135). Cambridge: Cambridge University Press.

Glenn, S., Heyes, A., and Moore, T. 1993. Carbon dioxide and methane fluxes from drained peat soils, southern Quebec. *Global Biogeochem. Cycles* 7:247–257.

Gorham, E. 1991. Northern peatlands: Role in the carbon cycle and probable responses to climate warming. *Ecol. App.* 1:182–195.

Graham, R.L., Turner, M.C., and Dale, V.H. 1991. How increasing CO_2 and climate change affect forests. *BioScience.* 40:575–587.

Groisman, P.Y., Karl, T.R., and Knight, R.W. 1994. Observed impact of snow cover on the heat balance and the rise of continental spring temperatures. *Science* 263:198–200.

Grulke, N.E., Riechers, G.H., Oechel, W.C., Hjelm, U., and Jaeger, C. 1990. Carbon balance in tussock tundra under ambient and elevated atmospheric CO_2. *Oecologia* 83:485–494.

Harden, J.W., Sunquist, E.T., Stallard, R.F., and Mark, R.K. 1992. Dynamics of soil carbon during deglaciation of the Laurentide ice sheet. *Science* 258:1921–1924.

Hengeveld, H. 1991. A State of the Environment Report. Atmospheric Environment Service, Environment Canada.

Hilbert, D.W., Prudhomme, T.I., and Oechel, W.C. 1987. Response of tussock tundra to elevated carbon dioxide regimes: Analysis of ecosystem flux through nonlinear modelling. *Oecologia* 72:466–472.

Hinzman, L.D., and Kane, D.L. 1992. Potential response of an arctic watershed during a period of global warming. *J. Geophys. Res.* 97:2811–2820.

Hogg, E.H., Lieffers, V.J., and Wein, R.W. 1992. Potential carbon losses from peat profiles: effects of temperature, drought cycles, and fire. *Ecol. App.* 2:298–306.

Idso, S.B., Kimball, B.A., and Allen, S.G. 1991. Net photosynthesis of sour orange trees maintained in atmospheres of ambient and elevated CO_2 concentration. *Agri. Forest Meteorol.* 54:95–101.

Kane, D.L., Gieck, R.E., and Hinzman, L.D. 1990. Evapotranspiration from a small arctic watershed. *Nord. Hydrol.* 21:253–272.

Kane, D.L., Hinzman, L.D., Woo, M.K., and Everett, K.R. 1992. Arctic hydrology and climate change. In: Chapin, F.S. III, Jefferies, R., Shaver, G., Reynolds, J., and Svoboda, J. (eds.), *Physiological Ecology of Arctic Plants: Implications For Climate Change*. New York: Academic Press.

Khalil, M.A.K., and Rasmussen, R.A. 1992. Decline in the trend of atmospheric methane. In: Ferguson, E.E., and Rosson, R.M. (eds.), U.S. Department of Commerce, National Oceanic and Atmospheric Administration, Climate Monitoring and Diagnostics Laboratory Summary Report No. 20 (pp. 105–107). Boulder, CO.

Khalil, M.A.K., Rasmussen, R.A., Shearer, M.J., Ge, S., and Rau, J.A. 1993. Methane from coal burning. *Chemosphere* 26:473–477.

Kling, G.W., Kipphut, G.W., and Miller, M.C. 1991. Arctic lakes and streams as conduits to the atmosphere: Implications for tundra carbon budgets. *Science* 251:298–301.

Körner, C. 1993. CO_2 fertilization: The great uncertainty in future vegetation development. Pages 53–70. In: Soloman, A.M., and Shugart, H.H. (eds.), *Vegetation Dynamics and Global Change*. New York: Chapman and Hall.

Körner, C., and Diemer, M. 1987. In situ photosynthetic responses to light, temperature, and carbon dioxide in herbaceous plants from low and high altitude. *Func. Ecol.* 1:179–194.

Kummerow, J., and Ellis, B.A. 1984. Temperature effect on biomass production and root/shoot biomass ratios in two arctic sedges under controlled environmental conditions. *Can. J. Bot.* 62:2150–2153.

Lachenbruch, A.H., and Marshall, B.V. 1986. Changing climate: geothermal evidence from permafrost in the Alaskan arctic. *Science* 234:689–696.

Landhäusser, S.M., and Wein, R.W. 1993. Postfire vegetation recovery and tree establishment at the arctic treeline: Climate-change-vegetation-regeneration hypotheses. *J. Ecol.* 81:665–672.

Larigauderie, A., Hilbert, D.W., and Oechel, W.C. 1988. Effect of CO_2 enrichment and nitrogen availability on resource acquisition and resource allocation processes in a grass, *Bromus mollis*. *Oecologia* 77:544–549.

Lashof, D.A., and Ahuja, D.R. 1990. Relative contributions of greenhouse gas emissions to global warming. *Nature* 344:529–531.

Limbach, W.E., Oechel, W.C., and Lowell, W. 1982. Photosynthetic and respiratory responses to temperature and light of three Alaskan tundra growth forms. *Holarctic Ecol.* 5:150–157.

Manabe, S., and Stouffer, R.J. 1993. Century-scale effects of increased atmospheric CO_2 on the ocean-atmosphere system. *Nature* 364:215–218.
Marion, G.M., and Black, C.H. 1987. The effect of time and temperature on nitrogen mineralization in arctic tundra soils. *Soil Sci. Soc. Am. J.* 51:1501–1507.
Marion, G.M., and Miller, P.C. 1982. Nitrogen mineralization in a tussock tundra soil. *Arctic Alpine Res.* 14:287–293.
Marion, G.M., and Oechel, W.C. 1993. Mid- to late-Holocene carbon balance in arctic Alaska and its implications for future global warming. *The Holocene* 3:193–200.
Mathews, E., and Fung, I. 1987. Methane emission from natural wetlands: Global distribution, area, and environmental characteristics of sources. *Global Biogeochem. Cycles* 1:61–86.
Meehl, G.A., Washington, W.M., and Karl, T.R. 1993. Low-frequency variability and CO_2 transient climate change: Part 1. Time averaged differences. *Clim. Dynam.* 8:117–133.
Melillo, J.M., Callaghan, T.V., Woodward, F.I., Salati, E., and Sinha, S.K. 1990. Effects on Ecosystems. In: Houghton, J.T., Jenkins, G.J., and Ephraums, J.J. (eds.), *Climate Change: The IPCC Scientific Assessment.* Cambridge: Cambridge University Press.
Melillo, J.M., McGuire, A.D., Kicklighter, D.W., Moore, B., III, Vorosmarty, C.J., and Schloss, A.L. 1993. Global climate change and terrestrial net primary production. *Nature* 363:234–240.
Miller, P.C., Kendall, R., and Oechel, W.C. 1983. Simulating carbon accumulation in northern ecosystems. *Simulation* 40:119–131.
Mitchell, J.F.B., Senior, C.A., and Ingram, W.J. 1989. CO_2 and climate: A missing feedback? *Nature* 341:132–134.
Moore, T.R., Heyes, A., and Roulet, N.T. 1994. Methane emissions from wetlands, southern Hudson Bay lowland. *J. Geophys. Res.* 99:1455–1467.
Moorhead, D.L., and Reynolds, J.F. 1993. Effects of climate change on decomposition in arctic tussock tundra: A modeling synthesis. *Arctic Alpine Res.* 25:403–412.
Moraes, F., and Khalil, M.A.K. 1993. Permafrost methane content: 2. Modeling theory and results. *Chemosphere* 26:595–607.
Morrissey, L.A., and Livingston, G.P. 1992. Methane emissions from Alaska arctic tundra: An assessment of local spatial variability. *J. Geophys. Res.* 97:16,661–16,670.
Nadelhoffer, K.J., Giblin, A.E., Shaver, G.R., and Laundre, J.A. 1991. Effects of temperature and substrate quality on element mineralization in six arctic soils. *Ecology* 72:242–253.
Neilson, R.P. 1993. Vegetation redistribution: A possible biosphere source of CO_2 during climatic change. *Water, Soil Air Pollut.* 70:659–673.
Oberbauer, S.F., Sionit, N., Hastings, S.J., and Oechel, W.C. 1986. Effects of CO_2 enrichment and nutrition on growth photosynthesis, and nutrient concentration of Alaskan tundra plant species. *Can. J. Bot.* 64:2993–2998.
Oechel, W.C., and Billings, W.D. 1992. Anticipated effects of global change on carbon balance of arctic plants and ecosystems. In: Chapin, F.S. III, Jefferies, R., Shaver, G., Reynolds, J., and Svoboda, J. (eds.), *Physiological Ecology of Arctic Plants: Implications for Climate Change* (pp. 139–168). New York: Academic Press.
Oechel, W.C., Cowles, S., Grulke, N., Hastings, S.J., Lawrence, B., Prudhomme, T., Riechers, G., Strain, B., Tissue, D., and Vourlitis, G. 1994. Transient nature of CO_2 fertilization in arctic tundra. *Nature* 371:500–503.
Oechel, W.C., Hastings, S.J., Vourlitis, G.L., Jenkins, M., Riechers, G., and Grulke, N. 1993. Recent change of arctic tundra ecosystems from a net carbon dioxide sink to a source. *Nature* 361:520–523.

Oechel, W.C., Riechers, G., Lawrence, W.T., Prudhomme, T.I., Grulke, N., and Hastings, S.J. 1992. "CO_2LT" an automated, null-balance system for studying the effects of elevated CO_2 and global climate change on unmanaged ecosystems. *Func. Ecol.* 6:86–100.

Oechel, W.C., and Vourlitis, G.L. 1994. The effects of climate change on land-atmosphere feedbacks in arctic tundra regions. *Trends Ecol. Evol.* 9:324–329.

Oechel, W.C., and Vourlitis, G.L. Effects of global change on carbon storage in cold soils. *Adv.* in *Soil Sci.* In press.

Oechel, W.C., Vourlitis, G.L., Hastings, S.J., and Bochkarev, S.A. Change in arctic CO_2 flux over two decades: Effects of climate change at Barrow, Alaska. *Ecol. Appl.* In press.

Paul, E.A., and Clark, F.E. 1989. *Soil Microbiology and Biochemistry*. San Diego, CA: Academic Press.

Prudhomme, T.I., Oechel, W.C., Hastings, S.J., and Lawrence, W.T. 1982. Net ecosystem gas exchange at ambient and elevated carbon dioxide concentrations in tussock tundra at Toolik Lake, Alaska: An evaluation of methods and initial results. The potential effects of carbon-dioxide induced climate change in Alaska: Proceedings of a Conference, April 7–8, 1982, School of Agriculture and Land Resources Management, University of Alaska, Fairbanks, Misc. Publication.

Ramanathan, V., Cicerone, H.B., Singh, H.B., and Kiehl, J.T. 1985. Trace gas trends and their potential role in climate change. *J. Geophys. Res.* 90:5547–5566.

Rastetter, E.B., McKane, R.B., Shaver, G.R., and Melillo, J.M. 1992. Changes in C storage by terrestrial ecosystems: How C-N interactions restrict responses to CO_2 and temperature. *Water, Air, Soil Pollut.* 64:327–344.

Reeburgh, W.S., Roulet, N.T., and Svensson, B.H. 1994. Terrestrial biosphere-atmosphere exchange in high latitudes. In: Prinn, R.G. (ed.), *Global Atmospheric-Biospheric Chemistry* (pp. 165–178). New York: Plenum Press.

Reeburgh, W.S., Whalen, S.C., and Alperin, M.J. 1993. The role of methyltrophy in the global methane budget. In: Murrell, J.C., and Kelley, D. (eds.), *Microbial Growth on C_i Compounds* (pp. 1–14). Andover: Intercept.

Roulet, N.T., Ash, R., and Moore, T.R. 1992. Low boreal wetlands as a source of atmospheric methane. *J. Geophys. Res.* 97:3739–3749.

Schell, D.M. 1983. Carbon-13 and carbon-14 abundances in Alaskan aquatic organisms: Delayed production from peat in arctic food webs. *Science*. 219:1068–1068.

Schlesinger, M.E., and Mitchell, J.F.B. 1987. Climate model simulations of the equilibrium climatic response to increased carbon dioxide. *Rev. Geophys.* 25:760–798.

Schlesinger, M.E., and Xingjian, J. 1991. Revised projection of future greenhouse warming. *Nature* 350:219–221.

Schlesinger, W.H. 1991. *Biogeochemistry: An Analysis of Global Change*. San Diego, CA: Academic Press Inc.

Sebacher, D.I., Harriss, R.C., and Bartlett, K.B. 1985. Methane emissions to the atmosphere through aquatic plants. *J. Environ. Qual.* 14:40–46.

Sebacher, D.I., Harriss, R.C., Bartlett, K.B., Sebacher, S.M., and Grice, S.S. 1986. Atmospheric methane sources: Alaskan tundra bogs, an alpine fen, and a subarctic boreal marsh. *Tellus* 38B:1–10.

Sharrat, B.S. 1992. Growing season trends in the Alaskan climate record. *Arctic* 45:124–127.

Shaver, G.R., Billings, W.D., Chapin, F.S. III, Giblin, A.E., Nadelhoffer, K.J., Oechel, W.C., and Rastetter, E.B. 1992. Global Change and the Carbon Balance of Arctic Ecosystems. *BioScience* 42:433–441.

Shaver, G.R., and Chapin, F.R. 1980. Response to fertilization by various plant growth forms in an Alaskan tundra: Nutrient accumulation and growth. *Ecology* 61:662–675.

Silvola, J. 1986. Carbon dioxide dynamics in mines reclaimed for forestry in eastern Finland. *Ann. Bot. Fenn.* 23:59–67.

Smith, T.M., and Shugart, H.H. 1993. The transient response of terrestrial carbon storage to a perturbed climate. *Nature* 361:523–526.

Stitt, M. 1991. Rising CO_2 levels and their potential significance for carbon flow in photosynthetic cells. *Plant Cell Envion.* 14:741–762.

Strain, B.R. 1992. Field measurements of CO_2 enhancement and change in natural vegetation. *Water, Soil Pollut. Air* 64:45–60.

Svensson, B.H., and Rosswall, T. 1984. In situ methane production from acid peat in plant communities with different moisture regimes in a subarctic mire. *Oikos* 43:341–350.

Tans, P., Fung, I., and Takahashi, T. 1990. Observational constraints on the global atmospheric CO_2 budget. *Science.* 247:1431–1439.

Tieszen, L.L. 1973. The seasonal course of aboveground production and chlorophyll distribution in a wet arctic tundra at Barrow, Alaska. *Arctic Alpine Res.* 4:307–324.

Tissue, D.T., and Oechel, W.C. 1987. Response of *Eriophorum vaginatum* to elevated CO_2 and temperature in the Alaskan tossock tundra. *Ecology* 68:401–410.

Torn, M.S., and Chapin, F.S. III. 1993. Environmental and biotic controls over methane flux from arctic tundra. *Chemosphere* 26:357–368.

Valentine, D.W., Holland, E.A., and Schimel, D.S. 1994. Ecosystem and physiological controls over methane production in northern wetlands. *J. Geophys. Res.* 99:1563–1571.

Vourlitis, G.L., Oechel, W.C., Hastings, S.J., and Jenkins, M.A. 1993. A system for measuring in situ CO_2 and CH_4 flux in unmanaged ecosystems: an arctic example. *Func. Ecol.* 7:369–379.

Vourlitis, G.L., Oechel, W.C., Hastings, S.J., and Jenkins, M.A. 1994. The effect of soil moisture and thaw depth on CH_4 flux from wet coastal tundra ecosystems on the North Slope of Alaska. *Chemosphere* 28:R1–R3.

Waelbroeck, C. 1993. Climate-soil processes in the presence of permafrost: a systems modeling approach. *Ecol. Model.* 69:185–225.

Watson, R.T., Meira Filho, L.G., Sanhueza, E., and Janetos, A. 1992. Sources and sinks. In: Houghton, J.T., Callander, B.A., and Varney, S.K. (eds.), *Climate Change 1992: The Supplemental Report to the IPCC Scientific Assessment.* Cambridge: Cambridge University Press.

Whiting, G.J., and Chanton, J.P. 1993. Primary production control of methane emission from wetlands. *Nature* 364:794–795.

Woodward, F.L. 1987. Stomatal numbers are sensitive to increases in CO_2 from preindustrial levels. *Nature* 327:617–618.

Zimov, S.A., Zimova, G.M., Davidov, S.P., Davidova, A.I., Voropaev, Y.V., Boropaeva, Z.V., Prosiannikov, S.F., Prosiannikova, O.V., Semiletova, I.V., and Semiletov, I.P. 1993. Winter biotic activity and production of CO_2 in siberian soils: A factor in the greenhouse effect. *J. Geophys. Res.* 98:5017–5023.

21. Phenomenological Models of the Primary Productivity of Zonal Arctic Ecosystems

Tagir G. Gilmanov

The primary productivity of an ecosystem, as measured by either the rate of gross photosynthesis of the photoautotrophs (called Gross Primary Productivity, *GPP*) or the rate of net photosynthesis of the autotrophs (as equal to the gross photosynthesis minus respiration, called Net Primary Productivity, *NPP*), belongs to the prime components of the gas exchange between the ecosystem and the atmosphere. For most terrestrial ecosystems, reliable empirical data and adequate mathematical models of production process are presently available; however, for arctic and alpine ecosystems the number and quality of related data remain very limited. The available publications and reviews are usually limited to pairwise linear correlative relationships (Bazilevich, 1992, 1994; Bazilevich et al., 1986; French, 1981; Wielgolaski et al., 1981), though the need for multivariate nonlinear analysis is generally acknowledged.

There is a growing consensus that under climate change the high latitude and high altitude ecosystems might experience particularly marked changes in their gas exchange regime, including the possible transformation from sink to source of carbon for the external atmosphere (Oechel et al., 1993). Under those conditions we have initiated the development of predictive models for productivity of arctic ecosystems. The goal is to obtain equations relating productivity estimates at the ecosystem level to easily available climatic and soil factors.

From the methodological point of view it is important to distinguish two extreme versions of predictive models, namely the phenomenological (e.g., empirical) and simulation (e.g., process or mechanistic) models, though a whole spectrum of intermediate variants between the two is possible. In this chapter the phenomenological models are discussed and analyzed.

Phenomenological and Simulation Models of Ecosystem Productivity

Let us consider one of the productivity characteristics of the ecosystem, for example, net primary productivity (*NPP*) measured in $t\,ha^{-1}\,yr^{-1}$, which will be denoted by the variable y, and let us list the ecological factors determining the ecosystem productivity as $x_1, x_2, \ldots, x_n$. Using this notation under the phenomenological model of *NPP* we will understand the multivariate function f, approximating the relationship between y and $x_1, x_2, \ldots, x_n$:

$$y = f(x_1, x_2, \ldots, x_n) + \varepsilon, \tag{1}$$

where ε is the error of approximation, usually described by a random variable with zero mean and a certain standard deviation σ. Usually it is a good idea to have at least an estimate of the value of σ, measured by the variable s, which is the statistical estimate of σ.

The models of this type are also known in the literature as regressional, statistical, or correlative. These terms are strictly associated with certain procedures and algorithms for identification of type 1 models, based on the definite assumptions about the statistical properties of the factor variables and the response variable, such as independence of the factors, normality of the error term ε, etc. (cf. Draper and Smith, 1981; Himmelblau, 1970). Because the data on ecological factors and productivity rarely satisfy those assumptions, and the methods that we apply are different from the classic methods of regression analysis, referring to the models of type 1 we prefer to use the term "phenomenological model." Phenomenological models do not require detailed knowledge of the mechanism of the production process, and stress quantitative generalization of the empirical information about the interrelation between productivity and the factors determining it.

In contrast to the phenomenological model 1, the dynamic (mechanistic) simulation model of *NPP* would be expressed by the mathematical functional F of the form:

$$y(t) = F\left[t, v_1(\cdot), v_2(\cdot), \ldots, v_k(\cdot), s_1(\cdot), s_2(\cdot), \ldots, s_n(\cdot)\right], \tag{2}$$

where the $y(t)$ is productivity *function* (not a single number!), determined at a given time interval $0 \le t \le T$, and $v_i(.) = v_i(\tau)$ $(0 \le \tau \le t, i = 1, \ldots, k)$ and $s_j(.) = s_j(\tau)$ $(0 \le \tau \le t, j = 1, \ldots, n)$ describe the input variables v_i (e.g., radiation, air temperature, precipitation, wind speed, etc.) and the state variables s_i (e.g., green phytomass, soil temperature and moisture, soil nutrients, etc.). The functional F is the solving operator of the system of simultaneous mathematical equations (usually difference or differential equations) that describe the temporary dynamics of the ecosystem.

There is no doubt that it is the dynamic simulation modeling approach that eventually would be the most reliable and efficient instrument of the ecological analysis, forecast, and optimization, as for example is demonstrated by the CENTURY grassland and ecosystem model (Parton et al., 1987; Parton et al., 1993; SCOPEGRAM, 1994). The available dynamic simulation models of arctic and alpine ecosystems (e.g., Bunnell and Scoullar, 1975; Collier et al., 1975; Miller et al., 1983; Reynolds and Acock, 1985; Reynolds and Leadley, 1992; Sandhaug et al., 1975; Timin et al., 1973; Waelbroeck, 1993) are less sophisticated than the existing grassland or forest models and require considerable development and improvement. On the other hand, the application to the arctic and alpine ecosystems of the generic ecosystem-level models, which may be tuned to simulate the dynamics of various ecosystem types, is not very much successful. For instance, the GEM model (General Ecosystem Model), which works quite successfully at low- and medium-latitude ecosystems, was found to be less satisfactory for the high-latitude ecosystems. For the polar deserts and alpine tundra ecosystems the GEM model provides the average estimates of the net primary productivity of $87\,g\,C\,m^{-2}\,yr^{-1}$ (Melillo et al., 1993), whereas the empirical values for the same ecosystem types are 5 to 10 times lower (Bliss and Matveyeva, 1992; Bliss et al., 1984; see also Table 21.2 below).

Taking into account the comparative shortcomings of the mechanistic simulation models of arctic ecosystems, for the purpose of obtaining the rapid and preliminary, but empirically based, estimates of the productivity of arctic ecosystems and its changes under climate change we have chosen the method of phenomenological modeling. This method still has a high potential in the field of comparative ecosystem analysis as is demonstrated by the works of H. Lieth and his associates (Esser, 1987; Esser, 1991; Esser et al., 1982; Lieth, 1975a; Lieth, 1976). Some of the principles of phenomenological modeling at the ecosystem level were discussed in our recent publication (Gilmanov, 1993) and they are elaborated in more detail in the following sketch of the history of phenomenological modeling of ecosystem productivity.

A Sketch of the History of Phenomenological Modeling of Ecosystem Productivity

The first phenomenological models of terrestrial ecosystem productivity were constructed by A. Mitscherlich and comprised the empirical formulae for prediction of the yield of various agricultural crops under the climatic and soil conditions of Germany (Baule, 1918; Mitscherlich, 1909). Mitscherlich was the first who proposed the formal approach to representation of the phenomenon of *interaction* of several ecological factors (namely, the multiplicative interaction, as opposed to the Liebig's type of alternative interaction—see below). He used the multivariate relationship of the form:

$$y = f(x_1, x_2, \ldots, x_n) = y_{\max}\left(1 - e^{-k_1 x_1}\right)\left(1 - e^{-k_2 x_2}\right) \ldots \left(1 - e^{-k_n x_n}\right). \quad (3)$$

Though the development of the theory of statistical experimental design by R. Fisher in the 1930s and the associated trend to polynomially describe the multivariate response functions dominated the papers on agricultural statistics, they were not able to substitute the intellectual fascination of the Mitscherlich approach.

Indeed, since with respect to each factor x_i the relationship (3) may be reformulated as a differential equation of the form:

$$\frac{dy}{dx_i} = k_i\left(A_i(x_1, x_2, \ldots, x_{i-1}, x_{i+1}, \ldots, x_n) - y\right), \quad (4)$$

stating that the growth rate of productivity with respect to a given factor x_i is proportional to the deviation of current productivity value from the maximum value $A_i(x_1, x_2, \ldots, x_{i-1}, x_{i+1}, \ldots x_n)$, determined by a combination of other factors, x_i. This interpretation gave Mitscherlich's equation undefeatable theoretical advantage compared to purely empirical polynomial approximations.

Grigoryev and Budyko (1956) made a notable contribution to the phenomenology of ecosystem productivity studies. They did not provide the final formulae for productivity calculations, but proposed the so-called "radiation index for dryness" of the landscapes and for the first time related it to the global pattern on geographical distribution of ecosystems and their productivity (Budyko, 1974; Grygoryev and Budyko, 1956). Rosenzweig (1968) proposed the formula for calculating the aboveground net primary productivity of ecosystems using climatic indices (evapotranspiration), and Ryabchikov (1968) suggested the biohydrothermal productivity index and used it as a basis for estimation of the total (above + belowground) net primary production of terrestrial ecosystems. The idea of using climatic parameters to obtain estimates of the productivity of terrestrial ecosystems was most extensively elaborated by H. Lieth and his colleagues (Lieth, 1972; Lieth, 1975b; Lieth and Box, 1977), which resulted in construction of the series of maps of global distribution of terrestrial productivity. In a sense these maps were the culmination of the first International Biological Program (IBP)-based phase of terrestrial productivity studies.

Nevertheless, as it was confirmed by later analysis, climatic factors alone (without taking into consideration the properties of soil and vegetation) are not sufficient to predict the productivity of terrestrial ecosystems (Esser et al., 1982). Introducing the so-called "soil factor" of productivity, F_{soil}, defined as a relation of the field estimate of productivity NPP_{field} to its "climatic" estimate NPP_{cli}:

$$F_{soil} = \frac{NPP_{field}}{NPP_{cli}} \tag{5}$$

Table 21.1 shows that, depending on the soil type, the climatically based estimate may be either four times greater (e.g., for Humic Acrisol F_{soil} = 0.22), or three times lower (e.g., for Gleyic Luvisol, F_{soil} = 2.37) than the actual field value (Esser et al., 1982).

The importance of the soil factor in predicting productivity of terrestrial ecosystems was known to ecologists long before this conclusion by Esser et al. (1982), at least since Liebig's and Boussengault's works in the 19th century. Therefore, a number of authors made attempts to explicitly include soil fertility factors into a set of predictors of terrestrial ecosystem productivity. Jenny (1941, 1980) was a pioneer of quantification of the soil factor relationships. The original book by Volobuev (1974) contained one of the first quantitative results directly relating *NPP* to edaphic factors. A number of soil fertility factors were taken into consideration in the models by Esser (1986, 1991) and Prentice et al. (1992). A number of significant empirical studies of the influence of soil nutrients on productivity of arctic vegetation were carried out by Khantimer (1970, 1974), Shaver and Chapin (1986), and Bowman et al. (1993). Nevertheless, the quantitative analysis of the significance of soil factors for improving the prediction of productivity in arctic

Table 21.1. Values of the Soil Factor F_{soil} Calculated for the Main Classification Units of the FAO/UNESCO Soil Map of the World

Soil Unit	F_{soil}	Soil Unit	F_{soil}
Gleyic acrisol	0.87	Lithosol	0.52
Humic acrisol	0.22	Lithosol-yermosol	1.14
Orthic acrisol	0.70	Fluvisol	0.49
Other acrisol	0.60	Eutric fluvisol	0.61
Dystric cambisol	0.94	Other fluvisol	0.55
Eutric cambisol	1.69	Albic luvisol	0.34
Humic cambisol	1.58	Chromic luvisol	1.04
Gelic cambisol	0.76	Ferric luvisol	1.65
Luvic chernozem	0.99	Gleyic luvisol	2.78
Dystric podzoluvisol	0.83	Orthic luvisol	0.85
Xanthic ferrasol	0.55	Distric histosol	1.29
Humic gleysol	0.47	Humic podzol	0.56
Gelic gleysol	0.57	Orthic podzol	0.61
Other gleysol	0.50	Other podzol	0.55

Adapted from Esser et al. (1982).

ecosystems was never performed. The present publication is an attempt to quantify and assess the role of soil factors in these ecosystems based on a comprehensive analysis of available empirical information.

The Methodology of Phenomenological Modeling of Terrestrial Ecosystem Productivity

A phenomenological model by definition constitutes a quantitative generalization of the available empirical information. The first question in considering the methodological foundation of a given model refers to the contents and quality of the data base in use. Our approach to the construction of the data base is described below. Concerning the other known-to-us data bases containing information on productivity and environmental and soil conditions of arctic ecosystems, for example, prepared at the initial (French, 1974; Webber, 1974; Wielgolaski, 1972, 1975) or at the final stages of the International Biological Programme (French, 1981; Tikhomirov et al., 1981; Wielgolaski et al., 1981), they are usually geographically biased and characterized by incomplete representation of either Eurasian (Russian) or North American sites. The most comprehensive from this point of view is the tundra section of the latest review by Bazilevich (1992, 1994), but her data base is mainly restricted to biomass and productivity parameters and contains no climatic and soil information.

Different investigators have taken different approaches in an attempt to describe the phenomenological model of the relationship between the ecosystem productivity and ecological factors that control it. These are discussed below.

1. The Mitscherlich-Forrester approach: The multivariate relationship $y = f(x_1, x_2, \ldots, x_n)$ is reconstructed as the product of the potential maximum productivity y_{max} and several normalized (bounded between 0 and 1) single-factor partial response functions (Baule, 1918; Forrester, 1961; Mitscherlich, 1909). This approach, known also as the hypothesis of multiplicative interaction of factors, may be summarized by the formula:

$$y = f(x_1, x_2, \ldots, x_n) = y_{max} f_1(x_1) f_2(x_2) \ldots f_n(x_n). \tag{6}$$

Despite the relatively weak theoretical foundation, this method has a wide and sufficiently successful application in both the phenomenological (Esser, 1991; Lieth, 1975a), and dynamic modeling of productivity of ecosystems including arctic and alpine (Bunnell and Scoullar, 1975; Melillo et al., 1993; Miller et al., 1978, 1983, 1984; Parton et al., 1987).

2. The Liebig-type approach: The desired multifactorial relationship is reconstructed using the concept of the limiting factor (Liebig, 1840). The theoretical foundation to this method was provided by Poletaev (1966, 1973):

$$y = y_{\max} \min\{f_1(x_1), f_2(x_2), \ldots, f_n(x_n)\}, \tag{7}$$

where $fi(x_i)$ are the normalized partial response functions of productivity to single factors separately.

3. The semiempirical approach: This method of identification of the multifactorial response function is applicable in those rare cases when the analytical form of the required relationship is known on the basis of certain theoretical assumptions and contains a number of unknown parameters that may be estimated from empirical data:

$$y = f(x_1, x_2, \ldots, x_n;\ a_1, a_2, \ldots, a_p). \tag{8}$$

The simplest example of this kind is the three-parametric nonrectangular hyperbolic equation relating net primary productivity y to photosynthetically active radiation x_1 and dark respiration x_2 (Johnson and Thornley, 1984):

$$y = \left(\frac{1}{2a_1}\left[a_2x_1 + a_3 - \sqrt{(a_2x_1 + a_3)^2 - 4a_1a_2a_3x_1}\,\right]\right) - x_2 \tag{9}$$

where a_3 means the maximum rate of gross photosynthesis, a_2 is the initial slope of the net productivity light curve, and a_1 is the curvature parameter.

4. Finally, as the natural generalization of the above, we can formulate the fourth, hybrid method for building the multivariate response functions combining the three mentioned approaches in a single formula:

$$\begin{aligned} y = y_{\max} f_1(x_1) \ldots f_k(x_k) \min\{f_{k+1}(x_{k+1}) \ldots f_m(x_m)\} \\ f(x_{m+1}, \ldots, x_n;\ a_1, \ldots, a_p). \end{aligned} \tag{10}$$

As can be seen from the formula, the first k factors interact multiplicatively with each other and the two next groups of $(m - k)$ and $(n - m)$ factors; the $(m - k)$ factors of the second group interact with each other alternatively, whereas the coaction of the factors of the third group is described by a semiempirical formula with parameters $(a_1, \ldots, a_p)$. This approach is similar to that taken by Pastor and Post (1986) and Parton et al. (1987).

The Global Climate/Soil/Productivity Data Base for Arctic Ecosystems

In the framework of solving the problem of building multifactorial predictive phenomenological models for productivity of arctic ecosystems, a need to construct a computerized data base on climate, soils, and productivity for the tundra biome worldwide became evident. As the starting point we used the recent compilations of literature data by N.I. Bazilevich on phytomass and productivity of the Eurasia and other regions (Bazilevich, 1992, 1994; Bazilevich et al., 1986). In addition, a number of new records describing the Russian, North American, and North European sites were added and the data collected through an extensive literature survey were added to the compilation.

The structure, composition, and the present time capacity of the global climate/soil/productivity data base for arctic ecosystems can be characterized by the following parameters. The basic unit of storage of information in the data base is the individual record, describing a specific ecosystem where the productivity measurements were performed and for which the original empirical data or, derived from them, estimates of phytomass and productivity are presented in the literature. Each individual record consists of several fields—elementary information cells, containing numerical (numbers) or symbolic (letters or words) data. The fields in each record are grouped according to their content into the following four divisions:

- The physicogeographical characteristics (geographical coordinates, location, zonal affiliation, ecosystem title, relief position, etc.; more than 20 fields in total);
- Climatic and meteorological parameters (temperature, precipitation, cloudiness, wind speed, radiation parameters, evaporation, and others derived from their thermal, hydrological, and phenological characteristics, such as temperature sums, length of the growing period, precipitation during the growing period, etc.; more than 80 parameters in total);
- Soil properties [content and reserve of C, N, P, K, Ca, Mg, bulk density and texture, hydrophysical, physicochemical (pH, cation exchange capacity, base saturation), and permafrost parameters; more than 50 characteristics in total];
- Phytomass and productivity characteristics [community title, dominant species, phytomass and productivity for the whole community and with various subdivisions (above/belowground, green/nongreen, vascular/cryptogamic, live/dead, etc.); more than 30 parameters totally].

The total number of fields in the record describing an ecosystem may exceed 160. Nevertheless, it should be acknowledged that only for a few of the most thoroughly studied ecosystems is it possible to ascertain such a detailed description. In more typical cases 50% of the needed characteristics were available and not for all ecosystems was it possible to describe the same fields. This problem of missing data obviously leads to serious difficulties in the process of identification of the multiparametric functions of several variables, but this is the reality that cannot be avoided.

The most important information from the data base, containing some key data on relations of productivity of zonal arctic ecosystems to climatic and soil factors at 145 sites, is presented in Table 21.2. Using the methods of

Table 21.2. Productivity, Climatic, and Soil Conditions of Zonal Tundra Ecosystems

Ecosystem	Zone/subzone	Location	Total Phytomass (B, t ha^{-1})	Net Primary Production (NPP, t ha^{-1} yr^{-1})	Green Phytomass (G, t ha^{-1})	Aboveground Phytomass (BA, t ha^{-1})	Aboveground Net Primary Production (ANPP, t ha yr^{-1})	Altitude (ALT, m)	Latitude (LAT, m)	Longitude (LON, m)
Polar desert	polar desert	Franz Jozef Land	1.580	0.24	1.29	1.29	0.15	50	81.00	51.50
Polar desert	polar desert	Chelyuskin, Mys	1.150	0.19	0.05	—	0.051	20	77.72	104.08
Moss/herb on plateau	polar desert	Devon Island	2.570	0.08	2.47	2.47	0.07	400	75.55	–84.67
Barrens	polar desert	Devon Island	0.155	0.0118	0.085	0.116	0.0083	250	75.00	–85.00
Barrens	polar desert	Devon Island	0.038	0.009	0.011	0.03	0.0061	—	75.07	–91.75
Snowflush	polar desert	Cornwallis Island	0.828	0.0165	0.439	0.688	0.014	—	74.75	–95.00
Barrens	polar desert	Somerset Island	0.416	0.0145	0.308	0.377	0.0125	225	74.07	–93.88
Barrens	polar desert	Somerset Island	0.030	0.0026	0.012	0.02	0.0017	225	74.07	–93.88
Cushion plant	polar desert	Somerset Island	0.393	0.0063	0.087	0.121	0.005	225	74.07	–93.88
Forb/moss polygonal	arctic	Novosibirskie Islands	3.790	1.24	1.16	1.19	0.52	20	73.33	142.67
Grass/dwarf shrub/moss spotty hummocky	arctic	Novosibirskie Islands	6.960	1.48	1.6	1.85	0.57	20	73.33	142.67
Cushion plant/moss	arctic	Maria Pronchshcheva Bay	2.500	0.86	0.3	0.9	0.46	35	75.57	113.27
Cushion plant/moss	arctic	Maria Pronchshcheva Bay	1.970	0.67	0.27	0.67	0.37	35	75.57	113.27
Forb/lichen spotty	arctic	Tiksi	2.130	0.46	1.06	1.06	0.15	11	72.17	128.33
Dwarf shrub/lichen spotty	arctic	Tiksi	3.040	0.66	1.69	2.33	0.62	11	72.17	128.33
Dwarf shrub/forb spotty	arctic	Tiksi	13.040	4.18	2.35	2.84	1.41	11	72.17	128.33
Dwarf shrub/grass/lichen spotty	arctic	Yamal	12.260	2.75	4.4	4.94	0.82	20	71.17	67.33
Dwarf shrub/grass/moss	arctic	Yamal	14.330	3.51	3.51	4.13	0.96	20	71.17	67.33
Dwarf shrub	arctic	Barrow	5.790	1.35	0.92	1.08	0.25	10	71.33	–156.67
Cushion plant/moss	arctic	Devon Island	7.990	0.54	6.71	7.49	0.49	50	75.55	–84.67
Dwarf shrub/moss	arctic	Spitzbergen	2.890	0.6	2.28	2.35	0.54	9	78.00	15.00
Raised beach	arctic	Devon Island	28.060	1.83	9.5	10.67	0.58	10	75.55	–84.67
Cushion plant/lichen	arctic	Disko	1.310	0.46	0.66	0.99	0.32	—	69.25	–53.50
Herb/moss	arctic	Queen Elizabeth Islands	8.486	0.518	6.884	7.367	0.293	80	75.00	–108.00
Tussock tundra	arctic	Toolik Lake	12.250	2.53	5.69	7.43	2.05	760	68.63	–159.57
Hilltop heath	arctic	Toolik Lake	4.389	0.5526	2.603	3.189	0.4326	760	68.63	–159.57
Riverside willow	arctic	Toolik Lake	23.560	4.96	6.2	13.94	4	760	68.63	–159.57
Birch/willow heath	subarctic	Tuktoyaktuk	—	4.9	1.5	2.85	1.25	18	69.45	–133.00
Dwarf shrub/moss	subarctic typical	Baydaratskii Bay	12.960	1.19	3.56	4.13	0.68	143	68.50	65.00
Shrub/moss	subarctic typical	Baydaratskii Bay	17.640	1.79	3.99	8.21	1.2	143	68.50	65.00
Shrub/moss	subarctic typical	Baydaratskii Bay	32.690	3.64	3.7	9.77	1.98	143	68.50	65.00
Moss	subarctic typical	Baydaratskii Bay	16.830	2.24	9.04	9.04	1.81	143	68.50	65.00
Polygonal sedge/lichen/moss tundra	subarctic typical	Marre Sale, Yamal	29.370	2.9	10.99	11.17	1.59	25	66.67	66.83
Dwarf shrub/sedge/lichen/moss	subarctic typical	Marre Sale, Yamal	35.310	3.68	13.18	13.41	2.1	25	66.67	66.83
Sedge	subarctic typical	Marre Sale, Yamal	29.994	3.25	11.334	11.394	1.91	25	66.67	66.83
Dwarf shrub/lichen	subarctic typical	Marre Sale, Yamal	5.010	0.66	2.72	3.11	0.52	25	66.67	66.83
Dwarf shrub/sedge/moss	subarctic typical	Tareya	17.330	3.84	5.02	5.33	1.14	35	72.67	91.50
Dwarf shrub/sedge/moss spotty hummocl	subarctic typical	Tareya	22.040	3.64	6.24	6.54	1.45	35	72.67	91.50
Dwarf shrub/moss spotty	subarctic typical	Agapa	7.440	0.81	0.94	1.44	0.33	120	71.50	88.83

Dwarf shrub/moss spotty	subarctic typical	Agapa	7.300	1.31	0.92	1.38	0.32	120	71.50	88.83
Dwarf shrub/moss spotty	subarctic typical	Agapa	7.080	0.73	0.74	1.08	0.25	120	71.50	88.83
Dwarf shrub/moss spotty	subarctic typical	Agapa	13.990	1.95	1.89	2.69	0.72	120	71.67	88.67
Eriophorum/sedge/moss spotty hummocl	subarctic typical	Agapa	24.750	4.91	3.72	4.75	1.58	115	71.50	88.83
Dwarf shrub/lichen/moss spotty	subarctic typical	Agapa	11.430	1.93	4.31	5.97	1.34	100	72.50	85.67
Dwarf shrub/herb/moss	subarctic typical	Agapa	20.190	3.49	6.06	6.41	1.42	100	72.50	85.67
Eriophorum/dw. shrub/sedge spotty	subarctic typical	Agapa	33.640	5.85	5.37	6.64	2.02	120	71.67	88.67
Dwarf shrub/*Eriophorum*/moss spotty humn	subarctic typical	Agapa	30.910	6.13	9.34	10.91	3.13	120	71.50	88.83
Dwarf shrub/*Eriophorum*/moss spotty humn	subarctic typical	Agapa	24.510	5.48	7.31	9.51	2.98	120	71.50	88.83
Eriophorum/dw. shrub/moss tussock	subarctic typical	Agapa	48.070	5.92	5.14	6.44	1.72	120	71.67	88.67
Shrub/sedge/moss spotty	subarctic typical	Agapa	30.860	5.73	5.47	6.86	1.73	120	71.67	88.67
Shrub/dwarf shrub/moss	subarctic typical	Pura	28.280	6.21	6.61	7.61	2.08	100	72.50	85.67
Shrub/dwarf shrub/moss	subarctic typical	Pura	21.790	3.31	6.54	8.65	2	100	72.50	85.67
Shrub/sedge/moss	subarctic typical	Agapa	13.010	3.17	3.02	3.91	1.47	120	71.50	88.83
Shrub/sedge/moss	subarctic typical	Agapa	22.760	5.39	6.08	9.03	2.76	120	71.50	88.83
Shrub/sedge/moss spotty	subarctic typical	Agapa	20.600	5.91	8.84	9.84	2.81	120	71.50	88.83
Shrub/*Eriophorum*/moss spotty tussock	subarctic typical	Agapa	34.280	6.89	11.15	13.55	3.21	120	71.50	88.83
Dwarf shrub/lichen/moss spotty	subarctic typical	Agapa	33.360	4.81	12.76	14.26	2.61	120	71.50	88.83
Dwarf shrub/sedge/moss spotty	subarctic typical	Agapa	27.030	4.78	7.07	8.33	1.93	120	71.67	88.67
Dwarf shrub/sedge/moss spotty	subarctic typical	Agapa	15.260	4.44	3.22	4.26	1.64	120	71.50	88.83
Dwarf shrub/lichen/moss spotty	subarctic typical	Agapa	10.240	2.51	2.39	3.24	0.91	120	71.50	88.83
Shrub/dwarf shrub/sedge/moss	subarctic typical	Agapa	17.900	4.21	3.95	4.9	1.56	120	71.67	88.67
Dwarf shrub/moss	subarctic typical	Agapa	33.850	4.07	4.61	5.85	1.62	120	71.67	88.67
Dwarf shrub/moss polygonal	subarctic typical	Agapa	12.360	2.54	3.28	4.36	1.04	120	71.50	88.83
Dwarf shrub/moss	subarctic typical	Agapa	13.250	2.32	3.55	4.25	0.96	120	71.67	88.67
Dwarf shrub/shrub/sedge/moss	subarctic typical	Agapa	23.640	4.51	4.17	6.9	1.79	120	71.50	88.83
Shrub/dwarf shrub/sedge/moss	subarctic typical	Pura	27.040	4.61	5.84	12.34	2.89	100	72.50	85.67
Sedge/dwarf shrub/moss spotty	subarctic typical	Taimyr	14.610	3.88	4.71	5.59	1.5	150	74.83	105.00
Shrub/sedge/moss hummocky	subarctic typical	Taimyr	14.620	2.76	6.57	7.77	1.57	140	74.83	105.00
Herb/dwarf shrub/moss hummocky	subarctic typical	Taimyr	20.800	3.81	6.04	8.04	1.68	140	74.83	105.00
Sedge/dwarf shrub/moss spotty	subarctic typical	Taimyr	15.690	4.8	5.09	5.69	1.85	150	74.83	105.00
Shrub/lichen	subarctic southern	Kola Peninsula	4.040	0.64	1.95	3.24	0.46	—	69.00	31.83
Shrub/lichen	subarctic southern	Kola Peninsula	6.220	0.75	2.99	4.94	0.55	150	68.17	37.00
Shrub/dwarf shrub	subarctic southern	Kola Peninsula	7.110	0.75	1.07	4.9	0.54	80	67.50	41.50
Lichen/dwarf shrub	subarctic southern	Kola Peninsula	12.260	1.54	5.69	9.37	1.11	200	67.50	41.50
Dwarf shrub/lichen	subarctic southern	Kola Peninsula	7.600	0.84	3.93	5.71	0.58	60	67.50	41.50
Shrub (*Betula*/Salix)	subarctic southern	Kola Peninsula	17.320	3.21	2.07	11.85	2.2	80	69.00	31.83
Moss/shrub	subarctic southern	Kola Peninsula	19.150	3.39	2	12.6	2.23	80	68.17	37.67
Dwarf shrub (*Empetrum*)	subarctic southern	Kola Peninsula	10.730	1.45	3.37	7.5	1.06	90	67.50	41.50
Dwarf shrub/shrub	subarctic southern	Kola Peninsula	15.150	2.64	2.93	10.74	1.92	60	67.50	41.50
Lichen/dwarf shrub	subarctic southern	Kola Peninsula	8.590	1.11	2.84	5.79	0.71	60	67.50	41.50
Dwarf shrub/lichen	subarctic southern	Kola Peninsula	7.990	1.02	4.09	6.2	0.75	250	69.00	41.50
Shrub/dwarf shrub	subarctic southern	Kola Peninsula	10.270	1.71	3.2	7.91	1.31	170	69.00	31.83
Shrub/moss/lichen hummocky	subarctic southern	Bol'shezemel'skaya tundra	20.600	1.77	3.33	3.75	0.63	200	67.67	64.50
Shrub hummocky	subarctic southern	Bol'shezemel'skaya tundra	11.990	1.67	3.38	4.77	1.17	200	67.67	64.50
Shrub/moss hummocky	subarctic southern	Bol'shezemel'skaya tundra	18.770	2.06	3.58	5.07	1.06	205	67.67	64.50
Shrub/moss	subarctic southern	Sivaya Maska	58.780	8.62	15.69	21.62	4.92	200	66.83	64.67

Continued

Table 21.2. *Continued*

Ecosystem	Zone/subzone	Location	Total Phytomass (B, t ha^{-1})	Net Primary Production (NPP, t $ha^{-1} yr^{-1}$)	Green Phytomass (G, t ha^{-1})	Aboveground Phytomass (BA, t ha^{-1})	Aboveground Net Primary Production (ANPP, t ha yr^{-1})	Altitude (ALT, m)	Latitude (LAT, m)	Longitude (LON, m)
Dwarf shrub/moss spotty	subarctic southern	Sivaya Maska	13.680	1.81	2.24	2.74	0.91	200	66.83	64.67
Dwarf shrub/lichen hummocky	subarctic southern	Sivaya Maska	17.310	2.84	4.5	5.77	1.22	200	66.83	64.67
Shrub/moss	subarctic southern	Sivaya Maska	31.800	4.21	9.12	12.47	2.1	200	66.83	64.67
Shrub/moss	subarctic southern	Sivaya Maska	16.850	3.84	3.9	5.05	1.15	200	67.67	64.50
Shrub/dwarf shrub/moss/lichen	subarctic southern	Bol'shezemel'skaya tundra	8.610	1.31	1.64	4.32	1.1	203	67.33	64.50
Shrub/dwarf shrub/moss	subarctic southern	Bol'shezemel'skaya tundra	17.800	1.74	4.52	7.84	1.24	190	67.33	64.50
Shrub/dwarf shrub/moss hummocky	subarctic southern	Bol'shezemel'skaya tundra	21.520	1.92	3.4	10.1	1.34	190	67.33	64.50
Shrub/moss	subarctic southern	Kharp	15.630	1.71	2.64	5.63	1.06	110	67.17	68.33
Dwarf shrub/*Eriophorum* hummocky	subarctic southern	Kharp	26.580	3.77	2.4	6.58	1.52	110	67.17	68.33
Dwarf shrub/spotty	subarctic southern	Kharp	5.960	0.92	1.07	2.56	0.62	110	67.17	68.33
Shrub/dwarf shrub hummocky	subarctic southern	Kharp	29.030	3.36	2.31	8.03	1.46	150	67.17	68.33
Dwarf shrub/moss spotty	subarctic southern	Kharp	9.440	1.26	3.42	6.03	1.06	110	67.17	68.33
Shrub/dwarf shrub/moss hummocky	subarctic southern	Kharp	23.580	2.94	5.56	9.77	1.64	110	67.17	68.33
Shrub/dwarf shrub/moss	aubarctic southern	Kharp	23.580	2.94	5.56	9.77	1.64	110	67.17	68.33
Shrub/dwarf shrub/moss	subarctic southern	Kharp	23.970	3.11	3.19	13.8	2.23	110	67.17	68.33
Dwarf shrub/moss	subarctic southern	Kharp	13.530	2.09	5.67	6.45	1.09	200	67.17	68.33
Dwarf shrub/moss	subarctic southern	Kharp	23.750	2.67	4.8	9.3	1.67	110	67.17	68.33
Dwarf shrub/*Eriophorum*/moss	subarctic southern	Kharp	12.190	2.51	2.49	4.61	0.9	110	67.17	68.33
Shrub/moss hummocky	subarctic southern	Kharp	15.670	2.08	4.85	6.35	1.44	110	67.17	68.33
Shrub/dwarf shrub/moss spotty	subarctic southern	Kharp	13.610	1.6	4.18	5.75	1.1	110	67.17	68.33
Shrub/moss	subarctic southern	Kharp	20.520	2	4.06	6.71	1.2	110	67.17	68.33
Dwarf/shrub/moss	subarctic southern	Kharp	16.130	1.79	4.13	6.13	1.19	110	67.17	68.33
Shrub/sedge/moss spotty	subarctic southern	Taimyr	40.210	6.71	6.21	9.21	2.66	50	69.00	86.50
Shrub/moss spotty	subarctic southern	Taimyr	51.720	6.29	9.62	31.72	4.24	50	69.00	86.50
Dwarf shrub/moss spotty	subarctic southern	Ary-Mas	20.900	3.1	4.95	6.35	1.4	31	73.00	102.00
Shrub/moss hummocky	subarctic southern	Kolyma	15.340	3.59	2.34	6.06	1.84	30	68.83	162.00
Dwarf shrub/hummocky	subarctic southern	Kolyma	12.780	1.02	0.72	2.63	0.42	30	68.83	162.00

Shrub (*Betula*/*Pinus*)	subarctic southern	Chukchi peninsula	46.000	5.1	3.3	6	1.1	150	67.00	173.00
Sedge/moss/dwarf shrub spotty	subarctic southern	Koryakskoe highland	23.760	3	3.48	5.02	1.12	200	62.45	166.47
Dwarf shrub/lichen hummocky	subarctic southern	Koryakskoe highland	25.280	2.29	3.08	5.01	0.89	200	62.45	166.47
Moss carpet	antarctic	Signi Island	31.100	6.7	31.1	31.1	6.7	30	–60.00	–45.00
Moss turf	antarctic	Signi Island	98.000	4.77	98	98	4.77	30	–60.00	–45.00
Fellfield	subantarctic	South Georgia	0.500	0.1	0.1	0.1	0.05	300	–54.28	–36.52
Dwarf shrub/shrub/moss	subantarctic	NE Sweden	7.810	1.15	3.78	4.28	0.8	—	68.37	19.05
Dwarf shrub/lichen birch woodland	forest tundra	Kola peninsula	23.420	2.14	2.86	18.7	1.56	330	69.00	31.83
Shrub/dwarf shrub/lichen birch woodland	forest tundra	Kola peninsula	18.780	1.87	3.94	14.96	1.35	330	69.00	31.83
Shrub/lichen birch woodland	forest tundra	Kola peninsula	27.110	2.95	3.07	19.61	1.98	330	69.00	31.83
Dwarf shrub birch woodland	forest tundra	Kola peninsula	41.300	4.32	4.24	30.93	3.07	140	69.00	31.83
Dwarf shrub birch woodland	forest tundra	Kola peninsula	22.180	2.23	3.46	16.07	1.56	140	69.00	31.83
Dwarf shrub/moss spruce woodland	forest tundra	Bolshezemel'skaya tundra	76.460	4.79	10.46	42.59	3.22	190	66.83	64.67
Dwarf shrub/moss spruce woodland	forest tundra	Bolshezemel'skaya tundra	39.450	2.11	4.66	21.02	1.43	190	66.83	64.67
Shrub/moss spruce woodland	forest tundra	Bolshezemel'skaya tundra	35.140	2.49	5.36	19.14	1.86	190	66.83	64.67
Moss/spruce woodland	forest tundra	Bolshezemel'skaya tundra	80.170	4.56	10.79	46.99	3.17	190	66.83	64.67
Shrub/spruce woodland	forest tundra	Bolshezemel'skaya tundra	90.760	4.5	8.85	54.04	2.8	190	66.83	64.67
Dwarf shrub/moss/birch woodland	forest tundra	Bolshezemel'skaya tundra	88.150	4.13	7.82	58.18	3.01	190	66.83	64.67
Dwarf shrub/larch woodland	forest tundra	Kharp, Yamal	34.940	3.72	4.09	18.12	2.11	40	67.17	68.33
Dwarf shrub/birch woodland	forest tundra	Lower Ob'	43.780	3.92	8.3	25.9	3.42	50	67.00	68.33
Dwarf shrub/moss/larch woodland	forest tundra	Ary-Mas	56.710	4.7	16.47	28.05	3.31	120	73.00	102.00
Dwarf shrub/moss/larch woodland	forest tundra	Ary-Mas	31.950	2.94	9.08	14.22	2.17	120	73.00	102.00
Larch woodland (redina)	forest tundra	Nadym-Pura watershed	28.660	1.643	10.34	12.56	0.943	150	66.10	73.70
Moss/larch woodland	forest tundra	Ary-Mas	71.460	6.26	25.45	42.8	5.05	120	73.00	102.00
Picea mariana woodland	forest tundra	Alaska	29.010	1.38	4.21	16.51	0.78	1671	64.75	–148.25
Betula pubescens woodland	forest tundra	Greenland	54.900	2.07	1.2	48.5	1.98	50	62.00	–43.00
Betula papyrifera woodland	forest tundra	N.-W.Canada	99.500	2.4	1.7	74.5	2.2	50	66.00	–128.00
Birch woodland	forest tundra	Kola peninsula	68.153	4.737	2.642	49.556	3.572	90	69.00	31.83
Salix thicket	subarctic typical	Baydaratskii Bay	42.450	4.26	1.31	16.66	2.29	—	68.50	80.17
Salix/sedge/moss	subarctic typical	Agapa	49.780	9.33	12.89	15.78	5.13	—	71.50	88.83
Salix/herb	subarctic typical	Agapa	19.500	2.49	1.44	2.5	0.74	—	71.67	88.67
Salix/herb	subarctic typical	Agapa	18.210	3.99	1.14	3.21	1.19	—	71.67	88.67
Dwarf shrub sward	subantarctic	South Georgia	50.210	16.05	5.83	15.21	11.05	10	–54.28	–36.52

Continued

Table 21.2. *Continued*

Mean Annual Temperature (T_{yr}, °C)	Mean June–August Temperature (*T*678, °C)	Annual Precipitation (Prc, mm)	Days with Temperature >0°C (*L*0)	Temperature sum >0°C (Sum0, °C*Days)	Soil Organic Matter (H, t ha^{-1})	Total Soil Nitrogen (N, t ha^{-1})	Exchangeable Cations (Cat, kg-eq. ha^{-1})	Cation exchange Capacity (CEC, kg-eq. ha^{-1})	Topsoil pH (pH)	Available Potassium K (t k_2O ha^{-1})	Available Phosphorus (P_2O_5) (t P_2O_5 ha^{-1})	Source(s)
−11.9	−0.37	98	45	—	45	—	100	—	6.25	—	—	Aleksandrova, 1971
−14.5	0.40	244	72	80	60.48	4.02	439.7	574.8	7	0.3	0.231	Matveyeva and Chernov, 1976; Chugunova, 1979
−17.5	3.23	185	84	304	9.94	1.49	—	—	7.8	—	—	Bliss, 1975
−17.5	3.23	185	84	320	6.51	—	585.92	—	8.4	—	—	Bliss, Svoboda and Bliss, 1984
−17.5	3.23	185	84	320	15.03	—	526.06	—	8.4	—	—	Bliss, Svoboda and Bliss, 1984
−17.1	1.30	127	75	—	11.46	—	509.36	—	8.4	—	—	Bliss, Svoboda and Bliss, 1984
−17.1	1.30	127	75	—	15.03	—	496	—	8.5	—	—	Bliss, Svoboda and Bliss, 1984
−17.1	1.30	127	75	—	11.46	—	407.49	—	8.5	—	—	Bliss, Svoboda and Bliss, 1984
−17.1	1.30	127	75	—	34.11	—	477.6	—	8.4	—	—	Bliss, Svoboda and Bliss, 1984
−14.4	1.70	175	94	120	274.34	11	96.44	314.2	3.93	—	—	Aleksandrova, 1958; Bazilevich, 1992
−14.4	1.70	175	94	120	218.74	9.58	1333.03	1354.0	5.7	—	—	Aleksandrova, 1958; Bazilevich, 1992
−14	2.30	219	90	283	144.75	5.58	807.51	—	6.73	1.08	0.360	Matveyeva et al., 1975; Matveyeva, 1979
−14	2.30	219	90	283	177.46	7.86	920.39	—	6.9	0.825	0.575	Matveyeva et al., 1975; Matveyeva, 1979
−13.4	5.67	313	121	—	—	—	—	—	—	—	—	Bazilevich et al., 1986
−13.4	5.67	313	121	—	64.53	—	407.48	2450.0	5.73	—	—	Bazilevich et al., 1986
−13.4	5.67	313	121	—	—	—	—	—	—	—	—	Bazilevich et al., 1986
−9.8	3.97	296	109	460	—	—	—	—	—	—	—	Vil'chek, 1984
−9.8	3.97	296	109	460	—	—	—	—	—	—	—	Vil'chek, 1984
−12.6	2.47	124	87	—	46.79	1.61	43.94	104.2	8.57	—	—	P.J. Webber, in: Wielgolaski et al. 1981
−17.5	3.23	185	84	362	—	—	—	—	—	—	—	Bliss, 1975
−4.39	3.47	378	120	—	—	—	—	—	—	—	—	Tishkov, 1982
−17.5	3.23	185	84	362	75.71	1.9	—	—	7.9	2.353	1.160	Bliss and Kerik, 1973
−3.6	5.90	200	90	800	—	—	—	—	6.8	—	—	M. Lewis, in: Wielgolaski et al., 1981
−17.6	1.23	62	50	210	33.22	0.21	128.12	—	6.2	—	—	Bliss and Svoboda, 1980b
−8	9.80	200	110	—	207.2	5.28	315.2	321.0	5.8	—	—	Shaver and Chapin, 1991
−8	9.80	200	110	—	94.28	1.79	22.45	62.1	4.55	—	—	Shaver and Chapin, 1991
−8	9.80	200	110	—	196.7	6.01	344.2	347.3	6.8	—	—	Shaver and Chapin, 1991
−10.7	7.93	130	109	993	331.45	10.62	—	2063.4	5.25	—	—	Haag, 1974; Haag and Bliss, 1974
−7.6	7.67	337	116	811	222.93	5.39	1444.36	1567.2	5.99	0.719	1.530	Ignatenko and Khakimzyanova, 1971; Bazilevich, 1992
−7.6	7.67	337	116	811	—	—	—	—	—	—	—	Ignatenko and Khakimzyanova, 1971; Bazilevich, 1992
−7.6	7.67	337	116	811	249.1	6.4	945.42	1451.7	4.99	0.848	0.760	Ignatenko and Khakimzyanova, 1971; Bazilevich, 1992
−7.6	7.67	337	116	811	—	—	—	—	—	—	—	Ignatenko and Khakimzyanova, 1971; Bazilevich, 1992
−8.3	4.83	325	116	558	173.16	—	493.65	842.1	4.8	0.397	0.322	Vasilyevskaya, Ivanov and Bogatyrev, 1986
−8.3	4.83	325	116	558	161.87	6.48	382.22	711.7	4.2	0.719	0.490	Vasilyevskaya, Ivanov and Bogatyrev, 1986
−8.3	4.83	325	116	558	—	—	—	—	—	—	—	Vasilyevskaya, Ivanov and Bogatyrev, 1986
−8.3	4.83	325	116	558	69.86	5.06	444.34	813.0	5.33	0.62	0.360	Vasilyevskaya, Ivanov and Bogatyrev, 1986

−14.1	6.57	297	98	657	261.96	7.58	2936.38	3269.3	6.33	1.411	0.722	Khodachek, 1969; Shamurin et al., 1972; Bazilevich, 1992
−13.4	6.57	297	98	657	389.08	9.56	579.48	712.4	5.8	0.75	0.156	Khodachek, 1969; Shamurin et al., 1972; Bazilevich, 1992
−12.3	7.80	344	115	—	144.16	3.95	2650.05	2937.3	6.35	—	—	Pospelova, 1973; Bazilevich, 1992
−12.3	7.80	344	115	—	37.47	1.23	273.97	315.9	6.7	0.093	0.024	Bogatyrev, 1976
−12.3	7.80	344	115	—	—	—	—	—	—	—	—	Pospelova and Zharkova, 1972; Bazilevich, 1992
−12.3	7.80	344	115	—	—	—	—	—	—	—	—	Pospelova, 1973; Bazilevich, 1992
−12.3	7.80	344	115	—	302.08	8.41	2052.16	2305.1	6.3	1.003	0.494	Pospelova, 1973; Bazilevich, 1992
−13.2	4.30	297	97	466	—	—	—	—	—	—	—	Vil'chek, 1987
−13.2	4.30	297	97	466	—	—	—	—	—	—	—	Vil'chek, 1987
−12.3	7.80	344	115	—	—	—	—	—	—	—	—	Pospelova, 1973; Bazilevich, 1992
−12.3	7.80	344	115	—	—	—	—	—	—	—	—	Pospelova and Zharkova, 1972; Pospelova, 1973; Bazilevich, 1992
−12.3	7.80	344	115	—	—	—	—	—	—	—	—	Pospelova and Zharkova, 1972; Pospelova, 1973; Bazilevich, 1992
−12.3	7.80	344	115	—	—	—	—	—	—	—	—	Pospelova and Zharkova, 1972; Bazilevich, 1992
−12.3	7.80	344	115	—	—	—	—	—	—	—	—	Pavlova, 1969; Bazilevich, 1992
−13.2	4.30	297	97	466	—	—	—	—	—	—	—	Vil'chek, 1987
−13.2	4.30	297	97	466	—	—	—	—	—	—	—	Vil'chek, 1987
−12.3	7.80	344	115	—	78.2	2.58	347.92	412.7	5.8	0.115	0.056	Bogatyrev, 1973, 1975
−12.3	7.80	344	115	—	—	—	—	—	—	—	—	Bogatyrev, 1973, 1975
−12.3	7.80	344	115	—	—	—	—	—	—	—	—	Pospelova and Zharkova, 1972; Bazilevich, 1992
−12.3	7.80	344	115	—	—	—	—	—	—	—	—	Pospelova and Zharkova, 1972; Bazilevich, 1992
−12.3	7.80	344	115	—	—	—	—	—	—	—	—	Pospelova and Zharkova, 1972; Bazilevich, 1992
−12.3	7.80	344	115	—	—	—	—	—	—	—	—	Pospelova and Zharkova, 1972; Bazilevich, 1992
−12.3	7.80	344	115	—	—	—	—	—	—	—	—	Pospelova, 1973; Bazilevich, 1992
−12.3	7.80	344	115	—	—	—	—	—	—	—	—	Pospelova, 1973; Bazilevich, 1992
−12.3	7.80	344	115	—	—	—	—	—	—	—	—	Pospelova, 1973; Bazilevich, 1992
−12.3	7.80	344	115	—	—	—	—	—	—	—	—	Pospelova and Zharkova, 1972; Bazilevich, 1992
−12.3	7.80	344	115	—	—	—	—	—	—	—	—	Pospelova, 1973; Bazilevich, 1992
−12.3	7.80	344	115	—	—	—	—	—	—	—	—	Pospelova, 1973; Bazilevich, 1992
−12.3	7.80	344	115	—	—	—	—	—	—	—	—	Bogatyrev, 1976
−13.2	4.30	297	97	466	—	—	—	—	—	—	—	Zyryanov and Kolpashchikov, 1981
−14.57	4.17	287	93	459	—	—	—	—	—	—	—	Pospelova and Orlov,1984
−14.57	4.17	287	93	459	—	—	—	—	—	—	—	Pospelova and Orlov, 1984; Bazilevich, 1992
−14.57	4.17	287	93	459	—	—	—	—	—	—	—	Pospelova and Orlov, 1984; Bazilevich, 1992
−14.57	4.17	287	93	459	—	—	—	—	—	—	—	Pospelova and Orlov, 1984; Bazilevich, 1992
−1.35	10.23	386	154	1191	77.51	—	84.38	202.3	4.93	—	—	Manakov, 1970, 1972
−1.7	10.57	399	159	1206	100.04	—	102.61	142.9	4.53	—	—	Manakov, 1970, 1972
−1.1	6.97	358	163	942	240.03	—	695.74	1003.5	4.8	—	—	Manakov, 1970, 1972
−1.1	6.97	358	163	942	327.46	—	42.05	352.9	4.55	—	—	Manakov, 1970, 1972
−1.1	6.97	358	163	942	106.88	—	61.83	175.2	4.73	—	—	Manakov, 1970, 1972
−1.35	10.23	386	154	1191	461.19	—	728.36	1634.2	4.08	—	—	Manakov, 1970, 1972
−1.1	6.97	397	163	942	207.33	—	96.57	361.4	4.47	—	—	Manakov, 1970, 1972
−1.1	6.97	358	163	942	283.07	—	680.71	1163.8	4.2	—	—	Manakov, 1970, 1972
−1.1	6.97	358	163	942	103.97	—	1362.6	1517.8	4.65	—	—	Manakov, 1970, 1972

Continued

Table 21.2. *Continued*

Mean Annual Temperature (T_{yr}, °C)	Mean June–August Temperature (*T678*, °C)	Annual Precipitation (Prc, mm)	Days with Temperature >0°C (*L0*)	Temperature sum >0°C (Sum0, °C*Days)	Soil Organic Matter (H, t ha^{-1})	Total Soil Nitrogen (N, t ha^{-1})	Exchangeable Cations (Cat, kg-eq. ha^{-1})	Cation exchange Capacity (CEC, kg-eq. ha^{-1})	Topsoil pH (pH)	Available Potassium K (t k_2O ha^{-1})	Available Phosphorus (P_2O_5) (t P_2O_5 ha^{-1})	Source(s)
−1.1	6.97	358	163	942	61.31	—	104.1	229.5	4.45	—	—	Manakov, 1970, 1972
−1.1	6.97	415	163	942	123.93	2.9	50.62	221.5	4.76	—	—	Manakov, 1970
−1.35	10.23	386	154	1191	48.8	1.3	54.83	143.2	4.84	—	—	Manakov, 1970
−7.6	7.67	337	116	811	—	—	1347.01	1732.4	4.6	—	—	Shamurin, 1970; Bazilevich, 1992
−7.6	7.67	337	116	811	99.92	—	1778.46	1967.6	5.1	—	—	Shamurin, 1970; Bazilevich, 1992
−7.6	7.67	337	116	811	219.73	—	747.8	3006.0	4.8	—	—	Shamurin, 1970; Bazilevich, 1992
−6	9.37	385	131	1004	252.2	7.59	1389.63	1705.4	3.97	0.675	0.471	Rakhmanina, 1971
−6	9.37	385	131	1004	—	—	—	—	—	—	—	Rakhmanina, 1971
−6	9.37	385	131	1004	—	—	1982.98	2299.1	5.4	—	—	Rakhmanina, 1971
−6	9.37	385	131	1004	344.93	10.96	2839.83	4168.3	4.27	1.555	0.521	Ignatenko et al., 1973
−7.6	7.67	337	116	811	—	—	—	—	—	—	—	Khantimer, 1970
−6.5	8.93	385	126	953	—	—	—	—	—	—	—	Vil'chek, 1986
−6.5	8.93	385	126	953	190.34	7.27	2844.3	3216.0	4.91	—	—	Vil'chek, 1986
−6.5	8.93	385	126	953	—	—	—	—	—	—	—	Vil'chek, 1986
−6.5	10.70	464	131	1160	210.86	—	2756.4	3098.2	4.65	1.562	0.807	Gasheva, 1974; Bazilevich, 1992
−6.5	10.70	464	131	1160	—	—	—	—	—	—	—	Gasheva, 1974; Bazilevich, 1992
−6.5	10.70	464	131	1160	84.05	—	5739.68	5773.9	5.51	1.828	2.198	Gasheva, 1974; Bazilevich, 1992
−6.5	10.70	464	131	1160	61.21	3.28	165.64	523.5	4.83	0.311	0.087	Gasheva, 1974; Bazilevich, 1992
−6.5	10.70	464	131	1160	—	—	—	—	—	—	—	Trotsenko, 1974; Bazilevich, 1992
−6.5	10.70	464	131	1160	—	—	—	—	—	—	—	Trotsenko, 1974; Bazilevich, 1992
−6.5	10.70	464	131	1160	—	—	—	—	—	—	—	Trotsenko, 1974; Bazilevich, 1992
−6.5	10.70	464	131	1160	109.43	—	—	—	—	—	—	Trotsenko, 1974; Bazilevich, 1992
−6.5	10.70	464	131	1160	—	—	—	—	—	—	—	Trotsenko, 1974; Bazilevich, 1992
−6.5	10.70	464	131	1160	—	—	—	—	—	—	—	Trotsenko, 1974; Bazilevich, 1992
−6.5	10.70	464	131	1160	—	—	—	—	—	—	—	Andreyashkina & Gorchakovsky, 1972; Bazilevich, 1992
−6.5	10.70	464	131	1160	53.21	—	1612.64	1631.7	4.94	0.799	0.789	Andreyashkina & Gorchakovsky, 1972; Bazilevich, 1992
−6.5	10.70	464	131	1160	—	—	—	—	—	—	—	Gorchakovskii and Andreyashkina, 1972; Bazilevich, 1992
−6.5	10.70	464	131	1160	—	—	—	—	—	—	—	Gorchakovskii and Andreyashkina, 1972; Bazilevich, 1992
−9.6	9.97	384	120	1037	—	—	—	—	—	—	—	Pospelova, 1973; Bazilevich, 1992
−9.6	9.97	384	120	1037	—	—	—	—	—	—	—	Pospelova, 1973; Bazilevich, 1992
−14.8	6.10	375	99	629	307.47	7.89	3615.25	3841.3	6.16	—	—	Ignatenko et al., 1972

−11.6	9.93	268	109	—	399.52	15.97	1179.37	1226.0	4.63	0.176	0.137	Andreev and Galaktionova, 1978
−11.6	9.93	268	109	—	—	—	—	—	—	—	—	Andreev and Galaktionova, 1978
−9.5	3.77	360	104	—	290.22	—	—	—	5.13	—	—	Bogdanov et al., 1979; Bazilevich, 1992
−8.6	9.03	346	118	—	—	—	—	—	—	—	—	Viakhireva-Vasil'kova et al., 1964; Bazilevich, 1992
−8.6	9.03	346	118	—	—	—	—	—	—	—	—	Viakhireva-Vasil'kova et al., 1964; Bazilevich, 1992
−3.3	0.87	400	120	210	274.97	10.83	—	—	4.9	0.122	0.500	Collins et al., 1975
−3.3	0.87	400	120	210	462.18	11.12	—	—	4.8	0.232	0.229	Collins et al., 1975
1.8	4.57	2105	170	800	15.74	0.72	—	—	5.1	0.001	0.002	Lewis Smith and Walton, 1975
−0.7	10.73	300	145	—	—	—	—	—	—	—	—	Bergman, 1980
−1.8	8.90	517	138	1005	—	—	—	—	—	—	—	Manakov, 1970, 1974
−1.8	8.90	517	138	1005	—	—	—	—	—	—	—	Manakov, 1970, 1974
−1.8	8.90	517	138	1005	—	—	—	—	—	—	—	Manakov, 1970, 1974
−1.8	8.90	517	138	1005	—	—	—	—	—	—	—	Manakov, 1970, 1974
−1.8	8.90	517	138	1005	208.84	5.06	152.96	346.1	4.31	—	—	Manakov, 1970, 1974
−6	9.37	433	131	1004	524.44	11.44	983.64	4044.7	3.83	—	—	Ignatenko et al., 1973; Ignatenko 1979
−6	9.37	433	131	1004	—	—	—	—	—	—	—	Chertovskoy et al., 1978; Bazilevich, 1992
−6	9.37	433	131	1004	—	—	—	—	—	—	—	Chertovskoy et al., 1978; Bazilevich, 1992
−6	9.37	433	131	1004	—	—	—	—	—	—	—	Chertovskoy et al., 1978; Bazilevich, 1992
−6	9.37	433	131	1004	—	—	—	—	—	—	—	Chertovskoy et al., 1978; Bazilevich, 1992
−6	9.37	433	131	1004	—	—	—	—	—	—	—	Chertovskoy et al., 1978; Bazilevich, 1992
−6.5	10.70	464	131	1160	—	—	—	—	—	—	—	Trotsenko, 1974; Bazilevich, 1992
−6.5	10.70	464	131	1160	—	—	—	—	—	—	—	Dyakonov and Retium, 1971; Bazilevich, 1992
−13.4	8.50	243	106	852	246.43	7.95	1686.55	1919.6	5.13	—	—	Ignatenko et al., 1972; Bazilevich, 1992
−13.4	8.50	243	106	852	—	—	—	—	—	—	—	Ignatenko et al., 1972; Bazilevich, 1992
−7.8	9.90	430	125	1080	67.17	—	617.82	890.1	4.66	—	—	Vasilevskaya, Ivanov and Bogatyrev, 1986
−13.4	8.50	243	106	852	—	—	—	—	—	—	—	Ignatenko et al., 1972; Bazilevich, 1992
−3.4	14.17	294	150	—	—	—	—	—	—	—	—	Barney and Van Cleve, 1973; Van Cleve, 1981; Bazilevich, 1992
−4	9.03	900	138	—	—	—	—	—	—	—	—	Elkington and Jones, 1974
−7.1	14.13	328	139	—	—	—	—	—	—	—	—	Sakai et al., 1979, cit Bazilevich, 1992
−1.35	10.23	386	154	1191	206.84	5.96	162.19	474.0	4.88	—	—	Manakov, 1970
−9.4	9.67	550	121	1015	—	—	—	—	—	—	—	Ignatenko and Khakimzyanova, 1971; Bazilevich, 1992
−12.3	7.80	344	115	—	—	—	—	—	—	—	—	Pospelova and Zharkova, 1972; Bazilevich, 1992
−12.3	7.80	344	115	—	—	—	—	—	—	—	—	Pospelova and Zharkova, 1972; Bazilevich, 1992
−12.3	7.80	344	115	—	—	—	—	—	—	—	—	Pospelova and Zharkova, 1972; Bazilevich, 1992
1.8	4.57	2105	170	800	646.85	35.22	—	1703.2	4.9	2.117	0.174	Lewis Smith and Walton, 1975

multivariate numerical analysis, a specific set of predictive equations was constructed describing relationships of productivity characteristics of zonal arctic ecosystems to climatic, soil, and vegetation factors–predictors. Because introduction of soil characteristic proved to be a significant step in the present work, before discussing the equations themselves it seems necessary to review some of the methods and approaches used to prepare the soil data included in the data base.

Introduction of Soil Characteristics into Phenomenological Models of Primary Productivity of Arctic and Alpine Ecosystems

The problem of incorporating soil characteristics into phenomenological models of ecosystem productivity deserves special attention because of the need to transform the traditional pedological data, expressed in the relative units (usually percentage of dry weight or volume) to quantifiable parameters. These parameters need to be expressed on the unit surface area basis. Therefore, the reserves of nutrients and water should be described per 1 ha (or 1 m^2) of the territory occupied by the corresponding ecosystem. This is because the absolute amounts of these resources, and not their relative concentrations, determine the more or less favorable conditions for plant growth. As an example, let us consider the task of estimation of the reserves of carbon of the soil organic matter.

Usually the soil organic matter data are presented as the percentage of carbon C_i on the absolute dry weight basis in the n consecutive soil horizons, labeled by the running index i ($i = 1, 2, \ldots, n$), C_1 referring to the first horizon from the surface, and C_n to the deepest horizon. Let h_i denote the thickness of the ith horizon (cm), and d_i as its average bulk density ($g\,cm^{-3}$). The total amount of carbon in the soil expressed in metric tons per hectare per whole depth of the profile would be given by:

$$C(t/ha) = \sum_{i=1}^{n} C_i d_i h_i . \tag{11}$$

In order to obtain the surface area-based estimates, the data on bulk density of soil horizons are required. Unfortunately, the bulk density data are seldom published along with other soil characteristics. To overcome this lack of available bulk density data, transformation of traditional soil data to reliable estimates of bulk density needs to be established.

An empirical formulation was determined using a large number of soil samples ($n = 307$) from different genetic horizons and depths retrieved from the literature. Each sample has an estimate of its bulk density d_v and some other traditional soil properties. Analysis of this data set has revealed that soil organic matter content H and the depth in the profile z where the sample was taken are the best estimators of bulk density. These results are consistent with findings by Gersper et al. (1980) and Post et al. (1982).

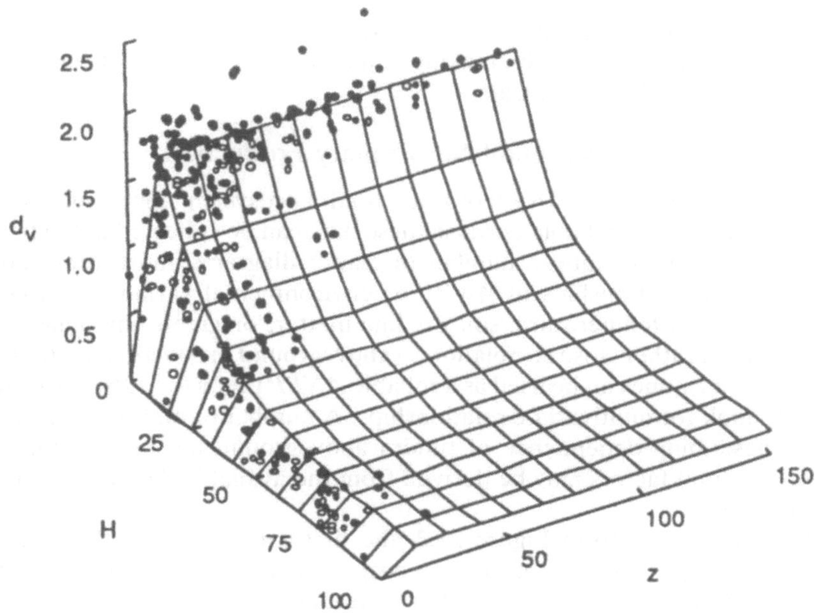

Figure 21.1. Comparison of the empirical data (points) on the bulk density of the arctic and alpine soils d_v ($t\,m^{-3}$) in relation to soil organic matter content H (% dry weight) and depth in the profile z (cm) with the model response surface, described by the approximating Eq. (12) ($n = 307$; $R^2 = 0.83$). The solid points lie above, and the open dots lie below the approximating surface.

Figure 21.1 shows the scatter diagram of the empirical points and the response surface, approximating the required relationship, described by the equation:

$$d_v = 0.166[1 - \exp(-0.529z)][1 + 8.733\exp(-0.057H)]. \tag{12}$$

The sufficiently high coefficient of determination $R^2 = 0.83$ indicates that the empirical relationship explains a large proportion of the data variability and may be used to estimate the bulk density of soil samples using depth and humus content, when the original data are not available.

Productivity of Zonal Arctic and Subarctic Ecosystems

Analysis of various relationships between the productivity characteristics of zonal tundra ecosystems and the associated ecological factors became available with the construction of the data base, including aboveground net primary production (*ANPP*), total ecosystem net primary production

(NPP), vascular and cryptogamic productivity components, etc. We shall consider them consecutively, making comments to problems of reliability and representability of the data.

Aboveground Net Primary Production

The information on the $ANPP$ of zonal tundra ecosystem apparently is the most extensive and reliable because these data can be collected in the field. Figure 21.2 demonstrates a number of scatter diagrams, illustrating some single-factor relationships of $ANPP$ to environmental variables. Though such factors as temperature, soil organic matter, or green phytomass reserve show a rather close correlation to aboveground production, neither of these factors alone can be used as a reliable $ANPP$ predictor. Obviously, we have to look for multivariate relationships. A number of multivariate functions, describing dependence of tundra aboveground net productivity to environmental factors may be derived from the available data base (Table 21.2).

As can be seen from Figure 21.2, the first candidates for the role of predictors of productivity of tundra ecosystems are the mean annual temperature T_{yr} and the amount of soil organic matter H. The predictive equation for the relationship $ANPP = f(T_{yr}, H)$, provided by the NonlinearFit algorithm (Novak, 1992), is given by the formula:

$$\mathrm{ANPP} = a_1 \exp\left(a_2 T_{yr} + a_3 H\right). \tag{13}$$

The parameter estimates are: $a_1 = 0.925$, $a_2 = 0.0526$, and $a_3 = 0.00358$. The coefficient of determination for this approximation is equal to $R^2 = 0.64$. To visualize the goodness of fit of Eq. (13), in Figure 21.3 we present the scatter diagram of the data and the graph of the model response surface, approximating these data with the mean square error $s = 1.08\,\mathrm{t\,ha^{-1}\,yr^{-1}}$. Mean air temperature and organic matter reserve are the most easily available parameters. Therefore, the formula (13) may be recommended for preliminary approximation of the aboveground productivity.

To obtain the more accurate description, consider the following model, expressing the productivity as a function of temperature T_{yr} and soil nitrogen reserve N:

$$\mathrm{ANPP} = \frac{a_1 \exp\left(a_2 T_{yr}\right)}{\left(1 + a_3 \exp\left(-a_4 \mathrm{N}\right)\right)}. \tag{14}$$

With parameters $a_1 = 8.49$, $a_2 = 0.136$, $a_3 = 86.87$, and $a_4 = 0.80$ this model explains 83% of the variation of empirical data subset with $n = 39$ relevant records ($R^2 = 0.83$) and the mean square error $s = 0.93\,\mathrm{t\,ha^{-1}\,yr^{-1}}$.

Introduction of additional factors may further improve the model. An obviously important predicting factor for productivity is the amount of

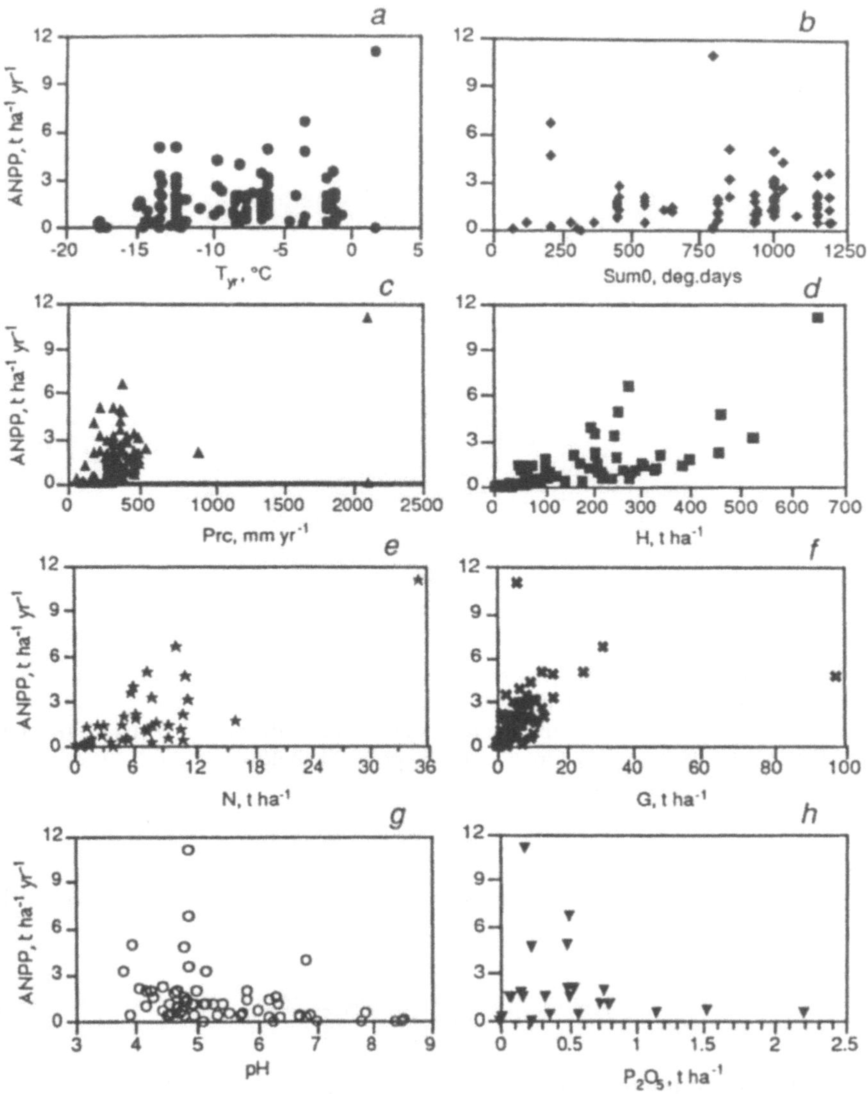

Figure 21.2. Scatter diagrams of relationships of the aboveground net primary production of zonal tundra ecosystems (*ANPP*, $t\,ha^{-1}\,yr^{-1}$) to some productivity factors: (*a*) mean annual temperature, T_{yr} (°C); (*b*) annual temperature sum ($T >$ 0°C), *Sum0* (degree days); (*c*) annual precipitation *Prc*, (mm); (*d*) total reserve of soil organic matter, *H* ($t\,ha^{-1}$); (*e*) total reserve of soil nitrogen, *N* ($t\,ha^{-1}$); (*f*) green phytomass (vascular + cryptogamic), *G* ($t\,ha^{-1}$); (*g*) topsoil pH; (*h*) available soil phosphorus, P_2O_5 ($t\,P_2O_5\,ha^{-1}$). Data adapted from Table 21.2.

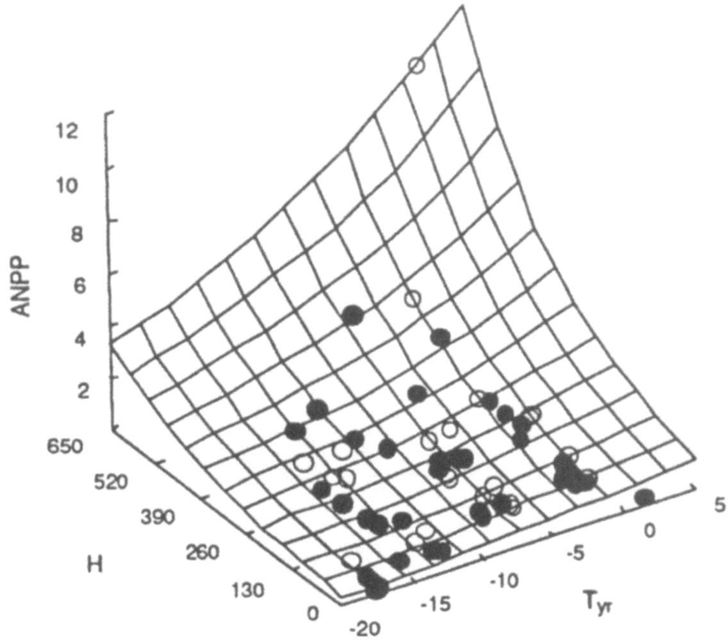

Figure 21.3. Empirical data subset (points) and the approximating response surface [Eq. (3)] describing the relationship of zonal tundra *ANPP* (t ha^{-1} yr^{-1}) to mean annual temperature T_{yr} (°C) and the reserve of soil organic matter H (t ha^{-1}) ($n = 66$; $R^2 = 0.64$; $s = 1.08$ t ha^{-1} yr^{-1}). Open and solid points lie below and above the approximating surface, respectively.

green phytomass G. A combination of amount of green phytomass with temperature and soil nitrogen leads to the model:

$$\text{ANPP} = a_1 \exp(a_2 T_{yr} + a_3 \text{N}) \frac{G}{(G + a_4)}. \tag{15}$$

With the parameter values $a_1 = 4.76$, $a_2 = 0.0522$, $a_3 = 0.0415$, and $a_4 = 6.04$, the determination coefficient increases to $R^2 = 0.87$ and the mean square deviation of the model from the $n = 39$ empirical points is as low as $s = 0.81$ t ha^{-1} yr^{-1} (Figure 21.4).

Another interesting model, providing close fitting to the data, is the following Liebig-type expression, relating aboveground productivity to temperature T_{yr}, soil nitrogen N, and available soil phosphorus P_2O_5:

$$\text{ANPP} = a_1 \exp(a_2 T_{yr}) \min\left\{ \frac{1}{1 + a_3 \exp(-a_4 \text{N})}, \ \exp\left(-\left(\frac{\text{P}_2\text{O}_5 - 0.57}{a_5} \right)^2 \right) \right\}. \tag{16}$$

The parameter values, bringing the best fit of the model to the data are: $a_1 = 10.49$, $a_2 = 0.17$, $a_3 = 64.09$, $a_4 = 0.87$, and $a_5 = 0.80$. The determination coefficient they allow is even greater than that in the previous case and equal to $R^2 = 0.93$. Nevertheless, due to the smaller number of data points for which data on temperature, nitrogen, phosphorus, and productivity are available ($n = 21$) in this and due to the greater number of parameters (k = 5) the estimate of mean square error of approximation in this case is somewhat higher: $s = 0.82\,\mathrm{t\,ha^{-1}\,yr^{-1}}$. The graphs in Figure 21.5 are given to illustrate the partial two-factorial response functions of the complex three-factorial relationship $ANPP = f(T_{yr}, \mathrm{N}, \mathrm{P_2O_5})$. Because of difficulties of visualization of points and response surface in a four-dimensional space to assess the goodness of fit of the model in Figure 21.5(d), a scatter diagram of the empirical *ANPP* data is plotted against the corresponding model values. One can see how the points concentrate near the diagonal, which would correspond to complete agreement between the model and the data.

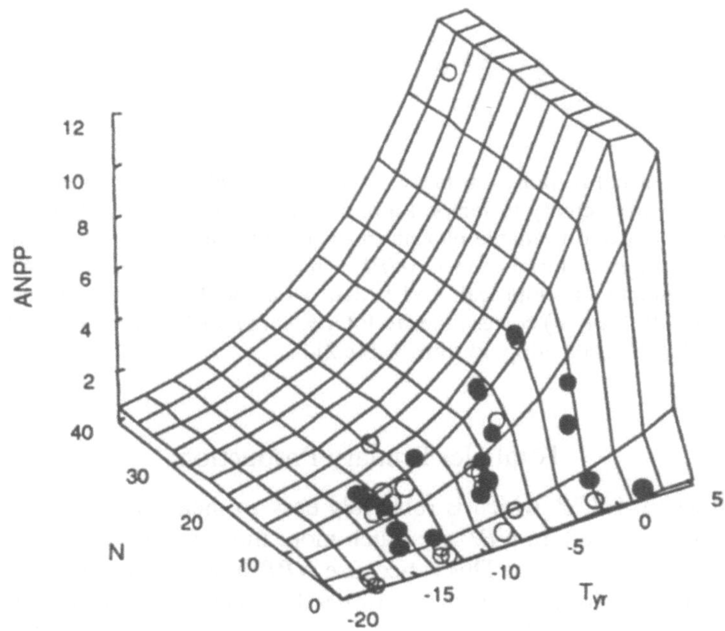

Figure 21.4. Original data subset (points) and the approximating response surface [Eq. (14)] for the relation of zonal tundra *ANPP* ($\mathrm{t\,ha^{-1}\,yr^{-1}}$) to mean annual temperature T_{yr} (°C) and total soil nitrogen N ($\mathrm{t\,ha^{-1}}$) ($n = 39$; $R^2 = 0.83$; $s = 0.93$ $\mathrm{t\,ha^{-1}\,yr^{-1}}$). Open and solid points lie below and above the approximating surface, respectively.

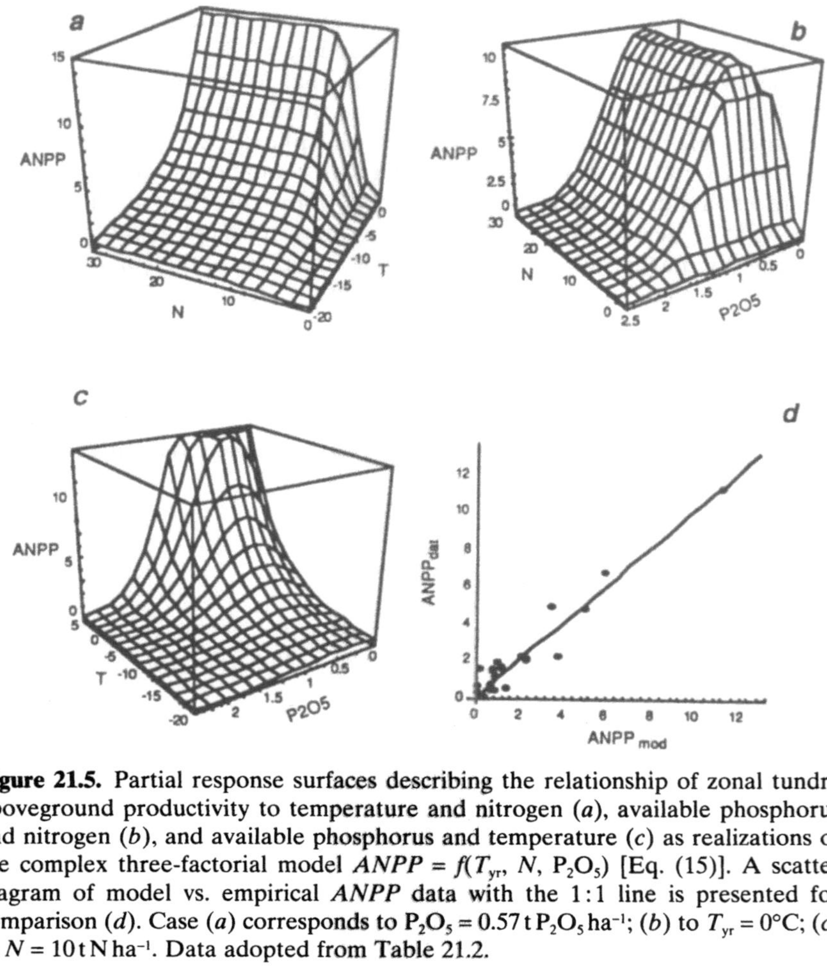

Figure 21.5. Partial response surfaces describing the relationship of zonal tundra aboveground productivity to temperature and nitrogen (*a*), available phosphorus and nitrogen (*b*), and available phosphorus and temperature (*c*) as realizations of the complex three-factorial model $ANPP = f(T_{yr}, N, P_2O_5)$ [Eq. (15)]. A scatter diagram of model vs. empirical *ANPP* data with the 1:1 line is presented for comparison (*d*). Case (*a*) corresponds to $P_2O_5 = 0.57\,t\,P_2O_5\,ha^{-1}$; (*b*) to $T_{yr} = 0°C$; (*c*) to $N = 10\,t\,N\,ha^{-1}$. Data adopted from Table 21.2.

Total Net Primary Production

In contrast to *ANPP*, which is relatively easy to measure in the field, the data on total ecosystem *NPP* are much more indirect and, sometimes, are based on approximations. This is because *NPP* should account for changes in belowground biomass, which is estimated with considerably lower precision (if estimated at all). It also should be noted that only half of the estimates in the *NPP* column of Table 21.2 are derived from the data of original authors (sometimes with minor recalculation), whereas the other half (marked by italics) are provided by Bazilevich (1992, 1994) on the basis of comparative extrapolations from the belowgound phytomass and shoot/root ratio for production at similar sites. Therefore, the following predictive equations for *NPP* should be considered as approximation. Nevertheless, they still provide interesting information on the current level of quantitative knowledge for an important functional characteristic of the ecosystem.

As can be seen from Figure 21.6, a fairly reasonable description of the general trend of response of *NPP* to climatic and soil conditions is provided by the mean annual temperature and the reserve of soil organic matter—the two factors that a long time ago were marked by Jenny (1941; 1980) as predictors for productivity. The equation describing the relationship of *NPP* to temperature T_{yr} and soil organic matter stock H is of a simple exponential type:

$$\mathrm{NPP} = a_1 \exp\left(a_2 T_{yr} + a_3 H\right). \quad (17)$$

The least-square estimates of its parameters on the $n = 66$ records subset of Table 21.2, delivering the highest value of $R^2 = 0.63$ and the lowest mean square error $s = 1.63\,\mathrm{t\,ha^{-1}\,yr^{-1}}$, are: $a_1 = 1.33$, $a_2 = 0.027$, and $a_3 = 0.0036$.

The scatter diagram of data versus corresponding model values [Figure 21.6(b)], allowing a more visible comparison, shows that several points are significantly far from the 1:1 line. Additional factor(s) are obviously needed to improve the model.

One of such factors, soil nitrogen *N*, helps to somewhat increase the R^2 and decrease the error, as can be seen from the model:

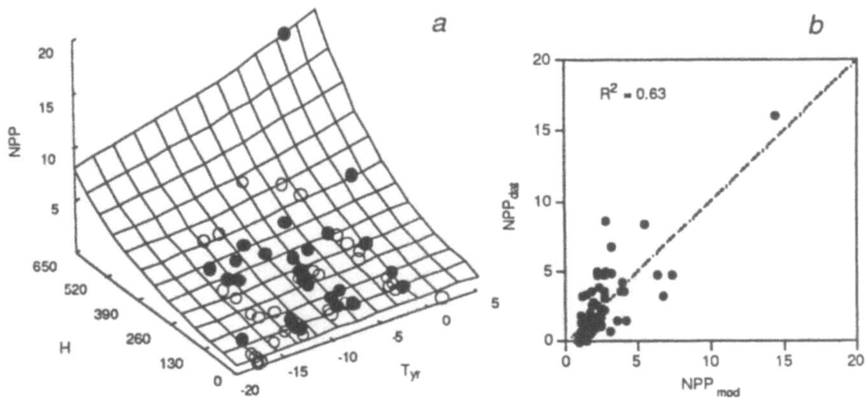

Figure 21.6. Net primary production of zonal tundra ecosystems *NPP* ($\mathrm{t\,ha^{-1}\,yr^{-1}}$) in relation to mean annual temperature *T* (°C) and soil organic matter stock *H* ($\mathrm{t\,ha^{-1}}$) ($n = 66$; $R^2 = 0.63$; $s = 1.63\,\mathrm{t\,ha^{-1}\,yr^{-1}}$): (*a*) data subset (points) and the approximating response surface [Eq. (17)]; open and solid points lie below and above the surface correspondingly; (*b*) scatter diagram of empirical data versus values calculated by Equation (17); the 1:1 line is drawn for comparison.

$$\mathrm{NPP} = \frac{a_1 \exp(a_2 T_{yr} + a_3 H)}{[1 + a_4 \exp(-a_5 N)]}, \tag{18}$$

($a_1 = 9.43$, $a_2 = 0.055$, $a_3 = 0.00064$, $a_4 = 4.70$, $a_5 = 0.182$, $n = 39$, $R^2 = 0.73$, $s = 1.60\,\mathrm{t\,ha^{-1}\,yr^{-1}}$).

Further improvement of the model is achieved with the introduction of the green phytomass variable G (Figure 21.7) leading to the equation:

$$\mathrm{NPP} = a_1 \exp(a_2 T_{yr} + a_3 N) \frac{G}{(G + a_4)} \tag{19}$$

with parameter estimates $a_1 = 3.83$, $a_2 = 0.012$, $a_3 = 0.050$, $a_4 = 2.45$, and statistical characteristics $n = 39$, $R^2 = 0.80$, and $s = 1.35\,\mathrm{t\,ha^{-1}\,yr^{-1}}$. Taking into account that a great part of *NPP* values in Table 21.1 are more or less rough estimates, the obtained level of agreement between "data" and the model might be considered as acceptable. Compared to Figure 21.6, the points on the scatter diagram for the model (19) lie closer to the diagonal line, indicating better agreement between the model and the data [Figure 21.7(b)]. Obviously, considering the green phytomass as the productivity predictor is not the best approach, considering difficulties of obtaining representative green phytomass data in the field. Probably for future use a kind of remotely measured index, correlating with the green phytomass, would be found and used for productivity predictions.

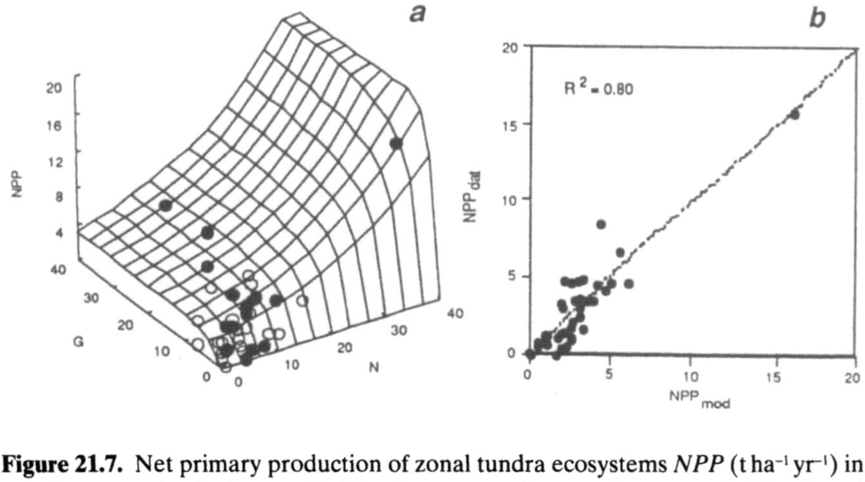

Figure 21.7. Net primary production of zonal tundra ecosystems *NPP* ($\mathrm{t\,ha^{-1}\,yr^{-1}}$) in relation to total soil nitrogen N ($\mathrm{t\,ha^{-1}}$), amount of green phytomass G ($\mathrm{t\,ha^{-1}}$), and mean annual temperature T_{yr} (°C) ($n = 39$; $R^2 = 0.80$; $s = 1.35\,\mathrm{t\,ha^{-1}\,yr^{-1}}$): (*a*) data subset (points) and the approximating response surface [Eq. (19), $T_{yr} = -9.2$°C]; open and solid points lie below and above the surface, respectively; (*b*) scatter diagram of empirical data NPP_{dat} versus values NPP_{mod} calculated by the model [Eq. (19)].

Vascular Versus Cryptogamic Productivity

In certain cases it is desirable to distinguish vascular and cryptogamic productivity. Because these two plant types greatly differ in their ecological and productivity characteristics, it is interesting to determine if those differences might be reflected in corresponding quantitative terms. The abovementioned arctic/alpine climate/soil/productivity data base contains some information to consider this problem. Some of the existing data in literature on tundra *ANPP* can be subdivided into vascular *ANPP* ($ANPP_v$) and cryptogamic *ANPP* ($ANPP_c$) components. We attempted to construct the formulae for $ANPP_v$ and $ANPP_c$ separately by correlating appropriate empirical estimates with relevant ecological factors.

As the data on vascular productivity are usually more accurate than the data on cryptogamic productivity, they allowed rather good quantitative models (Figure 21.8). For estimation of the *ANPP* of zonal tundra ecosystems the best fitting multivariate relationship was found to be the function of the temperature sum above 0°C (Sum0), amount of soil nitrogen *N*, and soil organic matter *H*:

$$\text{ANPP}_v = a_1 \exp(a_2\text{Sum0} + a_3 N)[1 - \exp(-a_4 H)] \tag{20}$$

were parameters are: $a_1 = 0.103$, $a_2 = 0.00184$, $a_3 = 0.0833$, and $a_4 = 0.110$, whereas the number of data points $n = 21$, $R^2 = 0.93$, and the mean square

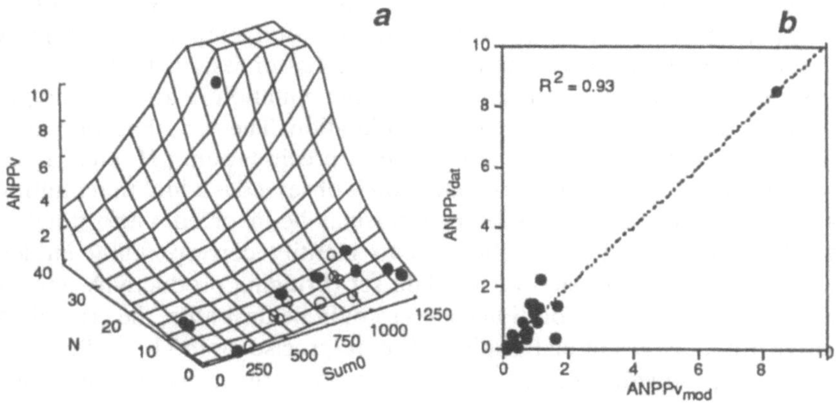

Figure 21.8. Vascular aboveground net primary production of zonal tundra ecosystems $ANPP_v$ (t ha^{-1} yr^{-1}) in relation to temperature sum above 0°C (Sum0) (degree × days) and total soil nitrogen *N* (t ha^{-1}): (*a*) empirical data (points) and partial response surface of $ANPP_v$ to temperature sum and amount of nitrogen generated by Eq. (20) with soil organic matter level $H = 500$ t ha^{-1}; (*b*) scatter diagram of data versus model (the 1:1 line is drawn for easy comparison).

error $s = 0.5\,\text{t ha}^{-1}\,\text{yr}^{-1}$. The temperature sum, Sum0, appears to be a more successful predictor of the aboveground productivity of vascular tundra plants than the mean annual temperature or the mean growing season temperature. This is due to the ability of temperature sum to accumulate information on the whole thermic history of the favorable periods during the growing season in contrast to the mean annual or mean seasonal temperatures, which are influenced also by occasional negative temperatures.

The combined effect of temperature sum and nitrogen reserve of the ecosystem on the vascular *ANPP* is illustrated by Figure 21.8, where the response surface corresponds to the third argument of Eq. (20)—the soil organic matter content H, taken at the level $H = 500\,\text{t ha}^{-1}$.

For the total vascular net primary production of zonal tundra ecosystems it was possible to obtain a predictive equation dependent on the temperature sum, and the amounts of nitrogen N and soil organic matter H:

$$NPP_v = a_1 \exp(a_2 \text{Sum0} + a_3 N)[1 - \exp(-a_4 H)] \quad (21)$$

which provides the best fit to data with parameters $a_1 = 0.64$, $a_2 = 0.0011$, $a_3 = 0.063$, and $a_4 = 0.027$ ($n = 21$; $R^2 = 0.87$; $s = 1.1\,\text{t ha}^{-1}\,\text{yr}^{-1}$). Note that despite the photosynthesizing phytomass (in this case the green vascular phytomass G_v) is not included in the list of arguments of Eq. (21), its coefficient of determination is greater and the error is smaller than those in Eq. (19) for vascular and cryptogamic *NPP*.

To evaluate the available data on cryptogamic production of zonal tundra ecosystems in relation to various ecological factors one can consider the scatter diagrams in Figure 21.9. Though there are some vague patterns there [e.g., positive correlation with growing season temperature (a), growing season precipitation (b), and soil organic matter reserve (d)], it was not possible to obtain any significant multivariate equation relating the cryptogamic production to factors under consideration. The fact, that for a given value of a single factor (e.g., temperature, precipitation, pH, etc.) a very different value of cryptogamic production is observed is a hint that multiplicative or alternative types of interactions may be involved. Unfortunately, the data available did not allow us to identify corresponding models. Whereas further expansion of the data set (bringing productivity and the related factors data for additional sites) would be desirable, the crux of the matter probably lies in elaboration of more precise methods for field estimation of production of such slowly growing organisms as lichens and mosses.

Summary

A set of phenomenological models was derived from the global climate/soil/productivity data base for zonal arctic/subarctic ecosystems. Such an attempt allows prediction of the aboveground and total net primary produc-

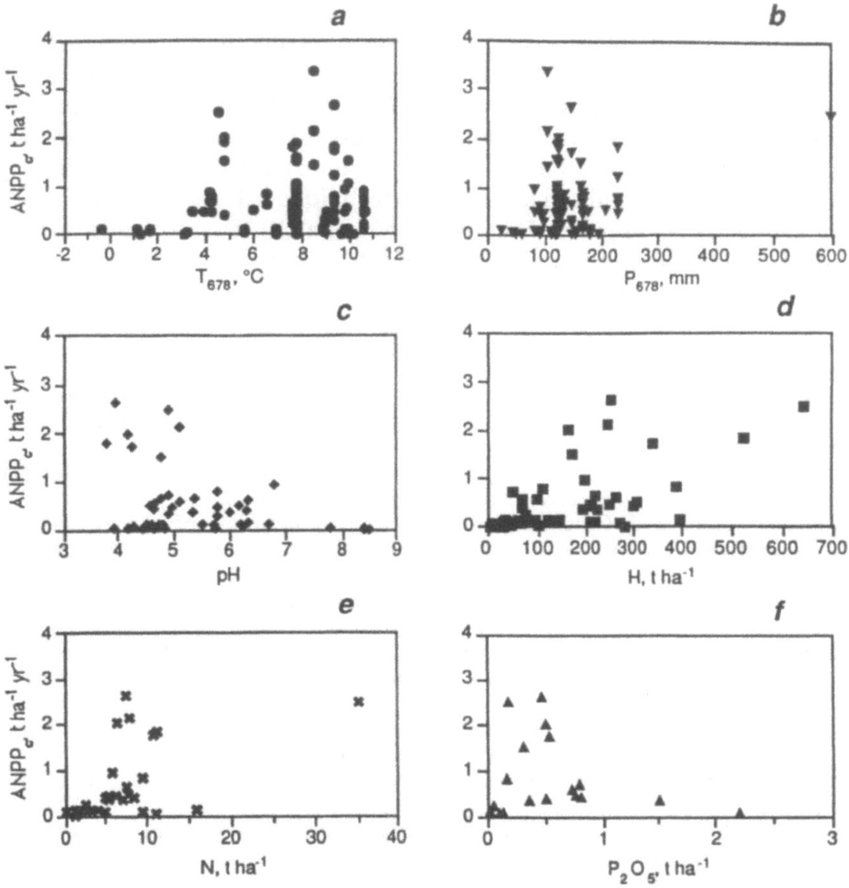

Figure 21.9. Projections of the multidimensional data set of cryptogamic productivity $ANPP_c$ (t $ha^{-1} yr^{-1}$) of zonal tundra ecosystems on the single factor coordinate planes: (*a*) mean June–August air temperature T_{678}, °C; (*b*) June–August precipitation P_{678}, mm; (*c*) topsoil *pH*; (*d*) soil organic matter reserve *H*, t ha^{-1}; (*e*) total soil nitrogen *N*, t ha^{-1}; (*f*) available soil phosphorus P_2O_5, t ha^{-1}.

tion of zonal tundra and related ecosystems based on such readily available climatic, soil, and vegetation characteristics as mean annual temperature, temperature sum, reserves on soil organic matter, nitrogen and phosphorus, and the amount of green plant phytomass. These equations explain a higher part of the variability of the available productivity data for zonal arctic ecosystem than the models based on climatic factors exclusively. The most obvious way to further increase the predictive power of phenomenological models of primary productivity of arctic ecosystems is to expand the data base by adding information on productivity of ecosystems with greater geographical representation and higher variability of soil characteristics (e.g., central and eastern parts of Canadian Arctic and western Alaska, currently not covered by the data base). This approach will expand the range of the relationships represented in the models of the multiple productivity factors. The described technique should also be applied to the analysis of data on productivity of azonal (meadow, mire) arctic ecosystems in their relation to climatic and soil factors. This technique will enable us to scale up the productivity estimates from ecosystem (patch) level to landscape, regional, and finally circumpolar levels of integration required in the context of global change studies.

References

Aleksandrova, V.D. 1958. An attempt to measure the areal and underground mass of plants in the arctic tundra. *Botanicheskii Zhurnal* (Leningrad) 43:1748–1762.

Aleksandrova, V.D. 1971. On the principles of the zonal division of the vegetation of the Arctic. *Botanicheskii Zhurnal* (Leningrad) 56:1–21.

Andreev, V.N., and Galaktionova, T.F. 1978. Formation of the reserves of aboveground phytomass in communities of subarctic tundra. In: *Sezonnaya i Pogodovaya Dinamika Fitomassy v Subarkticheskoy Tundre [Seasonal and Year-to-year Dynamics of Phytomass in Subarctic Tundra]* (pp. 164–177). Novosibirsk: Nauka.

Andreyashkina, N.I., and Gorchakovsky, P.L. 1972. Productivity of the shrub, dwarf shrub and herbaceous communities and a method of its investigation. *Ekologiya* (Sverdlovsk) 3:5–12.

Barney, R.J., and Van Cleve, K. 1973. Black spruce fuel weights and biomass in two interior Alaska stands. *Can. J. Forest Res.* 3:304–311.

Baule, B. 1918. Zu Mitscherlichs Gesetz der physiologischen Beziechungen. *Landw. Jahrb.* 51.

Bazilevich, N.I. 1992. *Computer Data Bases on Biological Productivity of the Ecosystems of the World.* Moscow: Scientific Technical Center "Ekologia," Russian Academy of Sciences.

Bazilevich, N.I. 1994. *Primary Productivity of the Natural Ecosystems of the North-East Eurasia.* Moscow: Nauka.

Bazilevich, N.I., Grenenshchikov, O.S., and Tishkov, A.A. 1986. *Geograficheskie Zakonomernosti Struktury I Funkcionirovaniya Ekosystem [Geographical Regularities of Structure and Functioning of Ecosystems].* Moscow: Nauka.

Bliss, L.C. 1975. Devon Island, Canada. In: Rosswall, T., and Heal, O.W. (eds.), *Structure and Function of Tundra Ecosystems* (pp. 17–60). Stockholm: Swedish Natural Sciences Research Council.

Bliss, L.C., and Kerik, J. 1973. Primary production of plant communities of the Truelove Lowland, Devon Island, Canada—rock outcrops. In: Bliss, L.C., and Wielgolaski, F.E. (eds.), *Primary Production and Production Processes, Tundra Biome* (pp. 27–36). Edmonton: IBP Tundra Biome Steering Committee.

Bliss, L.C., and Matveyeva, N.V. 1992. Circumpolar Arctic vegetation. In: Chapin, F.S. III et al. (eds.), *Arctic Ecosystems in a Changing Climate* (pp. 59–89). San Diego: Academic Press.

Bliss, L.C., and Svoboda, J. 1980. Plant communities and plant production on Banks and Victoria islands. Manuscript, cit. by: Wielgolaski et al., 1981.

Bliss, L.C., Svoboda, J., and Bliss, D.I. 1984. Polar deserts, their plant cover and plant production in the Canadian high arctic. *Holarctic Ecol.* 7:304–324.

Bogatyrev, L.G. 1973. Biological cycling of ash elements in a shrub-sedge-moss tundra. In: *Pochvy i Rastitel'nost Merzlotnykh Rayonov SSSR [Soils and Vegetation of the Permafrost Regions of the USSR]*. Magadan.
Bogatyrev, L.G. 1975. Transfer of plant litter in the tundra biogeocoenoses. *Vestnik Moskovskogo Universiteta*, Ser. 6. Biologiya, Pochvovedenie (Proceedings of the Moscow University, Ser. 6. Biology and Soil Science), N. 2.
Bogdanov, E.I., Ignatenko, I.V., and Pugachev, A.A. 1979. Soils, phytomass reserves and productivity of the tall shrub tundras of the south-eastern Chukotka. In: *Biologicheskaya Produktivnost' Pochv i Eye Uvelichenie v Interesakh Narodnogo Khozyaistva [Biological Productivity of Soils And its Increase for the People's Economy Purposes]* (pp. 26–27). Moscow: Moscow University Publishing House.
Bowman, W.D., Theodose, T.A., Schardt, J.C., and Conant, T. 1993. Constraints of nutrient availability on primary production in two alpine tundra communities. *Ecology* 74:2085–2097.
Budyko, M.I. 1974. *Climate and Life*. New York: Academic Press.
Bunnell, F.L. and Scoullar, K.A. 1975. ABISKO-II. A computer simulation model of carbon flux in tundra ecosystems. In: Rosswall, T., and Heal, O.W. (eds.), *Structure and Function of Tundra Ecosystems* (pp. 425–448). Stockholm: Swedish Natural Science Research Council.
Chertovskoy, V.G., Elizarov, F.P., Cemenov, B.A., and Kornyak, V.S. 1978. Forest growth conditions and productivity of near tundra forests. In: Chertovskoy, V.G. (ed.), *Ekologia Taezhnykh Lesov [Ecology of Taiga Forests]* (pp. 32–42). Archangelsk: Archangelsk Institute of Forest and Chemistry.
Chugunova, M.V. 1979. Some chemical properties of the soils of Mys Chelyuskin. In: Aleksandrova, V.D., and Matveyeva, N.V. (eds.), *Arctic Tundra and Polar Deserts of Taimyr* (pp. 74–77). Leningrad: Nauka.
Collier, B.D., Stenseth, N., Osborn, R., and Barkley, S. 1975. A simulation model of energy acquisition and utilization by the brown lemming (*Lemmus trimucronatus*) at Barrow, Alaska. *Oikos* 26:276–295.
Collins, N.J., Baker, J.H., and Tilbrook, P.J. 1975. Signy Island, maritime Antarctic. In: Rosswall, T., and Heal, O.W. (eds.), *Structure and Function of Tundra Ecosystems* (pp. 345–374). Stockholm: Swedish Natural Science Research Council.
Draper, N.R., and Smith, H. 1981. *Applied Regression Analysis*, 2nd ed. New York: Wiley.
Dyakonov, K.N., and Retium, A.Y. 1971. Data on productivity of tundra and forest-tundra in the region of the lower Ob' River. In: *Biological Productivity and Mineral Cycling in the Terrestrial Plant Communities* (pp. 43–46). Leningrad: Nauka.
Elkington, J.J., and Jones, B.M.G. 1974. Biomass and primary productivity of birch (*Betula pubescence* s. lat.) in south-west Greenland. *J. Ecol.* 62:821–830.
Esser, G. 1986. The carbon budget of the biosphere—structure and preliminary results of the Osnabruck Biosphere Model. *Veroff. Naturf. Ges. zu Emden von 1814* 7:160.
Esser, G. 1987. Sensitivity of global carbon pools and fluxes to human and potential climate impacts. *Tellus* 39B:245–260.
Esser, G. 1991. Osnabruck biosphere model: structure construction, results. In: Esser, G., and Overdieck, D. (eds.), *Modern Ecology: Basic and Applied Aspects* (pp. 679–709). Amsterdam: Elsevier.
Esser, G., Aselmann, I., and Lieth, H. 1982. Modelling the carbon reservoir in the system compartment "litter". In: *Mitteillungen aus dem Geologisch-Palaentologischen Institut der Universität Hamburg* 52:39–58. Hamburg: University of Hamburg Press.

Forrester, J.W. 1961. *Industrial Dynamics*. Cambridge: MIT Press.
French, D.D. 1974. Classification of IBP Tundra Biome sites based on climate and soil properties. In: Holding, A.J., Heal, O.W., MacLean, S.F., and Flanagan, P.W. (eds.), *Soil Organisms and Decomposition in Tundra: Proceedings of the Microbiology, Decomposition and Invertebrate Working Groups Meeting, Fairbanks, Alaska, August, 1973* (pp. 3–25). Stockholm: Tundra Biome Steering Committee.
French, D.D. 1981. Multivariate comparisons of IBP Tundra Biome site characteristics. In: Bliss, L.C., Heal, O.W., and Moore, J.J. (eds.), *Tundra Ecosystems: A Comparative Analysis* (pp. 47–51). Cambridge: Cambridge University Press.
Gasheva, A.F. 1974. Phytomass reserves in some communities of the "Kharp" field station. In: Gorchakovski, P.L., and Smirnov, V.S. (eds.), *Biomassa i Dinamika Rastitel'nogo Pokrova i Zhivotnogo Naseleniya v Lesotundre. Trudy Instituta Ekologii Rastenii i Zhivotnykh UNC AN SSSR, vyp. 88. [Biomass and the Dynamics of Plant Cover and Animal Population in Forest Tundra. Proc. Institute of Ecology of Plants and Animals of the Urals Scientific Center, Acad. Sci. USSR, Issue 88]* (pp. 106–107). Sverdlovsk: UNC AN SSSR.
Gersper, P.L., Alexander, V., Barkley, S.A., Barsdate, R.J., and Flint, P.S. 1980. The soils and their nutrients. In: Brown, J., Miller, P.C., Tieszen, L.L., and Bunnell, F.L. (eds.), *An Arctic Ecosystem: The Coastal Tundra at Barrow, Alaska* (pp. 219–254). Stroudsburg, PA: Dowden, Hutchinson, and Ross.
Gilmanov, T.G. 1993. Empirical and theoretical foundations for modelling the response of ecosystems to climatic change (with emphasis to boreal and arctic/alpine areas). In: Holten, J.I., Poulsen, G., and Oechel, W.C. (eds.), *Impact of Climatic Change on Natural Ecosystems* (pp. 43–74). Trondheim, Norway: Norwegian Institute for Nature research and Directorate for Nature Management.
Gorchakovsky, P.L., and Andreyaskina, N.I. 1972. Productivity of some shrub, dwarf shrub and herbaceous communities of forest-tundra. In: Wielgolaski, F.E., and Rosswall, T. (eds.), *Proceedings IVth International Meeting on the Biological Productivity of Tundra* (pp. 113–116). Stockholm: Tundra Biome Steering Committee.
Grigoryev, A.A., and Budyko, M.I. 1956. On the periodic law of geographical zonality. *Doklady Akademii Nauk SSSR* 110:129–132.
Haag, R.W. 1974. Nutrient limitations to plant production in two tundra communities. *Can. J. Bot.* 52:103–116.
Haag, R.W., and Bliss, L.C. 1974. Energy budget changes following surface disturbance to upland tundra. *J. App. Ecol.* 11:355–374.
Himmelblau, D.M. 1970. *Process Analysis by Statistical Methods*. New York: Wiley.
Ignatenko, I.V. 1979. *Pochvy Vostochnoevropeyskoy Tundry i Lesotundry [Soils of the East-European Tundra and Forest Tundra]*. Moscow: Nauka.
Ignatenko, I.V., and Khakimzyanova, F.I. 1971. Soils and total phytomass reserves in dwarf birch—white dryas and willow tundras of the East European Northlands, *Ekologia (Sverdlovsk)* 4:17–24.
Ignatenko, I.V., Norin, B.N., and Rakhmanina, A.T. 1973. Cycling of mineral elements and nitrogen in some biogeocoenoses of the East-European forest tundra. In: *Pochvy i Rastitel'nost' Merzlotnykh Rayonov SSSR [Soils and Vegetation of the Permafrost Regions of the USSR]* (pp. 335–350). Magadan: Institute of Biological Problems of the North.
Ignatenko, I.V., Knorre, I.V., Lovelius, A.V., and Norin, N.V. 1972. Standing crop in plant communities at the station Ary-Mas. In: Wielgolaski, F.E., and Rosswall, T. (eds.), *Proceedings IVth International Meeting on the Biological Productivity of Tundra* (pp. 140–149). Stockholm: Tundra Biome Steering Committee.
Jenny, H. 1941. *Factors of Soil Formation. A System of Quantitative Pedology*. New York: McGraw-Hill.

Jenny, H. 1980. The *Soil Resource: Origin and Behavior.* New York: Springer-Verlag.
Johnson, I.R., and Thornley, J.H.M. 1984. A model of instantaneous and daily canopy photosynthesis. *J. Theor. Biol.* 107:531–545.
Khantimer, I.S. 1970. Increment of the aboveground phytomass in the virgin and managed tundra. In: *Produktivnost' Biogeotsenozov Subarktiki [Productivity of the Biogeocoenoses of Subarctics]* (pp. 84–86). Sverdlovsk: Ural Branch of the Academy of Sciences of the USSR.
Khantimer, I.S. 1974. *Agricultural Management of Tundra.* Leningrad: Nauka.
Khodachek, E.A. 1969. Phytomass of the tundra phytocoenoses on the west Taimyr. *Bot. Zhurn. (Leningrad)* 54:1059–1075.
Lewis Smith, R.I., and Walton, D.W.H. (1975). South Georgia, subantarctic. In: Rosswall, T., and Heal, O.W. (eds.), *Structure and Function of Tundra Ecosystems. Ecol. Bull. (Stockholm)* (pp. 399–423). Stockholm: Swedish Natural Science Research Council.
von Liebig, J.F. 1840. *Die Organische Chemie in Ihrer Anwendung auf Agricultur und Physiologie.* Braunschweig: Verlag von Friedrich Vieweg und Sohn.
Lieth, H. 1972. Über die Primärproduktion der Pflanzendecke der Erde. *Z. Angewand. Botan.* 46:1–37.
Lieth, H. 1975a. Modeling the primary productivity of the world. In: Lieth, H., and Whittaker, R.H. (eds.), *Primary Productivity of the Biosphere* (pp. 237–263). New York: Springer-Verlag.
Lieth, H. 1975b. Primary productivity of the major vegetation units of the world. In: Lieth, H., and Whittaker, R.H. (eds.), *Primary Productivity of the Biosphere* (pp. 203–215). New York: Springer-Verlag.
Lieth, H. 1976. Biophysikalische Fragestellungen in der Okologie und Umweltforschung. Teil 2: Extremalprinzipien in Okosystemen. *Radiat. Environ. Biophys* 13:337–351.
Lieth, H., and Box, E.O. 1977. The gross primary productivity pattern of the land vegetation: A first attempt. *Tropical Ecol.* 18:109–115.
Manakov, K.N. 1970. *Elementy Biologicheskogo Krugovorota na Polyarnov Severe [Elements of the Biological Cycling at the Polar North].* Leningrad: Nauka.
Manakov, K.N. 1972. *Productivity and Biological Turn-over in the Tundra Biogeocoenoses of the Kola Peninsula.* Leningrad: Nauka.
Manakov, K.N. 1974. Mineral cycling in the birch woodlands of the forest tundra of Kola Peninsula. In: *Biologicheskie Processy i Mineral'nyi Obmen v Pochvakh Kol'skogo Poluostrova [Biological Processes and Mineral Cycling in Soils of the Kola Peninsula]* (pp. 110–137). Apatity: Kola branch, USSR Academy of Sciences.
Matveyeva, N.V. 1979. The structure of the plant cover in the polar deserts of the Taimyr Peninsula (Mys Chelyuskin) (In Russian). In: Aleksandrova, V.D., and Matveyeva, N.V. (eds.), *The arctic tundras and polar deserts of Taimyr* (pp. 5–27). Leningrad: Nauka.
Melillo, J.M., McGuire, A.D., D.W., K., Moore, III, B., Vorosmarty, C.J., and Schloss, A.L. 1993. Global climate change and terrestrial net primary production. *Nature*, 363:234–240.
Miller, P.C., Kendall, R., and Oechel, W.C. 1983. Simulating carbon accumulation in northern ecosystems. *Simulation* 40:119–131.
Miller, P.C., Oechel, W.C., Stoner, W.A., and Sveinbjornsson, B. 1978. Simulation of CO_2 uptake and water relations of four arctic bryophytes at Point Barrow, Alaska. *Photosynthetica* 12:7–20.
Miller, P.C., Miller, P.M., Blake-Jacobson, M., Chapin, F.S., Everett, K.R., Hilbert, D.W., Kummerow, J., Linkins, A.E., Marion, G.M., Oechel, W.C., Roberts, S.W.,

and Stuart, L. 1984. Plant-soil processes in Eriophorum vaginatum tussock tundra in Alaska: Systems modeling approach. *Ecol. Monogr.* 54:361–405.
Mitscherlich, L.A. 1909. Das Gesetz des Minimums und das gesetz des abnehmenden Bodenertrages. *Landw. Jahrb.* 38:537–552.
Novak, J.M. 1992. Nonlinear Fit program in the "Statistics" package provided by "Mathematica" software by Wolfram Research Institute.
Parton, W.J., Schimel, D.S., Cole, C.V., and Ojima, D.S. 1987. Analysis of factors controlling soil organic matter levels in great plains grasslands. *Soil Sci. Soc. Am. J.* 5:1173–1179.
Parton, W.J., Scurlock, J.M., Ojima, D.S., Gilmanov, T.G., et al. 1993. Observations and modeling of biomass and soil organic matter dynamics for the grassland biome worldwide. *Global Biogeochem. Cycles* 7:785–809.
Pastor, J., and Post, W.M. 1986. Influence of climate, soil moisture, and succession on forest carbon and nitrogen cycles. *Biogeochemistry* 2:3–28.
Pavlova, E.B. 1969. On the plant mass of tundras of the western Taimyr. *Vestnik Moskov. Univ. Ser. Biolog. [Bull. Moscow Univ., Biol. Ser.]* 5:62–27.
Poletaev, I.A. 1966. On mathematical models of biogeocoenoses (In Russian). *Problemy kibernetiki* 16:171–190.
Poletaev, I.A. 1973. "Predator-prey" models by Volterra and some of their generalizations using the Liebig's principle (In Russian). *Zhurnal Obshchei Biologii* 34:43–57.
Pospelova, E.B. 1973. Seasonal dynamics of increment of some tundra shrubs *Biologicheskie Nauki (Moscow)* 15:66–70.
Pospelova, E.B., and Orlov, M.V. 1984. Phytomass stock and structure of the spotty tundras of the eastern shore of Lake Taimyr [Russian SFSR, USSR]. *Ekologia (Sverdlovsk)* No. 1:14–21.
Pospelova, E.B., and Zharkova, Y.G. 1972. Plant cover and the phytomass of plant communities of the "Agapa" research station (In Russian). In: *Pochvy i Produktivnost' Rastitel'nykh Soobshchestv. Vyp. l [Soils and Productivity of Plant Communities. Issue l]* (pp. 55–78). Moscow: Moscow University Publishing House.
Post, W.M., Emanuel, W.R., Zinke, P.J., and Stangenberger, A.G. 1982. Soil carbon pools and world life zones. *Nature* 298:156–159.
Prentice, C., Cramer, W., Harrison, S.P., et al. 1992. A global biome model based on plant physiology and dominance, soil properties and climate. *J. Biogeogr.* 19:117–134.
Rakhmanina, A.T. 1971. Aboveground and belowground phytomass of some communities of the Easteuropean forest tundra. In: Bazilevich, N.I., and Rodin, L.E. (eds.), *Biological Productivity and mineral Cycling in Plant Communities* (pp. 37–42). Leningrad: Nauka.
Reynolds, J.F., and Acock, B. 1985. Modeling approaches for evaluating vegetation responses to carbon dioxide concentration. In: Strain, B.R., and Cure, J.D. (eds.), *Direct Effects of Increasing Carbon Dioxide on Vegetation* (pp. 33–51). Washington DC: U.S. Department of Energy.
Reynolds, J.F., and Leadley, P.W. 1992. Modeling the response of arctic plants to changing climate. In: Chapin, III, F.S. et al. (eds.), *Arctic Ecosystems in a Changing Climate. An Ecophysiological Perspective* (pp. 413–438). San Diego: Academic Press.
Rosenzweig, M.L. 1968. Net primary productivity of terrestrial communities: Prediction from climatological data. *Am. Natural.* 102:57–74.
Ryabchikov, A.M. 1968. Hydrothermal conditions and the productivity of plant mass in the principal landscape zones. *Vestnik Moskov. Univ. Ser. Geogr. [Bull. Moscow Univ. Ser. Geogr.]* No. 5:41–48.

Sandhaug, A., Kjelvik, S., and Wielgolaski, F.E. 1975. A mathematical simulation model for terrestrial tundra ecosystems. *Ecol. Studies* 17:251–266.

SCOPEGRAM (SCOPE Grassland Modeling Group): Parton, W.J., Coughenour, M.B., Ojima, D., Scurlock, J., Hall, D., Gilmanov, T.G. et al. 1994. In press. Grassland ecosystem modeling—temperate and tropical; Modeling global grassland production in relation to climatic change. In: Melillo, J., and Hall, D. (eds.), *Effects of Climate Change on Production and Decomposition in Coniferous Forests and Grasslands*. SCOPE volume. New York: John Wiley.

Shamurin, V.F. 1970. The main supply of phytomass in some tundra co-associations in the area of Vorkuta. In: *Biological Foundations of Utilization of Nature of the Far North* (pp. 25–29). Syktyvkar: Komi Publishing House.

Shamurin, V.T., Polozova, T.G., and Khodachek, E.A. 1972. Plant biomass of main plant communities at the Tareya station (Taimyr). In: *Tundra Biome (IBP). Proc. IV Intern. Meeting on Biol. Productivity of Tundra. Leningrad. Oct. 1971* (pp. 163–181). Stockholm: Tundra Biome Steering Committee.

Shaver, G.R., and Chapin, F.S. III. 1986. Effect of NPK fertilization on production and biomass of Alaskan tussock tundra. *Arctic Alpine Res.* 18:261–268.

Shaver, G.R., and Chapin, F.S.III. 1991. Production: biomass relationships and element cycling in contrasting arctic vegetation types. *Ecol. Monogr.* 61:1–31.

Tikhomirov, B.A., Shamurin, V.F., and Aleksandrova, V.D. 1981. Phytomass and primary production of tundra communities, USSR. In: Bliss, L.C., Heal, O.W., and Moore, J.J. (eds.), *Tundra Ecosystems: A Comparative Analysis* (pp. 227–237). Cambridge, New York: Cambridge University Press.

Timin, M.E., Collier, B.D., Zich, J., and Walter, D. 1973. *A Computer Simulation of the Arctic Tundra Ecosystem Near Barrow, Alaska*. U.S. Tundra Biome Rep. No. 73-1. San Diego: San Diego State University.

Tishkov, A.A. 1982. Phytomass of tundra plant communities on Spitzbergen. Manuscript.

Trotsenko, G.V. 1974. On the reserves of phytomass in main plant communities of forest tundra of the Priobye North. In: *Biologicheskie Problemy Severa. Vyp. 3. Botanika i Rastitel'nye Resursy [Biological Problems of the North. Issue 3. Botany and Plant Resources]* (pp. 188–192).

Van Cleve, K. 1981. Alaska data in woodland dataset. In: Reichle, D.E. (ed.), *Dynamic Properties of Forest Ecosystems* (pp. 648–650). Cambridge: Cambridge University Press.

Vasilyevskaya, V.D., Ivanov, V.V., and Bogatyrev, L.G. 1986. *Soils of the North of the West Siberia*. Moscow: Moscow State University Publishing House.

Vikhireva-Vasil'kova, V.V., Gavrilyuk, V.A., and Shamurin, V.F. 1964. Aboveground and belowground mass of some dwarf-shrub communities of Koryak Zemlya. In: *Problemy Severa, Vypusk 8 (Problems of the North, Issue No. 8)* (pp. 130–147).

Vil'chek, G.E. 1984. Productivity of tundra phytocoenoses of the environs of Cape Kharasavey [Russian SFSR, USSR]. *Biolog. Nauki* (*Moscow*) No. 7:67–71.

Vil'chek, G.E. 1986. Productivity of some phytocoenoses in Vorkuta tundra [Russian SFSR, USSR]. *Ekologiya (Sverdlovsk)* No. 2:8–13.

Vil'chek, G.E. (1987). Productivity of typical tundras of Taimyr [Russian SFSR, USSR]. *Ekologiya* (Sverdlovsk) No. 5:38–43.

Volobuev, V.R. 1974. *Introduction to the Energetics of Soil Formation*. Moscow: Nauka.

Waelbroeck, C. 1993. Climate-soil processes in the presence of permafrost: a systems modelling approach. *Ecol. Model.* 69:185–225.

Webber, P.J. 1974. Tundra primary productivity. In: Ives, J.D., and Barry, R.G. (eds.), *Arctic and Alpine Environment* (pp. 445–473). London: Methuen.

Wielgolaski, F.E. 1972. Vegetation types and plant biomass in tundra. *Arctic Alpine Res.* 4:291–305.

Wielgolaski, F.E. 1975. Productivity of tundra ecosystems. In: *Productivity of World Ecosystems: Proceedings of a Symposium. Presented August 31–September 1, 1972, at the V General Assembly of the Special Committee for the International Biological Program, Seattle, Washington* (pp. 1–12). Washington, DC: National Academy of Science.

Wielgolaski, F.E., Bliss, L.C., Svoboda, J., and Doyle, G. 1981. Primary production of tundra. In: Bliss, L.C., Heal, O.W., and Moore, J.J. (eds.), *Tundra Ecosystems: A Comparative Analysis* (pp. 187–225). Cambridge: Cambridge University Press.

Zyryanov, V.A., and Kolpashchikov, L.A. 1981. Plant resources and their use by the wild reindeer in western Taimyr. In: *Ekologiya i Khozyaistvennoe Ispol'zovanie Nazemnoy Fauny Eniseyskogo Severa [Ecology and Economical Use of the Terrestrial Fauna of the Enisey North]* (pp. 27–38). Novosibirsk.

22. Analysis of CO_2, Temperature, and Moisture Effects on Carbon Storage in Alaskan Arctic Tundra Using a General Ecosystem Model

Edward B. Rastetter, Robert B. McKane, Gaius R. Shaver, Knute J. Nadelhoffer, and Anne Giblin

Projected changes in global climate associated with accumulations of CO_2 and other greenhouse gases in the atmosphere might stimulate the release of a substantial portion of the estimated 4.3×10^{16} g of carbon stored in wet and moist arctic tundras (Oechel, 1989; Shaver et al., 1992). About 98% of this carbon is in soils and has accumulated over thousands of years as a result of cold, wet conditions that serve to slow decomposition. Increased temperatures will stimulate decomposition and hence release CO_2 from soils to the atmosphere. However, increased CO_2, increased temperature, and higher amounts of available nitrogen (as a result of faster decomposition) could stimulate rates of production in vegetation and thereby also stimulate CO_2 removal from the atmosphere. The key to which of these two opposing processes dominates appears to be closely tied to the nitrogen cycle (Billings et al., 1984; Rastetter et al., 1992; Shaver et al., 1992).

Three interactions between the nitrogen cycle and carbon storage are particularly important in this regard. First, in most ecosystems, including tundra, rates of nitrogen supply through rainfall, dry deposition, and nitrogen fixation are slow relative to internal nitrogen recycling rates. In mature ecosystems, slow inputs are roughly balanced by losses through leaching and gaseous exports. Changes in carbon storage depend in part upon the ability of plants and soils to scavenge additional nitrogen from this slow trickle passing through the ecosystem. If changes in climate or CO_2 affect nitrogen scavenging efficiency, then carbon storage should change in proportion to changes in total ecosystem nitrogen stocks. This ability to scavenge nitrogen depends not only on the rate of external nitrogen supply, but also on the efficiency of nitrogen retention by the ecosystem itself. For example, increased CO_2 concentrations would be expected to increase photosynthesis, which should in turn allow vegetation to allocate more resources to roots (Bazzaz, 1990; Norby et al., 1992) and thereby enhance the ability of plants to scavenge nitrogen passing through the ecosystem. Increased photosynthesis should also increase the amount and the C/N ratio of plant litter (Bazzaz, 1990). To decompose this larger flux of low-nitrogen litter, microbes must immobilize more inorganic nitrogen to meet their own metabolic needs. This higher demand for inorganic nitrogen by microbes should reduce the availability of nitrogen to plants.

Changing C/N ratios in vegetation and soils is the second carbon–nitrogen interaction that might influence ecosystem carbon storage under altered climate and CO_2. The relative abundances of woody versus nonwoody plant species, changes in the allocation of biomass to woody versus nonwoody tissues in existing vegetation, and changes in the C/N ratios of the plant tissues themselves are all potential mechanisms by which the vegetation C/N ratio might change. Long-term warming experiments suggest that warming will favor woody tundra species and a higher allocation to woody tissues in some species (Chapin et al., 1995). High CO_2 is known to increase the C/N ratios of tissues in many plants (Bazzaz, 1990).

The potential for changing the C/N ratios of soils is also high, especially in tundra. During decomposition, the C/N ratio of organic matter decreases from that of fresh litter (ca. 40) to that of humus (ca. 12). Because of the cold, wet conditions in tundra soils, decomposition is slow, organic matter is poorly decomposed, and the C/N ratio of tundra soils is therefore slightly higher than that of warmer, less waterlogged soils (Nadelhoffer et al., 1992). Any factor that stimulates decomposition rates, such as warming, will result in more highly decomposed soil organic matter, lower C/N ratios, and therefore less carbon storage in soils.

The redistribution of nitrogen between soils and vegetation is the final carbon–nitrogen interaction that will influence ecosystem carbon storage (Rastetter et al., 1991, 1992; Shaver et al., 1992). Soils have much lower C/N ratios than vegetation (about 18 vs. 40 in tundra). A redistribution of nitrogen between these two pools can have significant effects on carbon storage. For example, warmer conditions will stimulate the rate of decomposition, increasing CO_2 release, but also increasing soil–nitrogen turnover. If vegetation takes up the extra mineralized nitrogen and incorporates it into plant tissues, then there will be about twice as much carbon tied to that nitrogen as there was in the soil, thereby resulting in a net increase in carbon storage in the ecosystem as a whole. On the other hand, if the nitrogen released during decomposition is lost from the ecosystem, then there will be a net loss of carbon.

In the past, we have used the Marine Biological Laboratory General Ecosystem Model (MBL-GEM) to investigate changes in tundra carbon storage in response to increases in CO_2, temperature, nitrogen deposition, and ecosystem nitrogen throughput (McKane et al., 1995; submitted a, b; Rastetter et al., 1991, 1992). In this chapter we will investigate the additional effects of changes in soil moisture on changes in tundra carbon storage in response to increased CO_2 and temperature. Moisture is seldom limiting to photosynthesis on the tundra, but does affect decomposition (Flanagan and Veum, 1974; Nadelhoffer et al., 1992). Thus, its major effects are expected to act through the turnover of soil organic matter and particularly on the rate of nitrogen cycling. Oechel et al. (1993) conclude that losses of carbon they observed for Alaskan tundras was due predominantly to decreases in soil moisture and lower water tables rather than to a direct effect of increased temperature.

Changes in soil moisture associated with changes in climate are difficult to anticipate. Increases in temperature might be expected to enhance evapotranspiration, thereby decreasing soil moisture. Temperature might also increase the depth of thaw of tundra soils, which could enhance drainage and thereby dry at least the surface layers of the soil. However, decreased irradiance associated with higher cloudiness would tend to decrease moisture losses through evapotranspiration and increased precipitation could more than replace moisture lost by evapotranspiration or drainage. Climate model predictions of future changes in rainfall and cloudiness are still unreliable (Maxwell, 1992). We have therefore taken a sensitivity approach and will investigate both a 10% increase and 10% decrease in soil moisture.

MBL-GEM

The structure of MBL-GEM is described in detail in Rastetter et al. (1991). It is a process-based, lumped-parameter model of the carbon–nitrogen interactions in terrestrial ecosystems. It simulates stand-level photosynthesis and nitrogen uptake by plants, allocation of carbon and nitrogen to foliage, stems, and fine roots, respiration in these tissues, turnover of biomass through litter fall, and decomposition of that litter in soils. The model simulates responses to changes in atmospheric CO_2 concentration, temperature, soil moisture, irradiance, and inorganic nitrogen inputs from outside the ecosystem. The only losses from the ecosystem are CO_2 losses through plant and soil respiration and inorganic nitrogen losses, which are assumed to be proportional to inorganic nitrogen concentrations in soils.

A major feature of the model is that vegetation acclimates to changes in the environment to maintain a nutritional balance between carbon and nitrogen. Thus, environmental changes that stimulate photosynthesis (e.g., increased CO_2 or higher irradiance) result in an increase in allocation of carbon and nitrogen to fine roots, thereby stimulating nitrogen uptake. Similarly, environmental changes that stimulate nitrogen uptake (e.g., high inorganic nitrogen levels) increase allocation of carbon and nitrogen to foliage, thereby stimulating carbon uptake.

The model is implemented through a calibration to biomass and productivity data. The tundra calibration used in this chapter was based on data from a 9-year study by Chapin et al. (1995) in which experimental plots at Toolik Lake, Alaska, were manipulated with fertilizer, greenhouses, and shade houses. In addition, data from a two-times ambient CO_2 experiment at Toolik Lake by Oechel et al. (1986) were used. This calibration and an analysis of the application of MBL-GEM to these data are presented in detail in McKane et al. (submitted a). The calibration procedure was successful in finding a single parameter set that is consistent with all of these experimental manipulations.

Several caveats and clarifications must be made regarding the application of MBL-GEM to tundra ecosystems. First, because MBL-GEM is a lumped-parameter model, species composition is not explicitly simulated. This does not mean that the biogeochemical consequences of changes in species composition are not accommodated by the model. For example, in the 9-year experiment by Chapin et al. (submitted), the fertilizer treatment resulted in both a tripling of deciduous woody shrub biomass and an increase in the proportion of total biomass accounted for by deciduous shrubs (from about 30% to 70%). The biogeochemical result is a large increase in woody biomass, which is readily simulated by MBL-GEM (McKane et al., submitted a). Similarly, when the Toolik Lake, Moist Tundra calibration of MBL-GEM is run using climate data for Barrow, Alaska, the model simulates gross primary production, net primary production, and biomass consistent with that measured at Barrow despite the wide difference in community compositions between Toolik (dominated by *Eriophorum vaginatum* and *Betula nana*, Chapin et al., submitted) and Barrow (dominated by *Carex aquatilis*). Thus, our applications of MBL-GEM to tundra ecosystems have thus far been able to account for the biogeochemical consequences of changes in species composition. Nevertheless, we do not know how extreme these changes can be before this calibration of MBL-GEM is no longer valid.

Second, because the acclimation scheme in MBL-GEM is based on Thornley's (1972) root/shoot partitioning model, changes in nonvascular plant abundance are not well represented in our model. We have therefore based our calibration solely on the dynamics of the vascular component of the Toolik Lake moist tundra. Mosses and lichens at this site can make up 30% of the biomass. We have therefore left out a substantial component of the ecosystem from our analysis. In the Chapin et al. (1995) experiments, competition for light resulted in an inverse relation between the responses of vascular versus nonvascular plants to the experimental manipulations. That is, increases in vascular leaf area was accompanied by a decrease in the moss and lichen biomass and vice versa. If this result can be generalized, then changes in leaf area predicted by MBL-GEM will be at least partially matched by compensatory changes in moss and lichen biomass.

Third, MBL-GEM does not explicitly account for changes in depth of thaw in response to changes in climate. Changes in depth of thaw have two major effects on tundra biogeochemistry. The first is a change in soil volume and the accompanying change in the amount of active soil organic matter and of mineral nutrients deep in the soil. The second is a change in the drainage properties of the soils and the accompanying changes in the rates of soil processes, especially in shallow soil layers. Because the activity of these shallow soil layers tends to dominate the carbon and nitrogen dynamics of soils, the change in drainage will likely be the more important of these two effects of thaw depth. Our analysis of the responses to changes in soil moisture are designed to address this effect of a change in depth of thaw.

Finally, our experience with the model suggests that because of the slow turnover of soil organic matter, the response time of tundra to changes in climate and CO_2 can be long, rendering long-term validation of the model difficult. Long-term validation will have to wait for either reliable reconstructions of past soil moisture, temperature, nitrogen deposition, and cloudiness or for data from long-term future monitoring of the tundra to become available. In the interim, we present our results as a tentative analysis of the responses of tundra ecosystems to changes in CO_2 and climate, but urge intercomparison of the long-term results from many models. Intercomparison among models, developed from a variety of perspectives, is the only short-term way to evaluate long-term predications and is the best short-term substitute for long-term validation.

Simulations

MBL-GEM was used to examine 50-year responses of a moist tundra ecosystem dominated by the tussock forming sedge *Eriophorum vaginatum* to changes in CO_2, temperature, and moisture. A full-factorial analysis was performed on two levels of CO_2 (344 and 688 ppmv), two levels of temperature (8.3 and 13.3°C), and three levels of moisture (64%, 71%, and 78% water-filled pore space). In all simulations, the ecosystem was initially in equilibrium with present-day, average growing season conditions with CO_2 at 344 ppmv, temperature at 8.3°C, and moisture at 71% water-filled pore space. The simulations maintained these initial conditions for 2 years, then at the beginning of the third year, CO_2, temperature, and/or moisture were changed instantaneously to their manipulated levels and were maintained at these levels for the duration of the simulation (i.e., each factor was manipulated as a step function).

Changes in ecosystem carbon storage were analyzed by partitioning this storage among three factors relating to nitrogen dynamics in the ecosystem (Rastetter et al., 1992). This partitioning analysis associates changes in ecosystem carbon storage with (1) the net change in total ecosystem nitrogen, (2) the change in the C/N ratios of soils and vegetation, (3) the net exchange of nitrogen between soils (low C/N ratio) and vegetation (high C/N ratio), and (4) the interaction among these three factors. This analysis is designed to emphasize and clarify the importance of nitrogen, the limiting nutrient in these ecosystems, in regulating carbon storage.

Response to CO_2 and Temperature and Comparison to Past Results

Simulations for a doubling of CO_2 concentration, a 5°C increase in temperature, and the combination of these two treatments indicate that tundra should store carbon under all three treatments (Figures 22.1 to 22.3, Table 22.1). In none of the treatments is new nitrogen from outside the ecosystem a major contributor to carbon storage. Tundra is severely nitrogen limited with only about $0.03\,g\,N\,m^{-2}$ being deposited in precipitation each year (Barsdate and Alexander, 1975) into a total ecosystem pool of over $1500\,g\,N\,m^{-2}$. Dry deposition might double the total nitrogen deposition. Nitrogen fixation rates have been estimated at between 0.02 and $0.26\,g\,N\,m^{-2}\,yr^{-1}$ (Chapin and Bledsoe, 1992). Even if all this nitrogen were retained by the ecosystem (i.e., no nitrogen losses), it could amount to no more than

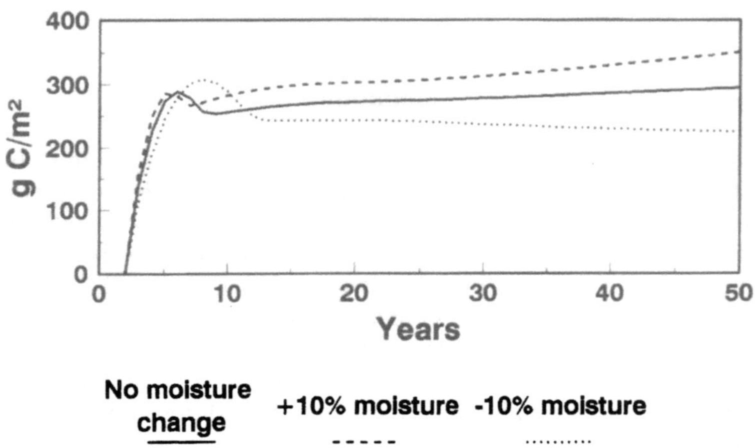

Figure 22.1. Cumulative changes in ecosystem carbon resulting from a doubling of CO_2 concentration (constant temperature) and the combination of a doubling of CO_2 concentration with either a 10% increase or 10% decrease in soil moisture.

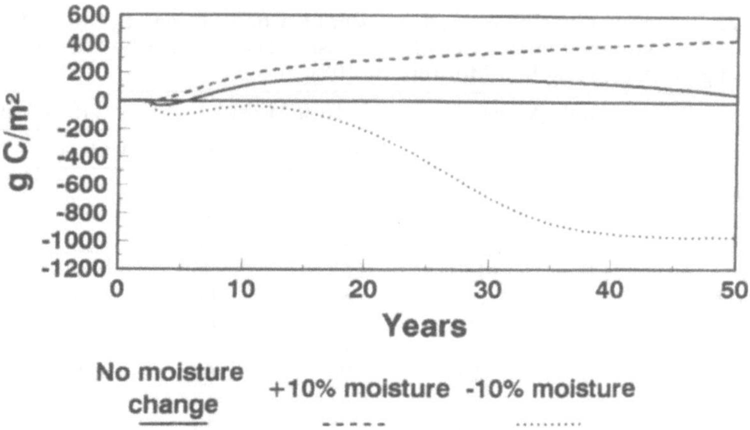

Figure 22.2. Cumulative changes in ecosystem carbon resulting from a 5°C increase in temperature (constant CO_2 concentration) and the combination of a 5°C increase in temperature with either a 10% increase or 10% decrease in soil moisture.

16 g N m^{-2} over 50 years, or about a 1% increase in total ecosystem nitrogen. If there were no accompanying change in C/N ratios of soils and vegetation and no redistribution of nitrogen between soils and vegetation, this change in total ecosystem nitrogen would translate into a proportional change in ecosystem carbon storage (i.e., about 1%). Due to the absence of major nitrogen fixers in the systems we simulated, we used a conservative nitrogen

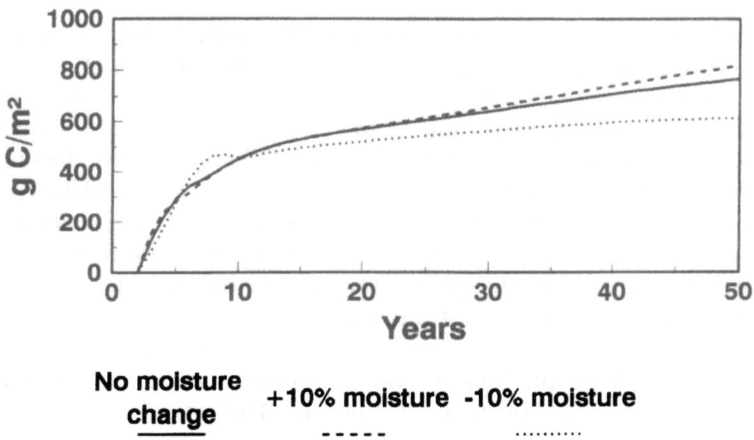

Figure 22.3. Cumulative changes in ecosystem carbon resulting from a doubling of CO_2 concentration and a 5°C increase in temperature and the combination of a doubling of CO_2 concentration and a 5°C increase in temperature with either a 10% increase or 10% decrease in soil moisture.

Table 22.1. Cumulative Change in Ecosystem Carbon Storage ($g C m^{-2}$) After 50 Years in Response to Doubling of CO_2, Increasing Temperature by 5°C, and Either Increasing or Decreasing Soil Moisture by 10%

	Amb. Moist		−10% Moist		+10% Moist	
	Amb. T	+5°C	Amb. T	+5°C	Amb. T	+5°C
Amb. CO_2						
total	0	53	−186	−967	130	435
New N	0	12	−14	−96	10	23
C/N	0	−140	−252	−996	199	334
N exch	0	247	87	306	−79	86
Inter	0	−66	−7	−181	0	−8
Soil	0	−166	−283	−1032	249	340
Veg	0	220	96	65	−119	95
$2 \times CO_2$						
Total	295	768	225	614	350	819
New N	6	24	−4	16	15	29
C/N	294	493	124	199	446	708
N exch	−3	196	72	343	−80	63
Inter	−1	56	34	56	−30	19
Soil	144	309	−83	−26	371	599
Veg	151	459	309	640	−21	220

Also shown is the amount of the total cumulative change in carbon that can be associated with each of three factors relating to the interactions between carbon and nitrogen and to changes in soil and vegetation carbon: New N: carbon storage associated with new nitrogen in the ecosystem, C/N: carbon storage associated with changes in vegetation and soil C/N ratios, N exch.: carbon storage associated with the exchange of nitrogen between soils and vegetation, intercarbon storage associated with the interactions among these three factors, Soil: change in soil carbon, Veg: change in vegetation carbon.

input of $0.06 g N m^{-2} yr^{-1}$ in our simulations. The contribution of new nitrogen to carbon storage is therefore less than 1% in our simulations.

For the CO_2 treatment, the increase in carbon storage (Figure 22.1) was due almost entirely to a widening of the C/N ratios of vegetation and soils (Table 22.1). The widening of the vegetation C/N ratio resulted from both a widening of the C/N ratios of all three tissues and particularly to a disproportionate increase in woody biomass relative to leaf and fine root biomass. Woody biomass increased by almost 45% in this treatment, whereas leaf biomass decreased by 10% and root biomass increased by 8%. The increase in root to leaf ratio is the expected acclimation response to increased CO_2 concentration. Increased productivity under elevated CO_2 also increased litter production and therefore increased the amount of young, high C/N ratio organic matter in soils. The total increase in ecosystem carbon storage after 50 years was about evenly distributed between soils and vegetation (Table 22.1).

For the temperature treatment, the increase in carbon storage is due almost entirely to a movement of nitrogen from soils to vegetation (Table 22.1, Figure 22.2). The temperature increase stimulated decomposition and therefore nitrogen mineralization. Because of the severe nitrogen limitation, most of this nitrogen was taken up and incorporated into plant tissues and very little was lost from the ecosystem. After about 20 years, the respiratory costs of the accumulated plant biomass caused the rate of carbon gain by vegetation to decrease below the rate of carbon loss by soils. The ecosystem therefore began to reverse the gains in carbon made during the first 20 years (Figure 22.2). Over the 50 years, soils lost about 166 g C m^{-2} and vegetation gained about 220 g V m^{-2} (Table 22.1). The stimulation of decomposition by higher temperatures has a disproportionate effect on the younger, more labile soil organic matter with higher C/N ratios. Thus, the gain in carbon due to the movement of nitrogen from soils to vegetation is somewhat smaller than would be predicted from the C/N ratio of the whole soil.

In the combined CO_2 and temperature treatment, increases in soil and vegetation C/N ratios and movement of nitrogen from soils to vegetation contributed to the increase in carbon storage (Table 22.1, Figure 22.3). However, the response to the combined treatment was more than twice the sum of the independent responses to CO_2 and temperature. This synergism was due to the combination of a stimulation of photosynthesis by increased CO_2 and a release from nitrogen limitation resulting from the temperature-induced increase in nitrogen mineralization. Because the vegetation was exposed to higher nitrogen levels at the same time that photosynthesis was stimulated by CO_2, the resulting increase in biomass could be allocated to tissues with lower C/N ratios like leaves and fine roots. Because an increase in the amounts of these tissues enhances the ability to capture carbon and take up nitrogen, a positive feedback developed that enhanced ecosystem carbon storage. Litter production resulting from the higher productivity more than replaced the carbon lost through higher decomposition rates. Thus, carbon stocks in both vegetation and soils increased over the 50-year simulation (Table 22.1).

These results are qualitatively similar to those reported earlier (Rastetter et al., 1991, 1992), but there are some important differences. Calibration to data from a 9-year, multiple-factor experiment (McKane et al., submitted a) placed additional, longer-term constraints on the model that were not available for the earlier simulations. The major differences between earlier results and those reported here involve mostly the photosynthetic response to temperature. The earlier simulations were based on a fit of the Lassiter (1975) temperature–response function to data such as those reported by Larcher (1980, p. 115). These data give fairly broad ranges for the temperature minima, optima, and maxima for photosynthesis of general classes of vegetation and are based on short-term metabolic measurements. Therefore, they do not sufficiently constrain the model to match the longer-term data from the Chapin et al. (1995) 9-year experiment. Even subtle changes in the relative responses of photosynthesis and respiration to temperature can make a large difference in long-term ecosystem carbon storage. More long-term temperature experiments are needed if reliable projections of ecosystem responses to changing climate are to be made.

A second difference between earlier simulations and those reported here is in the rate of acclimation of vegetation to changes in CO_2 and temperature. For example, in earlier simulations of the response of tundra to a doubling of CO_2, net ecosystem productivity showed a large increase initially, decreased to zero by year 6, was slightly negative for another 6 years, and then remained virtually zero thereafter. The new simulations show the same pattern of ecosystem productivity, but the duration of the dynamic is about 7, rather than 12 years (Figure 22.1). These new results are more consistent with the data reported by Oechel et al. (1986).

A final difference between earlier and present simulations is in the interaction between the CO_2 and temperature responses. In earlier simulations (Rastetter et al., 1991, 1992), there was no obvious synergism in the combined CO_2–temperature response; the combined response was virtually equal to the sum of the independent responses to the two factors. This lack of a synergism may be due to the slower response time of vegetation in the earlier calibration. Because of the slower response in the earlier calibration, the vegetation could not take full advantage of the transient increase in nitrogen availability and the positive feedback resulting in the synergism never fully developed.

The Effects of Moisture Changes

In moist arctic tundras, moisture levels are typically high enough to inhibit decomposition and seldom low enough to inhibit photosynthesis. Thus, the major effect of a change in soil moisture is likely to be on soil processes. We used a moisture-response function for decomposition derived from an experiment by Flanagan and Veum (1974). These data are for short-term manipulations of moisture and may not reflect longer-term effects of prolonged changes in soil moisture. However, long-term data are not available. The short-term data should give at least a qualitative assessment of the effects of soil moisture, especially given the uncertainty in the projections of changes in soil moisture itself. Nevertheless, long-term manipulations of soil moisture should be done to improve our understanding of tundra carbon storage.

When moisture was increased by 10% in the simulations, all three key ecosystem properties changed along with the decrease in decomposition rate (Table 22.1). The most important of these was an increase in the C/N ratio of soils. The soil C/N ratio increased because slower rates of decomposition resulted in an accumulation of fresh plant litter, which has a higher C/N ratio than the soil as a whole. This change alone accounted for an increase of nearly 200 g C m^{-2} in the ecosystem over the 50-year simulation. The second property to change was the total amount of nitrogen in the ecosystem. This increase in nitrogen resulted from the slower rate of nitrogen turnover in the soil, which reduced the amount of mineral nitrogen available to be exported from the ecosystem. Thus, nitrogen inputs to the ecosystem were not balanced by commensurate losses. This increase in ecosystem nitrogen accounted for an accumulation of another $10\,g\,C\,m^{-2}$ in the ecosystem. Opposing these increases in C/N ratios and total ecosystem nitrogen was a tendency to lose carbon associated with a net movement of nitrogen from vegetation (high C/N ratio) to soils (low C/N ratio). Nitrogen lost from the vegetation as litter could not be recovered by uptake because of the decreased mineral-nitrogen availability resulting from the slower turnover of organic nitrogen. Thus, there was a net movement of nitrogen from vegetation to soils that accounted for a loss of $79\,g\,C\,m^{-2}$ from the ecosystem. The effect of all three of these properties was a decrease in vegetation carbon, but a larger increase in soil carbon resulting in a net increase in ecosystem carbon of about $130\,g\,m^{-2}$ (Table 22.1, Figure 22.4).

When moisture was decreased by 10%, the same three ecosystem properties changed by roughly the same amount, but in opposite directions (Table 22.1, Figure 22.4). Thus, higher decomposition rates resulted in the loss of younger, high C/N ratio organic matter in the soil, thereby decreasing the overall C/N ratio of the soil and accounting for a $252\,g\,C\,m^{-2}$ loss from the ecosystem. The increased nitrogen turnover increased the export of mineral nitrogen from the ecosystem, which could account for a loss of another $14\,g\,m^{-2}$ of carbon. However, changes in these two properties were counteracted to some degree by the net movement of nitrogen from soils to vegetation, which accounted for an $87\,g\,m^{-2}$ increase in carbon storage by the ecosystem. The effect of the dryer conditions was a small gain in vegetation carbon, a three times greater loss of soil carbon resulting in a net loss of $186\,g\,C\,m^{-2}$ loss from the ecosystem (Table 22.1, Figure 22.4).

The interactions between changes in soil moisture and changes in CO_2 and temperature are, for the most part, straightforward. Over the 50-year simulation, increased soil moisture further increased carbon storage in the CO_2, temperature, and combined treatments (Figures 22.1 to 22.3, Table 22.1). However, the moisture effect was not simply additive. An increase in moisture alone increased carbon storage by $130\,g\,m^{-2}$ over 50 years. Coupled with the CO_2 treatment or the combined CO_2 and temperature treatment, increased moisture only accounted for an additional 50 to $55\,g\,C\,m^{-2}$. However, increased moisture coupled to increased temperature had a marked synergistic effect on carbon storage; the combined effect was over two times as great as the sum of their individual effects. This pattern of nonadditive effects on carbon storage suggests that increased moisture effectively counteracts the stimulation of decomposition by increased temperature, but also counteracts the stimulation of production by CO_2 by decreasing nitrogen turnover in the soil.

Decreased soil moisture decreased the amount of carbon stored in the CO_2, temperature, and combined treatments (Figures 22.1 to 22.3, Table

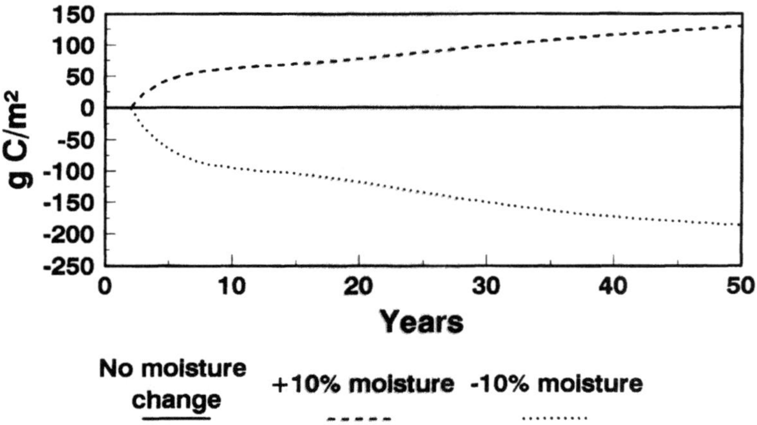

Figure 22.4. Cumulative changes in ecosystem carbon resulting from a 10% increase and a 10% decrease in soil moisture with no changes in CO_2 concentration or temperature.

22.1). Again the effects of moisture were not additive. The moisture decrease alone decreased carbon storage by 186 g m^{-2} over 50 years. The effect of decreased moisture on the CO_2 treatment and the combined CO_2 and temperature treatment was to reduce carbon storage below that of either treatment alone, but by less than 186 g C m^{-2}. However, the effect of reduced moisture on the temperature treatment was again synergistic and resulted in a loss of nearly 1000 g C m^{-2} from the ecosystem.

The synergism between increased temperature and decreased moisture is particularly striking and may represent a perturbation that takes the simulation beyond the limits of extrapolation for this particular calibration. In particular, the increase in nitrogen mineralization in this simulation resulted in an extreme allocation of carbon and nitrogen to leaves and away from fine roots. Before such an extreme allocation pattern ever materialized in nature, it is likely that the species composition of the vegetation would change sufficiently that the present calibration would no longer be valid and another calibration to a different vegetation (e.g., a shrub tundra) would be justified.

Nevertheless, the qualitative results of this simulation have important ramifications, regardless of their quantitative validity. The joint effect of increased temperature and decreased moisture on nitrogen mineralization was strong enough to release the ecosystem from nitrogen limitation. The vegetation was able to utilize this increased supply of nitrogen for about 25 years, but eventually saturated its uptake capacity. At that point, nitrogen began to leak out of the ecosystem. This loss of nitrogen is a real possibility for tundra ecosystems. These ecosystems have large amounts of nitrogen tied up in soils. There is more nitrogen in tundra soils than in both soils and vegetation combined for many temperate forests (Cole and Rapp, 1981; Giblin et al., 1991). If changes in temperature and moisture are sufficient to mineralize a large portion of this nitrogen, it is unlikely that any vegetation likely to develop on the tundra will be able to utilize all of it. The nitrogen will therefore be exported to streams and lakes and possibly in gaseous forms to the atmosphere. In either case, the effect of this nitrogen leakage on adjacent ecosystems, on the global climate system, and on tundra carbon storage could be important. However, the long response times of tundra ecosystems to this particular perturbation indicate that the important effects would not be observed in even decade-long experimental manipulations.

Summary

Carbon storage in tundra ecosystems is closely tied to the nitrogen cycle. To change carbon storage in the arctic, either the amounts of nitrogen in ecosystems must change, C/N ratios of soils and vegetation must change, or nitrogen must be redistributed between soils and vegetation. Because nitrogen inputs into tundra are small, gains in carbon associated with an increase in nitrogen capital are unlikely. On the other hand, losses in carbon associated with decreases in nitrogen capital are a real possibility, especially if changes in climate stimulate rates of nitrogen turnover in tundra soils.

Changes in carbon storage associated with changes in C/N ratios of vegetation and soils are constrained by the metabolic limits of plants and soil microbes. Nevertheless, our simulations indicate that constraints on C/N ratios are an important mechanism by which changes in climate and CO_2 will affect carbon storage. This is especially true for soils where even small changes in the C/N ratio have large effects on carbon storage.

Changes in carbon storage associated with the redistribution of nitrogen between soils and vegetation are potentially large, especially if nitrogen is redistributed from soils to vegetation. Any net movement of nitrogen from soils to vegetation should increase carbon storage. However, it is not likely that any type of tundra vegetation will be able to absorb more than a fraction of the nitrogen that might be mobilized from soils. The excess nitrogen would be lost to adjacent ecosystems or to the atmosphere with as yet undetermined effects.

Because of the large areal extent of tundra, the changes in carbon storage predicted by our model can have important implications for the global carbon budget. However, these predicted changes are small relative to the very large amount of carbon stored in tundras (about 27,000 $g\,C\,m^{-2}$). For example, a 1000 $g\,m^{-2}$ change of carbon storage represents less than a 5% change in the amount of carbon in the ecosystem. All of our simulations indicate that changes in carbon storage over 50 years will be smaller than this amount. A change of this magnitude would be well below the detection limits if it were exclusively from soils. On the other hand, if a large portion of this change in carbon storage were in vegetation, it would be easily detected. All of our simulations except the one for a doubling of CO_2 alone indicate a detectable change in vegetation carbon. However, this change in vegetation carbon is almost always overshadowed by a larger, but undetectable, change in soil carbon that is often in the opposite direction.

Our simulations indicate that the full effect of changes in CO_2, temperature, and moisture may not be fully manifested for a decade or more and that the initial direction of change may even be opposite to the direction of the long-term change (e.g., increased temperature alone, Figure 22.2). Furthermore, there are important nonadditive interactions among these three factors that would not be obvious without a full-factorial analysis. Unfortunately, long-term, full-factorial experiments are simply too expensive within the present funding climate. Even if these experiments were to be performed, the final results would not be available for several decades. The interim solution to this problem is to couple experimental and observational data to modeling. In this way the model can be used to fill gaps in the factorial analysis. However, because of the lack of long-term data, the long-term projections by these models will have to be accepted without rigorous validation. Reliance on such projections from a single model would be ill advised. Therefore, these long-term projections should be evaluated against similar projections from other models developed from broadly different perspectives. Consensus among a variety of model predictions may be the only available substitute for long-term validation.

Acknowledgments. This research was supported by funds from the National Science Foundation and the U.S. Environmental Protection Agency.

References

Barsdate, K.J., and Alexander, V. 1975. The nitrogen balance of arctic tundra: Pathways, rates, and environmental implications. *J. Environ. Qual.* 4:111–117.

Bazzaz, F.A. 1990. The response of natural ecosystems to the rising global CO_2 levels. *Annu. Rev. Ecol. Syst.* 21:167–196.

Billings, W.D., Peterson, K.M., Luken, O.J., and Mortensen, D.A. 1984. Interaction of increasing atmospheric carbon dioxide and soil nitrogen on the carbon balance of tundra microcosms. *Oecologia* 65:26–29.

Chapin, D.M., and Bledsoe, C.S. 1992. Nitrogen fixation in arctic plant communities. In: Chapin, F.E., Jefferies, R.L., Reynolds, J.F., Shaver, G.R., and Svoboda, J. (eds.), *Arctic Ecosystems in a Changing Climate* (pp. 301–319). San Diego: Academic Press.

Chapin, F.S., III, Shaver, G.R., Giblin, A.E., Nadelhoffer, K.J., and Laundre, J.A. 1995. Responses of arctic tundra to experimental and observed changes in climate. *Ecology* 76:694–711.

Cole, D.W., and Rapp, M. 1981. Element cycling in forest ecosystems. In: Reichle, D.E. (ed.), *Dynamic Properties of Forest Ecosystems* (pp. 341–409). London: Cambridge University Press.

Flanagan, P.W., and Veum, A.K. 1974. Relationships between respiration, weight loss, temperature and moisture in organic residues on tundra. In: Holding, A.J.,

Heal, O.W., MacLean, S.F., and Flanagan, P.W. (eds.), *Soil Organism and Decomposition in Tundra* (pp. 249–277). Stockholm: Tundra Biome Steering Committee.

Giblin, A.E., Nadelhoffer, K.J., Shaver, G.R., Laundre, J.A., and McKerrow, A.J. 1991. Biogeochemical diversity along a riverside toposequence in arctic Alaska. *Ecol. Monogr.* 61:415–435.

Larcher, W. 1980. *Physiological Plant Ecology*. Berlin: Springer-Verlag.

Lassiter, R.R. 1975. Modeling the dynamic of biological and chemical components of aquatic ecosystems. Technical report EPA-660/3-75-012, United States Environmental Protection Agency.

Maxwell, B. 1992. Arctic climate: Potential for change under global warming. In: Chapin, F.E., Jefferies, R.L., Reynolds, J.F., Shaver, G.R., and Svoboda, J. (eds.), *Arctic Ecosystems in a Changing Climate* (pp. 11–34). San Diego: Academic Press.

McKane, R.B., Rastetter, E.B., Melillo, J.M., Shaver, G.R., Hopkinson, C.S., Fernandes, D.N., Skole, D.L., and Chomentowski, W.H. 1995. Effects of global change on carbon storage in tropical forests of South America. *Global Biogeochemical Cycle* 9:329–350.

McKane, R.B., Rastetter, E.B., Shaver, G.R., Nadelhoffer, K.J., Giblin, A.E., Laundre, J.A., and Chapin, F.S., III. Submitted a. Effects of experimental changes in CO_2 and climate on carbon storage in Arctic tundra. *Ecology*.

McKane, R.B., Rastetter, E.B., Shaver, G.R., Nadelhoffer, K.J., Giblin, A.E., Laundre, J.A., and Chapin, F.S., III. Submitted b. Reconstruction and analysis of historical changes in carbon storage in Arctic tundra. *Ecology*.

Nadelhoffer, K.J., Giblin, A.E., Shaver, G.R., and Linkins, A.E. 1992. Microbial processes and plant nutrient availability in arctic soils. In: Chapin, F.E., Jefferies, R.L., Reynolds, J.F., Shaver, G.R., and Svoboda, J. (eds.), *Arctic Ecosystems in a Changing Climate* (pp. 281–300). San Diego: Academic Press.

Norby, R.J., Gunderson, C.A., Wullschieger, S.D., O'Neill, E.G., and McCracken, M.K. 1992. Productivity and compensatory responses of yellow-poplar trees in elevated CO_2. *Nature* 357:322–324.

Oechel, W.C. 1989. Nutrient and water flux in a small arctic watershed: An overview. *Holarctic Ecol.* 12:229–237.

Oechel, W.C., Riechers, G.H., Beyers, J., Cowles, S., Grulke, N., Hastings, S., Oberbaur, S., Prudhomme, T., and Sionit, N. 1986. Response of tundra ecosystems to elevated carbon dioxide. Progress report 037, U.S. Department of Energy, Carbon Dioxide Research Division.

Oechel, W.C., Hastings, S.J., Vourlitis, G., Jenkins, M., Riechers, G., and Grulke, N. 1993. Recent change of arctic tundra ecosystems from net carbon dioxide sink to source. *Nature* 361:520–523.

Rastetter, E.B., Ryan, M.G., Shaver, G.R., Melillo, J.M., Nadelhoffer, K.J., Hobbie, J.E., and Aber, J.D. 1991. A general biogeochemical model describing the responses of the C and N cycles in terrestrial ecosystems to changes in CO_2, climate, and N deposition. *Tree Physiol.* 9:101–126.

Rastetter, E.B., McKane, R.B., Shaver, G.R., and Melillo, J.M. 1992. Changes in C storage by terrestrial ecosystems: How C–N interactions restrict responses to CO_2 and temperature. *Water, Air, Soil Pollution* 64:327–344.

Shaver, G.R., Billings, W.D., Chapin, F.S., Giblin, A.E., Nadelhoffer, K.J., Oechel, W.C., and Rastetter, E.B. 1992. Global change and the carbon balance of arctic ecosystems. *Bioscience* 42:433–441.

Thornley, J.H.M. 1972. A model to describe the partitioning of photosynthate during vegetative growth. *Ann. Bot.* 36:419–430.

23. Century Trends in the Volume Balance of Boreal Forests: Implications for Global CO_2 Balance

Allan N.D. Auclair and Julie A. Bedford

Considerable uncertainty exists on the actual carbon budget of the boreal forest, and of its response to changes in global climate. These features have assumed considerable importance in light of ongoing, rapid changes in temperature at high latitudes (Jones and Wigley, 1990; Mitchell et al., 1990). North and south limits of the boreal forest are separated by 2 to 4°C, suggesting that the existing trees could be adversely impacted. Loss of boreal forest in Canada and the Commonwealth of Independent States (CIS) would be significant to their economies. In addition, large-scale fires (Auclair and Carter, 1993) and dieback (Neilson, 1990; Turner et al., 1993) may feedback C to the atmosphere through the release of transient pulses of CO_2.

Tans et al. (1990), using deconvolution techniques, estimated the boreal *plus* arctic regions to be a net source of 0 to $+0.7\,\mathrm{Pg\,C\,yr^{-1}}$ of C to the atmosphere. The most recent evidence based on $^{13}C/^{12}C$ ratios of atmospheric CO_2 suggests that the temperate *plus* boreal forests can be a large

* Disclaimer: Although the information in this document has been funded wholly or in part by the United States Environmental Protection Agency, Office of Environmental Processes and Effects Research (Washington, DC) through EPA Contract No. 68-D1-0109 to SPA, Inc., it has not been subjected to Agency peer review process and therefore does not necessarily reflect the views of the Agency and no official endorsement should be inferred.

net CO_2 sink in some years (e.g., 3.5 $Pg\,C\,yr^{-1}$ in 1992 and 1993, Ciais et al., 1995).

There are few complete vegetation *plus* soil-based estimates of the net C flux in boreal forest across the Northern Hemisphere. Apps et al. (1993) estimated the net C flux to be 0.7 $Pg\,C\,yr^{-1}$ based on a literature synthesis that included fluxes in peatlands and forest products. Sedjo (1992) also estimated a rate of 0.7 $Pg\,C\,yr^{-1}$ but in boreal *plus* temperate forests in the Northern Hemisphere based on forest volume changes.

C balance of high latitude ecosystems represents a moving target particularly subject to climatic variations. Recent modeling studies have demonstrated the potential for shifts from a net C sink to a net C source in both boreal forest (Dai and Fung, 1993) and arctic tundra (Oechel et al., 1993). The national forest inventory assessments of the United States and Canada indicate a net increase in wood volume and a potential carbon sink, at least since 1950. Substantial changes have occurred in the past four decades in both the losses and additions (Table 23.1). These forest records suggest that it is important to look historically and at individual processes to adequately understand (and predict) the forces driving these changes.

The objective of this chapter is to estimate changes in the annual net volume balance of boreal forests in the Northern Hemisphere this century (1890 to 1990). This is a preliminary but important step in developing a complete C budget over the period of historical forestry records.

Table 23.1. The Estimated Volume Budgets in Canada[a] and the United States[b] for Different Inventory Periods

	Average Annual GMV Depletion ($m^3 \times 10^6$/yr)	Average Annual GMV Accrual ($m^3 \times 10^6$/yr)	GMV Balance ($m^3 \times 10^6$)
Canada			
1977–1981	296.00	365.60	−69.60
1982–1986	230.00	364.60	−134.60
1987–1991	308.40	363.60	−55.20
United States			
1952	110.88	504.47	−393.59
1962	123.03	595.77	−472.74
1970	495.38	675.16	−179.78
1976	517.83	736.01	−218.18
1986	592.13	764.17	−172.04
1991	616.99	767.61	−150.62

The GMV depletion includes annual removals plus annual mortality (fire, pests, weather); GMV accrual is the total addition due to growth in standing forest, to new tree recruitment by regeneration, and to planting/reforestation.

[a] Canadian Forest Service (1994).

[b] Powell et al. (1993, Table 35) and Waddell et al. (1989, Tables 27, 28, 29).

Data and Methods

Approach

We compiled the national forestry data bases of the USDA Forest Service and Canadian Forest Service into a standard format of annual total area (ha) and GMV or gross merchantable volume (m^3) of forest depletions in Alaska and Canada. Gross merchantable volume is the gross volume in cubic meters less deductions for rot, roughness, and poor form. Volume is computed for the central stem from a 28-cm (1 foot) stump to a minimum 10 cm (4 inch) top diameter outside bark, or to the point where the central stem breaks into limbs (Waddell et al., 1989).

Depletion records consisted of annual or periodic records of losses to wildfire, pests, and harvesting. Most records were consistent from 1950 to present. Several depletions were recorded annually (e.g., annual extent of fire in the US, from 1918 onwards), or in the case of harvesting, had been previously reconstructed from the literature (e.g., Clawson, 1979). Where records early in the century were unavailable, we extrapolated from the first decade of available depletion records (usually 1920 to 1929) to 1990.

Accrual terms included forest volume conserved by fire suppression, gains by forests regrowing following depletion, and gains due to increased tree growth rate over the century. Fire suppression and regrowth were each estimated indirectly from depletion terms; increased tree growth was estimated from literature on tree chronologies (see details below).

All initial sources of the data on area and volume of the forest depletions, and final estimates of depletions and accruals that involved any modifying assumptions to amend gaps or inconsistencies in the data were comprehensively documented (Auclair et al., 1994a, 1994b).

The annual balance between forest volume depletions and accruals, or "net forest volume" was calculated each year, 1890 to 1990, by subtracting from the sum of forest volume depleted (wildfire, pests, harvest), the sum of the forest volume accrued (regrowth, conservation by fire suppression, increased tree growth).

Net Forest Volume = total of forest depletions – total of forest accruals

pest-kill	increased tree growth
wildfire	fire suppression*
harvesting	regrowth

There were two exceptions in our calculation of net forest volume. First, the volume conserved by fire suppression was *not* included in our estimate of net forest volume on the consideration that the statistics of volume loss to wildfire already included the decreased extent of loss to wildfire. Second, the increased tree growth was excluded in an initial set of graphic displays on the consideration that the volume attributed to this factor was due to other than land cover changes. Increased tree growth was included in a modified graph to demonstrate what the combined effects of regrowth *plus* long-term changes in tree growth rate may have in accounting for "missing" carbon in the global budget (Tans et al., 1990).

Boreal Forest included "main taiga," and Forest Tundra "northern and maritime taiga" plus "wooded tundra" as defined by Olson et al. (1983). Boreal Forest occupied 49% and Forest Tundra occupied 51% of the 1170 million ha total area in the two biome types. Regions analyzed included Alaska, Canada, Scandinavia (Norway, Sweden, Finland), the Commonwealth of Independent States (CIS), and China.

Estimates of Forest Depletion

Loss to Forest Pests

The GMV of trees lost to insects and diseases in Canada had been summarized for the 1977 to 1981 (Sterner and Davidson, 1982), 1982 to 1987 (Power, 1991), and 1988 to 1990 (Forestry Canada, 1988, 1989, 1990) periods. In the absence of a national summary of volume loss to pests prior to 1977, we would have extrapolated the 1977 to 1990 average level of pest depletion back to 1890.

Such an extrapolation would not have been satisfactory. The most prominent pest, namely the spruce budworm [*Choristoneura fumiferana* (Clem.)] had been previously monitored (Brown, 1970) and observed to be highly cyclical over the century. Spruce budworm caused two-thirds of the 1977 to 1987 insect loss and two-fifths of the total loss to all pests (Power, 1991). Historical trends (1909 to 1966) in the *area* of infestation summarized by Brown (1970) were updated from 1967 to 1976 (Canadian Forestry Service, 1967 to 1976). The 1910 level was then used to approximate the levels from 1909 to 1990. Forest volume losses to spruce budworm were estimated by multiplying the annual areal estimates above by the average GMV loss/area of infestation based on the 1977 to 1987 period (Power, 1991).

By differentiating spruce budworm from diseases and "other" insects we were able to obtain a partial but more accurate annual estimate of total losses to pests in Canada. To approximate the total GMV losses to all pests the GMV loss to spruce budworm was added to the GMV loss to diseases and "other" insects. The latter were estimated using the average 1977 to 1987 GMV loss of each pest category to represent the 1890 to 1990 period (Power, 1991).

The biome distribution of the total volume of pest-kill in Canada was based on the magnitude of losses to the 10 most common pest species, combined with information on the tree species most typically infected by each pest (Martineau, 1984). The GMV loss in Boreal Forest was estimated to have been 65% of total mortality to pests. There were no records of pest infestation in Forest Tundra.

The GMV losses to pests in Alaska from 1943 to 1990 were estimated using the USDA Forest Service annual softwood/hardwood mortality assessments (Powell et al., 1993) assuming that 65% of the annual mortality was due to pests (Lewis, personal comm.). The GMV loss from 1942 to 1890 was assumed to be the average GMV loss over the 1943 to 1990 period. The fraction of loss to pests in Boreal Forest was assumed to be half of the "softwood" loss to pests in the State. In the absence of factual information we arbitrarily assumed no loss to pests in Forest Tundra.

Loss to Wildfires

The annual extent of forest wildfires in Canada 1918 to 1990 (Forestry Canada, 1992a) was determined from annual fire records. The annual extent from 1890 to 1917 was approximated using the average annual extent, 1920 to 1929.

Records in Canada prior to 1946 did not include the Yukon, the Northwest Territories, or Newfoundland. The average annual extent of forest fires in these areas, 1946 to 1990, was computed as a fraction of the total in the other Canadian provinces over the same period. This fraction was multiplied by the annual area burned, 1918 to 1946, to approximate the full extent of burning in Canada prior to 1946. Stocks (personal comm.) noted that all fires, including many large fires in remote areas of Canada, were likely not detected in the early years of record. To adjust for the lack of full reporting in early observation years, the areal extent of burning was increased by 25% over the 1918 to 1935 period. The level in 1935 (i.e., 125% of actual level) was decreased linearly to 100% of actual level in 1960 based on the assumption that more or less complete detection was in place by the 1960 decade.

To estimate forest volume killed by wildfire, empirically determined levels of GMV/ha mortality in each biome were multiplied by the areal extent of burning. Biomass surveys (Bonner, 1985; Botkin and Simpson, 1990) indicated 27 m^3/ha in Forest Tundra and 80 m^3/ha in Boreal Forest were representative of *unburned* stands and we used these density estimates. The latter compared to a density of 83.7 m^3/ha in *fire-killed* Boreal Forests of Canada on "productive nonreserved forest land" (Honer and Bickerstaff, 1985, pp. 80, 81).

The distribution of wildfires in Canadian Boreal Forest (76.5%) and Forest Tundra (22.5%) was based on the incidence of wildfires in the 1980 to 1989 period (Kurz et al., 1992, pp. 66, 91).

The areal extent of forest fires in Alaska was based on the US Department of Interior, Bureau of Land Management statistics for the 1944 to 1990 period (Alaska Fire Service, 1992). To approximate the extent of burning in the State over the pre-1944 period, we computed the areal extent of burning in Alaska as a fraction of the total area of burns in Canada over the 1944 to 1990 period. This fraction was then multiplied by the annual extent of burning in Canada to roughly approximate the extent of wildfires in Alaska 1890 to 1943. Because 88% of the forest land burned in Canada 1980 to 1989 was in subarctic and western regions (Kurz et al., 1992, p. 91), this approach was considered a reasonable extrapolation.

To estimate volume of annual mortality, the annual area of burn was multiplied by the same volume densities as in Canada, namely $27 m^3$/ha in Forest Tundra and $80 m^3$/ha in Boreal Forest. USDA Forest Service (1944–1990) reports annually the fraction of wildfires in "nonforested watershed" in Alaska. These areas contain scattered trees, shrubs, or rangeland and were arbitrarily assumed to contain an additional forest volume equivalent to 10% of that in closed forest.

Discussion with fire scientists in Alaska suggested that about 40% of the area burned in the State was in Boreal Forest, and 55% in Forest Tundra.

Forest Harvesting

The total GMV harvested in Canada 1926 to 1990 was based on the sum of the roundwood equivalents in primary products (Forestry Canada, 1992b). The annual wood volume harvested in Canada from 1890 to 1925, was assumed to follow the same trend as in the United States (Clawson, 1979); US values over this period were adjusted downward to coincide with the harvest level in Canada in 1926. The area harvested annually was estimated by dividing the total GMV harvested in Boreal Forest by the mean harvest GMV/ha over the 1977 to 1981 period (Honer and Bickerstaff, 1985, p. 70).

In the absence of biome-specific summaries for Canada, we had to approximate indirectly the fraction of the total Canadian harvest originating in Boreal Forest. The levels of "softwood in primary products" 1926 to 1969 (Rotherham, 1991) and 1970 to 1990 (Forestry Canada, 1992) were used to estimate the percent distribution in coniferous forest types. Boreal Forest was estimated to be the source of 75% of the total softwood GMV; this corresponded closely with the Forestry Canada estimates of harvesting in Boreal and Cordilleran regions in 1986. There was no commercial harvesting in Forest Tundra (Kurz et al., 1992, p. 92).

The GMV harvested in Alaska Boreal Forest was obtained from forest inventory assessments for the period from 1952 to 1990 (Powell et al., 1993). Boreal Forest was assumed to be 1% of the total GMV harvested in the State based on the fraction harvested in white spruce (*Picea glauca* [Moench.] Voss.) and paper birch (*Betula papyrifera* Marsh.) (Hutchison, 1968, Table 23). No commercial harvesting was assumed to have occurred prior to 1942; harvest levels from 1943 to 1952 were increased linearly from 0 to the observed 1952 level. Area of Boreal Forest harvested annually was approximated by dividing by the same GMV/ha as in Canadian Boreal Forest. There was assumed to have been no commercial harvesting in Alaska Forest Tundra.

Estimates of Forest Accrual

Volume Conserved by Fire Suppression

Forest volume saved by fire fighting was the difference between the regression trends of volume burned from 1920 to 1976, and a line of no difference projected from 1935 to 1990. The procedure assumed that increasingly effective fire suppression capability in Canadian Boreal Forest began about 1935.

Linear regression equations were calculated using annual values of the volume burned over the 1920 to 1976 period, inclusive. The regression values were then extrapolated from 1977 to 1990. The rationale for using a regression (and not the highly variable annual values) was that the capability to suppress wildfires was probably progressive, and more or less consistent over time. The 1977 to 1990 period was not included in the regression because the levels of burning in Canada had increased markedly (Auclair and Carter, 1993) and masked the real efficacy of fire fighting. There was no indication of a sudden change in policy or in resources to fight forest fires in and after 1977; it was assumed that the increasing level of suppression typical of 1935 to 1976 prevailed over the 1977 to 1990 period.

There was no significant regression trend towards decreased volume of mortality due to wildfire in Alaska over the period of record from 1944 to 1990. In fact, the trend increased very slightly. On the basis of the procedure used, we assumed that the volume saved from burning was zero. This observation was consistent with perceptions of fire scientists in the State.

Regrowth Following Depletion

Forest volume in regrowth was an estimate of the cumulative volume depleted over time that had not reached age of maturity. It was assumed that all forests lost to pests, wildfire, and harvesting began to regenerate the year of disturbance and would reach full volume by linear growth at the biome-specific age to maturity.

To achieve this we backed our computer algorithm to year 1765 and used the average rates of depletion from 1890 to 1900. Age to maturity was estimated to be 100 years in Forest Tundra and 125 years in Boreal Forest. By initiating the areal calculations in 1765, any lag in depletions 125 years hence (i.e., by 1890) were accounted for. Failure to use this algorithm resulted in an artifact, namely a sharp decrease in volume of regrowth by the 1930 decade. The volume of regrowth in any year was the product of the cumulative area of depletion (within each depletion process, biome, and year) multiplied by the rates of regrowth. Rates of regrowth used were the forest GMV volume at maturity determined for each process and forest biome, divided by the age to maturity.

Increased Tree Growth Rate

There was particularly convincing evidence of a severalfold increase in the rate of tree growth in Boreal Forest and Forest Tundra in Canada since 1890 (Jacoby and D'Arrigo, 1989; Jozsa and Powell, 1987; Payette et al., 1985).

Evidence of increased tree growth was based on a literature synthesis of dendrochronologies standardized for tree age and stand density. For purposes of comparison, we stratified the synthesis by major forest biomes across North America, namely Forest Tundra, Boreal Forest (west), Boreal Forest (east), western Temperate Conifer Forest, northern Temperate Hardwood Forest, and southeastern US Oak-Pine Forest. The factor increase in growth rate in each biome was estimated by dividing the growth rate in 1990 by the growth rate in 1890 (Table 23.2).

To estimate the increased forest volume annually, the biome factor increase of Forest Tundra and of Boreal Forest in Canada and Alaska were multiplied by "standard tree chronology annual increments" of Jacoby and D'Arrigo (1989). These standard increments were available for 1890 to 1973 only. We assumed no tree growth rate change over the 1974 to 1990 period. This was consistent with evidence in Alaska of decreased growth rate index as a result of decreased precipitation and increased moisture stress in the 1970 and 1980 decades (Jacoby and D'Arrigo, 1995).

Table 23.2. Average Increase in Tree Growth Rate in Major Forest Biomes of North America This Century Based on an Extensive Overview of the Published Dendrochronology Literature

Forest Biome	Area (% Total)	Factor Increase	Reference
Forest Tundra	22	2.45	Jacoby (1983), Jacoby and D'Arrigo (1989), Payette and Filion (1986), Payette et al. (1985), Scott et al. (1989)
Boreal (West)	28	3.20	Jozsa and Powell (1987)
Boreal (East)	16	1.65	Dendron (1988)
Temperate Conifer	16	1.70	Brubaker and Graumlich (1987), Graybill and Idso (1993), Lamarche (1974), Lamarche et al. (1984), Parker (1987)
Temperate Hardwood	18	1.15	National Geophysical Data Center (1993)
AVERAGE (Weighted)		2.18	

The factor increase is the tree annual growth rate in 1990 (or the most recent year of record) divided by that in 1890. Approximate biome distributions are from Olson et al. (1983).

Forest Volume Estimates for Eurasia

Rates of forest harvest in Scandinavia (United Nations, 1986, 1992), the CIS and northern China (Houghton and Hackler, 1994; Melillo et al., 1988; United Nations, 1989, 1992) were reconstructed yearly from literature and government summaries, 1890 to 1990.

In the absence of readily available statistics on losses to pests and wildfire in Eurasia, we used a series of steps to approximate the level and pattern of loss across the century. First, we selected a surrogate biome from North America that we judged to be a reasonable ecological equivalent. For example, main taiga in the CIS was assumed to be reasonably represented by main taiga in Canada. Second, pest and wildfire rates were determined from literature sources. The rate of loss to pests in Scandinavia was $0.10\,m^3 ha^{-1} yr^{-1}$ (United Nations, 1986) and this was assumed to apply to the CIS and northern China. This rate compared to $0.07\,m^3 ha^{-1} yr^{-1}$ in Alaska and $0.08\,m^3 ha^{-1} yr^{-1}$ in Canada Boreal Forest. Wildfire in Eurasian forests (i.e., 0.3 million ha burned per yr per 100 million ha of forest land, Auclair and Carter, 1993) were about half the burn rate in Alaska and Canada. Third, the areal extent of each the North American and Eurasian biomes was estimated based on Olson et al. (1983). Areal coefficients (i.e., area of Eurasian biome /area of surrogate North American biome) were multiplied by the rate of forest loss in the Eurasian biome and by the rate of loss in the surrogate North American biome to obtain an approximate rate of loss. Studies currently in progress (Nabuurs, 1994; Nilsson et al., 1994) may soon make available accurate statistics for comparison and verification of our approximations.

Results and Discussion

Shifts in Tree Growth Rate This Century

Black spruce [*Picea mariana* (Mill.) BSP.] at forest limits in the eastern subarctic showed about a 2.5-fold increase between 1890 and 1980 (Payette et al., 1985). Tree chronologies of white spruce along the northern tree line across Alaska and Canada showed a similar level of increase between 1890 and 1973 (Jacoby and D'Arrigo, 1989). The annual biomass increment in the stemwood of white spruce in Boreal Forest of western Canada increased threefold between the previous and present centuries (Jozsa and Powell, 1987). The change in radial increment of white spruce in Boreal Forest of southeastern Canada was only 1.6-fold by comparison (Dendron, 1988). Growth rate increase of eastern Temperate Hardwood Forest, western Temperate Coniferous Forest, and southeastern Oak-Pine Forest of the US ranged from 1.8- to 1.3-fold. Overall the average increase across the forest biomes of North America (weighted by area) increased 2.18-fold (Table 23.2).

The empirical observations in Table 23.2 indicate larger changes in the more northern biomes suggesting a temperature-driven response to warming. This was consistent with statistical tests of D'Arrigo and Jacoby (1993). These authors could find no evidence of a CO_2 fertilization effect on the growth rate increase in white spruce at tree line, but cautioned that their conclusions may not apply to the Boreal Forest as a whole or to specific conditions such as high elevations and strip bark morphology (Graybill and Idso, 1993).

Shifts in Forest Volume This Century

Wildfire decreased from 1920 to notably low levels by 1965, followed by a surge in the 1980s. Losses to pests remained more or less constant until the mid-1970s when levels doubled. Natural mortality (fire, pests) dominated depletions until about 1945 after which harvesting was increasingly dominant. Total depletions in Boreal Forest across the Northern Hemisphere about doubled from 1890 to 1990 [Figure 23.1(A)].

Forest volume accrual due to increased tree growth rate was significantly greater by 1920 than either the volume added annually by regrowth on formerly depleted sites or by the volume conserved in live forest by fire suppression. Total accruals of volume in Boreal Forest across the Northern Hemisphere increased fourfold from 1890 to 1990 [Figure 23.1(B)].

Net forest volume due to landcover changes alone (i.e., total depletions minus regrowth on depleted sites) showed little trend over the 1920 to 1965 period at the level of the hemisphere. After 1965, wildfire and accelerated harvesting combined to shift the budget sharply toward a net loss of forest volume [Figure 23.2(A)]. In North America, the net forest volume showed a similar reversal about 1970 [Figure 23.2(B)]. The Eurasian forest volume budget showed a net depletion across all decades [Figure 23.2(C)]. This difference between North America and Eurasia (and specifically the USSR) had been noted previously (Melillo et al., 1988). Our analysis suggested that the low wildfire fire rates in Eurasia, about half that of North America (Auclair and Carter, 1993), resulted in low regrowth volume.

Net forest volume due to landcover *plus* increased tree growth rate changes showed a strong net volume accrual from 1890 through about 1965. From 1965 to 1990 the trend was toward lower rates of net accrual (Figure 23.3). The ostensible reason for this was an increase in losses to pests, wildfire and harvesting [Figure 23.1(A)], and at the same time, an arrest in the tree growth rates over the 1970 and 1980 decades [Figure 23.1(B)]. Despite the decreased accrual in recent decades, the Boreal Forest remains a significant potential sink of CO_2 as implied from the still positive net volume accrual. Also evident is the possibility of a diminished C sink and the likelihood of a net C source should the increased losses and slowed tree growth continue.

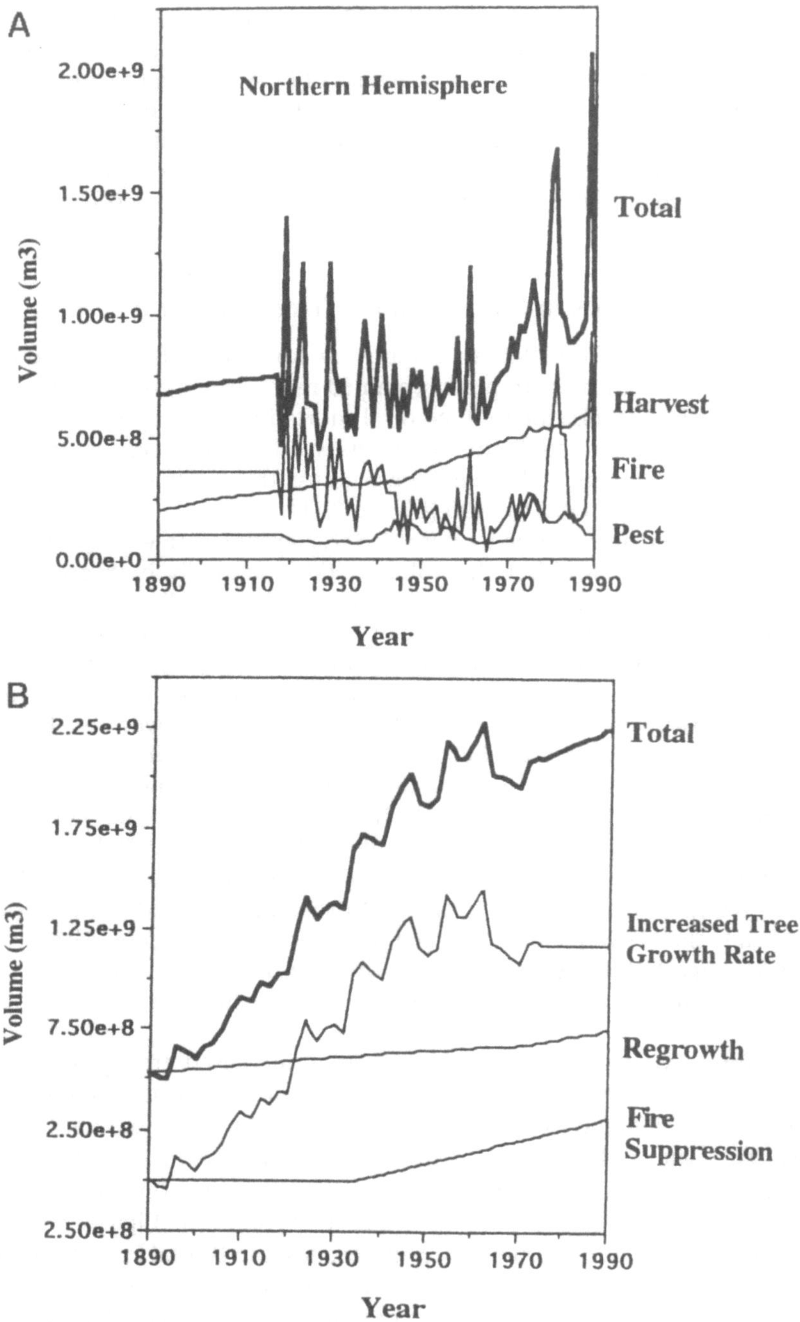

Figure 23.1. Total volume of (A) depletions, and (B) accruals in Boreal Forest (*plus* Forest Tundra) of the Northern Hemisphere.

Net Biospheric Flux Versus Net Forest Volume Balance

Estimates of the Boreal Forest net volume balance paralleled the trends in global net biospheric CO_2 flux (Sarmiento et al., 1992) [Figure 23.4(A)].

Ideally, net CO_2 flux estimates in Boreal Forest would have been used in the regression. The analysis of Figure 23.4 compares biome-scale "fluxes" in wood volume (i.e., depletion/accrual balance) and CO_2 fluxes (i.e., source/sink balance) of Sarmiento et al. (1992).

Regression analysis indicated a particularly high R^2 (=0.80) over the 1890 to 1960 period [Figure 23.4(B)], and an R^2 (=0.52) over the entire century, 1890 to 1990. The lack of any R^2 over the 1961 to 1990 period ($R^2 = 0.001$) resulted from the accelerated global net biospheric CO_2 sink at a time of increasing forest net depletions [Figure 23.4(A)]. The marked net CO_2 sink from 1962 to 1979 may be partly explained by an unusually wet period in the tropic and subtropic forest zones (Dai and Fung, 1993).

Some probable reasons for the high R^2 variation accounted for from 1890 to 1960 included the following: (1) The 1.2 billion ha in boreal forest represents a significant fraction (30%) of the world's productive forest, (2) approximately 96% of boreal forest (based on results of the Canadian 1986 forest inventory of trees in <100 year class) is currently in regrowth, (3) relatively long lag times (tenfold that of the tropical forests) resulting from slow postdisturbance decomposition and regeneration of biomass and soil C (Auclair and Carter, 1993) suggests that much if not most of the landscape may not be in C equilibrium (i.e., highly interactive with the atmosphere), (4) a strong link ($R^2 = 0.64$) is known to exist between the annual tree ring width of white spruce on climate-sensitive sites at tree line across Alaska and Canada and seasonal amplitude of atmospheric CO_2 drawdown, 1971 to 1981, at Point Barrow, Alaska (D'Arrigo et al., 1987), and (5) large changes in growth and forest cover have occurred within the biome in response to shifts in management (i.e., harvesting, fire suppression) and climate over the century. Dai and Fung (1993) concluded that climatic changes in the last five decades have induced significant variations in biospheric fluxes at the northern middle latitudes (30°N to 60°N). By contrast, the large precipitation variations at low latitudes (15°S to 25°N) caused compensating changes between net primary production and soil respiration, and resulted in insignificant changes in net carbon flux.

General Discussion

Benefits and Limitations of Approach

There is an urgent need to synthesize historical information on long-term changes in forest cover and productivity. In most countries the national forestry data is the only consistent record of forest changes available, and are now being used in C flux estimates (Birdsey, 1992; Kurz et al., 1992).

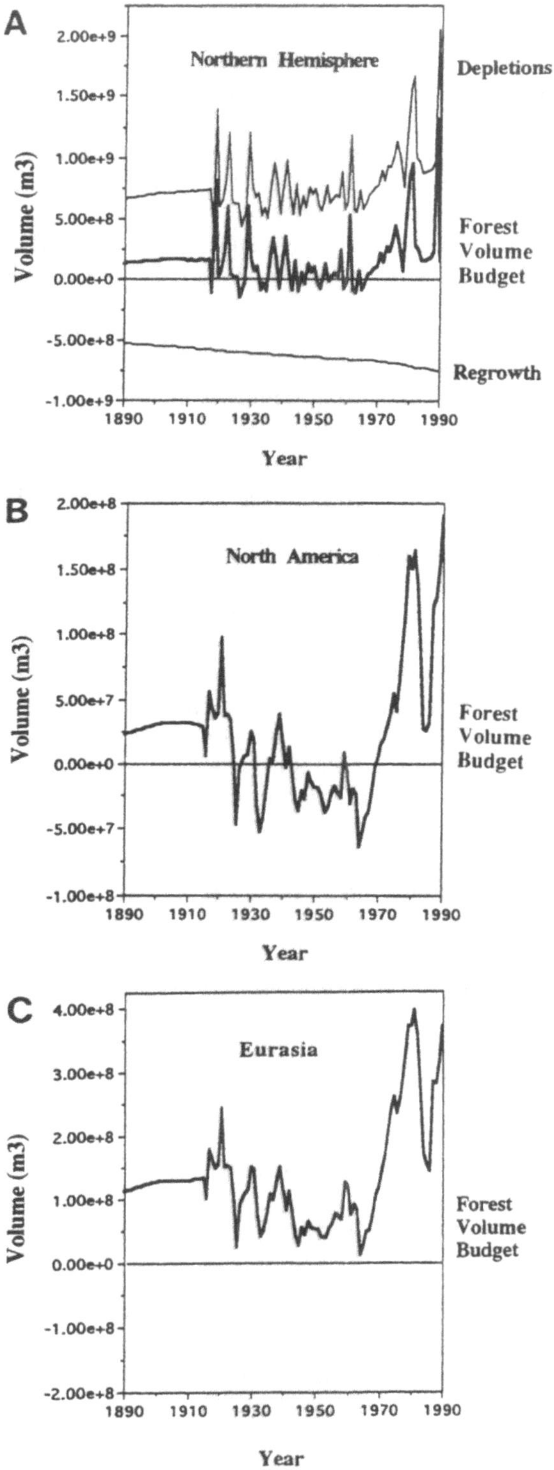
A
Northern Hemisphere
Depletions
Forest Volume Budget
Regrowth
Volume (m3)
2.00e+9
1.50e+9
1.00e+9
5.00e+8
0.00e+0
-5.00e+8
-1.00e+9
1890 1910 1930 1950 1970 1990
Year
B
North America
Forest Volume Budget
Volume (m3)
2.00e+8
1.50e+8
1.00e+8
5.00e+7
0.00e+0
-5.00e+7
-1.00e+8
1890 1910 1930 1950 1970 1990
Year
C
Eurasia
Forest Volume Budget
Volume (m3)
4.00e+8
3.00e+8
2.00e+8
1.00e+8
0.00e+0
-1.00e+8
-2.00e+8
1890 1910 1930 1950 1970 1990
Year

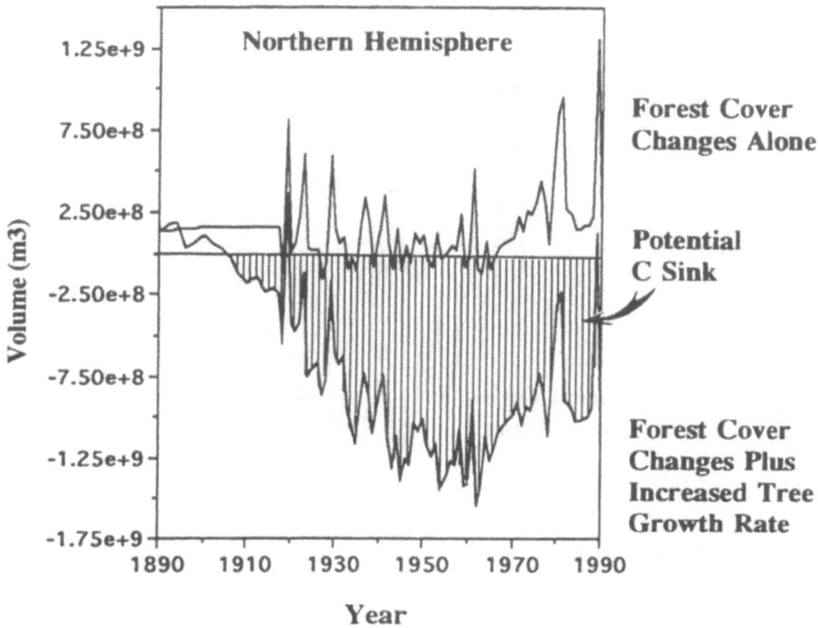

Figure 23.3. Changes in net forest volume this century in Boreal Forests (*plus* Forest Tundra) of the Northern Hemisphere. The forest volume budgets are for forest cover changes alone (total depletions – regrowth) and for forest cover changes plus increased tree growth rate (total depletions – total regrowth *plus* increased tree growth rate). The shaded area represents the net forest accrual and a potential C sink.

The records provide biome-scale, empirical data that are preferable to site studies that extrapolate from one or few locales to the level of the hemisphere (Bonan, 1991). Forestry volume statistics are on-line (i.e., computed periodically or annually). The few industrial countries involved makes feasible a complete synthesis on circumboreal forests. We considered the Eurasian extrapolation to be a temporary and provisional measure, and

---

Figure 23.2. (A) Changes in volume budget of Boreal Forest (*plus* Forest Tundra) of the Northern Hemisphere based on total forest depletions minus regrowth. (B) Five-year running mean of the same budget for North America and (C) for Eurasia showing the sudden, persistent reversal of the sink trend about 1970. Depletions represent a potential SOURCE of carbon, and are the total volume of depletions of Figure 23.1(A). Regrowth is the regrowth term of Figure 23.1(B); by convention the positive *sign has been inverted* to represent a potential carbon SINK. The Forest Volume Budget of each figure is the total depletion minus the total regrowth term and represents changes over the century in the potential for a carbon source (positive values) or potential carbon sink (negative values).

A

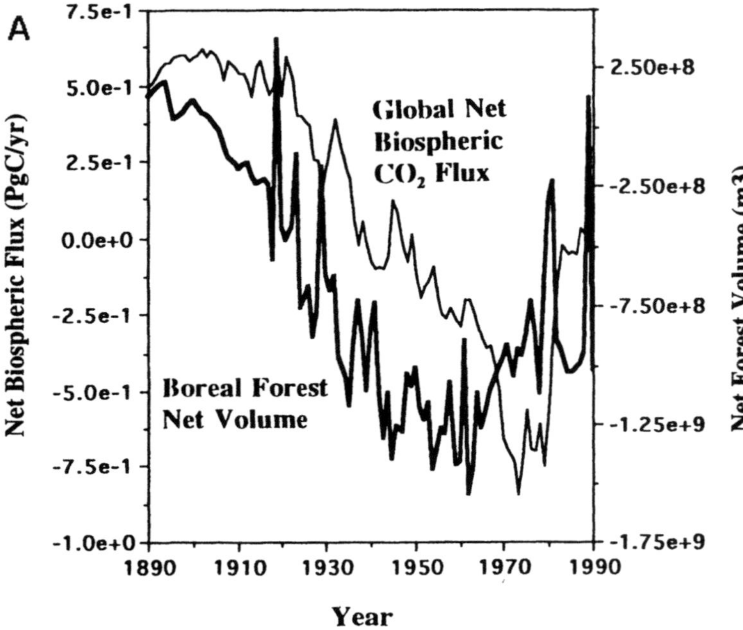

B

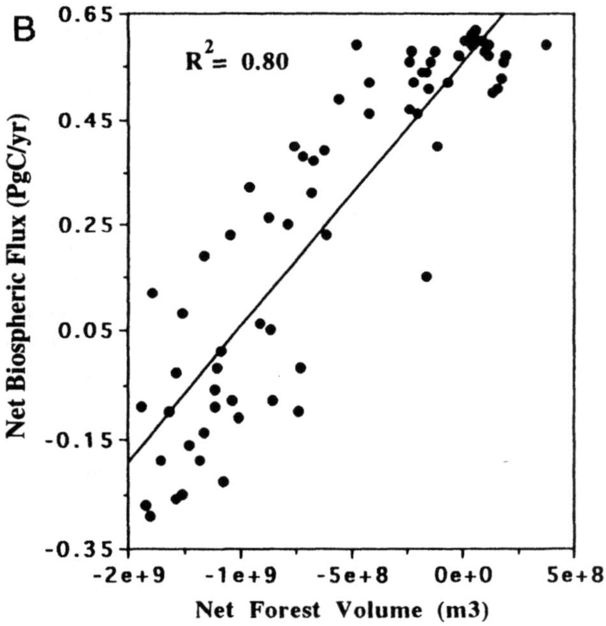

Figure 23.4. (A) Comparison of the balance of net forest volume in Boreal Forest (*plus* Forest Tundra) of the Northern Hemisphere with the global net biospheric CO_2 flux (Sarmiento et al., 1992). (B) First order polynomial regression of the net forest volume × net biospheric flux over the 1890 to 1960 interval.

recognize the possibility of including comprehensive forestry statistics from Scandinavia and the CIS (Nabuurs, 1994; Nilsson et al., 1994).

The range of variation and error in the data is of paramount concern because it will determine the inferences that can be made from the analysis. Forest volume inventories are considered to be very reliable. The accuracy reported for the provincial forest volume inventories in Canada was ±2% and ±6% nationally (Honer and Bickerstaff, 1985). The accuracy of forest volume inventories in Alaska was within ±3% and within ±1% in the US as a whole (Powell et al., 1993; Powell, personal comm.).

Conclusions on Importance of Boreal Forests in Global Change

The Boreal Forest, including adjacent areas of Forest Tundra, is a significant net sink of CO_2. The peak net volume accrual of about 1.25 billion m^3/yr was observed in the 1950 decade. The current (1970 to 1990) net volume accrual is about 0.75 billion m^3/yr (Figure 23.3). This is equivalent to a net C flux of 0.8 to 1.3 $Pg\,C\,yr^{-1}$ assuming that about 1 ton of C per 4 m^3 of wood (Birdsey, 1992), and 4 to 7 tons of C occur in the total biomass and soil of boreal and mixed forest ecosystems per 1 ton of C in GMV of trees (Moulton and Richards, 1990, Table 7). This estimate is slightly larger than the 0.7 $Pg\,C\,yr^{-1}$ estimated by Apps et al. (1993).

The net C sink in the Boreal Forests is a moving target and has changed significantly over this century. The increase in the size of the net C sink (i.e., net volume accrual) from 1920 through 1960 was due largely to an increase in tree growth rate combined with extensive forest regrowth and rigorous fire suppression. A decrease in the sink size since 1965 was due to increased fire and pest activity, increased harvesting, and no or little increase in tree growth rate. The trend toward an increasing sink early in the century has shifted in the last two decades to increasingly large forest volume losses and a potential source of CO_2 to the atmosphere.

The depletion/accrual balance in Boreal Forest is climate sensitive, a conclusion consistent with observations of Apps et al. (1993) and Sampson et al. (1993). The Boreal Forest was highly responsive to the "climatic jump" of 1976 (Ebbesmeyer et al., 1991) and to continued warming since 1965 (Hansen and Lebedeff, 1987). This is indicated by the fivefold increase in losses to wildfire (Auclair and Carter, 1993), the doubling of pest-related mortality (Powell et al., 1993), and decreased tree growth rate (Jacoby and D'Arrigo, 1995). It is the influence of climate that provides the strong link between changes in the biome and the global net CO_2 flux in the biosphere.

One result of the combination of potentially volatile sources and responsive sinks is the need to carefully assess the risks and tradeoffs in any attempts to manage biologically the boreal forests to "stabilize the atmosphere" (Lashof and Tirpak, 1990; Winjum et al., 1991).

Priorities of Future GCRP Research and Policy Studies

We have developed a comprehensive model to estimate sources, sinks, and net fluxes of C using the area and volume statistics of the national forestry (and related soil) data bases (Auclair et al., 1994a, 1994b). There is a need to apply the model across all regions of boreal forest, and the results to a series of statistical tests that address specific policy-relevant questions of priority to the IGBP Global Change Research Program (IGBP, 1990). These include the contribution of boreal forests to the biospheric CO_2 flux and the sensitivity the forest C flux to shifts in climate (IPCC, 1990, p. 92; IGBP, 1990, Focus 2.4 and Focus 2.7). The model may also be used to assess practical options for silvicultural management of trace gas emissions (IPCC, 1990, p. 93) and C sink protection (FFI, 1993).

Acknowledgements. We are especially grateful to Dr. Jorge Sarmiento of Princeton University, Princeton, NJ, USA, for providing the most recent estimates of the global biospheric CO_2 flux, and to Dr. David Turner, U.S. Environmental Protection Agency, Corvallis, OR, USA, for the ORNL Raster Projections of forest areas.

References

Alaska Fire Service. 1992. Annual forest fire reports, 1944–1990. Alaska Fire Service, Bureau of Land Management, Department of Interior, P.O. Box 35005, Fort Wainwright, Alaska, USA.

Apps, M.J., Kurz, W.A., Luxmoore, R.L., Nilsson, L.O., Sedjo, R.A., Schmidt, R., Simpson, L.G., and Vinson, T.S. 1993. Boreal forests and tundra. *Water, Air, Soil Pollution* 70:39–53.

Auclair, A.N.D., and Carter, T.B. 1993. Forest wildfires as a recent source of CO_2 at northern latitudes. *Can. J. Forest Res.* 23:1528–1536.

Auclair, A.N.D., Bedford, J.A., and Revenga, C. 1994a. Area and volume trends in forest depletions and accruals. Task Report 2 to the U.S. Environmental Protection Agency, Office of Research and Development, Washington, DC. Washington, DC: Science and Policy Associates, Inc.

Auclair, A.N.D., Bedford, J.A., and Revenga, C. 1994b. The missing carbon sink in the terrestrial biosphere: Task Report 8 to the U.S. Environmental Protection Agency, Office of Research and Development, Washington, DC. Washington, DC: Science and Policy Associates, Inc.

Birdsey, R.A. 1992. Carbon storage and accumulation in United States forest ecosystems. General Technical Report WO-59, U.S.D.A. Radnor, PA: Forest Service.

Bonan, G.B. 1991. Seasonal and annual carbon fluxes in a boreal forest landscape. *J. Geophys. Res.* 96(D9):17329–17338.

Bonner, G.M. 1985. Inventory of forest biomass in Canada. FORSTATS Program, Petawawa National Institute of Forestry, Canadian Forestry Service, Chalk River, Ontario. Canada.

Botkin, D.B., and Simpson, L.G. 1990. Biomass of the North American boreal forest. *Biogeochemistry* 9:161–174.

Brown, C.E. 1970. A cartographic representation of spruce budworm *Choristoneura fumiferana* (Clem.) infestation in eastern Canada, 1909–1966. Canadian Forestry

Service Publication No. 1263, Department of Fisheries and Forestry, Ottawa, Ontario, Canada.
Brubaker, L.B., and Graumlich, L.J. 1987. 100-year records of forest productivity at high elevations in western Washington, USA. In: Kairiukstis, L., Bednarz, Z., and Feliksik, E. (eds.), *Methods of Dendrochronology I.* (pp. 277–286). Proceedings of the Task Force Meeting on Methodology of Dendrochronology: East/West Approaches. 2–6 June 1986, Krakow, Poland. Warsaw, Poland: International Institute of Applied Systems Analysis, and the Polish Academy of Sciences-Systems Research Institute.
Canadian Forestry Service. 1967–1976. Forest insect and disease survey. Annual Reports, 1967–1976. Canadian Forestry Service, Environment Canada. Ottawa, Ontario, Canada.
Canadian Forestry Service. 1994. The state of Canada's forest 1993. Fourth Report to Parliament. Canadian Forestry Service, Department of Natural Resources. Ottawa, Canada.
Ciais, P., Tans, P.P., Trolier, M., White, J.W.C., and Francey, R.J. 1995. A large northern hemisphere terrestrial CO_2 sink indicated by the $^{13}C/^{12}C$ ratio of atmospheric CO_2.
Clawson, M. 1979. Forests in the long sweep of American history. *Science* 204:1168–1174.
Dai, A., and Fung, I.Y. 1993. Can climatic variability contribute to the "missing" CO_2 sink? *Global Biogeochem. Cycles* 7:599–610.
D'Arrigo, R.D., and Jacoby, G.C. 1993. Tree growth–climate relationships at the northern boreal forest tree line of North America: Evaluation of potential response to increasing carbon dioxide. *Global Biogeochem. Cycles* 7:525–536.
D'Arrigo, R., Jacoby, G.C., and Fung, I.Y. 1987. Boreal forests and atmosphere–biosphere exchange of carbon dioxide. *Nature* 329:321–323.
Dendron. 1988. An investigation of the possible impact of acid rain on forest growth in Canada. Vol. I. Main Report. Dendron Resource Surveys Ltd., Ottawa, Ontario, Canada.
Ebbesmeyer, C.C., Cayan, D.R., McLain, D.R., Nichols, F.H., Peterson, D.H., and Redmond, K.T. 1991. 1976 step in the Pacific climate: Forty environmental changes between 1968–1975 and 1977–1984. In: Betancourt, J.L., and Tharp, V.L. (eds.), Proc. 7th Annual Pacific Climate (PACLIM) Workshop (pp. 115–126). April 1990. Interagency Ecological Studies Program Tech. Rpt. 26. San Francisco, CA: California Dept. Water Resources.
FFI. 1993. Forests for the Future Initiative: Cooperative partnerships among U.S. Agencies and interested countries and organizations. January 15, 1993 Report of an Interagency Task Force, Council to the President, Washington, DC.
Forestry Canada. 1988–1990. Forest insect and disease survey. Annual Reports, 1988–1990. Forestry Canada. Ottawa, Ontario, Canada.
Forestry Canada. 1992a. Annual forest fire statistics on number of fires and area burned in Canada, 1918–1990. Petawawa National Institute of Forestry, Canadian Forestry Service, Chalk River, Ontario, Canada.
Forestry Canada. 1992b. National forest harvest database, 1926–1992. Forestry Canada. Ottawa, Ontario, Canada.
Graybill, D.A., and Idso, S.B. 1993. Detecting the aerial fertilization effect of atmospheric CO_2 enrichment in tree-ring chronologies. *Global Biogeochem. Cycles* 7(1):81–95.
Houghton, R.A., and Hackler, J. 1994. The Houghton global land use dataset: Updated database and documentation notes. Unpublished report. The Woods Hole Research Center, Woods Hole, MA. [Currently under review for publica-

tion by the Carbon Dioxide Information Analysis Center (CDIAC), Oak Ridge National Laboratory, Oak Ridge, TN.]

Hansen, J., and Lebedeff, S. 1987. Global trends of measured surface air temperature. *J. Geophys. Res.* 92(D11):13345–13372.

Honer, T.G., and Bickerstaff, A. 1985. Canada's forest area and wood volume balance 1977–1981: An appraisal of change under present levels of management. Information Report BC-X-272. Pacific Forestry Centre, Canadian Forestry Service, Victoria, British Columbia, Canada.

Honer, T.G., Clark, W.R., and Gray, S.L. 1991. Determining Canada's forest area and wood volume balance, 1977–1986. In: Brand D.G. (ed.), *Canada's Timber Resources* (pp. 17–25). Info. Rpt. PI-X-101, Forestry Canada, Chalk River, ON, Canada.

Hutchison, O.K. 1968. Alaska's forest resource. Research Bulletin PNW-19. Pacific Northwest Range and Experiment Station, U.S.D.A. Forest Service, Portland, OR, USA.

IGBP. 1990. The International Geosphere Biosphere Programme: A study of global change—the initial core projects. International Geosphere Biosphere Programme Secretariat, Royal Swedish Academy of Sciences, Stockholm.

IPCC. 1990. Formulation of response strategies. Working Group III, Intergovernmental Panel on Climate Change. World Meteorological Organization and United Nations Environment Programme.

Jacoby, G.C. 1983. A dendroclimatic study in the forest-tundra ecotone on the east shore of Hudson Bay. In: Morisset, P., and Payette, S. (eds.), *Tree-line Ecology* (pp. 95–100). Collection Nordicana No. 47, Proceedings of the Northern Quebec Tree-line Conference. Centre for Northern Studies, Laval University, Quebec, Quebec, Canada.

Jacoby, G.C., and D'Arrigo, R.D. 1989. Reconstructed northern hemisphere annual temperature since 1671 based on high-latitude tree-ring data from North America. *Climatic Change* 14:39–59.

Jacoby, G.C., and D'Arrigo, R.D. 1995. Tree ring width and density of climatic and potential forest change in Alaska. *Global Biogeochem. Cycles* 9(2):227–234.

Jones, P.D., and Wigley, T.M.L. 1990. Global warming trends. *Sci. Am.* 263(2):84–91.

Jozsa, L.A., and Powell, J.M. 1987. Some climatic aspects of biomass productivity of white spruce stem wood. *Can. J. Forest Res.* 17:1075–1079,

Kurz, W.A., Apps, M.J., Webb, T.M., and McNamee, P.J. 1992. The carbon budget of the Canadian forest sector: Phase I. Rpt. NOR-X-326, Forestry Canada, Edmonton, AB, Canada.

Lamarche, V.C. 1974. Paleoclimatic inferences from long tree-ring records. *Science* 183:1043–1048.

Lamarche, V.C., Graybill, D.A., Fritts, H.C., and Rose, M.R. 1984. Increasing atmospheric carbon dioxide: Tree ring evidence for growth enhancement in natural vegetation. *Science* 225:1019–1021.

Lashof, D.A., and Tirpak, D.A. (eds.) 1990. Policy options for stabilizing global climate. Rpt. 21P-2003.1, Washington, DC: U.S. Environmental Protection Agency.

Martineau, R. 1984. Insects harmful to forest trees. Multiscience Publications Ltd., Montreal, Quebec, Canada.

Melillo, J.M., Fruci, J.R., Houghton, R.A., Moore, B., and Skole, D.L. 1988. Land-use change in the Soviet Union between 1850 and 1980: Causes of a net release of CO_2 to the atmosphere. *Tellus* 40B:116–128.

Mitchell, J.F.B., Manabe, S., Tokioka, T., and Meleshko, V. 1990. Equilibrium climate change. In: Houghton, J.T., Jenkins, G.J., and Ephraums, J.J. (eds.),

Climate Change: The IPCC Scientific Assessment (pp. 131–172). Cambridge, UK: Cambridge University Press.

Moulton, R.J., and Richards, K.R. 1990. Costs of sequestering carbon through tree planting and forest management in the United States. Gen. Tech. Rpt. WO-58. U.S. Department of Agriculture, Forest Service, Washington DC.

Nabuurs, G.J. 1994. State-of-the-art in the field of forest sector carbon balance studies, with reference to the European situation. Working Paper No. 2. European Forest Institute, Joensuu, Finland.

National Geophysical Data Center. 1993. The International Tree-Ring Data Bank. World Data Center-A for Paleoclimatology, N.O.A.A., U.S. Department of Commerce, Boulder, CO.

Neilson, R.P. 1990. Biosphere feedback during climate change. In: King, G.A., Winjum, J.K., Dixon, R.K., and Arnaut, L.Y. (eds.), *Response and Feedbacks of Forest Systems to Global Climate Change*. Corvallis, OR: U.S. Environmental Protection Agency.

Nilsson, S., Shvidenko, A., Bondarev, A., and Danilin, I. 1994. Siberian forestry. Working Paper WP-94-08. International Institute for Applied Systems Analysis, Laxenburg, Austria.

Oechel, W.C., Hastings, S.J., Vourlitis, G., Jenkins, M., Riechers, G., and Grulke, N. 1993. Recent change of Arctic tundra ecosystems from a net carbon dioxide sink to a source. *Nature* 361:520–523.

Olson, J.S., Watts, J.A., and Allison, L.J. 1983. Carbon in live vegetation of major world ecosystems. Rpt. ORNL-5862. Oak Ridge National Laboratory, Oak Ridge, TN.

Parker, M.L. 1987. X-ray densitometry and image analysis as methods in dendrochronology in Canada and the United States. In: Kairiukstis, L., Bednarz, Z., and Feliksik, E. (eds.), *Methods of Dendrochronology I* (pp. 57–68). Proceedings of the Task Force Meeting on Methodology of Dendrochronology: East/West Approaches. 2–6 June 1986, Krakow, Poland. Warsaw, Poland: International Institute of Applied Systems Analysis, and the Polish Academy of Sciences-Systems Research Institute.

Payette, S., and Filion, L. 1986. White spruce expansion at the tree line and recent climatic change. *Can. J. Forest Res.* 15:241–251.

Payette, S., Filion, L., Gauthier, L., and Boutin, Y. 1985. Secular climate change in old-growth tree-line vegetation of northern Quebec. *Nature* 315:135–138.

Powell, D.S., Faulkner, J.L., Darr, D.R., Zhu, Z., and MacCleery, D.W. 1993. Forest statistics of the United States, 1992. Pre-publication Draft Tables. FIERR Staff Report, Forest Service, U.S. Department of Agriculture, Washington, DC.

Power, J.M. 1991. National data on forest pest damage. In: Brand, D.G. (ed.), *Canada's Timber Resources* (pp. 119–129). Info. Report PI-X-101, Petawawa National Forestry Institute, Forestry Canada, Chalk River, Ontario, Canada.

Rotherham, T. 1991. Timber harvest statistics: Past practice and present needs. In: Brand, D.G. (ed.), *Canada's Timber Resources* (pp. 105–118). Information Report PI-X-101. Petawawa National Forestry Institute, Canadian Forestry Service, Chalk River, Ontario, Canada.

Sampson, R.N., Apps, M., Brown, S., Cole, C.V., Downing, J., Heath, L.S., Ojima, D.S., Smith, T.M., Solomon, A.M., and Wisniewski, J. 1993. Workshop summary statement: Terrestrial biospheric carbon fluxes—quantification of sinks and sources of CO_2. *Water, Air, Soil Pollution* 70:3–15.

Sarmiento, J.L., Orr, J.C., and Siegenthaler, U. 1992. A perturbation simulation of CO_2 uptake in an ocean general circulation model. *J. Geophys. Res.* 97:3621–3645.

Scott, P.A., Hansell, R.I.C., and Fayle, D.C.F. 1988. Establishment of white spruce populations and responses to climatic change at the treeline, Churchill, Manitoba, Canada. *Arctic Alpine Res.* 19(1):45–51.

Sedjo, R.A. 1992. Temperate forest ecosystems in the global carbon cycle. *Ambio* 21:274–277.

Sterner, T.E., and Davidson, A.G. 1982. Forest insect and disease conditions in Canada 1981. Canadian Forestry Service, Ottawa, Ontario, Canada. [Also cited on p. 126 in Power, J.M. 1991. National data on forest pest damage. pp. 119–129 in Brand, D.G. (ed.), *Canada's Timber Resources*. Forestry Canada, Petawawa National Forestry Institute, Information Report PI-X-101, Chalk River, Ontario. Canada.]

Tans, P.P., Fung, I.Y., and Takahashi, T. 1990. Observational constraints on the global atmospheric CO_2 budget. *Science* 247:1431–1438.

Turner, D.P., Lee, J.J., Koerper, G.J., and Barker, J.R. (eds.) 1993. The forest sector carbon budget of the United States: Carbon pools and flux under alternative policy options. U.S. Environmental Protection Agency, Rpt, EPA/600/3-93/093, Corvallis, OR.

United Nations. 1986. European timber trends and prospects to the year 2000 and beyond. Volumes I and II. Report ECE/TIM/30. New York: United Nations.

United Nations. 1989. Outlook for the forest and forest products sector of the USSR. Report ECE/TIM/48. New York: United Nations.

United Nations. 1992. Forest products, 1979–1990. FAO Forestry Series No. 25/ FAO Statistics Series No. 103, FAO Yearbook 1990. Rome: United Nations.

USDA Forest Service. 1944–1990. Annual forest fire reports. Fire and Aviation Management Division, USDA Forest Service, Independence at 14th Street, Washington, DC.

Waddell, K.L., Oswald, D.D., and Powell, D.S. 1989. Forest statistics of the United States, 1987. Resource Bulletin PNW-RB-168, Pacific Northwest Research Station, U.S.D.A. Forest Service. Portland, OR.

Winjum, J.K., Schroeder, P.E., and Kenady, M.J. (eds.) 1991. Large-scale reforestation. Proceedings of the International Workshop on Large-Scale Reforestation, May 1990. Rpt. EPA/600/9-91/014. Corvallis, OR, U.S. Environmental Protection Agency.

5. Relationships Between Ecosystem Change and Land Use

24. Aboriginal Participation in Global Change Research in Northwest Territories of Canada

Ellen Bielawski

This chapter is about people. People in Canada's North, who are neither statistics nor passive recipients of policies developed by governments on the basis of scientific research, nor, most importantly, people disconnected from the Arctic and Subarctic terrestrial, marine, or aquatic ecosystems. In contrast, Inuit (formerly called Eskimos) and Dene (Athapaskan Indians), the aboriginal peoples of Canada's Northwest Territories, are active participants in these ecosystems. They react to the research process and to the policies that are developed from research results. Unlike scientists, they do not traditionally separate their land, sea, and freshwater worlds into discrete ecosystems. Both conceptual and practical problems with global change research occur whether aboriginal people are advisers to scientific research projects, subjects in such research, participants in research, or affected by policies for response to global change. The following touches upon some human aspects of global change research in the circumpolar region.

I will first, discuss differences between traditional aboriginal world views and scientific views of Canada's Arctic and Subarctic. This discussion will center on the differences between scientific knowledge and what is called aboriginal, indigenous, or traditional knowledge. Martha Johnson (1992a, p. 4) summarizes the strengths and weaknesses of different terms, deriving this definition for "Traditional Environmental Knowledge" (TEK):

Traditional environmental knowledge, or TEK, can generally be defined as a body of knowledge built up by a group of people through generations of living in close contact with nature. It includes a system of classification, a set of empirical observations about the local environment, and a system of self-management that governs resource use. The quantity and quality of traditional environmental knowledge varies among community members, depending upon gender, age, social status, intellectual capability, and profession (Hunter, spiritual leader, healer, etc.). With its roots firmly in the past, traditional environmental knowledge is both cumulative and dynamic, building upon the experience of earlier generations and adapting to the new technological and socioeconomic changes of the present.

The Traditional Knowledge Working Group, Government of the Northwest Territories (1991, p. 11) produced the following definition of traditional knowledge after 2 years of discussion:

The lack of common understanding about the meaning of traditional knowledge is frustrating for those who advocate or attempt in practical ways to recognize and use traditional knowledge. For some, traditional knowledge is simply information which aboriginal peoples have about the land and animals with which they have a special relationship. But for aboriginal people, traditional knowledge is much more. One elder calls it a common understanding of what life is about. Knowledge is the condition of knowing something with familiarity gained through experience or association. The traditional knowledge of northern aboriginal peoples has roots based firmly in the northern landscape and a land-based life experience of thousands of years. Traditional knowledge offers a view of the world, aspirations, and an avenue to "truth" different from those held by non-aboriginal people whose knowledge is based largely on European philosophies.

Both definitions describe the knowledge that Inuit and Dene hold and live on their lands and waters. So also does Sayer's (1984) description of knowledge as labor as well as communicative interaction.

Differences between aboriginal and scientific world view are the context for conceptual and practical problems with global change research involving, or affecting, aboriginal people. I argue that most global change research in Canada's North has some effect on aboriginal people, because either (1) research takes place on aboriginal lands, and/or (2) aboriginal people observe and are knowledgeable about environmental indicators significant to global change research.

I will attempt to illustrate differences in aboriginal and scientific world views with brief description about research comparing aboriginal knowledge and scientific knowledge, with recommendations for incorporating aboriginal knowledge in global change research developed through conference workshops in May 1993.

I describe conducting research in partnership with aboriginal people. This description raises numerous questions, ranging from developing interview guides in aboriginal languages that reflect aboriginal rather than scientific categories; to the level and content of training for local, aboriginal people to do research; to the difficulties of working with communities in the process of recovering from the assaults of colonialism.

Lastly, I briefly describe some of the indications of changing climate that aboriginal people are observing at present and the impacts on aboriginal land use. I illustrate the difficulty of isolating causal variables in changing land use, and hence of attempting to relate changes in land use to global change or climate change.

Aboriginal People in Canada's Northwest Territories

The Northwest Territories (NWT) is about one-third of Canada's land mass, but of a national population around 23 million, only 47,000 live in the NWT. The aboriginal people are Inuit and Dene. Both groups include smaller regional groups with their own specific names. Inuit and Inuvialuit inhabit the Eastern, Central, and Western Arctic, primarily along the coast. One Inuit group lives on the Barrengrounds west of Hudson Bay. Human occupancy of Inuit land began around 4000 years ago and all of it is directly or indirectly ancestral to Inuit Culture (for a synthesis of recent data, see Bielawski, 1988). Dene inhabit the Subarctic forest and tree line areas of the western NWT. All Dene in the NWT traditionally ranged beyond the tree line for caribou hunting on the Barrengrounds, and many still do. Thus, it is impossible to exclude them from consideration of Arctic ecosystems. Human occupancy of Dene lands began around 8000 years ago and is ancestral to Dene culture. The boundary between Dene and Inuit has, by all evidence, always been fluid and contentious, and is so today. This is significant in the context of ecosystems, because each group needs and needed resources on the far side of the boundary, and acquired them either through travel, occupancy, trade, conflict, or, as at present, politically negotiated settlements. Soon the NWT will be divided into an Inuit territory, Nunavut, and a Western Territory. Inuit will form the majority population in Nunavut. Dene will become a minority in the Eurocanadian and mixed population of the west.

Dene and Inuit maintain practical access to much of their traditional land, although renewable and nonrenewable resources are all regulated by the Crown and the Northwest Territories government. Comprehensive claims settlements and Treaty Entitlements under negotiation will change this substantially, as we enter the 502nd year since Europeans came in contact with North American aboriginal people. For Inuit and Dene, the most substantial change in their lifeways has occurred in the last 40 years when they moved from nomadic lives on the land to life in permanent settlements.

Comparing Aboriginal Knowledge and Scientific Knowledge

Neither Inuit nor Dene see themselves as separate from nature. To the contrary, they see themselves as intimately related to the natural world and all of its other inhabitants. Their place in this world is spiritual as well as practical; sustenance from the land comes in calories and other supplies from animals, trees, rocks, berries, and so on. Sustenance also comes through maintaining one's role as custodian of lands and waters. Aboriginal knowledge and worldview requires that people know the land (in its broadest sense; land in the sense that we might use the term "environment"), learn from it, pay attention to the messages it gives spiritually and practically, and care for it.

The traditional scientific view opposes the aboriginal view. This is clear in the organization of science and academia, where social and natural research are commonly separated. Also, disciplinary and subdisciplinary specialty is the norm in research and in training for research.

This contrast—the removal of people from nature in the scientific view, and the inclusion of people with nature in the aboriginal view—is a factor inhibiting meaningful environmental research in and on the Arctic environment. Because people are both part of this environment and significant actors in it, research that ignores people is at best specific, and at worst incomplete. In the aboriginal view, much scientific research is incomplete, because it ignores much of reality (Cruikshank, 1984). It is overly specific, where aboriginal knowledge is holistic and integrative.

A second contrast is also significant. Science tends to see the world in terms of causality and linearity; in aboriginal world views, cycles and analogy provide greater explanation and are more readily perceived and understood.

Until recently, aboriginal knowledge was regarded as culturally specific and locally constrained. In the past decade, overdue recognition of aboriginal knowledge has yielded abundant, well-documented, and scientifically sound evidence of a rich body of indigenous knowledge globally (Colorado, 1988; Cruikshank, 1981, 1984; Feit, 1988; Freeman, 1988, 1992, Freeman and Carbyn, 1988; Government of the Northwest Territories, 1991; Hobson, 1992; Johnson, 1992a; Johnson and Ruttan, 1992; Merculieff, 1990; Salmond, 1985; Waldram, 1986, and many others). Most of this work concentrates on TEK.

Evaluated in terms of commonly recognized parameters of science, Inuit knowledge compares well with science (Bielawski, 1992a, 1992b; Denny, 1986, and others). Like science, Inuit knowledge is consensual, replicable, generalizable, incorporating, and, to some extent, predictive. I have not done comparable epistemological research on Dene knowledge, but I expect the same characteristics will be evident in a formal study comparing Dene knowledge with science, for they are evident when Dene are on the land. In contrast to science, however, there is little evidence that Inuit controlled conditions for experiment, nor sought increasing accuracy in measurement of time and space.

Conducting Research in Partnership with Aboriginal People

Much of this topic is general to conducting any research with Inuit or Dene. Conceptual problems are exacerbated as one moves away from topics that are easily matched with aboriginal categories. Weather, as opposed to climate, is a good example of this. In discussing climate variability research with Dene Sonline (Chipewyan), "climate" is interpreted as "weather"; the Dene yati language does not include a word for "climate" (personal communication, Archie Catholique, April 1993, Lutsel k'e, NWT). Three kinds of difficulties exist: conceptual problems; funding structures; and practical problems in aboriginal communities. Problems with funding structures are too complex to be discussed here. In contrast, the rewards are significant. These include, especially, data enrichment through collection strategies and interpretation drawn from two worldviews, and the trust that develops between aboriginal community and researcher.

The conceptual problem is to discover categories for data collection that match the aboriginal and scientific worldviews. Research utilizing traditional knowledge thus far has relied primarily on extracting aboriginal knowledge from its context so that it matches categories of information determined by the needs of scientists. This approach serves the needs of science, but does not do justice to aboriginal knowledge. Aboriginal knowledge is both context embedded and implies correct, spiritually based relationships in the environment. We are only beginning to deal with this conceptual problem, and I have no data to report. This is really *the key* intellectual problem for research on indigenous knowledge.

The practical problems of conducting research in partnership with aboriginal communities are essentially the legacy of colonialism: low level of education and training appropriate to research; poorly developed employment skills; alcoholism; chronic poor nutrition that denies people health, initiative, and energy; poverty that reflects itself in people's desire for jobs on research projects and leads them to work out on the land with neither the traditional nor modern gear required for efficient labor. Our work includes substantive research training that transfers basic employment and research skills to individuals and communities (see Ryan and Robinson, 1990, for detailed description of Participatory Action Research methodology). Communities control the data through participation, verification, review of all reports and other research products, and through holding copyright to records of their knowledge. Exercising copyright is a practical problem, and some of our work is in the negotiation and exercising of intellectual property rights. Individuals and communities have learned a long and deep mistrust of researchers. Conducting research in partnership with them requires facing this mistrust, sometimes daily, and demonstrating the integrity of one's commitment to their knowledge, in combination with good research that does justice to aboriginal knowledge and to science.

Climate Change and Aboriginal Land Use

In the aboriginal view, land *use* is a very limited concept. Dene and Inuit relationships with land and waters, as I have tried to describe briefly above, include custodianship and sustenance far beyond simple "use." I use the concept only to describe some impacts of climate change. Also I must caution that in large-scale climate change studies, land use may be taken to mean industrial use, and aboriginal use is ignored.

Recent climatic change in Canada's NWT includes marked warming in the western territories (Dene and Inuvialuit regions) contrasted with cooling in the Eastern Arctic. Cooler conditions in the East seem to be requiring less change in land use than are warmer conditions in the West. Higher temperatures in the Dene and Inuvialuit regions have been coupled with greater instability in weather patterns; hotter, drier summers during which forest fires are significant in restricting human land use directly, and subsequently indirectly as animal populations adjust to the habitat changes caused by major fires. Extreme heat is uncomfortable for people and on at least one occasion caused health problems for Elders; smoke also exacerbates respiratory illnesses. Water levels, the location of fish, and the length of portages for boats are all significant to land use. Melting permafrost and/or rising water levels that require communities to relocate, will, if they occur, be direct social costs of climate warming. The most consistent significant impact to date in the Subarctic is thinner ice coupled with severe overflows that are very dangerous for winter travellers; and heavier, wetter snow that is much more difficult for travellers on the land than the lighter, drier snow mature trappers remember.

It is very difficult to isolate the effects of these conditions on aboriginal land users from the effects of other variables in changing land use. In the Subarctic, for example, trapping has declined markedly during the past several years. The warming trend has made winter travel more difficult and dangerous. Two other factors are significant in the decline in trapping. Trappers describe low fur prices, and their dislike of the new, humane traps as the major reasons for not trapping. Low fur prices are related to the animal rights movement, which has stalled trapping in the NWT. The rising cost of supplies required for modern trapping, notably gas for snow machines and airplane charters from aboriginal communities to trapping camps, taken together with low fur prices, mean that people do not currently profit in cash from trapping. They profit in the well-being that comes from working on the land and in providing fur and land food to their communities, but trapping as a way of life is declining and may not grow again in the future. Fewer young men work as trappers each year, as educational and other employment opportunities become available and require their steady participation in community jobs. All of these factors, combined with the difficult land travel caused by the greater snowfall and less predictable ice conditions, work against trapping. It is not possible to isolate the effect of climatic conditions alone on this aspect of aboriginal land use.

One important effect of climate variability is people's concern for the land and the animals. People talk about the indicators of environmental change and ponder the effects on other species. They are very concerned about the health of land and animals, and are aware, for example, of the potential effects of increased UVB radiation on animals even more so than on humans.

In hunting, fishing, and gathering, people are adapting to changing conditions thus far. Environmental changes are discussed and adaptations to them developed and shared in constantly honed local knowledge. Significant changes in animal populations will be felt. Again, however, the significance of these changes will be embedded in other social and economic changes. These vary widely among regions, and especially among communities. For example, nearby industrial development that provides wage employment may significantly change community and individual participation in hunting, reducing the overall impact of climate change. Policy for responding to global change, it if is developed in isolation from aboriginal people, may be more detrimental than direct environmental change.

Policy Recommendations

These recommendations come from the Human Dimensions of Global Change Research Conference (Whitehorse, Canada, May 1993) organized by the Association of Canadian Universities for Northern Studies. I quote from the "Key Concerns and Recommendations" of the report:

1. "Apart from cultural survival (but also intractably linked to it) Traditional Knowledge has an important role to play in dealing with questions of global change. Monitoring climate change at the micro-level; monitoring environmental change; identifying associations and indices of environmental quality; identifying appropriate resource management models; measuring and monitoring population change are all of great importance."

2. Relatively few southern researchers have an understanding of the scope, complexity or utility of Traditional Knowledge. . . ."

3. "There is need for reciprocity between southern researchers and northern communities whereby the benefits from research utilizing Traditional Knowledge flow back into the communities. . . ."

4. "Application of Traditional Knowledge to local and global environmental problems provides a basis for cooperative research with reciprocal benefits manifested in a better understanding of global systems on one hand and enhancement of community 'health' on the other. Successful cooperative frameworks that would facilitate such research should be identified." (Duerden and Masuzumi, n.d.)

If as scientists we are concerned with research applications and policy for responses to global change, we have work to do in addressing these recommendations.

Summary

The human context of global change research includes people as part of the Arctic terrestrial ecosystem. Inuit and Dene are observing and adapting to current climate variability, which is one of several variables driving changing use of the Arctic and Subarctic environment. Inuit and Dene knowledge encompasses observation and interpretation that is a valid source of information about global change. Their knowledge contrasts significantly with scientific knowledge. Matching categories of aboriginal knowledge with scientific categories is the key intellectual problem requiring research. Aboriginal people need to be part of the research process and the policy development for response to global change. Without their participation, policy development for the North might be more detrimental to them than the direct effects of global change.

References

Bielawski, E. 1988. Paleoeskimo variability: The early Arctic Small Tool tradition in the central Canadian Arctic. *Am. Antiq.* 53:52–74.

Bielawski, E. 1992a. Inuit indigenous knowledge and science in the North. *North. Perspect.* 20(1):5–8.

Bielawski, E. 1992b. Cross-cultural epistemology: Cultural readaptation through the pursuit of knowledge. Looking to the Future, Inuit Studies Occasional Papers 4:59–69. Association Inuksiutiit Katimajiit, University Laval, Ste-Foy, Quebec.

Colorado, P. 1988. Bridging native and western science. *Convergence* Vol. XXI, No. 2 and 3, pp. 49–69.

Cruikshank, J. 1981. Legend and Landscape: Convergence of oral and scientific traditions in the Yukon Territory. *Arctic Anthropol.* XVIII-2:67–93.

Cruikshank, J. 1984. Oral tradition and scientific research: approaches to knowledge in the North. Social science in the North: communicating northern values. Occasional Publication 9, Association of Canadian Universities for Northern Studies.

Denny, J.P. 1986. Cultural ecology of mathematics: Ojibway and Inuit hunters. In: Close, M.P. (ed.), *Native American Mathematics* (pp. 129–180). Austin: University of Texas Press.

Duerden, F., and Masuzumi, B. n.d. Global change and traditional knowledge. Draft report from Global Change Workshop, Association of Canadian Universities for Northern Studies, Whitehorse, Yukon, Canada, May 1993.

Feit, H.A. 1988. Self-management and state management: forms of knowing and managing northern wildlife. Traditional knowledge and renewable resource management in northern regions. Boreal Institute for Northern Studies Occasional Publication 23, pp. 72–91.

Freeman, M.M.R. 1988. Ethnoscience, prevailing science and Arctic co-operation. Paper presented to the International Conference on Arctic Co-operation, Toronto.

Freeman, M.M.R. 1992. The nature and utility of traditional ecological knowledge. *North. Perspect.* 20(1):9–12.

Freeman, M.M.R., and Carbyn, L.N. (eds.) 1988. Traditional knowledge and renewable resource management. Boreal Institute for Northern Studies Occasional Publication 23.

Government of the Northwest Territories. 1991. Report of the Traditional Knowledge Working Group. Yellowknife, Department of Culture and Communications.

Johnson, M. (ed.) 1992a. Lore: Capturing Traditional Environmental Knowledge. Dene Cultural Institute and International Development Research Centre, Ottawa.

Johnson, M., and Ruttan, R. 1992. Dene traditional environmental knowledge: A pilot project. In: Johnson, M. (ed.), *Lore: Capturing Traditional Environmental Knowledge*. Ottawa: Dene Cultural Institute and International Development Research Centre.

Merculieff, L. 1990. Western society's linear systems and aboriginal cultures: The need for two-way exchanges for the survival. Presented at the Sixth International Conference of Hunting and Gathering Societies, Fairbanks, Alaska. Commissioner's Office, State of Alaska.

Ryan, J., and Robinson, M.P. 1990. Implementing participatory action research in the Canadian North: A case of the Gwich'in language and cultural project. *Culture* X(2):57–71.

Salmond, A. 1985. Maori epistemologies. In: Overing, J. (ed.), *Reason and Morality*. London: Tavistock Publication.

Sayer, A. 1984. *Method in Social Science: A Realist Approach*. London: Hutchinson.

Waldram, J. 1986. Traditional knowledge systems: The recognition of indigenous history and science. *Saskatchewan Indian Federated Coll. J.* 2(2):115–124.

Index

Ecological Studies

Volume 68
Stable Isotopes in Ecological Research (1989)
P.W. Rundel, J.R. Ehleringer, and K.A. Nagy (Eds.)

Volume 69
Vertebrates in Complex Tropical Systems (1989)
M.L. Harmelin-Vivien and F. Bourlière (Eds.)

Volume 70
The Northern Forest Border in Canada and Alaska (1989)
J.A. Larsen

Volume 71
Tidal Flat Estuaries: Simulation and Analysis of the Ems Estuary (1988)
J. Baretta and P. Ruardij (Eds.)

Volume 72
Acidic Deposition and Forest Soils (1989)
D. Binkley, C.T. Driscoll, H.L. Allen, P. Schoeneberger, and D. McAvoy

Volume 73
Toxic Organic Chemicals in Porous Media (1989)
Z. Gerstl, Y. Chen, U. Mingelgrin, and B. Yaron (Eds.)

Volume 74
Inorganic Contaminants in the Vadose Zone (1989)
B. Bar-Yosef, N.J. Barrow, and J. Goldshmid (Eds.)

Volume 75
The Grazing Land Ecosystems of the African Sahel (1989)
H.N. Le Houérou

Volume 76
Vascular Plants as Epiphytes: Evolution and Ecophysiology (1989)
U. Lüttge (Ed.)

Volume 77
Air Pollution and Forest Decline: A Study of Spruce (*Picea abies*) on Acid Soils (1989)
E.-D. Schulze, O.L. Lange, and R. Oren (Eds.)

Volume 78
Agroecology: Researching the Ecological Basis for Sustainable Agriculture (1990)
S.R. Gliessman (Ed.)

Volume 79
Remote Sensing of Biosphere Functioning (1990)
R.J. Hobbs and H.A. Mooney (Eds.)

Volume 80
Plant Biology of the Basin and Range (1990)
C.B. Osmond, G.M. Hidy, and L.F. Pitelka (Eds.)

Volume 81
Nitrogen in Terrestrial Ecosystem: Questions of Productivity, Vegetational Changes, and Ecosystem Stability (1990)
C.O. Tamm

Volume 82
Quantitative Methods in Landscape Ecology: The Analysis and Interpretation of Landscape Heterogeneity (1990)
M.G. Turner and R.H. Gardner (Eds.)

Volume 83
The Rivers of Florida (1990)
R.J. Livingston (Ed.)

Volume 84
Fire in the Tropical Biota: Ecosystem Processes and Global Challenges (1990)
J.G. Goldammer (Ed.)

Volume 85
The Mosaic-Cycle Concept of Ecosystems (1991)
H. Remmert (Ed.)

Volume 86
Ecological Heterogeneity (1991)
J. Kolasa and S.T.A. Pickett (Eds.)

Volume 87
Horses and Grasses: The Nutritional Ecology of Equids and Their Impact on the Camargue (1992)
P. Duncan

Volume 88
Pinnipeds and El Niño: Responses to Environmental Stress (1992)
F. Trillmich and K.A. Ono (Eds.)

Ecological Studies

Volume 89
Plantago: A Multidisciplinary Study (1992)
P.J.C. Kuiper and M. Bos (Eds.)

Volume 90
Biogeochemistry of a Subalpine Ecosystem: Loch Vale Watershed (1992)
J. Baron (Ed.)

Volume 91
Atmospheric Deposition and Forest Nutrient Cycling (1992)
D.W. Johnson and S.E. Lindberg (Eds.)

Volume 92
Landscape Boundaries: Consequences for Biotic Diversity and Ecological Flows (1992)
A.J. Hansen and F. di Castri (Eds.)

Volume 93
Fire in South African Mountain Fynbos: Ecosystem, Community, and Species Response at Swartboskloof (1992)
B.W. van Wilgen et al. (Eds.)

Volume 94
The Ecology of Aquatic Hyphomycetes (1992)
F. Bärlocher (Ed.)

Volume 95
Palms in Forest-Ecosystems of Amazonia (1992)
F. Kahn and J.-J. DeGranville

Volume 96
Ecology and Decline of Red Spruce in the Eastern United States (1992)
C. Eagar and M.B. Adams (Eds.)

Volume 97
The Response of Western Forests to Air Pollution (1992)
R.K. Olson, D. Binkley, and M. Böhm (Eds.)

Volume 98
Plankton Regulation Dynamics (1993)
N. Walz (Ed.)

Volume 99
Biodiversity and Ecosystem Function (1993)
E.-D. Schulze and H.A. Mooney (Eds.)

Volume 100
Ecophysiology of Photosynthesis (1994)
E.-D. Schulze and M.M. Caldwell (Eds.)

Volume 101
Effects of Land-Use Change on Atmospheric CO_2 Concentrations: South and South East Asia as a Case Study (1993)
V.H. Dale (Ed.)

Volume 102
Coral Reef Ecology (1993)
Y.I. Sorokin (Ed.)

Volume 103
Rocky Shores: Exploitation in Chile and South Africa (1993)
W.R. Siegfried (Ed.)

Volume 104
Long-Term Experiments With Acid Rain in Norwegian Forest Ecosystems (1993)
G. Abrahamsen et al. (Eds.)

Volume 105
Microbial Ecology of Lake Plußsee (1993)
J. Overbeck and R.J. Chrost (Eds.)

Volume 106
Minimum Animal Populations (1994)
H. Remmert (Ed.)

Volume 107
The Role of Fire in Mediterranean-Type Ecosystems (1994)
J.M. Moreno and W.C. Oechel (Eds.)

Volume 108
Ecology and Biogeography of Mediterranean Ecosystems in Chile, California, and Australia (1994)
M.T.K. Arroyo, P.H. Zedler, and M.D. Fox (Eds.)

Ecological Studies

Volume 109
Mediterranean Type Ecosystems: The Function of Biodiversity (1994)
G.W. Davis and D.M. Richardson (Eds.)

Volume 110
Tropical Montane Cloud Forests (1994)
L.S. Hamilton, J.O. Juvik, and F.N. Scatena (Eds.)

Volume 111
Peatland Forestry (1995)
E. Paavilainen and J. Päivänen

Volume 112
Tropical Forests: Management and Ecology (1995)
A.E. Lugo and C. Lowe (Eds.)

Volume 113
Arctic and Alpine Biodiversity: Patterns, Causes and Ecosystem Consequences (1995)
F.S. Chapin III and C. Körner (Eds.)

Volume 114
Crassulacean Acid Metabolism: Biochemistry, Ecophysiology and Evolution (1995)
K. Winter and J.A.C. Smith (Eds.)

Volume 115
Islands: Biological Diversity and Ecosystem Function (1995)
P.M. Vitousek, H. Andersen, and L. Loope (Eds.)

Volume 116
High-Latitude Rainforests and Associate Ecosystems of the West Coast of the Americas (1995)
R.G. Lawford, P.B. Alaback, and E. Fuentes (Eds.)

Volume 117
Global Change and Mediterranean-Type Ecosystems (1995)
J.M. Moreno and W.C. Oechel (Eds.)

Volume 118
Impact of Air Pollutants on Southern Pine Forests (1995)
S. Fox and R.A. Mickler (Eds.)

Volume 119
Freshwater Ecosystems of Alaska (1996)
A.M. Milner and M.W. Oswood (Eds.)

Volume 120
Landscape Function and Disturbance in Arctic Tundra (1996)
J.F. Reynolds and J.D. Tenhunen (Eds.)

Volume 121
Biodiversity and Savanna Ecosystem Processes. A Global Perspective (1996)
O.T. Solbrig, E. Medina, and J.F. Silva (Eds.)

Volume 122
Biodiversity and Ecosystem Processes in Tropical Forests (1996)
G.H. Orians, R. Dirzo, and J.H. Cushman (Eds.)

Volume 123
Marine Benthic Vegetation. Recent Changes and the Effects of Eutrophication (1996)
W. Schramm and P.H. Nienhuis (Eds.)

Volume 124
Global Change and Arctic Terrestrial Ecosystems (1996)
W.C. Oechel (Ed.)